Advances in Antifungal Development: Discovery of New Drugs and Drug Repurposing

Editors

Jong Heon Kim
Luisa W. Cheng
Kirkwood Land

MDPI • Basel • Beijing • Wuhan • Barcelona • Belgrade • Manchester • Tokyo • Cluj • Tianjin

Editors
Jong Heon Kim
Foodborne Toxin Detection
and Prevention Research Unit
Western Regional Research
Center, USDA-ARS
Albany, CA 94710
United States

Luisa W. Cheng
Foodborne Toxin Detection
and Prevention Research Unit
Western Regional Research
Center, USDA-ARS
Albany, CA 94710
United States

Kirkwood Land
Department of Biological
Sciences
University of the Pacific
Stockton, CA 95211
United States

Editorial Office
MDPI
St. Alban-Anlage 66
4052 Basel, Switzerland

This is a reprint of articles from the Special Issue published online in the open access journal *Pharmaceuticals* (ISSN 1424-8247) (available at: www.mdpi.com/journal/pharmaceuticals/special_issues/antifungal_ph).

For citation purposes, cite each article independently as indicated on the article page online and as indicated below:

LastName, A.A.; LastName, B.B.; LastName, C.C. Article Title. *Journal Name* **Year**, *Volume Number*, Page Range.

ISBN 978-3-0365-4768-8 (Hbk)
ISBN 978-3-0365-4767-1 (PDF)

Contents

About the Editors

Jong Heon Kim

Jong H. Kim earned his Ph.D. in Molecular, Cellular and Developmental Biology at Ohio State University in 1997. After conducting his post-doctoral research in the field of oxidative stress responses in eukaryotic cells and site-specific gene integration into eukaryotic chromosomes at the University of California, Berkeley, CA, USA, and the United States Department of Agriculture (USDA), he joined USDA Western Regional Research Center (WRRC), Albany, CA, as a Research Molecular Biologist in 2003. Dr. Kim investigated the development of new, highly effective intervention strategies for the control of fungal pathogens, especially those producing mycotoxins. The prevention of fungal pathogens is often very difficult since potent antifungal agents for eliminating resistant pathogens are limited. To address this, he validated the use of safe natural products or their derivatives in debilitating fungal defense systems to environmental stressors, identifying fungal antioxidants or cell wall systems as the effective targets of natural products. He also identified multiple factors, such as differences in fungal physiology and species, ingredients in formulations, and the antioxidant and pro-oxidant potential of compounds, that can positively or negatively influence the antifungal or antimycotoxigenic activity of natural products. Dr. Kim's expertise in antifungal chemosensitization as an intervention strategy is widely recognized by fellow researchers at the USDA, in academia, at research institutes, and at corporations. His research has resulted in several grants, participation as a PI/co-PI in the USDA innovation fund and a US Centers for Disease Control and Prevention (CDC)-USDA interagency agreement. He has authored works on invention disclosures, a patent, news citations, and awards in scientific meetings. Dr. Kim consults regularly with industries, who adopt his recommendations and fund his research.

Luisa W. Cheng

Luisa W. Cheng earned her Ph.D. in Microbiology and Immunology at the University of California Los Angeles in 2001. She conducted post-doctoral research at the University of California, Berkeley, and then joined the United States Department of Agriculture, Western Regional Research Center (WRRC), Albany, CA, as a Research Biologist in 2006. She developed oral and systemic mouse models for the toxicological study of ricin, botulinum neurotoxins (BoNT), shiga toxins and abrin. Using these models, she determined the toxicity and bioavailability of pure and crude toxins in the presence or absence of complex food matrices and under different food processing conditions (different pH and temperature) in complex matrices. Dr. Cheng developed mouse models for the antibody or small-molecule neutralization of BoNTs and Shiga toxins, which provided a valuable proof of concept for the use of antibodies or small molecules for animal rescue, including timing and dosage. She also developed in vitro intestinal cell models to study oral botulinum neurotoxin absorption, the mechanism of toxin translocation through the intestines, and the contribution of neurotoxin-associated proteins to botulinum intestinal translocation and tested the use of probiotics in the inhibition of the toxin binding of epithelial cells. In collaboration with other scientists, she provided many other small-molecule drugs or natural chemicals for testing in BoNT epithelial cell binding inhibition.

Kirkwood Land

Kirkwood M. Land earned his Ph.D. in Microbiology and Immunology at the University of California Los Angeles in 2001. After pursuing his post-doctoral research in the field of drug discovery against protozoal parasites at the University of California San Francisco School of Medicine as Giannini Family Foundation Postdoctoral Fellow in Medical Research, Dr. Land joined University of the Pacific as an Assistant Professor in 2003, where he focused on anti-protozoan drug discovery. He has been an Associate Professor since 2010 and Director of Student Learning Assessment since 2017. His research program focuses on the control of microbial growth in biological and environmental matrices, including humans, animals, and food. His group leverage collaborations with other scientific investigators and agencies to identify new lead compounds that could serve as scaffolds for new drug discovery against human and animal pathogens. Dr. Land is also interested in the repurposing of US Food and Drug Administration (FDA)-approved drugs as alternative treatments for infectious diseases. Recently, in light of his drug discovery studies, he has also begun to examine zoonotic diseases—with a focus on avian and feline pathogens potentially transmissible to humans, as well as the environmental factors that contribute to wild and domesticated animal infections. The goal is to identify new pathogens which may pose a threat to human health.

Preface to "Advances in Antifungal Development: Discovery of New Drugs and Drug Repurposing"

This reprint summarizes the latest advances in antifungal development, with an emphasis on the discovery of new drugs, development of new intervention strategies, and drug repurposing.

While fungal infections caused by yeast or filamentous fungal pathogens are persistent problems, the current antifungal drugs or intervention methods have exhibited limited efficacy in curing fungal diseases, mainly due to the narrow spectrum of the activity of conventional drugs, the stagnation in the development of new antifungal agents, increased incidence of fungal resistance to the conventional interventions, and toxicities caused by antifungal drugs.

The thirteen published articles reported in this Special Issue provide an overview of the current management approach to fungal diseases, highlighting measures that help develop new antifungal drugs and drug-repurposing strategies. The articles underline further the new antifungal drugs from natural sources, design, synthesis and antifungal evaluation of newly synthesized compounds, nanotechnology systems for drug delivery, and drug repurposing for overcoming multidrug-resistant fungal pathogens.

We strongly encourage the readers in this field to explore the articles in this reprint, which could help to develop new strategies for the treatment of fungal diseases, overcoming multi-drug resistance and target-specific drug delivery.

We are extremely grateful to all the authors for their valuable contributions to this Special Issue. We also thank MDPI and the Editor-in-Chief of *Pharmaceuticals*, J.J. Vanden Eynde, for the decision to publish this reprint and Ms. Evelyn Du for her kind assistance and technical support.

Jong Heon Kim, Luisa W. Cheng, and Kirkwood Land
Editors

pharmaceuticals

MDPI

Editorial

Advances in Antifungal Development: Discovery of New Drugs and Drug Repurposing

Jong H. Kim [1],*, Luisa W. Cheng [1],* and Kirkwood M. Land [2],*

[1] Foodborne Toxin Detection and Prevention Research Unit, Western Regional Research Center, USDA-ARS, 800 Buchanan St., Albany, CA 94710, USA

[2] Department of Biological Sciences, University of the Pacific, 3601 Pacific Avenue, Stockton, CA 95211, USA

* Correspondence: jongheon.kim@usda.gov (J.H.K.); luisa.cheng@usda.gov (L.W.C.); kland@pacific.edu (K.M.L.); Tel.: +1-510-559-5841 (J.H.K.); +1-510-559-5600 (L.W.C.); +1-209-946-7608 (K.M.L.)

Citation: Kim, J.H.; Cheng, L.W.; Land, K.M. Advances in Antifungal Development: Discovery of New Drugs and Drug Repurposing. *Pharmaceuticals* **2022**, *15*, 787. https://doi.org/10.3390/ph15070787

Received: 14 June 2022
Accepted: 23 June 2022
Published: 24 June 2022

Publisher's Note: MDPI stays neutral with regard to jurisdictional claims in published maps and institutional affiliations.

This Special Issue of *Pharmaceuticals* describes recent advances accomplished in the field of antifungal development, especially the discovery of new drugs and drug repurposing. The subjects of the published articles include: new drugs from natural or synthetic sources, design, synthesis, and antifungal evaluation of newly-synthesized compounds, novel nanotechnology systems for drug delivery, drug repurposing for fungal control and/or overcoming the multidrug-resistant fungi, repurposing antifungal drugs for antiviral therapy, evaluation of differential antifungal efficacy of echinocandins, among others.

Infectious diseases caused by fungal pathogens, such as aspergillosis, candidiasis, or cryptococcus, are recurring problems. Current antifungal interventions often exhibited very limited efficacy in treating fungal infections, partly because the spectrum of the activity of conventional systemic antifungal drugs is narrow while the development of new antifungal drugs has become stagnant; azole and polyene drugs were introduced before 1980, whereas the echinocandin drug CAS was approved for the clinical uses since 2000 [1]. Besides 5-flucytosine (5FC), only three classes of antifungal drugs are used in clinical settings, namely, azoles (fluconazole (FLU), itraconazole (ITR), voriconazole (VOR), posaconazole (POS), isavuconazole (ISA)), polyenes (amphotericin B (AMB)), and echinocandins (caspofungin (CAS), micafungin (MICA), anidulafungin (ANID)).

Regarding the antifungal spectrum, for example, the yeast pathogens *Candida albicans*, *Candida glabrata*, *Candida parapsilosis,* and *Candida tropicalis* showed susceptibility to the above-mentioned antifungal drugs (AMB, 5FC, FLU, ITR, VOR, POS, ISA, CAS, MICA, and ANID), whereas the other two *Candida* species, namely, *Candida krusei* and *Candida lusitaniae*, did not show sensitivity to FLU or AMB, respectively, indicating differential susceptibilities of pathogens to the antifungal drugs exist depending on the types of fungi treated or drugs administered [2–4].

Also, increased incidences of pathogen resistance to conventional antifungal interventions make fungal diseases a global human health concern. For instance, sequential combination therapy is one of the strategies for effective control of fungi. However, prior exposure of fungal pathogens, such as *Candida* species or *Aspergillus fumigatus*, to azole drugs (FLU, ITR, or VOR) lowered the susceptibility of the fungal biofilms or germlings to polyenes such as AMB [5,6]. It has been determined that azole drugs could potentiate fungal tolerance to AMB via heat-shock protein 90 (Hsp90)-mediated oxidative stress defenses in pathogens [7].

Azoles are also applied in crop fields as fungicides to control plant fungal pathogens; for instance, more than 25% of current fungicide sales are azoles [8]. From the clinical perspective, increased application of agricultural fungicides provides environmental selection pressure for the development of azole-resistant fungal pathogens such as *A. fumigatus* possessing the TR34/L98H mutation, which resulted in the recent placement of azole-resistant *A. fumigatus* on the microorganism "watchlist (antibiotic resistance threat)" by the United States Centers for Disease Control and Prevention (CDC) [9].

Candidiasis is an infectious disease caused by the genus *Candida* which has been the second-highest cause of superficial/mucosal human infections and the fourth highest cause of bloodstream infections [10,11]. *Candida* species are opportunistic pathogens that change from harmless/normal inhabitants to pathogenic depending on the fluctuations in the host environment [11,12]. According to the CDC, around 95% of invasive infections by *Candida* in the United States are caused by *C. albicans*, *C. glabrata*, *C. parapsilosis*, *C. tropicalis*, and *C. krusei* [13]. While *C. albicans* is the leading cause of candidemia, increasing numbers of infections have been reported to be triggered by non-*albicans* species, which often exhibited resistance to antifungal drugs [13]. For example, around 7% of *Candida* bloodstream isolates examined at CDC showed resistance to FLU where >70% of the resistant isolates are the non-*albicans* species *C. glabrata* or *C. krusei* [13]. Therefore, the non-albicans *C. auris* or other *Candida* species have also been classified as an "urgent" or "serious" threat pathogens by CDC [9], which exhibited severe human infections (oral, vaginal, bloodstream infections) and multidrug resistance.

The toxicity triggered by conventional antifungal drugs also hampered the effectiveness of antifungal therapy. For example, the polyene AMB was the first antifungal drug introduced to clinical settings more than fifty years ago. However, AMB also triggered varying toxicity to the hosts, such as infusion-related reactions, nephrotoxicity, etc. Therefore, different types of formulations have been developed, for instance, lipid-associated AMB formulations such as the AMB lipid complex (AMB-LC), liposomal AMB (L-AMB), and colloidal dispersion of AMB (AMB-CD) [14,15].

In this Special Issue, thirteen works (seven original research articles, six reviews) were published on the recent advances in antifungal development, providing current tools, methods, strategies, or insights on antifungal therapy.

Sousa et al. [16] provided a summary of the state-of-the-art strategies for novel nanotechnology approaches for drug delivery systems and new drugs derived from natural sources. The authors provided an overview of the new drug delivery systems (types of nanoparticles) which are applicable for various antifungal agents such as azoles, amphotericin B, nystatin, etc., the drug chemical group, the route of administration (oral, topical, transdermal, etc.), and their dosage forms. The strategies not only counteract emerging fungal infections but also overcome the increase in drug-resistant strains. Of note, several mechanisms were discussed how nanoparticles overcome the development of fungal resistance [16], which include: (1) The chemical characteristics and concurrent multiple mechanisms caused by nitric oxide, chitosan, and metallic nanoparticles that destabilize fungal cell structures, (2) Packaging multiple antimicrobial drugs within the same nanoparticle, hence disrupting the resistance mechanisms, (3) Liposomes/dendrimers-based drug formulation that overcomes decreased uptake and increased efflux of drugs, and (4) Targeting antifungal drugs to the specific site of infection, enabling the localized release of high doses of drugs while keeping the total concentrations of the drugs administered low. The review [16] also focused on the identification of new natural antifungal compounds from the marine environment; the approaches provide platforms to discover possible new drug leads while highlighting the challenges of the translation of the identified natural compounds into the clinical pipeline.

Bouz and Doležal [17] reviewed recent advances in antifungal drug development in clinical settings. The review summarized novel antifungal agents in clinical development including first-in-class agents, new chemical structures developed against established molecular targets, modifications of formulation to marketed antifungal drugs, repurposing of non-antifungal agents to treat fungal infections, and membrane interacting peptides as antifungals, among others. Of note, the review described further the immunotherapy using the antifungal antibodies MAb 2G8 and efungumab (mycograb) advancing into clinical trials. MAb 2G8 targets laminarin (consisting mainly of β-glucans) and thus binds to the cell walls of *Candida albicans* and *Cryptococcus neoformans*, inhibiting fungal growth and capsule formation [18], whereas the new, humanized version of monoclonal MAb 2G8 precisely targets β-1,3 glucans of *Candida* spp. including *Candida auris* [19]. Novel antifungal targets are also discussed, such as heme biosynthesis and sphingolipid synthesis which are unique fungal target pathways, thus enabling the designing of novel drugs and overcoming the pathogen resistance to conventional antifungal interventions.

Recently, β-glucans, a subgroup of polysaccharides, have been implicated in medicinal applications, such as the treatment of diabetes and immunomodulation [20]. Since β-glucans often exhibit very limited water solubility, the cationization, namely, the addition of a positive charge to the polymer structures using "glycidyl trimethyl ammonium chloride (GTMAC)" has been a tool to enhance water solubility. Meanwhile, the interaction between the positively charged polymers, namely, polycations, and a biological membrane could be the cause of the toxicity exerted by the developed macromolecules, hence it became a very attractive strategy to improve the antifungal activity of the candidate polymers. In their research article, Kaminski et al. [21] investigated cationic derivatives of natural β-glucan polymers that were synthesized by reacting the polysaccharides from *Saccharomyces boulardii* (SB) and *Cetraria islandica* with GTMAC. In the study, three different polymer structures were obtained exhibiting selective antifungal activities against *Scopulariopsis brevicaulis*, *Aspergillus brasiliensis*, and *Fusarium solani*. In an in vivo model study of *Aspergillus brasiliensis* infection in *Galleria mellonella* (insect) using SB derivative, positive antifungal results were also observed while toxicity assay via fibroblast 3T3-L1 cell line revealed negligible toxicity of the compounds at various concentrations.

Using the fission yeast *Schizosaccharomyces pombe* (a model fungal system), Yagüe et al. [22] determined that various echinocandin drugs induce differential effects in cytokinesis progression and cell integrity in fungi. *S. pombe* contains three β(1,3)-D-glucan synthases (GSs), Bgs1, Bgs3, and Bgs4, which exert essential functions with non-overlapping roles involved in maintaining cell integrity and morphogenesis. Results by Yagüe et al. demonstrated that MICA exerts fungicidal activity by solely inhibiting Bgs4, whereas lethal doses of other echinocandin drugs ANID and CAS trigger an early cytokinesis arrest during fungal growth possibly by the concerted inhibition of several GSs.

Clotrimazole has been used for treating the yeast pathogen *C. albicans* and other fungi. The antifungal mechanism of clotrimazole is to interfere with ergosterol biosynthesis via the inhibition of the fungal cytochrome 14α-demethylase, which eventually triggers the disruption of the structure/function of the fungal cell wall including cell wall leakage. Since clotrimazole has been applied for treating vulvovaginal candidiasis for around 50 years, Mendling et al. [23], via analyzing thirty-seven randomized controlled studies, tried to estimate the current effectiveness of topical clotrimazole under different disease severity of populations infected. In patients with uncomplicated vulvovaginal candidiasis, application of single intravaginal doses of 500 mg clotrimazole (vaginal tablets) resulted in high cure rates, thus it was concluded that a single dose of clotrimazole 500 mg was equally effective as multiple doses of lower clotrimazole strengths. Moreover, prolonged treatment with the drug showed effectiveness in severe, recurrent infections and further in pregnant women. Therefore, Mendling et al. concluded that, despite its long-term usage, clotrimazole resistance in vaginal candidosis is scarce and hence, clotrimazole can continuously be used for vaginal health in the future [23].

Cryptococcus neoformans is also an opportunistic yeast pathogen causing cryptococcus, especially in the immune-compromised patients; around 73% of 223,100 annual cases of cryptococcal meningitis are diagnosed in sub-Saharan Africa with a 75% mortality rate [24]. Since photodynamic treatment (PDT) is a method of curing aerobic microbes susceptible to oxidative damage [25], Ogundeji et al. [24] investigated the photodynamic activity of acetylsalicylic acid (ASA; aspirin) against respiring cryptococcal cells. Cryptococcal cells were exposed to 0.5 or 1 mM of ASA with ultraviolet light (UVL) for 10 min, which resulted in a significant reduction in the fungal growth compared to the non-treated control cells. The treated cells exhibited: (a) a significant loss of mitochondrial membrane potential, (b) cellular accumulation of toxic reactive oxygen species, (c) altered ultrastructural appearance, and (d) limited expression levels of the capsular-associated gene, *CAP64*, leading to an acapsular phenotype.

Retinoids are chemical compounds derived from vitamin A or its structural analogs including all-*trans* retinoic acid (ATRA; tretinoin), a vitamin A metabolite. Retinoids possess a potent antimicrobial activity against a broad spectrum of pathogens, including bacteria, fungi and viruses. In their review, Cosio et al. [26] evaluated/summarized the results of thirty-nine in vitro/in vivo antifungal studies that demonstrated the antifungal efficacy of retinoids against human opportunistic fungal pathogens, such as yeast fungi (*Candida* spp., *Rhodotorula mucilaginosa*, *Malassezia furfur*) colonizing on the skin/mucosal surfaces of humans, environmental molds (*Aspergillus* spp., *Fonsecae monofora*) and several species of dermatophytes infecting humans and animals. The data from retinoid studies indicated that vitamin A or ATRA lowers the incidence and severity of fungal diseases via the immunoadjuvant properties of these compounds, where the antifungal efficacy, tolerability, and safety profile of test compounds have been proven against localized and systemic fungal infections [26].

In the study by Rodríguez-Villar et al. [27], a set of chemical compounds were synthesized and their antifungal activities were examined against *C. albicans*, *C. glabrata*, and *C. tropicalis* strains. They found that chemical derivatives containing "3-phenyl-1H-indazole" moiety exhibited the highest, broad anticandidal activity. In particular, *N,N*-diethylcarboxamide substituent showed the highest activity against *C. albicans* and *C. glabrata* (both miconazole susceptible and resistant species). Collectively, results indicated that the 3-phenyl-1H-indazole moiety can be a useful scaffold for the development of new anticandidal drugs.

Azole fungicides bind to and inhibit lanosterol 14α-demethylase, which is a fungal cytochrome P450 (CYP) enzyme involved in the ergosterol biosynthesis in fungal pathogens. While azoles are considered the most successful class of drugs for fungal control, extensive use of azole-class fungicides and prophylactic antifungal therapy resulted in the development of fungal resistance to clinical azole drugs. In their review, Kane and Carter [28] documented the enhancement of the potency of azole drugs with drug synergy, where significant antifungal synergism between azoles (FLU, ITR, VOR, ISA, POS, Ketoconazole, Miconazole) and known chemicals such as other antifungals (polyenes, echinocandins, etc.), anti-bacterials, anti-parasitics, anti-virals, calcium inhibitors, statins, bisphosphonates, immunomodulators, psychoactives, calcineurine inhibitors, essential oil extracts, etc., has been evaluated. Furthermore, azole synergy with novel compounds was also presented where the synergism with heat shock protein 90 (Hsp90) inhibitors, isoquinolone, phthalazine, natural metabolites berberine, piperidol, caffeic acid, guttiferone, the anti-inflammatory celecoxib, and phenylpentanol β-glucan synthase inhibitor SCY-078, ion chelator DIBI, target of rapamycin (TOR) inhibitor AZD8055, efflux modulators, as well as novel antifungal chalcones effectively control fungal pathogens including azole resistant *C. albicans*, *C. glabrata* and other *Candida* species. Considering no azole-based antifungal combinations are currently applied to treat fungal infections, co-application of secondary compounds exhibiting azole synergy, as shown above, could function as new fungal control regimens. It was concluded that "azole synergy" with "secondary compounds" will be a promising strategy for treating intrinsic/acquired resistance to the conventional drugs exerted by certain fungal pathogens [28].

The wide distribution of feline sporotrichosis in Brazil has been problematic, requiring new therapeutic alternatives [29]. In the in vitro and in silico analyses of new synthetic derivatives against wild type and non-wild type *Sporothrix brasiliensis*, de Souza et al. [29] demonstrated the antifungal efficacy of derivatives of three novel hydrazone and eleven novel quinone using the broth microdilution method, where the minimum inhibitory and fungicidal concentrations of test compounds ranged from 1 to >128 μg/mL. Particularly, the three hydrazone derivatives were determined as promising antifungal candidates against ITR-resistant *S. brasiliensis*, which is the most virulent and prevalent species causing feline sporotrichosis [29].

In view of the difficulties to treat invasive fungal infections, the development of new, alternative antifungal strategies that are efficient and appropriate measures are urgently needed. Antifungal drug repurposing is the repositioning process of already marketed medicinal drugs developed for treating human diseases to cure fungal infections [30]. The key merit of drug repurposing is that the mechanisms of action, cellular targets, or safety of the marketed drugs are already known, thus enabling accelerated regulatory approval once antifungal efficacy is determined. In their review, Peyclit et al. [31] provided a comprehensive summary of current antifungal drugs, their mechanisms of resistance, as well as an overview of the potent antifungal activity of non-traditional antifungal drugs. The review also presented mechanisms of action and the synergistic drug/compound combinations that enhance the efficacy of current antifungal interventions. Notably, drug repurposing for treating the emerging multidrug resistant (MDR) fungus such as *Candida auris*, *Aspergillus* or *Cryptococcus* species has been discussed, addressing the potential of antifungal drug repurposing to cure the difficult human fungal pathogens [31].

Conversely, certain antifungal drugs have been repurposed to treat viral or amoeboidal infections. Schloer et al. in their review [32] postulated that repurposing marketed antifungal drugs with well-known safety profiles can additionally target "host cell factors" necessary for viral proliferation, such as virus entry, replication, etc. Hence, these host cell factors could be a promising cellular target for the development of novel prophylaxis and treatment approaches against viral infections. ITR is one of the azoles inhibiting fungal ergosterol biosynthesis. Studies have shown that, in host cells, ITR directly interacts with the endolysosomal cholesterol transporter "Niemann–Pick disease, type c1 (npc1)" protein. A dysfunctional NPC1 interferes with the intracellular lipid transport, thus resulting in the excessive accumulation of lipids such as cholesterol in the endolysosomal compartment [33]. Importantly, NPC1 is recently explored as an antiviral drug target that can be interfered with ITR [34,35]. Other host cell factors to be interfered with ITR for the treatment of viral infections includes oxysterolbinding protein 1 (OSBP) and OSBP-Related Proteins (ORP) involved in lipid homeostasis [36], mammalian target of rapamycin (mTOR), hedgehog, and Wnt signaling pathways that regulate apoptosis and/or counteracts stress-induced autophagy [37,38], etc.

Acanthamoeba castellanii is a causative agent (amoeboid) of *Acanthamoeba* keratitis (AK), a serious eye infection resulting in an inflammation in the cornea, which could trigger permanent visual impairment/blindness [39]. Isavuconazonium sulfate is an FDA-approved antifungal drug for treating invasive fungal diseases such as aspergillosis and mucormycosis. The prodrug isavuconazonium sulfate is metabolized further into the active molecule ISA which previously exhibited amebicidal and cysticidal activity against *Acanthamoeba* T4 strains. However, the activity of the prodrug isavuconazonium sulfate against trophozoites and cysts was not determined. Shing et al. [39] recently demonstrated that isavuconazonium also possessed potent amoebicidal activity at as low as 1.4 nM and prevented excystation of cysts at as low as 136 µM, thus suggesting that, like ISA, the prodrug isavuconazonium maintains activity against *Acanthamoeba* T4 strains, which could be adapted for *Acanthamoeba* keratitis treatment.

In summary, the research articles and reviews presented in this Special Issue provide useful information and insight for the development of new antifungal drugs or intervention strategies. Identification of new, safe molecules, and cellular targets, as well as elucidation of their antifungal mechanisms of action will further the effective control of fungal pathogens, especially those resistant to current therapeutic agents.

Author Contributions: J.H.K., L.W.C. and K.M.L. contributed equally to this Editorial. All authors have read and agreed to the published version of the manuscript.

Funding: This research received no external funding.

Institutional Review Board Statement: Not applicable.

Informed Consent Statement: Not applicable.

Acknowledgments: The editors would like to acknowledge and thank the authors for their contributions, and all the reviewers for their effort, expertise and constructive suggestions that significantly contributed to the quality of this Special Issue. We would also like to thank Kathleen L. Chan, Food-borne Toxin Detection and Prevention Research Unit, Western Regional Research Center, USDA-ARS, for technical assistance.

Conflicts of Interest: The authors declare no conflict of interest.

References

1. Roemer, T.; Krysan, D.J. Antifungal Drug Development: Challenges, Unmet Clinical Needs, and New Approaches. *Cold Spring Harb. Perspect. Med.* **2014**, *4*, a019703. [CrossRef] [PubMed]
2. Houšť, J.; Spížek, J.; Havlíček, V. Antifungal Drugs. *Metabolites* **2020**, *10*, 106. [CrossRef] [PubMed]
3. Nami, S.; Aghebati-Maleki, A.; Morovati, H.; Aghebati-Maleki, L. Current antifungal drugs and immunotherapeutic approaches as promising strategies to treatment of fungal diseases. *Biomed. Pharmacother.* **2019**, *110*, 857–868. [CrossRef]
4. Nett, J.E.; Andes, D.R. Antifungal Agents: Spectrum of Activity, Pharmacology, and Clinical Indications. *Infect. Dis. Clin. N. Am.* **2016**, *30*, 51–83. [CrossRef] [PubMed]
5. Rajendran, R.; Mowat, E.; Jones, B.; Williams, C.; Ramage, G. Prior in vitro exposure to voriconazole confers resistance to amphotericin B in *Aspergillus fumigatus* biofilms. *Int. J. Antimicrob. Agents* **2015**, *46*, 342–345. [CrossRef]
6. Vazquez, J.A.; Arganoza, M.T.; Boikov, D.; Yoon, S.; Sobel, J.D.; Akins, R.A. Stable Phenotypic Resistance of *Candida* Species to Amphotericin B Conferred by Preexposure to Subinhibitory Levels of Azoles. *J. Clin. Microbiol.* **1998**, *36*, 2690–2695. [CrossRef]
7. Delattin, N.; Cammue, B.P.; Thevissen, K. Reactive oxygen species-inducing antifungal agents and their activity against fungal biofilms. *Future Med. Chem.* **2014**, *6*, 77–90. [CrossRef]
8. Bowyer, P.; Denning, D.W. Environmental fungicides and triazole resistance in *Aspergillus*. *Pest Manag. Sci.* **2014**, *70*, 173–178. [CrossRef]
9. Centers for Disease Control and Prevention (U.S.). *Antibiotic Resistance Threats in the United States, 2019*; Department of Health and Human Services, CDC: Atlanta, GA, USA, 2019. Available online: https://stacks.cdc.gov/view/cdc/82532 (accessed on 3 June 2022).
10. Bongomin, F.; Gago, S.; Oladele, R.O.; Denning, D.W. Global and Multi-National Prevalence of Fungal Diseases—Estimate Precision. *J. Fungi* **2017**, *3*, 57. [CrossRef]
11. Brown, G.D.; Denning, D.W.; Gow, N.A.R.; Levitz, S.M.; Netea, M.G.; White, T.C. Hidden Killers: Human Fungal Infections. *Sci. Transl. Med.* **2012**, *4*, 165rv13. [CrossRef]
12. Gonçalves, B.; Ferreira, C.; Alves, C.T.; Henriques, M.; Azeredo, J.; Silva, S. Vulvovaginal candidiasis: Epidemiology, microbiology and risk factors. *Crit. Rev. Microbiol.* **2016**, *42*, 905–927. [CrossRef]
13. Centers for Disease Control and Prevention (U.S.). Invasive Candidiasis Statistics. Available online: https://www.cdc.gov/fungal/diseases/candidiasis/invasive/statistics.html (accessed on 3 June 2022).
14. Hamill, R.J. Amphotericin B Formulations: A Comparative Review of Efficacy and Toxicity. *Drugs* **2013**, *73*, 919–934. [CrossRef] [PubMed]
15. Marena, G.D.; Ramos, M.A.d.S.; Bauab, T.M.; Chorilli, M. A Critical Review of Analytical Methods for Quantification of Amphotericin B in Biological Samples and Pharmaceutical Formulations. *Crit. Rev. Anal. Chem.* **2022**, *52*, 555–576. [CrossRef] [PubMed]
16. Sousa, F.; Ferreira, D.; Reis, S.; Costa, P. Current Insights on Antifungal Therapy: Novel Nanotechnology Approaches for Drug Delivery Systems and New Drugs from Natural Sources. *Pharmaceuticals* **2020**, *13*, 248. [CrossRef] [PubMed]
17. Bouz, G.; Doležal, M. Advances in Antifungal Drug Development: An Up-to-Date Mini Review. *Pharmaceuticals* **2021**, *14*, 1312. [CrossRef]
18. Rachini, A.; Pietrella, D.; Lupo, P.; Torosantucci, A.; Chiani, P.; Bromuro, C.; Proietti, C.; Bistoni, F.; Cassone, A.; Vecchiarelli, A. An Anti-Beta-Glucan Monoclonal Antibody Inhibits Growth and Capsule Formation of *Cryptococcus neoformans* In Vitro and Exerts Therapeutic, Anticryptococcal Activity In Vivo. *Infect. Immun.* **2007**, *75*, 5085–5094. [CrossRef]
19. Di Mambro, T.; Vanzolini, T.; Bruscolini, P.; Perez-Gaviro, S.; Marra, E.; Roscilli, G.; Bianchi, M.; Fraternale, A.; Schiavano, G.F.; Canonico, B.; et al. A new humanized antibody is effective against pathogenic fungi in vitro. *Sci. Rep.* **2021**, *11*, 19500. [CrossRef]
20. Vetvicka, V.; Vannucci, L.; Sima, P.; Richter, J. Beta Glucan: Supplement or Drug? From Laboratory to Clinical Trials. *Molecules* **2019**, *24*, 1251. [CrossRef]
21. Kaminski, K.; Skora, M.; Krzyściak, P.; Stączek, S.; Zdybicka-Barabas, A.; Cytryńska, M. Synthesis and Study of Antifungal Properties of New Cationic Beta-Glucan Derivatives. *Pharmaceuticals* **2021**, *14*, 838. [CrossRef]
22. Yagüe, N.; Gómez-Delgado, L.; Curto, M.Á.; Carvalho, V.S.D.; Moreno, M.B.; Pérez, P.; Ribas, J.C.; Cortés, J.C.G. Echinocandin Drugs Induce Differential Effects in Cytokinesis Progression and Cell Integrity. *Pharmaceuticals* **2021**, *14*, 1332. [CrossRef]
23. Mendling, W.; Atef El Shazly, M.; Zhang, L. Clotrimazole for Vulvovaginal Candidosis: More Than 45 Years of Clinical Experience. *Pharmaceuticals* **2020**, *13*, 274. [CrossRef] [PubMed]
24. Ogundeji, A.O.; Mjokane, N.; Folorunso, O.S.; Pohl, C.H.; Nyaga, M.M.; Sebolai, O.M. The Repurposing of Acetylsalicylic Acid as a Photosensitiser to Inactivate the Growth of Cryptococcal Cells. *Pharmaceuticals* **2021**, *14*, 404. [CrossRef]
25. Stapleton, M.; Rhodes, L.E. Photosensitizers for photodynamic therapy of cutaneous disease. *J. Dermatol. Treat.* **2003**, *14*, 107–112. [CrossRef] [PubMed]

26. Cosio, T.; Gaziano, R.; Zuccari, G.; Costanza, G.; Grelli, S.; Di Francesco, P.; Bianchi, L.; Campione, E. Retinoids in Fungal Infections: From Bench to Bedside. *Pharmaceuticals* **2021**, *14*, 962. [CrossRef] [PubMed]

27. Rodríguez-Villar, K.; Hernández-Campos, A.; Yépez-Mulia, L.; Sainz-Espuñes, T.d.R.; Soria-Arteche, O.; Palacios-Espinosa, J.F.; Cortés-Benítez, F.; Leyte-Lugo, M.; Varela-Petrissans, B.; Quintana-Salazar, E.A.; et al. Design, Synthesis and Anticandidal Evaluation of Indazole and Pyrazole Derivatives. *Pharmaceuticals* **2021**, *14*, 176. [CrossRef] [PubMed]

28. Kane, A.; Carter, D.A. Augmenting Azoles with Drug Synergy to Expand the Antifungal Toolbox. *Pharmaceuticals* **2022**, *15*, 482. [CrossRef] [PubMed]

29. de Souza, L.C.d.S.V.; Alcântara, L.M.; de Macêdo-Sales, P.A.; Reis, N.F.; de Oliveira, D.S.; Machado, R.L.D.; Geraldo, R.B.; dos Santos, A.L.S.; Ferreira, V.F.; Gonzaga, D.T.G.; et al. Synthetic Derivatives against Wild-Type and Non-Wild-Type *Sporothrix brasiliensis*: In Vitro and In Silico Analyses. *Pharmaceuticals* **2022**, *15*, 55. [CrossRef]

30. Cha, Y.; Erez, T.; Reynolds, I.J.; Kumar, D.; Ross, J.; Koytiger, G.; Kusko, R.; Zeskind, B.; Risso, S.; Kagan, E.; et al. Drug repurposing from the perspective of pharmaceutical companies. *Br. J. Pharmacol.* **2018**, *175*, 168–180. [CrossRef]

31. Peyclit, L.; Yousfi, H.; Rolain, J.-M.; Bittar, F. Drug Repurposing in Medical Mycology: Identification of Compounds as Potential Antifungals to Overcome the Emergence of Multidrug-Resistant Fungi. *Pharmaceuticals* **2021**, *14*, 488. [CrossRef]

32. Schloer, S.; Goretzko, J.; Rescher, U. Repurposing Antifungals for Host-Directed Antiviral Therapy? *Pharmaceuticals* **2022**, *15*, 212. [CrossRef]

33. Vanier, M.; Millat, G. Niemann–Pick disease type C. *Clin. Genet.* **2003**, *64*, 269–281. [CrossRef] [PubMed]

34. Côté, M.; Misasi, J.; Ren, T.; Bruchez, A.; Lee, K.; Filone, C.M.; Hensley, L.; Li, Q.; Ory, D.; Chandran, K.; et al. Small molecule inhibitors reveal Niemann–Pick C1 is essential for Ebola virus infection. *Nature* **2011**, *477*, 344–348. [CrossRef] [PubMed]

35. Herbert, A.S.; Davidson, C.; Kuehne, A.I.; Bakken, R.; Braigen, S.Z.; Gunn, K.E.; Whelan, S.P.; Brummelkamp, T.R.; Twenhafel, N.A.; Chandran, K.; et al. Niemann-Pick C1 Is Essential for Ebolavirus Replication and Pathogenesis In Vivo. *mBio* **2015**, *6*, e00565-00515. [CrossRef] [PubMed]

36. Schloer, S.; Goretzko, J.; Kühnl, A.; Brunotte, L.; Ludwig, S.; Rescher, U. The clinically licensed antifungal drug itraconazole inhibits influenza virus in vitro and in vivo. *Emerg. Microbes Infect.* **2019**, *8*, 80–93. [CrossRef]

37. Karam, B.S.; Morris, R.S.; Bramante, C.T.; Puskarich, M.; Zolfaghari, E.J.; Lotfi-Emran, S.; Ingraham, N.E.; Charles, A.; Odde, D.J.; Tignanelli, C.J. mTOR inhibition in COVID-19: A commentary and review of efficacy in RNA viruses. *J. Med. Virol.* **2021**, *93*, 1843–1846. [CrossRef]

38. Kuss-Duerkop, S.K.; Wang, J.; Mena, I.; White, K.; Metreveli, G.; Sakthivel, R.; Mata, M.A.; Muñoz-Moreno, R.; Chen, X.; Krammer, F.; et al. Influenza virus differentially activates mTORC1 and mTORC2 signaling to maximize late stage replication. *PLoS Pathog.* **2017**, *13*, e1006635. [CrossRef]

39. Shing, B.; Balen, M.; Debnath, A. Evaluation of Amebicidal and Cysticidal Activities of Antifungal Drug Isavuconazonium Sulfate against Acanthamoeba T4 Strains. *Pharmaceuticals* **2021**, *14*, 1294. [CrossRef]

 *pharmaceuticals*

Review

Current Insights on Antifungal Therapy: Novel Nanotechnology Approaches for Drug Delivery Systems and New Drugs from Natural Sources

Filipa Sousa [1,*], **Domingos Ferreira** [1], **Salette Reis** [2] **and Paulo Costa** [1,*]

[1] UCIBIO, REQUIMTE, Laboratory of Pharmaceutical Technology, Department of Drug Sciences, Faculty of Pharmacy, University of Porto, Rua Jorge Viterbo Ferreira n° 228, 4050-313 Porto, Portugal; domingos@ff.up.pt

[2] LAQV, REQUIMTE, Department of Chemical Sciences, Faculty of Pharmacy, University of Porto, Rua Jorge Viterbo Ferreira n° 228, 4050-313 Porto, Portugal; shreis@ff.up.pt

* Correspondence: filipambtsousa@gmail.com (F.S.); pccosta@ff.up.pt (P.C.)

Received: 24 August 2020; Accepted: 13 September 2020; Published: 15 September 2020

Abstract: The high incidence of fungal infections has become a worrisome public health issue, having been aggravated by an increase in host predisposition factors. Despite all the drugs available on the market to treat these diseases, their efficiency is questionable, and their side effects cannot be neglected. Bearing that in mind, it is of upmost importance to synthetize new and innovative carriers for these medicines not only to fight emerging fungal infections but also to avert the increase in drug-resistant strains. Although it has revealed to be a difficult job, new nano-based drug delivery systems and even new cellular targets and compounds with antifungal potential are now being investigated. This article will provide a summary of the state-of-the-art strategies that have been studied in order to improve antifungal therapy and reduce adverse effects of conventional drugs. The bidirectional relationship between Mycology and Nanotechnology will be also explained. Furthermore, the article will focus on new compounds from the marine environment which have a proven antifungal potential and may act as platforms to discover drug-like characteristics, highlighting the challenges of the translation of these natural compounds into the clinical pipeline.

Keywords: nanoparticles; fungi; drug delivery systems; marine; biological synthesis; myconanotechnology

---

1. Introduction

There is a wide range of fungal infections, from superficial, affecting skin, to systemic infections with invasion of internal organs [1]. Fungal infections affect millions of people every year worldwide. Of these, more or less 1.5 million are invasive fungal infections therefore requiring advanced treatment and hospitalization. Most of these disseminated infections are caused by *Candida, Cryptococcus, Aspergillus,* and *Pneumocystis* species, being the cause of cryptococcosis, candidiasis, aspergillosis, and pneumocystis pneumonia, respectively [2].

Superficial fungal infections are rather common and, despite rarely being life threatening, they can spread to other skin regions and even become widespread. Furthermore, they can be transmitted to other people and may cause secondary bacterial skin infections, harming the quality of a person's life. Skin mycoses are classified according to the causative fungal agents into dermatophytosis, yeast infections, and mold infections [1].

Invasive fungal infections represent a significant burden to healthcare systems, having high morbidity and mortality rates. These rates are most worrisome among immunocompromised patients that are more prone to opportunistic infections, such as patients with Acquired Immune Deficiency Syndrome (AIDS), transplant patients whose immune systems are suppressed to prevent organ rejection, patients with cancer who are taking immunosuppressive chemotherapy or autoimmune patients undergoing immunosuppressive therapy [2,3].

The currently major available agents to treat invasive fungal infections can be grouped into four main classes according to their mechanism of action: polyenes, azoles, allylamines, and echinocandins (Table 1) [4]. They all present drawbacks when it comes to spectrum of activity, drug–drug interactions, pharmacokinetics and pharmacodynamics, resistance mechanisms, and the toxicity of the compounds themselves. Furthermore, there are some limitations in terms of clinical efficacy and efficiency, mainly because of their physical-chemical properties, like their hydrophobic character that leads to a low solubility in water and also selectivity problems deriving from the similarities between fungi and human cells [3,5].

Table 1. Targets of each group of antifungals [6,7].

Class	Target (Mechanism of Action)		Antifungal
Azoles	Ergosterol (inhibition of lanosterol 14-α-demethylase)	Imidazoles	Miconazole
			Econazole
			Ketoconazole
			Clotrimazole
		Triazoles	Itraconazole
			Fluconazole
			Voriconazole
Allylamines	Ergosterol (inhibition of squalene epoxidase)		Terbinafine
			Naftifine
			Butenafine
Polyenes	Cell membrane (production of ROS)		Amphotericin B
	Ergosterol (inhibition of lanosterol 14-α-demethylase)		Nystatin
Echinocandines	Cell wall (block of β-1,3 glucan synthesis)		Caspofungin, Micafungin, Anidulafungin
Other antifungals	Chelation of polyvalent metal cations		Ciclopirox
	Microtubules (prevention of the formation of the mitotic spindle)		Griseofulvin
	Ergosterol (inhibition of D14 reductase and D7-D8 isomerase)		Amorolfine

Nevertheless, the design and development of new drug delivery systems or even new antifungals is an emerging need, owing to the following facts [8]:

- There are 20–40% mortality rates with invasive mycoses, therefore these figures need to be improved;
- The increase in patients undergoing prolonged antifungal therapies reflects the need to develop better fungicidal drugs and thus reduce the length of the treatments and the costs associated;
- There is still space for improvement in pharmacokinetics and pharmacodynamics, in order to reduce the frequency of drug use;
- More attention needs to be given to the host toxicities and drug–drug interactions of current therapy so that their effects can be eliminated or, at least, minimized;
- New therapy groups with different mechanisms of action are needed; this way, these new drugs may synergize with present ones and allow better responses;
- There is an alarming growth in antifungal resistance in all therapeutic groups [8].

Nanotechnology is an emerging field of science that has shown an undeniable versatility and has boosted a revolution when it comes to medical treatments, quicker diagnosis, cellular regeneration, and drug delivery [9,10]. The material to produce nanoparticles can be divided into three main groups: polymers, lipids, or metals, each one giving rise to a different type of nanoparticle [11]. The main representatives of each of these three different groups of nanoparticles are mentioned in Figure 1 below.

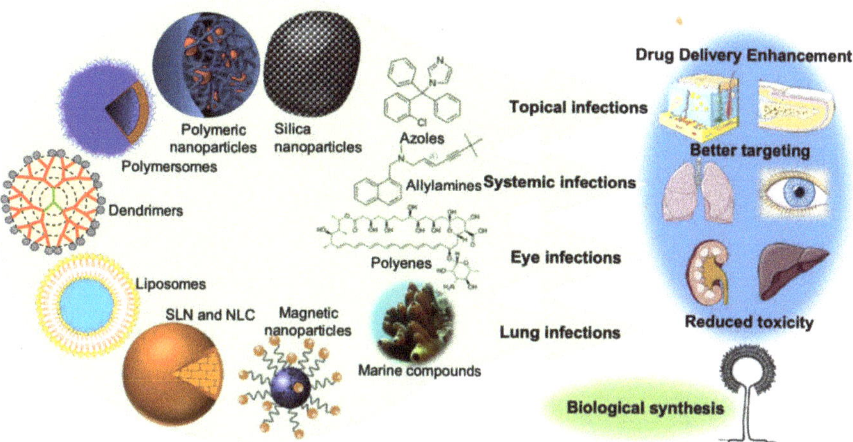

Figure 1. The new drug delivery systems based on nanotechnology that are currently being employed in order to enhance drug delivery, promote a better targeting, and reduce the toxicity of conventional antifungal drugs. It is also important to point out the importance of the production of nanoparticles by fungi (biological synthesis) and the undeniable potential of the sea as a source of new molecules with antifungal activity.

Nanoparticles have been employed in pharmaceutical formulations because of their ability to alter and improve the pharmacokinetic and pharmacodynamic properties of the drugs. This is given to their capability to increase the solubility and stability of the drugs, to allow a controlled release and to exhibit biocompatibility with tissues and cells, which is reflected in an overall improvement on therapeutic efficiency [11,12]. In addition, its subcellular size is compatible with an intravascular injection and its high surface area is amenable to modification so that the drug is released in a specific target, thus reducing the systemic adverse effects and increasing the therapeutic compliance, by decreasing the usual dose and the frequency of administration [13,14]. This targeted-specific action is possible since, at a nanomolecular level, it is possible to incorporate target ligands that allow a preferential binding of certain types of cells, by conjugation with antibodies and peptides on the surface of the transporters [15–17]. Hence, the development of new biopharmaceutical systems, especially nanoparticulate carriers, is a good strategy to improve the therapeutic efficacy, safety, and compliance of conventional antifungal drugs.

In Table 2 an overview of the new antifungal drug delivery systems is presented, and the drug chemical group, their route of administration, and their dosage form provided.

Table 2. Some of the novel drug delivery systems already developed for each antifungal drug.

Antifungal Drugs	Novel Drug Delivery Systems	Routes of Administration	Dosage Forms	References
Miconazole	Niosomes	Transdermal	Gel	[18]
	SLN	Oral	N.A.	[19]
		Topical	Gel	[20]
	Microemulsion	Topical	N.A.	[21]
	Liposomes	Topical	Gel	[22]
	Nanoemulsion	Topical	N.A.	[23]
	Nanosponges	Vaginal	Gel	[24]
	Transfersomes	Topical	Gel	[25]
Econazole	Microemulsion	Percutaneous	N.A.	[26]
		Topical	Gel	[27]
	SLN	Topical	Gel	[28]
	NLC	Topical	Gel	[29]
	Liposomes	Topical	Gel	[30]
	Ethosomes	Topical	Gel	[31]
	Transethosomes	Transdermal	Gel	[32]
	Nanosponges	Topical	Hydrogel	[33]
	Niosomes	Transdermal	Gel	[34]
	Polymeric micelles	Topical	N.A.	[35]
	Nanoemulsion	Topical	N.A.	[36]
Ketoconazole	SLN/NLC	Topical	Gel	[37]
	Niosomes	Topical	Gel	[38]
	Microemulsion	Oral	N.A.	[39]
	Spanlastics	Ocular	N.A.	[40]
	Dendrimers	Topical	Hydrogel	[41]
	Liposomes	Topical	N.A.	[42]
Clotrimazole	Liposomes	Topical	Gel	[43]
	Nanosponges	Topical	Hydrogel	[44]
	Ethosomes	Topical	Gel	[45]
	Niosomes	Topical	Gel	[46]
	Polymeric emulgel	Topical	Gel	[47]
	Polymeric micelles	Topical	N.A.	[35]
	SLN/NLC	Topical	N.A.	[48]
	Microemulsion	Buccal	Gel	[49]
		Vaginal	Gel	[50]
	Transfersomes	Transdermal/Topical	N.A.	[51]
Itraconazole	Transfersomes	Transdermal	N.A.	[52]
	SLN	Ocular	N.A.	[53]
	NLC	Inhalation	N.A.	[54]
	Niosomes	Topical	N.A.	[55]

Table 2. *Cont.*

Antifungal Drugs	Novel Drug Delivery Systems	Routes of Administration	Dosage Forms	References
	Microemulsion	Transdermal	N.A.	[56]
	Liposomes	Topical	N.A.	[57]
	Polymeric nanoparticles	Oral	N.A.	[58]
	Polymersome	Intravenous	N.A.	[54]
	Spanlastics	Ocular	N.A.	[59]
	Silica nanoparticles	Oral	N.A.	[60]
Fluconazole	Microemulsion	Vaginal	Gel	[61]
	Niosomes	Ocular	Gel	[62]
	Liposomes	Intravitral	N.A.	[63]
	SLN	Topical	Gel	[64]
	NLC	Oral	N.A.	[65]
	Microsponges	Topical	Gel	[66]
	Ethosomes	Topical	Gel	[67]
	Spanlastics	Ocular	N.A.	[68]
	Polymeric amphiphilogel	Topical	Gel	[69]
	Polymeric micelles	Topical	N.A.	[35]
Voriconazole	Microemulsion	Ocular	N.A.	[70]
	Polymeric nanoparticles	Ocular	N.A.	[71]
		Pulmonar	N.A.	[72]
	SLN	Topical	Gel	[73]
	Transethosome	Topical	N.A.	[74]
	Ethosome	Topical	N.A.	[75]
Terbinafine	Liposomes	Topical	Gel	[76]
	SLN	Topical	N.A.	[77]
	Transfersomes	Topical	N.A.	[78]
	Spanlastics	Transungual	N.A.	[79]
	Polymeric chitosan nanoparticles	Topical	Hydrogel	[80]
Naftifine	Microemulsion	Topical	N.A.	[81]
	Niosomes	Topical	Gel	[82]
Butenafine	Microemulsion	Topical	Hydrogel	[83]
Amphotericin B	Liposomes	Intravenous	N.A.	[84]
	SLN/NLC	Oral	N.A.	[85]
		Topical	N.A.	[86]
	Magnetic nanoparticles	Nasal instilation	N.A.	[87]
	Nanoemulsion	Topical	N.A.	[88]
	Polymeric nanoparticles	Intravenous	N.A.	[89]
		Oral	N.A.	[90]

Table 2. *Cont.*

Antifungal Drugs	Novel Drug Delivery Systems	Routes of Administration	Dosage Forms	References
	Polymersomes	Oral	N.A.	[91,92]
	Transfersomes	Topical	N.A.	[93]
	Micelles	Intravenous	N.A.	[94]
	Silica nanoparticles	Intravenous	N.A.	[95]
	SLN	Topical	N.A.	[96]
	Nanoemulsion	Topical	N.A.	[97]
Nystatin	Liposomes	Intravenous	N.A.	[98]
	Niosomes	Parenteral	N.A.	[99]
Griseofulvin	Niosomes	Oral	N.A.	[100]
Ciclopirox	Niosomes	Topical	Gel	[101]
Caspofungin, Micafungin, Anidulafungin, Amorolfine	No nano-tech studies yet released			

N.A.: the dosage form is not mentioned in the reference cited; SLN: Solid Lipid Nanoparticles; NLC: Nanostructured Lipid Carriers.

However, the efficacy and human safety of these new therapies remain uncertain in most of the articles found in literature. They generally lack controlled clinical trials and sometimes the suggested routes of administration are less practical, or the production cost may hinder the replacement of the conventional treatment. Nevertheless, in other cases, the opposite is verified, and some options have potential to become a viable first line treatment [102]. Moreover, given the widespread use of antifungal agents and the limited therapeutic offer, fungi have developed resistance mechanisms, like overexpression of efflux pump proteins and formation of biofilms. These mechanisms can mean not only a decrease in a drug's effective concentration, but also changes and subexpression of drug targets and metabolic bypass [6]. It is important to add that resistance is a cross-cutting issue to all of the currently available classes of antifungal agents, therefore overcoming antifungal resistance can be considered as the mainstay for improving therapeutic strategies to treat antifungal infections [2,103].

Despite the uprising of these issues in antifungal therapy, there are several mechanisms by which nanoparticles overcome the development of resistance mechanisms:

- The chemical features and simultaneous multiple mechanisms used by nitric oxide, chitosan, and metallic nanoparticles make the likelihood of resistance development unviable (for example, through the direct reaction of reactive nitrogen oxide intermediates with DNA structure) [104,105];
- The resistance mechanisms can be prevented by packaging multiple antimicrobial drugs within the same nanoparticle, because the likelihood of multiple simultaneous gene mutations in the same cell is low. The most striking examples are the encapsulation of antifungal drugs in chitosan or silver nanoparticles, combining the antifungal properties of both and decreasing the possibility of drug resistance [104,106];
- Some nanoparticles, such as liposomes and dendrimers, are able to overcome the resistance mechanisms of decreased uptake and increased efflux of drug from the microbial cell. Liposomes are able to quickly fuse with the plasma membrane of the microbial cell and release a high concentration of drug into its plasma membrane or cytoplasm, thereby circumventing the decreased uptake mechanism of resistance. This means a faster delivery and avoidance of the transmembrane pumps that catalyze increased efflux of drugs. Dendrimers, on the other hand, are extensively branched molecules, whose surface can be filled with positively charged quaternary ammonium compounds, which bind to negatively charged microbial cell envelopes and increase membrane permeability. This allows the entrance of more dendrimers to the

microbial cell, the flow of its cytoplasmic contents to the exterior, and the ultimate destruction of the microbial cell membrane. This goes to show that dendrimers are also able to surpass the resistance mechanism of decreased uptake of drug [107]. Other nanoparticles, specifically nitric oxide nanoparticles made of silica and zinc oxide nanoparticles are able to overcome biofilm formation by killing the microbes present in already formed biofilms or by inhibiting biofilm formation through the generation of reactive oxygen species, respectively [108,109];

- Nanoparticles have been used to target antifungal drugs to the specific site of infection, allowing the local release of high concentrations of drug, while keeping the total dose of drug administered low. This high local dose is able to destroy the infecting fungi before they can develop resistance, thereby overcoming this worrisome issue and translating into fewer side effects upon the patient [104].

That being said, it is also important that the research done, not only focuses on formulating these systems, but also in overcoming the major challenges that their placing on the market faces: the physical instability of nanoparticles, their small capacity of drug loading, the cytotoxicity/immunogenicity, and the high cost of production and standardization, given the complexity of the formulations. Besides that, there is almost a complete lack of studies in vivo as reaching the therapeutic range needed to perform these studies has proven to be an arduous job. That lies in the fact that, in many cases, there is an anticipated release of the drug, aggregation and precipitation of the nanoparticles, and the accumulation in non-target tissues.

2. Nanotechnology and Mycology

Mycology and Nanotechnology have created a bidirectional relationship throughout the years. This dynamic interface between mycology and nanotechnology led to the creation of the term "myconanotechnology" (Figure 2) [110]. Nanotechnology has proven to be an interesting strategy to increase the potency and efficiency of conventional antifungals, to enable a decrease in toxicity and cost, to avoid an anticipated degradation, to ameliorate the drug distribution, by increasing the circulation time and improving pharmacokinetics, and also to improve drug targeting, with promising in vitro and in vivo results [5]. Furthermore, many metallic nanoparticles have been used against human and plant pathogenic fungi in the light of their intrinsic antifungal activity and a wide spectrum of fungi are able to biosynthesize nanoelemental particles [110].

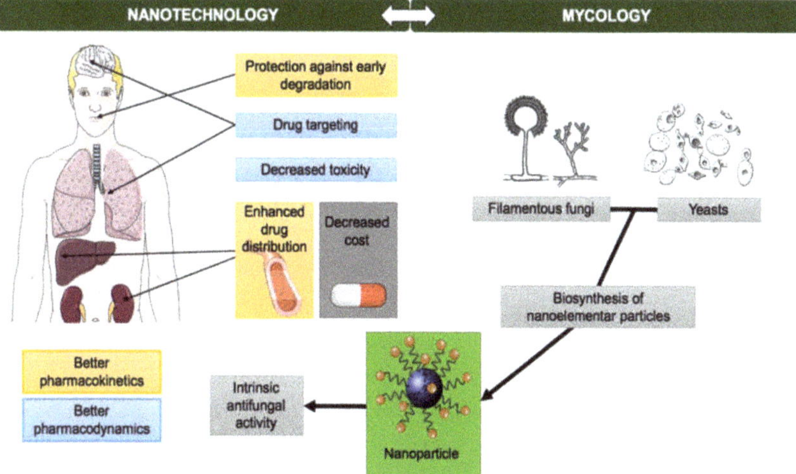

Figure 2. Bidirectional relationship of Nanoparticles and Mycology: nanotechnology has proven to be useful in improving antifungals pharmacokinetics and pharmacodynamics and many fungi have been used to biologically synthetize nanoparticles.

2.1. Antifungal Potential of Nanoparticles

Metallic nanoparticles have been used to eliminate fungi that are pathogenic to Man and to plants, because of their intrinsic antimicrobial activity [110]. The exact mechanisms this activity occurs through are only hypothesized and can be explained through three main pathways: (1) direct uptake of nanoparticles, (2) indirect activity of nanoparticles by production of reactive oxygen species (ROS), (3) impairment of cell wall/membrane through accumulation [111]. It is highly probable that it is the combination of these multiple pathways that is responsible for antimicrobial activity [112].

Nanoparticles undergo dissolution processes thanks to their electrochemical potential [113]. This leads to their separation into ions within the microbial fluid or in the culture medium. These ions also accumulate in the interior or exterior causing an inhibitory answer against microtubules. The accumulation of nanoparticles outside the microtubules causes the formation of layers that block cellular respiratory chain and destroy the microtubules [111].

The electrical charge of the nanoparticle is vital for the interaction that occurs between it and the carried drug. The electrostatic mechanism justifies why the antimicrobial activity was firstly described in silver nanoparticles. It is widely accepted that the positive charge of the silver ion is crucial for the antimicrobial activity of these nanoparticles through electrostatic attraction between the negatively charged cellular membrane of microorganisms and the positively charged membrane of nanoparticles [111]. Ag^+ has high affinity to thiol groups in cysteine of respiratory chain enzymes, therefore it uncouples the synthesis of adenosine triphosphate (ATP). Ag^+ also binds to proteins of transport from the respiratory chain, causing the leak out of protons and thus the collapse of the proton motive force. Furthermore, Ag^+ obstructs the uptake of phosphate and so promotes the efflux of intracellular phosphate [113].

Silver nanoparticles exhibit potent antifungal activity against clinical isolates and ATCC strains of *Trichophyton mentagrophytes* with concentrations of 1–7 μg/mL and a MIC (minimum inhibitory concentration) of 25 μg/mL against *Candida albicans* [114]. Silver nanoparticles also reveal good antifungal activity against *Aspergillus niger*, by inhibiting spore germination and preventing biofilm formation; when combined with simvastatin, there is an additive and synergistic effect that increases the antifungal effect, perhaps because simvastatin, as an ergosterol synthesis inhibitor (see Table 1), disrupts fungal cell membrane, which allows the entry of the nanoparticles [115].

Metals present in nanoparticles can act as catalysts, reacting with biomolecules thanks to their high specific surface area, inducing the direct production of free radicals when exposed to the acidic environment of lysosomes or interacting with oxidative organelles [112,113].

ROS, like superoxide anions, hydroxyl radicals, and hydrogen peroxide, are oxygen-derived by-products formed when a material is exposed to an oxygenated environment, allowing their interaction with biomolecules. This way, they can cause an imbalance between the production of reactive species and the capacity of the biological system to detoxify reactive intermediates or repair damage [111]. Although the antioxidant cell defense prevents the effects of ROS to some extent, excessive ROS production may cause oxidative stress and lipid peroxidation, leading to membrane impairment, mitochondrial dysfunction, and DNA damage. That being a toxic mechanism to human cells, cytotoxicity tests should be performed on nanoparticles that owe their antimicrobial effects to ROS production, therefore avoiding interactions and toxic reactions in human beings [111].

Chitosan and its chemical derivates have been used as building blocks for drug delivery nanoformulations in light of their biocompatibility, biodegradability, and mucoadhesivity, presenting some advantages such as in situ gelling performance, mucoadhesive properties, and ability to prolong the release of low-molecular-weight compounds to macromolecular drugs [116]. Chitosan nanoparticles have proven to exhibit great antimicrobial activity against Candida infections [1]. According to the literature and research already undertaken, this antimicrobial activity is attributed to positively charged amino groups that react with negatively charged groups of lipopolysaccharides and proteins on the surface of the microbial cells, resulting in disintegration of the cell membrane. By doing this, the nanoparticles are able to bind with DNA molecules and inhibit mRNA and protein synthesis.

In the specific case of fungi, chitosan acts by inhibiting the sporulation and germination of spores, by interfering with the activity of the growth-promoting enzymes [1,117].

Zinc oxide nanoparticles (ZnONPs) have also proved antifungal activity against dermatophyte infections and other pathogenic fungi, such as Candida and Aspergillus [1]. Meanwhile, the synergistic antifungal activity of ZnONPs was evaluated along with common antifungal drugs, which revealed that their inhibitory efficiency can be increased in combination with ZnONPs, which could possibly reduce the overuse of these drugs, decrease their toxicity, and increase their antifungal activity [118]. In addition, these nanoparticles might be an interesting and promising alternative to conventional preservatives in cosmetics in the future [119].

In addition to these nanoparticles mentioned above, dendrimers also exhibit antifungal activity and provide the opportunity for complex therapy in which dendrimers are both the drug carrier and the adjunctive component of the dosage form [41].

2.2. Synthesis of Nanoparticles by Fungi

Thanks to their tolerance and ability to bioaccumulate metals, fungi now occupy a central role in the biological production of metallic nanoparticles (Table 3) [10]. They can not only be used to produce the nanomaterials that will make up the nanosystem coating, but also be, themselves, carried by the nanosystem. This way, new ecological processes are developed, which imply a reduced waste of solvents and chemical substances. The biosynthesis methods are simpler and allow size and shape control of nanoparticles. Besides fungi, other organisms are used to synthesize nanoparticles, for instance, bacteria, plants, or plant extracts [110]. Compared to bacteria, fungi produce higher quantity of enzymes, which is translated into a higher yield in the nanoparticles production. Furthermore, their growth, both in laboratory and at an industrial scale, is easier to control [110].

Table 3. Some examples of metallic nanoparticles produced by fungi and their method of synthesis [10,120].

Fungal Species	Nanoparticles Type	Method of Synthesis
Phoma sp.	Silver	Extracellular
Fusarium oxysporum	Gold; Magnetite	Extracellular
Verticillium sp.	Silver	Intracellular
Aspergillus fumigatus	Silver	Extracellular
Aspergillus niger	Silver	Extracellular
Fusarium semitectum	Silver	Extracellular
Trichoderma asperellum	Silver	Extracellular
Phaenerochaete chrysosporium	Silver	Extracellular

The biological synthesis can either be intracellular or extracellular according to the nanoparticles' locations. In intracellular synthesis, the ions are transported to the inner part of microbial cells to form nanoparticles in the presence of an enzyme. The nanoparticles formed inside the organism are of shorter size, when compared to the extracellular ones, because there is nucleation of particles inside the organisms. Extracellular synthesis has more applications than intracellular, as there is no need to join cellular components from the cell. The majority of the fungi produce nanoparticles extracellularly as a result of their secretory components that participate in nanoparticles reduction and capping [10].

Both yeasts and filamentous fungi can be used to synthetize nanoparticles. The process of synthetizing filamentous fungi-mediated nanoparticles is easy and cost-effective, since the mycelia in the biomass has high surface area and high intracellular metal absorption. Moreover, the fungi cell wall has many functional groups that facilitate the absorption to the metals [110].

Silver nanoparticles are the most fundamental among metallic nanoparticles that are involved in biomedical applications, specially cancer diagnosis and antimicrobial therapy [114]. One of the most remarkable examples is the extracellular synthesis of silver nanoparticles from filamentous fungi, like *Fusarium solani*, a pathogenic fungus isolated from infected onions. Ingle et al. [121] believe that, in these cases, the fungus does the "phagocytosis" of the nanoparticles, depositing them in their cellular wall, by binding to their functional groups. Afterwards, the fungus carries the nanoparticles and excretes them through exocytosis. The silver nanoparticles obtained were quite stable in solution because fungus-secreted proteins capped the nanoparticles. The authors argued that the procedure with this fungus could work for other metal nanomaterials such as gold and platinum with countless applications in the medical field [121].

Afterwards, many other scientists were able to optimize the biological production of silver nanoparticles using other fungi, such as *Fusarium oxysporum* [122], *Cochliobolus lunatus*, *Beauveria bassiana* [123], *Bipolaris maydis* [124], and many others. It was also concluded that some of these biologically synthesized silver nanoparticles showed enhanced antifungal activity with fluconazole against *Phoma glomerata*, *Phoma herbarum*, *Fusarium semitectum*, *Trichoderma* sp., *Phoma glomerata*, *Phoma herbarum*, and *Candida albicans* [114]. Surprisingly, it was additionally observed that the potential of silver nanoparticles is much wider than the inhibition of human and plant pathogenic fungi, as they also inhibit indoor fungal species such as *Penicillium brevicompactum*, *Aspergillus fumigatus*, *Cladosporium cladosporoides*, *Chaetomium globosum*, and *Stachybotrys chartarum* [125].

Likewise, yeasts can be useful for nanoparticle synthesis, as they produce enzymes responsible for the reduction of metallic salts and their conversion into elementary nanoparticles. Some examples are the biosynthesis of cadmium nanoparticles by *Candida glabrata* and *Schizosaccharomyces pombe* [110] and of selenium and silver nanocompounds by *Saccharomyces cerevisiae* in aerobic conditions [126,127].

2.3. Antifungal Drug Administration

The administration of antifungal drugs is not restricted to oral or parenteral routes. Other routes such as transungual, pulmonary, and ocular have also acquired great importance in the treatment of certain fungal infections. The development of a drug delivery system for attaining therapeutic concentration at these target organs is a challenge that requires a comprehensive understanding of the dynamics and specific features of the nails, lungs, and eyes, respectively.

2.3.1. The Transungual Route

This special topical route acquires singular importance in antifungal therapy, mainly when onychomycosis (*Tinea ungium*) caused by dermatophytes (such as *Trichophyton rubrum*) or yeasts are concerned [128]. These infections affect the nail plate and/or the nail bed and are very frequent [129].

Drug delivery through nails has its own challenges and the use of nail penetration enhancers is compulsory in formulations. The nail plate is composed by cross linked keratin linkages, an extensive bonding network responsible for its rigidity. Although there has been considerable research regarding new approaches for transungual drug delivery, topical permeability was limited by its barrier properties. Therefore, the lookout for novel approaches is important to enhance treatment efficacy and reduce treatment time and relapse rate [128]. There are still no nanotech solutions for this purpose, but medicated nail lacquers loaded with ciclopirox or amorolfine are the most feasible delivery vehicles [130]. The main hurdle in the development of nail lacquers for nail disorders is delivering the therapeutically effective concentrations to the site of infection, which is often under the nail [129].

2.3.2. Pulmonary Delivery

Drug delivery to the lungs is challenging because the availability of therapeutic quantities of antifungal drug at the site of infection is often inadequate on account of high blood flow turnover in the tissue. This can be problematic since a local and prolonged drug release is desirable when targeting this organ.

Developing a nanoparticle-based system and delivering it to the lungs in case of a pulmonary infection may help retain the antifungal in lungs over a prolonged period and may reduce the toxicity compared to the parenteral route [72,131].

A study in which voriconazole was encapsulated within a PLGA nanoparticle revealed that this is a better option to deliver the drug in deep lung tissue at a high concentration for a prolonged period of time and assumedly to provide greater antifungal effect in fungal infections. This was concluded because these nanoparticles were retained for a prolonged time in lungs and showed higher biodistribution when compared to voriconazole alone [72].

2.3.3. The Ocular Route

The eye is a particularly puzzling organ for drug delivery systems. Physical barriers that hinder drug access into the eye limit the drug bioavailability. In addition, physiological processes like blinking and tear drainage reduce the residence time of ocular drug delivery systems and drastically reduce the amount of trans-corneal drug absorption [131,132].

Fungal ocular infections are much less prevalent when compared to bacteria or virus. However, fungal keratitis is considered the second cause of blindness in developing countries and is an important cause of morbidity [133,134]. People with prolonged use of corticosteroids or antibiotics, with diabetes mellitus, with corneal trauma or surgery, and even people wearing contact lenses are at a higher risk of having fungal keratitis [134].

New formulations such as polymer-based nanoparticles, liposomes, dendrimers, SLNs, spanlastics, and niosomes were developed to enhance drug bioavailability to the eye and to minimize antifungal adverse effects [135]. They also have the ability to overcome the disadvantages of conventional eye drops, like short residence time, by prolonging the contact time at the corneal surface and achieving a sustained release of the drug. In addition, the encapsulation into these systems protects antifungals from degradation promoted by the metabolic enzymes on conjunctival and corneal surfaces and in tear fluids [134].

A study assessed the efficacy of a fluconazole liposomal formulation for candidal keratitis that aimed at prolonging the antifungal action by increasing the contact time. The authors reported 86.4% healing observed in rabbits treated with fluconazole encapsulated liposomes, presumably because of higher viscosity and lipid solubility of fluconazole-loaded liposomes [136].

Furthermore, a fluconazole-loaded niosomal gel can also be successfully used as a topical drug delivery system for corneal fungal infections. Ethosomes are known to improve both transcorneal permeability and ocular bioavailability of poorly water-soluble drugs, which is a tremendous improvement in topical ocular drug delivery [137].

Spanlastics also enhance permeability and bioavailability of antifungal drugs and their nano-size makes them able to reach the posterior segment of the eye, when eye drops are used [138]. For instance, an itraconazole-loaded spanlastic proved to be safe and non-irritant to the eyes; moreover, the elasticity of these vesicles serves as a drug deliver for both anterior and posterior eye diseases [59].

2.4. An Overview of Nanoparticle Types and Their Applicability on Antifungal Therapy

2.4.1. Lipid Nanoparticles

Albeit the safety and tolerability of systemic antifungal therapy has improved considerably, a rising proportion of immunocompromised patients are receiving systemic antifungal agents for progressively longer treatment courses and that increases the probability of longer-term risks, drug interaction, and other toxicity-related reactions in different organs [139].

Ambisome® was the first successful example of a nanotech antifungal drug and it was manufactured by the Nexatar Company USA in 1990 [84,140]. In this formulation, amphotericin B was incorporated in an unilamellar liposomal bilayer of approximately 45–80 nm. Compared to the conventional formulation, this one showed less toxicity and prolonged circulation time, which suggests a higher distribution rate [84]. Even though Ambisome® circumvented the initial toxicity issues, its daily-basis usage is still limited by its cost [5]. Thus, nanosomal formulations of amphotericin B were latter developed, with a special focus on lipid nanoparticles for intravenous delivery, which showed lower cytotoxicity against human kidney cells and much less hematotoxicity compared to the ones already marketed [141].

Besides amphotericin B, some other drugs have also been incorporated into liposomes with promising results. One good example of that is nystatin, a drug used mainly topically because of its poor oral bioavailability, thus excluded for the treatment of systemic fungal infections [134]. A daily injection of liposomal nystatin, however, showed great efficacy in the treatment of invasive aspergillosis, although with some mild renal toxicity and some infusion-related events [98]. Additionally, a multilamellar liposome gel system containing clotrimazole revealed to be useful in the treatment of vaginal candidiasis, providing a sustained release in the vaginal fluid and a reduced dosing interval [43].

Econazole is a topical antifungal that may be an irritant to the skin when conventional vehicles are employed. Applying a liposomal formulation has revealed to be a good strategy to minimize the irritating potential and to increase the compliance. Bioavailability studies of the liposomal gel revealed a seven-fold increase in the drug concentration in the epidermis, when compared to a control. Thus, the authors proved that it is possible to reduce the amount of drug applied and keep the therapeutic efficacy, minimizing the cutaneous irritation [142].

Transfersomes are generally regarded as an upgrade of liposomes, because they overcome their poor penetration through the stratum corneum, by modifications in the bilayer composition [134]. They are elastic nanovesicles made up from phospholipids and edge activators, most often surfactants or hydrophilic detergents with high mobility.

These particles are essentially studied as carriers for dermal and transdermal delivery, but have also shown to be effective carriers for genetic material and vaccines [143]. Ultradeformable liposomes or transfersomes were tested in a topical delivery system for amphotericin B and were found to enhance the drug penetration towards deep skin layers in a scale 40 times higher than Ambisome®, making this system clinically significant [93].

Ethosomes are essentially a type of transfersome that employs ethanol instead of the edge activator molecules; the main difference lies in the fact that ethanol evaporates once applied on the skin, whereas the edge activator molecules remain on the skin surface after water evaporation from the formulations [144].

Ethanol also plays an essential role in enhancing the delivery of both hydrophilic and lipophilic drugs without impairing their deformability and elastic characteristics, which was a significant disadvantage of transfersomes [143]. Henceforth ethosomes show great potential as topical delivery systems for antifungal drugs [134].

The formulation of an ethosomal gel containing econazole nitrate pointed out the outstanding potential of this system to function as a topical delivery system, since it allowed controlled drug release, increased antifungal activity, and good stability after storage [31].

Fluconazole is used both for local and systemic fungal infections. Despite its worldwide usage, it shows low penetration rate because of its high solubility and low permeability. The preparation of an ethosomal gel for topical delivery of fluconazole has proved to be an efficient way to overcome the issues associated with bioavailability, degradation, stability, and side effects of this therapeutic agent [67].

Although the great potential of ethosomes, when the concentration of ethanol used is above 30–40% w/w, the vesicular membrane tends to become more permeable and leaky, leaking out the entrapped drug, specially hydrophilic/ionized drugs [45,145]. Furthermore, this high ethanolic content may affect the skin causing irritation or contact dermatitis. To solve that, Cavamax W7, a permeation enhancer, was developed so as to improve topical delivery and to reduce the amount of ethanol used therefore reducing the risk of adverse effects [45]. The Cavamax W7 ethosomes were able to reach the last layer of epidermis (stratum basale) which turned them into a very valuable tool in the treatment of deep-seated fungal infections [134]. A clotrimazole encapsulated Cavamax W7 composite ethosome gel presented a more stable and more efficient vesicular system than the conventional ethosomal formulation along with a better antifungal activity against *Candida albicans* and *Aspergillus niger* [45].

Transethosomes are lipid vesicles with irregular shapes that represent a combination of the concepts applied to transfersomes and ethosomes, having a high content of ethanol (up to 30%) along with an edge activator molecule. This combination causes a rearrangement in the lipid bilayer of transethosomes, allowing them to have both higher deformability and skin permeation/penetration than the other deformable molecules. The formulation of these vesicles also presents the advantage of being quite easy to scale up, which constitutes an enormous advantage for industrial purposes [144]. Contrary to other deformable vesicles, transethosomes improve skin delivery of drugs both under occlusive and non-occlusive condition [146]. For these reasons, they are considered the most promising lipid nanoparticles in antifungal therapy.

An econazole nitrate loaded transethosome gel for transdermal delivery has been developed. Comparing it with the marketed cream of econazole nitrate, the authors found out that the former presented less ex vivo penetration, higher ex vivo skin retention, and higher in vitro antifungal activity [32]. Transethosomes containing terbinafine, amphotericin B, ketoconazole also disclosed enhanced permeation [146]. In addition, voriconazole transethosomes significantly enhanced skin permeation and deposition of the drug when compared to conventional liposomes, transfersomes, and ethosomes [74].

Regardless of the fact that transethosomes represent the state-of-the-art in ultradeformable vesicles, they still have some drawbacks that are important to mention such as the solubility that the drugs need to fulfil in both lipophilic and aqueous environment so that they can reach the dermal microcirculation and access the systemic circulation. In addition, the molecular size of the drug needs to be reasonable to make the percutaneous absorption possible (40–200 nm) [146].

Solid Lipid Nanoparticles (SLN) and Nanostructured Lipid Carriers (NLC) have become very attractive options to the development of drugs for cutaneous application and as an alternative to overcome the drawbacks of liposomes. They provide an occlusive effect by virtue of their ability to form a surface film that reduces the transepidermal water loss, which can reduce the atopic eczema symptoms and improve the skin appearance [147]. Besides that, they exhibit an excellent tolerability profile and its reduced size promotes a closer contact with the stratum corneum, intensifying the skin penetration of drugs. These particles are also able to increase the chemical stability of compounds sensitive to light, oxidation, and hydrolysis [148].

One application of SLN and NLC is the incorporation of azoles to treat cutaneous fungal infections. The most common azoles (clotrimazole, miconazole, econazole) are extremely water insoluble, being very difficult to administrate and to release those drugs in infection sites. However, compounds with lipophilic character can be efficiently encapsulated in SLNs [107].

In 2010, a formulation was developed in which the SLN nanoparticles were dispersed in a hydrogel to carry miconazole nitrate (MN), an azole frequently used in cutaneous fungal infections [149]. The studies indicated that this MN-loaded SLN-bearing hydrogel formulation was less prone to cause cutaneous irritation in comparison to the MN hydrogel and MN suspension formulation, showing an even greater efficacy in animal models (male albino rabbits). Furthermore, SLN-bearing hydrogel provided sustained release of miconazole (slow initial release with a lag time of 1 h and a sustained drug release over a 24 h period) with greater drug deposition on the skin. This happens because gels help disperse the matrix carriers evenly, increasing the contact time and the deposition of carriers on the skin, resulting in a higher cutaneous penetration of the drug [16,149].

2.4.2. Polymeric Nanoparticles

Polymeric nanoparticles are produced either from natural or synthetic polymers, in which the drug is dissolved, trapped, or bound to the nanoparticle matrix. Depending upon the composition of the organic phase and the preparation method, nanocapsules (matrix-like structure) or nanospheres (core-shell morphology) can be obtained [110,150].

These nanoparticles are capable of carrying proteins, DNA, and drugs, such as antifungals, for cells and specific target organs, which increases the safety profile. Their nanometric size promotes an effective permeation through the cellular membranes and increases their stability in blood flow.

It is also important to point out the gelling systems that can be used to formulate polymeric carriers: microsponges and nanosponges, amphiphilic gels and emulgels or gellified emulsions, which play an important role in cutaneous drug delivery [143].

Microsponge and nanosponge systems are polymer-based spheres that can encapsulate or suspend many substances and then be incorporated into a dosage form (hydrogel) or be used for oral delivery [44,66,151].

A polymeric microsponge based gel system was successfully produced for the topical delivery of fluconazole, with especially great entrapment efficiency, production yield, and extended drug release, which allows a reduction in application frequency, a feature with great importance in recidivist and prolonged fungal infections [66].

A nanosponge based gel formulation loaded with clotrimazole showed controlled release with reduced side effects, indicating the safe and effective profile of these colloidal carriers for topical use [46]. An econazole nanosponge hydrogel was developed and was found to solubilize poorly soluble drugs and also had the ability of forming a local depot for sustained drug release [33].

Amphiphilic gels are semisolid systems of nonionic nature that can be used as topical/transdermal carriers without promoting irritation of the skin. They aim at delivering the antifungal in a level within the therapeutic window for as long as possible and to avoid fluctuations in plasma drug level [69].

An amphiphilogel of fluconazole was successfully formulated for topical application, presenting overall stability and cumulative drug release within the range expected [69].

Emulgels have dual release control system (emulsion/microemulsion and gel), which increases the overall stability of the antifungal formulations [143]. In comparison to conventional creams or ointments, they exhibit better application properties, faster, and more complete drug release profile and they lack greasiness and residues upon application [47].

Emulgels are suitable solutions to incorporate hydrophobic drugs in water soluble gel bases, for instance the incorporation of clotrimazole in an emulgel formulation prepared using either Carbopol 934 or HMPC 2910 showed good physical properties, stability, and antifungal activity [47].

A formulation with amphotericin B encapsulated in poly(lactic-co-glycolic acid) PLGA in association with ε-caprolactone was developed and proved the potential of polymeric nanoparticles in preventing cytotoxicity, yet preserving the fungicidal efficacy. With this polymeric combination, the authors described an amphotericin B encapsulation efficacy of 84% and a fungicidal effect for *Candida albicans* equal to the free drug, with the advantage of presenting less toxicity and less mortality associated [91].

A nanoformulation of the same drug in PLGA was also developed but, this time, it was functionalized with dimercaptosuccinic acid (DMSA). This acid exhibits tropism to lungs, being appropriate to be included in a formulation in which it is desired a release at this site. In this study an intraperitoneal administration was also tested and verified that the therapeutic effect on paracoccidioidomycosis (PCM) was equivalent to the free drug. The greatest advantage of this formulation was that it only required to be taken every three days, given the slow release of amphotericin B from the nanoparticles, and did not require a daily administration as the currently commercialized formulation [152].

Chitosan is a very versatile polymer owing to the ease presented by their groups to be modified or deacetylated [153]. Some studies have shown that this polymer also possesses antioxidant, antimicrobial, and anti-inflammatory properties, becoming very attractive from a biopharmaceutical viewpoint [154]. Some other studies pointed out that chitosan nanoparticles themselves have a huge potential to become safe and effective antifungal agents [155].

Chitosan nanoparticles containing amphotericin B were initially developed with the aim of avoiding the drug gastrointestinal degradation, increasing the stability and the bioavailability in target organs like the lung, the liver, and the spleen, whilst decreasing renal exposition. The researchers determined that when administered by an oral route, these nanoparticles were effective in treating visceral leishmaniosis, aspergillosis, and candidiasis, showing an efficacy comparable to the parenteral

Ambisome® formulation. Moreover, this natural polysaccharide was also employed in the cutaneous release of amphotericin B in wounds caused by infected burns with *Candida albicans*, promoting better tissue healing and increased antifungal activity, when compared to the conventional formulation [156].

Polymeric micelles present a unique architecture where the hydrophobic core can incorporate hydrophobic drugs, such as antifungals, which leads to a very significant improvement in their aqueous solubility [134]. These systems have attracted attention in drug delivery because their critical micelle concentration (CMC) is several 1000-fold lower (<10 mg/L) than classic micelles, this means that ionic surfactants have greater ease in self-assembling in water to form micelles [157].

Amphotericin B has been incorporated in polymeric micelles to treat brain fungal infections. The system was intended to improve drug solubility and permeability through biological membranes as well as obviate issues like high toxicity and low efficacy against *Candida meningoencephalitis* [94].

Polymersomes are spherical structures composed of an aqueous core surrounded by a polymeric bilayer membrane, being viewed as synthetic analogues to liposomes. They are very versatile structures, which spontaneously self-assemble from amphiphilic diblock copolymers. This capacity increases drug efficacy and enables them to encapsulate both hydrophilic and hydrophobic drugs, including antifungals [158,159]. Furthermore, polymersomes present stimuli-responsive drug release, which means that their physical and chemical properties are mutable in response to certain features (pH, temperature, redox conditions, light, magnetic field, ionic strength, or even concentration). This ability is promising in drug-controlled release, an important feature in antifungal drug therapy [159,160]. In Figure 3 the process by which a polymersome is formed is represented as well as its ability to carry different biological molecules.

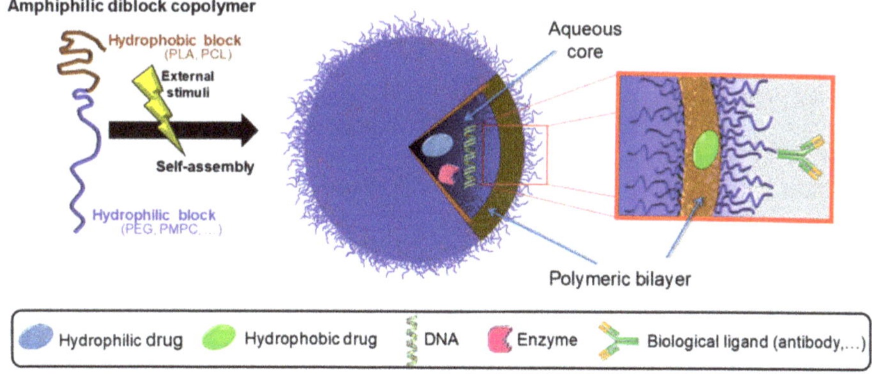

Figure 3. Schematic representation of the formation of a polymersome and its versatile properties. Polymersomes are generally self-assembled from block copolymers, presenting a unique structure that is able to encapsulate different biological molecules.

A research group developed an amphotericin B loaded polymersome by solvent injection method, using (PEG)$_3$-PLA as co-polymer. This formulation was compared with two marketed formulations (Fungizone® and Ambisome®) in terms of release, molecular organization of amphotericin B, and hemolysis. The results were similar to the marketed formulations, which indicated the potential for further in vivo development [92].

Dendrimers are homogenous polymeric tridimensional nanoarchitectures characterized by a highly-branched and symmetrical structure consisting of a central core (a single atom or a group of atoms), building blocks of repeating units emanating from the core (generations), and a high density of water-soluble functional groups on the surface (terminal group) [161–163]. The elements are added through sequential interactive chemical reactions to the central core, the spaces within the voids facilitate the encapsulation of active substances within the dendritic structure, and the terminal functional groups dictate the efficacy of nucleic acid complexation or drug entrapment [164]. Their nano and uniform size structure (2–10 nm in diameter), high degree of branching, water solubility, multivalence, well-defined molecular weight, and available internal cavities makes them extremely valuable as drug delivery systems and as carrier systems for antifungal agents [161,165].

Albeit different types of dendrimers have been described, the most frequently used for antifungal therapy are polyamidoamine (PAMAM) and polypropylene imine (PPI) dendrimers [163]. PAMAM dendrimers containing ketoconazole were shown to improve the solubility and the in vitro release of the drug and also to enhance the antifungal activity of ketoconazole [41].

2.4.3. Metallic Nanoparticles

Three main groups of metallic nanoparticles, gold, silver, and magnetic, can be used for the vectorization of antifungal drugs.

Metallic nanoparticles can be synthetized in three different ways: chemically, physically, and biologically. Chemical synthesis has been associated with many side effects related with the absorption of toxic chemical particles on the surface of nanoparticles [10], thereupon the biological synthesis is acquiring some expression [166].

Gold nanoparticles are used in immunochemical studies to identify protein interactions and in DNA fingerprinting to detect DNA in a sample. They are also used to detect aminoglycosides like streptomycin, gentamicin, and neomycin [10].

Silver nanoparticles are the most efficient on account of their antimicrobial efficacy against bacteria, viruses, and other eukaryotic microorganisms. In fact, they are the most used materials, being applied as antimicrobial agents in textile industry, for water treatment, in solar protectors, etc. [10].

Albeit nanostructures of iron, cobalt, and nickel exhibit superparamagnetic properties and high magnetic susceptibility, superparamagnetic iron oxide nanoparticles such as magnetite (Fe_3O_4), hematite (α-Fe_2O_3), and maghemite (γ-Fe_2O_3), are the most studied types of magnetic nanoparticles. This type of nanoparticle has received special attention by virtue of their capacity to be influenced by magnetic fields, therefore being easily directed and released in a specific site of the organism [10,167].

Superparamagnetic iron oxide nanoparticles present other unique properties, for instance: low toxicity, biocompatibility, potent magnetic targeting capacity, and chemical inertia, thus, they have many biomedical applications, for example in the cancer research field, steam cells, tissue repair, drug release, genetic therapy, DNA analysis, and clinical diagnosis through magnetic resonance [10,167].

Nowadays, nanoparticles of superparamagnetic iron oxide are the ones that answer more efficiently to an external magnetic field, being the ones with more potential to become drug carriers [5]. In order to bypass cytotoxic effects, these Fe_3O_4 magnetic nanoparticles need to be coated, usually with DMSA (meso-2,3-dimercaptosuccinic acid); this is a procedure that is important not only to increase cell internalization and biocompatibility, but also to carry active molecules to the nanoparticle's surface, essential to drug delivery [168].

2.4.4. Other Drug Delivery Systems

Niosomes are non-phospholipid vesicles made of non-ionic surfactants that serve as drug depots in the body as they release the drug in a sustained fashion through its bilayers. They are also able to improve oral bioavailability of poorly soluble drugs, enhance the skin permeability of topical drugs, protect the enclosed active substance from deleterious factors, and increase the stability of the entrapped drug [169].

Nistatin was successfully encapsulated in niosomes and a safe and effective formula for parenteral administration was obtained. This formulation provided reduced nephrotoxicity and hepatoxicity in female Wistar rats and showed pronounced efficacy against *Candida albicans* with a higher level of drug in vital organs [99]. The encapsulation of ciclopirox in niosomes increased the half-life, promoted a prolonged drug release, and minimized the side effects [103]. Given the poor water solubility of griseofulvin and its slow absorption from oral route, a niosome was produced in order to enhance bioavailability of this active substance, to accelerate its absorption and to obtain a sustained release, acting as a depot inside the body [100].

Ketoconazole niosomal gel showed more prolonged action than conventional ketoconazole formulations, hence it can be developed in order to improve the antifungal activity [38].

Spanlastics are termed as modified niosomes which present better permeability because they have edge activators in their composition, like transfersomes and transethosomes [134,145].

These systems show great flexibility allowing them to pass through fenestrations smaller than their own radius in order to enter the cell, simultaneously minimizing the possibility of damaging the vesicles while squashing [138].

Firstly, there were ketoconazole-loaded spanlastics for ocular drug delivery [40] and then spanlastics were loaded with terbinafine hydrochloride for treatment of onychomycosis [79].

Microemulsions are colloidal carriers that consist of a liquid dispersion of oil and water stabilized by an interfacial film of surfactant. Due to this, they are able to incorporate drugs of different lipophilicity [170]. They are promising colloidal carriers because of their transparency, ease of preparation, and long-term stability [171].

When it comes to drug delivery through stratum corneum, microemulsions are very resourceful considering the ability of the oils and surfactants included in their composition to act as skin penetration enhancers [170,172]. Furthermore, they are also good candidates for oral delivery of poorly water-soluble drugs as they can improve their solubilization.

Terbinafine hydrochloride is an example of a slightly water-soluble drug, which presented higher solubility, an increase in the dissolution rate, and better efficacy when incorporated within a microemulsion [170,171]. Likewise, a microemulsion system of voriconazole also showed a significant increase in the antifungal activity against *Candida albicans*, along with better drug skin penetration [173].

Nanoemulsions are thermodynamically and kinetically more stable than emulsions, so they have been evaluated as colloidal carriers to improve the efficacy and tolerability of some antifungal drugs. The capacity of nanoemulsions to dissolve large quantities of drugs with low solubility, their compatibility, their ability to protect drugs from enzymatic degradation and hydrolysis as well as their capacity to penetrate the deeper skin layers, turns them into ideal drug delivery vectors [143,174].

A nystatin nanoemulsion for topical application was developed and was found to have a higher antifungal effect than nystatin itself, representing a therapeutic improvement [97].

Silicon dioxide nanoparticles have been studied as drug carriers to enhance the antimicrobial activity of some drugs, because of their biodegradability, low toxicity, capacity to stimulate macrophages, and the ease by which they are synthetized and modified. The foremost advantage of these nanoparticles is that they can be loaded with large amounts of drug [95,175].

There are four main types of silica nanoparticles, but their antifungal potential is yet to be studied in some cases:

- Nitric oxide-silica nanoparticles with proven anti-biofilm activity [108];
- Metal modified silica nanoparticles, which can include silver or copper, metals that have a very well documented antimicrobial effect, derived from the cell membrane and DNA damages, interaction with enzymes from thiol groups or are associated with generating hydrogen peroxide [176];
- Surface-modified silica nanoparticles by quaternary ammonium compounds loaded with antifungal agents [177,178];
- Bioglasses and bioceramics [179].

Amongst them, mesoporous silica nanoparticles are the ones who have become the most promising candidates for many biomedical applications, specifically for antifungal therapy, because of their uniformed mesoporous tunnels and narrow pore size distribution. Besides that, they present outstanding biocompatibility and chemical stability and can be degraded and metabolized over a relatively short term. These aspects allow high drug loading and reduce the probability of particle-induced toxicity [95,180].

In a study, one group of mice with candidiasis induced by *Candida albicans* was treated with silica nanoparticles functionalized with amphotericin B and the other group was treated with officinal amphotericin B. It was possible to observe that the first group showed a significant increase in survival rates [95]. This confirms the fact that silica nanoparticles coated with quaternary ammonium surfactants have higher fungicidal and fungistatic effect against *Candida albicans* than colloidal silver [177,178].

In order to increase the oral bioavailability of itraconazole, a poorly water-soluble antifungal drug, mesoporous silica particles formulation was developed. It was concluded that ordered mesoporous silica can, in fact, be considered a promising carrier seeing that it enhances the oral bioavailability of extremely low water-soluble drugs [60].

It has been demonstrated that metals can easily accumulate in soil and then enter the food chain, thus the combination of Ag+ ions with silica prevents the formation of these ions and reduces its toxicity [176]. That is the reason why silica nanoparticles have also been studied for agricultural purposes as safe and effective alternative fungicides, for example to manage tomato early blight [180].

3. Hidden Potential and Challenges of Natural Antifungal Compounds

From 1981 to 2014, 32 new chemical entities were placed on the market to treat fungal infections: one from a biological source (interferon gamma-n1), that is a peptide produced by a biotechnological procedure; three derived from a natural product that suffered a semisynthetic modification (anidulafungin, caspofungin, micafungin); 25 totally synthetic discovered by random screening or through the modification of an existing agent (fluconazole, itraconazole, ketoconazole, amorolfine, ciclopirox; three synthetically synthesized with the molecule or the pharmacophore mimicking a natural product (butenafine, liranaftate, terbinafine) [181].

In fact, almost 90% of the antifungal agents approved within this period were from a synthetic origin. This paucity of natural products in modern treatments remains a reality and 1950s agents like amphotericin and griseofulvin are still widely used [181].

One good example of an efficient employment of natural products to nanoparticle synthesis happened with *Mentha pulegium* L., commonly known as pennyroyal, a flowering herb with antitussive, carminative, and antiseptic effects. This herb was successfully used to synthesize stable colloidal silver nanoparticles with promising antifungal effects against *Candida albicans* [182].

On the other hand, the marine environment represents a valuable and unexplored platform to the discovery of new compounds [183]. In Table 4, there is an overview of the natural products that have shown in vitro or in vivo potential as antifungal agents isolated from diverse marine organisms: microorganisms (bacteria and fungi), invertebrates (sponges, corals and sea cucumbers), and marine algae are presented.

Table 4. Overview of antifungal natural compounds produced by marine organisms [183,184].

Marine Organism	Source Organism	Type of Compound	Compound Name	Spectrum of Activity
Bacteria (30% of total)	*Bacillus licheniformis*	Glycolipid	Ledoglucomide C, Iedoglycolipid	*Aspergillus niger, Rhizoctonia solani, Botrytis cinerea, and Colletotrichum acutatum, Candida albicans*
	Bacillus subtilis	Lipopeptide	Gageopeptides A-D	*R. solani, P. capsici, B. cinerea, C. acutatum*
	Actinoalloteichus sp. NPS702	Macrolide	Neomaclafungins A-I	*Trichophyton mentagrophytes*
	Streptomyces sp.	Peptide	Mohangamide A	*C. albicans*
	Bacillus marinus	Macrolide	Macrolactins T and B	*Pyricularia oryzae, A. solani*
	Tolypothrix	Lipopeptide	Hassallidin A	*A. fumigatus and C. albicans*
	Chondromyces pediculatus	Peptide	Pedein A	*Rhodotorula glutinis*
Fungi (15% of total)	*Stagonosporopsis cucurbitacearum*	Alkaloid	Didymellamide A	*C. neoformans, C. albicans, C. glabrata*
	Aspergillus sclerotiorum	Peptide	Sclerotide B	*C. albicans*
	Penicillium bilaiae MA-267	Sesquiterpene	Penicibilaenes A and B	*C. gloeosporioides*

Table 4. *Cont.*

Marine Organism	Source Organism	Type of Compound	Compound Name	Spectrum of Activity
Sponge (35%)	*Theonella swinhoei*	Peptide	Theonegramide, Theonellamide G, Cyclolithistide A	*C. albicans*
	Halichondria cylindrata	Peptide	Halicylindramide D and E	*Mortierella ramanniana*
	Siliquariaspongia mirabilis, Theonella swinhoei	Peptide	Theopapuamide A; B and C	*C. albicans*
	Jaspis johnstoni	Peptide	Jasplakinolide	*C. albicans, C. pseidrotropicalis, C. parapsilosis*
	Monanchora arbuscular	Alkaloid	Batzelladine L	*A. flavus*
	Xestospongia muta	Furan	Mutafuran D	*Cryptococcus neoformans var.grubii*
Corals (5%)	*Clavelina oblonga*	Alkanol	(2S,3R)-2-aminododecan-3-ol	*C. albicans* ATCC 10231, *C. glabrata*
Sea cucumbers (6%)	*Stichopus variegates*	Triterpene glycoside	Variegatuside D	*C. albicans, C. pseudo- tropicalis, C. parapsilosis, and M. gypseum*
Algae (9%)	*Caulerpa racemos*	Xylene	Caulerprenylol B	*T. rubrum*

Sponges have always interested pharmacologists, chemists, and biologists as a rich source of antimicrobial compounds with peculiar activities. These colonial organisms have a sessile nature, hence the necessity of producing compounds as a way of protecting themselves, communicating, or modulating their cellular functions [184].

Peptides have high relevance as potential drugs, given their large spectrum of bioactivity. Sponges that belong to the Theonella gender are a recognized source of uncommon peptides with antibacterial, antifungal, and anti-HIV properties. The antifungal activity of some of these peptides is even surplus to other commercial formulations, as seen through their diffusion zone in the agar diffusion method [184].

Although some formulations containing marine compounds are already being subjected to clinical trials, none of them target fungal infections [183].

Theopapuamide A is a cyclic depsipeptide (a peptide in which one or more amides were replaced by an ester group) being firstly isolated from the sponge *Theonella swinhoei* from Papua New Guinea [183]. Its isolation, structural elucidation, and stereochemical analysis are well described in literature [185]. It is capable of inhibiting the growth of *Candida albicans* wild strains and also strains resistant to amphotericin B [186]. In addition, it presents anti-HIV activity, thanks to the 3,4-dimethyl-L-glutamine residue in its chemical structure [183,184].

Besides the marine environment, a considerable number of studies have been conducted on medicinal plants and alternative compounds, such as secondary metabolites, phenolic compounds, essential oils, and extracts. Some plant extracts (*Curcuma zedoaria, Psidium guajava, Plectranthus amboinicus, Lippia alba*) have shown activity against *Candida* spp. and others (*Asteraceae, Euphorbiaceae, Rubiaceae,* and *Solanaceae*) against filamentous fungi. Despite the streamline nature of the discovery process, the biological activity found during the screening of plant extracts, may not be experimentally reproducible. This can be due to many reasons: the chemical constituents in the crude extracts may be different, the solvent used to extract may destroy some compounds and chemical composition may vary according to the growth stage or geographic origin. Some components of essential oils, mainly terpenes or terpenoids, such as eugenol, camphor, curcumin, geraniol, and linalol show activity against a wide variety of Candida species and others like clemateol or citral are active against *Trichophyton* spp. or *Malassezia* spp. respectively. Propolis, which represents the resinous substances collected from plants by bees, demonstrated activity against *Candida* spp., some dermatophytes, and onychomycosis. Ajoene, a compound derived from garlic, has shown effectiveness in the treatment of paracoccidioidomycosis and *Fusarium* spp. infections. Other compounds such as saponins, alkaloids, flavonoids, coumarins, xanthones, lignans, and tannins also presented antifungal activity [187].

There is, in fact, a growing awareness of the limited structural diversity in existing compound libraries and that there are many benefits in exploiting the huge chemical diversity and high biological activity of natural compounds. Furthermore, it is possible to use these natural compounds as platforms to discover drug-like characteristics and produce new substances [188]. Therefore, natural compounds with antifungal properties represent a valuable alternative to current antifungals and their incorporation in nanocarriers is a significant step ahead in nanoformulation, besides being a more sustainable option, because it employs marine resources [183].

Since the discovery of penicillin, the pharmaceutical industry has extensively relied on natural products as sources of structural templates for drug discovery and development [188,189].

Regardless of this fact, the discovery and development of natural-derived medicines has been on a steady decline over the last years, possibly due to high throughput screening. This tool has changed the paradigm of drug discovery in the pharmaceutical industry since it allows the automatic screening of thousands to millions of drug candidates within vast libraries. There is every likelihood that these libraries contain a significant representation of all existing compounds, complicating the discovery of a new natural chemical entity itself [190].

Although a valuable and precious resource, natural products brought about their fair share of challenges in a wide variety of features, compromising their inclusion in the clinical pipeline.

The main issue lies in the fact that natural substances are not always abundant and they generally require laborious extraction and purification steps through complex, expensive, and time-consuming processes, often to obtain just few quantities of extract [190]. Moreover, there are no standard protocols and research groups modify the existing procedures, making the studies not reproducible and not comparable [187].

In addition, environmental aspects pose significant hurdles for drug discovery and development because the substance may come from an endangered species and its overexploitation can lead to habitat destruction [188].

The regulatory requirements for substances containing natural substances (medicines, nutraceuticals, and cosmeceuticals) range from rather strict to non-existent and vary between regions, being under the surveillance of different authorities [188].

4. Ongoing Clinical Trials on Myconanotechnology

Several pharmaceutical industries and authors have been performing clinical trials on antifungal nanomedicines. Despite the growing improvement on applying nanotechnology for drug development, just a few of those nanomedicines have been approved for their clinical applications. The major issues with clinical trials are their time-consuming process and the difficulty while accessing and researching information on previous nanomedicine trends. In the mycology field, amphotericin-B is the active substance that represents the majority of the nanomedicines already available in the market: Amphotec® (Three Rivers Pharmaceuticals, approved in 1992), Abelcet® (Sigma-Tau, approved in 1995) and Ambisome® (Gilead Sciences, approved in 1997), all lipid-based formulations [191,192]. An econazole liposomal formulation (Pevaryllipogel®) is also available on the market [193]. In Table 5 some examples of recent clinical trials are presented.

Table 5. Some examples of ongoing clinical trials on myconanotechnology [194].

Trade Name/Sponsor	ClinicalTrials.gov Identifier	Antifungal	Nanoformulation	Clinical Phase	Disease
Sara Botros, Minia University	NCT04110834	Itraconazole	Nanoemulsion gel	II	*Tinea versicolor*
Sara Botros, Minia University	NCT04110860	Voriconazol	Nanoemulsion gel	II	*Tinea versicolor*
Matinas BioPharma	NCT02971007	Amphotericin B	Cochleate lipid-crystal nanoparticle	II	Vulvovaginal candidiasis
Matinas BioPharma	NCT02629419	Amphotericin B	Cochleate lipid-crystal nanoparticle	II	Mucocutaneous candidiasis

Table 5. *Cont.*

Trade Name/Sponsor	ClinicalTrials.gov Identifier	Antifungal	Nanoformulation	Clinical Phase	Disease
Ahmed Abdellatif, Al-Azhar University	NCT03752424	-	Silver nanoparticle gel	I	Mycosis
Mona Badran, Cairo University	NCT03666195	-	Titanium dioxide nanoparticles	Recruiting	Candidiasis
Rasha Hamed, Assiut University	NCT04431804	-	Silver nanoparticle	Recruiting	Invasive aspergillosis
Celtic Pharma Development Services	NCT01145807	Terbinafine (TDT067)	Transfersome	III	Onychomycosis

5. Conclusions

The present pace of antifungal drug development is highly unlikely to keep up with the clinical needs, especially with the uprising of resistance to current agents. Therefore, more therapeutic answers to fungal illnesses are urgently needed.

The subject of sustainability and environmental impact has gained considerable relevance in today's societies and scientific research, motivating a heated debate on whether there should be an active research for new natural compounds with biological activity or if the pillar of today's research for new compounds should be high throughput screening. The fact is that two out of the three major classes of antifungal drugs (polyenes and echinocandins) were screened from natural products. The utilization of natural compounds, such as chitosan is, irrefutably, a more sustainable option and brings both environmental and pharmacokinetic advantages to nanotechnological approaches. Compounds from a marine origin exhibit an undeniable potential, but their activity and toxicity mechanisms are yet to be clarified.

Nanoparticles have been presented as promising solutions, mainly due to their ability to target specific sites where fungi are harbored, their capacity to enhance the pharmacological effect of drugs, optimizing their physiochemical characteristics, thereby allowing an administration through a more comfortable route. All these features can enable lower dosing, more comfortable regimens, increased bioavailability, and less serious adverse effects.

Special attention has been given to certain kinds of nanoparticles nowadays mainly because of the outstanding features they exhibit: (a) magnetic nanoparticles and their capacity to directly restrict fungal growth, (b) ultradeformable vesicles (transethosomes) and their ease on scaling up, (c) mesoporous silica nanoparticles and their high drug loading, (d) polymersomes and their ability to carry both hydrophobic and hydrophilic substances and to respond to external stimuli, (e) PAMAM dendrimers and their versatile and biocompatible structure. We speculate that future antifungal therapies will mostly lie in these five types of nanoparticles and will take advantage of the current knowledge of using them for other purposes (for instance, as anticancer agents).

These new nanotechnological systems should be able to surpass the issues already mentioned and should also mean a significant upgrade when comparing to the conventional treatments, fighting antifungal resistance, presenting a broad spectrum of activity, with an emphasis on potency increase and little host toxicity, thus having the potential to be industrially produced.

Interdisciplinary cooperation may be the key to a striving success of nanomedicine and nanotechnology, not to mention a proper exploitation of natural medicinal products. Physicists, health-care researchers, and clinical researchers have complementary knowledge that could be put together (for example, innovative screening strategies and novel chemical libraries) so that more effective, practical, and safe nanoparticles are designed. That would increase the chances of an industry funding and, ultimately, a rethinking of current antifungal arsenal, while providing a better quality of life for patients.

Author Contributions: Conceptualization, F.S., P.C. and S.R.; writing—original draft preparation, F.S., S.R.; writing—review and editing, F.S., P.C.; supervision, P.C., D.F. and S.R.; project administration, P.C. All authors have read and agreed to the published version of the manuscript.

Funding: This research received no external funding.

Conflicts of Interest: The authors declare no conflict of interest.

References

1. Rai, M.; Ingle, A.P.; Pandit, R.; Paralikar, P.; Gupta, I.; Anasane, N.; Dolenc-Voljč, M. Nanotechnology for the Treatment of Fungal Infections on Human Skin. In *The Microbiology of Skin, Soft Tissue, Bone and Joint Infections*; Academic Press: Cambridge, MA, USA, 2017; pp. 169–184.
2. Pianalto, K.M.; Alspaugh, J.A. New Horizons in Antifungal Therapy. *J. Fungi* **2016**, *2*, 26. [CrossRef]
3. Chang, Y.L.; Yu, S.J.; Heitman, J.; Wellington, M.; Chen, Y.L. New facets of antifungal therapy. *Virulence* **2017**, *8*, 222–236. [CrossRef] [PubMed]
4. Nami, S.; Aghebati-Maleki, A.; Morovati, H.; Aghebati-Maleki, L. Current antifungal drugs and immunotherapeutic approaches as promising strategies to treatment of fungal diseases. *Biomed. Pharm.* **2019**, *110*, 857–868. [CrossRef] [PubMed]
5. Souza, A.C.; Amaral, A.C. Antifungal Therapy for Systemic Mycosis and the Nanobiotechnology Era: Improving Efficacy, Biodistribution and Toxicity. *Front. Microbiol.* **2017**, *8*, 336. [CrossRef] [PubMed]
6. Scorzoni, L.; de Paula e Silva, A.C.A.; Marcos, C.M.; Assato, P.A.; de Melo, W.C.M.A.; de Oliveira, H.C.; Costa-Orlandi, C.B.; Mendes-Giannini, M.J.S.; Fusco-Almeida, A.M. Antifungal Therapy: New Advances in the Understanding and Treatment of Mycosis. *Front. Microbiol.* **2017**, *8*, 36. [CrossRef] [PubMed]
7. Sirish Sadhna Khatry, S.N.; Sadanandan, M. Novel Drug Delivery Systems for Antifungal Therapy. *Int. J. Pharm. Pharm. Sci.* **2010**, *2*, 6–9.
8. Perfect, J.R. Is there an emerging need for new antifungals? *Expert Opin Emerg Drugs* **2016**, *21*, 129–131. [CrossRef] [PubMed]
9. Roli Jain, A.P. A Review of Kinetics of Nanoparticulated Delayed Release Formulations. *J. Nanomed. Nanotechnol.* **2015**, *6*, 2.
10. Hasan, S. A review on Nanoparticles: Their Synthesis and Types. *Res. J. Recent Sci.* **2015**, *4*, 1–3.
11. Nagavarma, B.V.N.; Ayaz, A.H.K.S.Y.; Vasudha, L.S.; Shivahumar, H.G. Different Techniques for Preparation of Polymeric Nanoparticles—A review. *Asian J. Pharm. Clin. Res.* **2012**, *5*, 16–23.
12. Bhatt, P.; Lalani, R.; Vhora, I.; Patil, S.; Amrutiya, J.; Misra, A.; Mashru, R. Liposomes encapsulating native and cyclodextrin enclosed Paclitaxel: Enhanced loading efficiency and its pharmacokinetic evaluation. *Int. J. Pharm.* **2017**, *536*, 95–107. [CrossRef]
13. Jinhyun Hannah Lee, Y.Y. Controlled drug release from pharmaceutical nanocarriers. *Chem. Eng. Sci.* **2015**, *125*, 75–84.
14. D'Souza, S. A Review of In Vitro Drug Release Test Methods for Nano-Sized Dosage Forms. *Adv. Pharm.* **2014**, *2014*, 12.
15. Goyal, R.; Macri, L.K.; Kaplan, H.M.; Kohn, J. Nanoparticles and nanofibers for topical drug delivery. *J. Control Release* **2016**, *240*, 77–92. [CrossRef]
16. Rangari, A.T. Polymeric Nanoparticles Based Topical Drug Delivery: An Overview. *Asian J. Biomed. Pharm. Sci.* **2015**, *5*, 5–12. [CrossRef]
17. Siegel, R.A.; Rathbone, M.J. Chapter 2—Overview of Controlled Release Mechanisms. In *Fundamentals and Applications of Controlled Release Drug Delivery, Advances in Delivery Science and Technology*; Society, C.R., Ed.; Springer: New York, NY, USA, 2012.
18. Firthouse, P.U.M.; Halith, S.M.; Wahab, S.U.; Sirajudeen, M.; Mohideen, S.K. Formulation and Evaluation of Miconazole Niosomes. *Int. J. Pharmtech Res.* **2011**, *3*, 1019–1022.
19. Aljaeid, B.M.; Hosny, K.M. Miconazole-loaded solid lipid nanoparticles: Formulation and evaluation of a novel formula with high bioavailability and antifungal activity. *Int. J. Nanomed.* **2016**, *11*, 441–447. [CrossRef]
20. Bhalekar, M.R.; Pokharkar, V.; Madgulkar, A.; Patil, N.; Patil, N. Preparation and evaluation of miconazole nitrate-loaded solid lipid nanoparticles for topical delivery. *AAPS Pharmscitech* **2009**, *10*, 289–296. [CrossRef]

21. Shahzadi, I.; Masood, M.I.; Chowdhary, F.; Anjum, A.A.; Nawaz, M.A.; Maqsood, I.; Zaman, M.Q. Microemulsion Formulation for Topical Delivery of Miconazole Nitrate. *Int. J. Pharm. Sci. Rev. Res.* **2014**, *24*, 30–36.

22. Elmoslemany, R.M.; Abdallah, O.Y.; El-Khordagui, L.K.; Khalafallah, N.M. Propylene glycol liposomes as a topical delivery system for miconazole nitrate: Comparison with conventional liposomes. *AAPS Pharmscitech* **2012**, *13*, 723–731. [CrossRef]

23. Maha, H.L.; Sinaga, K.R.; Masfria, M. Formulation and evaluation of miconazole nitrate nanoemulsion and cream. *Asian J. Pharm. Clin. Res.* **2018**, *11*, 319–321. [CrossRef]

24. Kumar, P.S.; Hematheerthani, N.; Ratna, J.V.; Saikishore, V. Design and characterization of miconazole nitrate loaded nanosponges containing vaginal gels. *Int. J. Pharm. Anal. Res.* **2016**, *5*, 410–417.

25. Qushawy, M.; Nasr, A.; Abd-Alhaseeb, M.; Swidan, S. Design, Optimization and Characterization of a Transfersomal Gel Using Miconazole Nitrate for the Treatment of Candida Skin Infections. *Pharmaceutics* **2018**, *10*, 26. [CrossRef]

26. Ge, S.; Lin, Y.; Lu, H.; Li, Q.; He, J.; Chen, B.; Wu, C.; Xu, Y. Percutaneous delivery of econazole using microemulsion as vehicle: Formulation, evaluation and vesicle-skin interaction. *Int. J. Pharm.* **2014**, *465*, 120–131. [CrossRef] [PubMed]

27. Evelyn, D.; Wooi, C.C.; Kumar, J.R.; Muralidharan, S.; Dhanaraj, S.A. Development and evaluation of microemulsion based gel (MBGs) containing econazole nitrate for nail fungal infection. *J. Pharm. Res.* **2012**, *5*, 2385–2390.

28. Sanna, V.; Gavini, E.; Cossu, M.; Rassu, G.; Giunchedi, P. Solid lipid nanoparticles (SLN) as carriers for the topical delivery of econazole nitrate: In-vitro characterization, ex-vivo and in-vivo studies. *J. Pharm. Pharm.* **2007**, *59*, 1057–1064. [CrossRef]

29. Keshri, L.; Pathak, K. Development of thermodynamically stable nanostructured lipid carrier system using central composite design for zero order permeation of econazole nitrate through epidermis. *Pharm. Dev. Technol.* **2013**, *18*, 634–644. [CrossRef]

30. Xianrong, Q.; Liu, M.H.; Liu, H.Y.; Maitani, Y.; Nagai, T. Topical econazole delivery using liposomal gel. *S.T.P. Pharma Sci.* **2003**, *13*, 241–245.

31. Verma, P.; Pathak, K. Nanosized ethanolic vesicles loaded with econazole nitrate for the treatment of deep fungal infections through topical gel formulation. *Nanomedicine* **2012**, *8*, 489–496. [CrossRef]

32. Verma, S.; Utreja, P. Transethosomes of Econazole Nitrate for Transdermal Delivery: Development, In-vitro Characterization, and Ex-vivo Assessment. *Pharm. Nanotechnol.* **2018**, *6*, 171–179. [CrossRef]

33. Sharma, R.; Walker, R.B.; Pathak, K. Evaluation of the Kinetics and Mechanism of Drug Release from Econazole nitrate Nanosponge Loaded Carbapol Hydrogel. *Indian J. Pharm. Educ. Res.* **2011**, *45*, 25–31.

34. Kumar, Y.P.; Kumar, K.V.; Shekar, R.R.; Ravi, M.; Kishore, V.S. Formulation and Evaluation of Econazole Niosomes. *Sch. Acad. J. Pharm.* **2013**, *2*, 315–318.

35. Bachhav, Y.G.; Mondon, K.; Kalia, Y.N.; Gurny, R.; Möller, M. Novel micelle formulations to increase cutaneous bioavailability of azole antifungals. *J. Control. Release* **2011**, *153*, 126–132. [CrossRef] [PubMed]

36. Youenang Piemi, M.P.; Korner, D.; Benita, S.; Jean-Paul, M. Positively and negatively charged submicron emulsions for enhanced topical delivery of antifungal drugs. *J. Control. Release* **1999**, *58*, 177–187. [CrossRef]

37. Souto, E.B.; Muller, R.H. SLN and NLC for topical delivery of ketoconazole. *J. Microencapsul.* **2005**, *22*, 501–510. [CrossRef]

38. Shirsand, S.; Para, M.; Nagendrakumar, D.; Kanani, K.; Keerthy, D. Formulation and evaluation of Ketoconazole niosomal gel drug delivery system. *Int. J. Pharm. Investig.* **2012**, *2*, 201–207. [CrossRef]

39. Tiwari, N.; Sivakumar, A.; Mukherjee, A.; Chandrasekaran, N. Enhanced antifungal activity of Ketoconazole using rose oil based novel microemulsion formulation. *J. Drug Deliv. Sci. Technol.* **2018**, *47*, 434–444. [CrossRef]

40. Kakkar, S.; Kaur, I.P. Spanlastics—A novel nanovesicular carrier system for ocular delivery. *Int. J. Pharm.* **2011**, *413*, 202–210. [CrossRef]

41. Winnicka, K.; Wroblewska, M.; Wieczorek, P.; Sacha, P.T.; Tryniszewska, E. Hydrogel of ketoconazole and PAMAM dendrimers: Formulation and antifungal activity. *Molecules* **2012**, *17*, 4612–4624. [CrossRef]

42. Ashe, S.; Nayak, D.; Tiwari, G.; Rauta, P.R.; Nayak, B. Development of liposome-encapsulated ketoconazole: Formulation, characterisation and evaluation of pharmacological therapeutic efficacy. *Micro. Nano Lett.* **2015**, *10*, 126–129. [CrossRef]

43. Ning, M.; Guo, Y.; Pan, H.; Chen, X.; Gu, Z. Preparation, in vitro and in vivo evaluation of liposomal/niosomal gel delivery systems for clotrimazole. *Drug Dev. Ind. Pharm.* **2005**, *31*, 375–383. [CrossRef] [PubMed]

44. Kumar, A.S.; Sheri, P.S.; Kuriachan, M.A. Formulation and Evaluation of Antifungal Nanosponge Loaded Hydrogel for Topical Delivery. *Int. J. Pharm. Pharm. Res.* **2018**, *13*, 362–379.

45. Akhtar, N.; Pathak, K. Cavamax W7 Composite Ethosomal Gel of Clotrimazole for Improved Topical Delivery: Development and Comparison with Ethosomal Gel. *Am. Assoc. Pharm. Sci.* **2012**, *13*, 344–355. [CrossRef] [PubMed]

46. Shirsand, S.; Kumar, R.; Keshavshetti, G.; Bushetti, S.S.; Padivala, V.S. Formulation and Evaluation of Clotrimazole Niosomal Gel for Topical Application. *Rajiv Gandhi Univ. Health Sci. J. Pharm. Sci.* **2015**, *5*, 32–38. [CrossRef]

47. Yassin, G.E. Formulation and Evaluation of Optimized Clotrimazole Emulgel Formulations. *Br. J. Pharm. Res.* **2014**, *4*, 1014–1030. [CrossRef]

48. Souto, E.B.; Wissing, S.A.; Barbosa, C.M.; Müller, R.H. Development of a controlled release formulation based on SLN and NLC for topical clotrimazole delivery. *Int. J. Pharm.* **2004**, *278*, 71–77. [CrossRef]

49. Kaewbanjong, J.; Heng, P.W.S.; Boonme, P. Clotrimazole microemulsion and microemulsion-based gel: Evaluation of buccal drug delivery and irritancy using chick chorioallantoic membrane as the model. *J. Pharm. Pharm.* **2017**, *69*, 1716–1723. [CrossRef]

50. Bachhav, Y.G.; Patravale, V.B. Microemulsion-based vaginal gel of clotrimazole: Formulation, in vitro evaluation, and stability studies. *AAPS Pharmscitech* **2009**, *10*, 476–481. [CrossRef]

51. Maheshwari, R.G.S.; Tekade, R.K.; Sharma, P.A.; Darwhekar, G.; Tyagi, A.; Patel, R.P.; Jain, D.K. Ethosomes and ultradeformable liposomes for transdermal delivery of clotrimazole: A comparative assessment. *Saudi Pharm. J. Spj. Off. Publ. Saudi Pharm. Soc.* **2012**, *20*, 161–170. [CrossRef]

52. Zheng, W.S.; Fang, X.Q.; Wang, L.L.; Zhang, Y.J. Preparation and quality assessment of itraconazole transfersomes. *Int. J. Pharm.* **2012**, *436*, 291–298. [CrossRef]

53. Mohanty, B.; Majumdar, D.K.; Mishra, S.K.; Panda, A.K.; Patnaik, S. Development and characterization of itraconazole-loaded solid lipid nanoparticles for ocular delivery. *Pharm. Dev. Technol.* **2015**, *20*, 458–464. [CrossRef] [PubMed]

54. Pardeike, J.; Weber, S.; Haber, T.; Wagner, J.; Zarfl, H.P.; Plank, H.; Zimmer, A. Development of an itraconazole-loaded nanostructured lipid carrier (NLC) formulation for pulmonary application. *Int. J. Pharm.* **2011**, *419*, 329–338. [CrossRef] [PubMed]

55. Wagh, V.D.; Deshmukh, O.J. Itraconazole Niosomes Drug Delivery System and Its Antimycotic Activity against Candida albicans. *Isrn. Pharm.* **2012**, *2012*, 653465. [CrossRef] [PubMed]

56. Chudasama, A.; Patel, V.; Nivsarkar, M.; Vasu, K.; Shishoo, C. Investigation of microemulsion system for transdermal delivery of itraconazole. *J. Adv. Pharm. Technol. Res.* **2011**, *2*, 30–38.

57. Leal, A.F.; Leite, M.C.; Medeiros, C.S.; Cavalcanti, I.M.; Wanderley, A.G.; Magalhaes, N.S.; Neves, R.P. Antifungal activity of a liposomal itraconazole formulation in experimental Aspergillus flavus keratitis with endophthalmitis. *Mycopathologia* **2015**, *179*, 225–229. [CrossRef]

58. Leal, A.F.; Leite, M.C.; Medeiros, C.S.; Cavalcanti, I.M.; Wanderley, A.G.; Magalhaes, N.S.; Neves, R.P. Development of an itraconazole encapsulated polymeric nanoparticle platform for effective antifungal therapy. *J. Mater. Chem. B* **2016**, *4*, 1787–1796.

59. ElMeshad, A.N.; Mohsen, A.M. Enhanced corneal permeation and antimycotic activity of itraconazole against Candida albicans via a novel nanosystem vesicle. *Drug Deliv.* **2016**, *23*, 2115–2123. [CrossRef]

60. Mellaerts, R.; Mols, R.; Jammaer, J.A.G.; Aerts, C.A.; Annaert, P.; Van Humbeeck, J.; Van den Mooter, G.; Augustijns, P.; Martens, J.A. Increasing the oral bioavailability of the poorly water soluble drug itraconazole with ordered mesoporous silica. *Eur. J. Pharm. Biopharm.* **2008**, *69*, 223–230. [CrossRef]

61. Bachhav, Y.G.; Patravale, V.B. Microemulsion based vaginal gel of fluconazole: Formulation, in vitro and in vivo evaluation. *Int. J. Pharm.* **2009**, *365*, 175–179. [CrossRef]

62. Soliman, O.A.E.; Mohamed, E.A.; Khatera, N.A.A. Enhanced ocular bioavailability of fluconazole from niosomal gels and microemulsions: Formulation, optimization, and in vitro-in vivo evaluation. *Pharm. Dev. Technol.* **2017**, *24*, 1–52. [CrossRef]

63. Gupta, S.K.; Dhingra, N.; Velpandian, T.; Jaiswal, J. Efficacy of fluconazole and liposome entrapped fluconazole for C. albicans induced experimental mycotic endophthalmitis in rabbit eyes. *Acta Ophthalmol. Scand* **2000**, *78*, 448–450. [CrossRef] [PubMed]

64. El-Housiny, S.; Shams Eldeen, M.A.; El-Attar, Y.A.; Salem, H.A.; Attia, D.; Bendas, E.R.; El-Nabarawi, M.A. Fluconazole-loaded solid lipid nanoparticles topical gel for treatment of pityriasis versicolor: Formulation and clinical study. *Drug Deliv.* **2017**, *25*, 78–90. [CrossRef] [PubMed]

65. Kelidari, H.R.; Moazeni, M.; Babaei, R.; Saeedi, M.; Akbari, J.; Parkoohi, P.I.; Nabili, M.; Morteza-Semnani, A.A.G.K.; Nokhodchi, A. Improved Yeast Delivery of Fluconazole with a Nanostructured Lipid Carrier System. *Biomed. Pharmacother.* **2017**, *89*, 83–88. [CrossRef] [PubMed]

66. Moin, A.; Deb, T.K.; Osmani, R.A.M.; Bhosale, R.R.; Hani, U. Fabrication, characterization, and evaluation of microsponge delivery system for facilitated fungal therapy. *J. Basic Clin. Pharm.* **2016**, *7*, 39–48. [PubMed]

67. Indora, N.; Kaushik, D. Design, development and evaluation of ethosomal gel of fluconazole for topical fungal infection. *Int. J. Eng. Sci. Invent. Res. Dev.* **2015**, *1*, 280–306.

68. Kaur, I.P.; Rana, C.; Singh, M.; Bhushan, S.; Singh, H.; Kakkar, S. Development and evaluation of novel surfactant-based elastic vesicular system for ocular delivery of fluconazole. *J. Ocul. Pharm.* **2012**, *28*, 484–496. [CrossRef]

69. Lalit, S.K.; Panwar, A.S.; Darwhekar, G.; Jain, D.K. Formulation and Evaluation of Fluconazole Amphiphilogel. *Der. Pharm. Lett.* **2011**, *3*, 125–131.

70. Kumar, R.; Sinha, V.R. Preparation and optimization of voriconazole microemulsion for ocular delivery. *Colloids Surf. B Biointerfaces* **2014**, *117*, 82–88. [CrossRef]

71. Basaran, E.; Karaca, H.; Yenilmez, E.; Guven, U. Voriconazole incorporated polymeric nanoparticles for ocular application. *Lat. Am. J. Pharm.* **2017**, *36*, 1983–1994.

72. Das, P.J.; Paul, P.; Mukherjee, B.; Mazumder, B.; Mondal, L.; Baishya, R.; Debnath, M.C.; Dey, K.S. Pulmonary Delivery of Voriconazole Loaded Nanoparticles Providing a Prolonged Drug Level in Lungs: A Promise for Treating Fungal Infection. *Mol. Pharm.* **2015**, *12*, 2651–2664. [CrossRef]

73. Pandurangan, D.K.; Bodagala, P.; Palanirajan, V.K.; Govindaraj, S. Formulation and evaluation of voriconazole ophthalmic solid lipid nanoparticles in situ gel. *Int. J. Pharm. Investig.* **2016**, *6*, 56–62. [CrossRef] [PubMed]

74. Song, C.K.; Balakrishnan, P.; Shim, C.K.; Chung, S.-J.; Chong, S.; Kim, D.-D. A novel vesicular carrier, transethosome, for enhanced skin delivery of voriconazole: Characterization and in vitro/in vivo evaluation. *Colloids Surf. B Biointerfaces* **2011**, *92*, 299–304. [CrossRef] [PubMed]

75. Faisal, W.; Soliman, G.M.; Hamdan, A.M. Enhanced skin deposition and delivery of voriconazole using ethosomal preparations. *J. Liposome. Res.* **2018**, *28*, 14–21. [CrossRef] [PubMed]

76. Tanriverdi, S.T.; Hilmioglu Polat, S.; Yesim Metin, D.; Kandiloglu, G.; Ozer, O. Terbinafine hydrochloride loaded liposome film formulation for treatment of onychomycosis: In vitro and in vivo evaluation. *J. Liposome Res.* **2016**, *26*, 163–173. [CrossRef]

77. Chen, Y.-C.; Liu, D.-Z.; Liu, J.-J.; Chang, T.-W.; Ho, H.-O.; Sheu, M.-T. Development of terbinafine solid lipid nanoparticles as a topical delivery system. *Int. J. Nanomed.* **2012**, *7*, 4409–4418.

78. Ghannoum, M.; Isham, N.; Henry, W.; Kroon, H.A.; Yurdakul, S. Evaluation of the morphological effects of TDT 067 (terbinafine in Transfersome) and conventional terbinafine on dermatophyte hyphae in vitro and in vivo. *Antimicrob. Agents Chemother.* **2012**, *56*, 2530–2534. [CrossRef]

79. Elsherif, N.I.; Shamma, R.N.; Abdelbary, G. Terbinafine Hydrochloride Trans-ungual Delivery via Nanovesicular Systems: In Vitro Characterization and Ex Vivo Evaluation. *Aaps Pharmscitech* **2017**, *18*, 551–562. [CrossRef]

80. Ozcan, I.; Abaci, O.; Uztan, A.H.; Aksu, B.; Boyacioglu, H.; Guneri, T.; Ozer, O. Enhanced topical delivery of terbinafine hydrochloride with chitosan hydrogels. *Aaps Pharmscitech* **2009**, *10*, 1024–1031. [CrossRef]

81. Erdal, M.S.; Ozhan, G.; Mat, M.C.; Ozsoy, Y.; Gungor, S. Colloidal nanocarriers for the enhanced cutaneous delivery of naftifine: Characterization studies and in vitro and in vivo evaluations. *Int. J. Nanomed.* **2016**, *11*, 1027–1037. [CrossRef]

82. Barakat, H.S.; Darwish, I.A.; El-Khordagui, L.K.; Khalafallah, N.M. Development of naftifine hydrochloride alcohol-free niosome gel. *Drug Dev. Ind. Pharm.* **2009**, *35*, 631–637. [CrossRef]

83. Pillai, A.B.; Nair, J.V.; Gupta, N.K.; Gupta, S. Microemulsion-loaded hydrogel formulation of butenafine hydrochloride for improved topical delivery. *Arch Derm. Res.* **2015**, *307*, 625–633. [CrossRef] [PubMed]

84. Hay, R.J. Liposomal amphotericin B, AmBisome. *J. Infect.* **1994**, *28* (Suppl. 1), 35–43. [CrossRef]

85. Jansook, P.; Pichayakorn, W.; Ritthidej, G.C. Amphotericin B-loaded solid lipid nanoparticles (SLNs) and nanostructured lipid carrier (NLCs): Effect of drug loading and biopharmaceutical characterizations. *Drug Dev. Ind. Pharm.* **2018**, *44*, 1693–1700. [CrossRef] [PubMed]

86. Butani, D.; Yewale, C.; Misra, A. Topical Amphotericin B solid lipid nanoparticles: Design and development. *Colloids Surf. B Biointerfaces* **2016**, *139*, 17–24. [CrossRef]

87. Saldanha, C.A.; Garcia, M.P.; Iocca, D.C.; Rebelo, L.G.; Souza, A.C.; Bocca, A.L.; Almeida Santos Mde, F.; Morais, P.C.; Azevedo, R.B. Antifungal Activity of Amphotericin B Conjugated to Nanosized Magnetite in the Treatment of Paracoccidioidomycosis. *PLoS Negl. Trop Dis.* **2016**, *10*, e0004754. [CrossRef]

88. Sosa, L.; Clares, B.; Alvarado, H.L.; Bozal, N.; Domenech, O.; Calpena, A.C. Amphotericin B releasing topical nanoemulsion for the treatment of candidiasis and aspergillosis. *Nanomedicine* **2017**, *13*, 2303–2312. [CrossRef]

89. Souza, A.C.; Nascimento, A.L.; de Vasconcelos, N.M.; Jeronimo, M.S.; Siqueira, I.M.; R-Santos, L.; Cintra, D.O.; Fuscaldi, L.L.; Pires Junior, O.R.; Titze-de-Almeida, R.; et al. Activity and in vivo tracking of Amphotericin B loaded PLGA nanoparticles. *Eur. J. Med. Chem.* **2015**, *95*, 267–276. [CrossRef]

90. Italia, J.L.; Kumar, M.N.; Carter, K.C. Evaluating the potential of polyester nanoparticles for per oral delivery of amphotericin B in treating visceral leishmaniasis. *J. Biomed. Nanotechnol.* **2012**, *8*, 695–702. [CrossRef]

91. Tang, X.; Zhu, H.; Sun, L.; Hou, W.; Cai, S.; Zhang, R.; Liu, F. Enhanced antifungal effects of amphotericin B-TPGS-b-(PCL-ran-PGA) nanoparticles in vitro and in vivo. *Int. J. Nanomed.* **2014**, *9*, 5403–5413.

92. Jay Prakash Jain, N.K. Development of amphotericin B loaded polymersomes based on (PEG)3-PLA co-polymers: Factors affecting size and in vitro evaluation. *Eur. J. Pharm. Sci.* **2010**, *40*, 456–465. [CrossRef]

93. Perez, A.P.; Altube, M.J.; Schilrreff, P.; Apezteguia, G.; Celes, F.S.; Zacchino, S.; de Oliveira, C.I.; Romero, E.L.; Morilla, M.J. Topical amphotericin B in ultradeformable liposomes: Formulation, skin penetration study, antifungal and antileishmanial activity in vitro. *Colloids Surf. B Biointerfaces* **2016**, *139*, 190–198. [CrossRef] [PubMed]

94. Moreno-Rodriguez, A.C.; Torrado-Duran, S.; Molero, G.; Garcia-Rodriguez, J.J.; Torrado-Santiago, S. Efficacy and toxicity evaluation of new amphotericin B micelle systems for brain fungal infections. *Int. J. Pharm.* **2015**, *494*, 17–22. [CrossRef] [PubMed]

95. Lykov, A.; Gaidul, K.; Goldina, I.; Konenkov, V.; Kozlov, V.; Lyakhov, N.; Dushkin, A. Silica Nanoparticles as a Basis for Efficacy of Antimicrobial Drugs. In *Nanostructures for Antimicrobial Therapy*; Elsevier: Amsterdam, The Netherlands, 2017; pp. 551–575.

96. Khalil, R.M.; Rahman, A.A.A.E.; Kassem, M.A.; Ridi, M.S.E.; Samra, M.M.A.; Awad, G.E.A.; Mansy, S.S. Preparation and in vivo Assessment of Nystatin-Loaded Solid Lipid Nanoparticles for Topical Delivery against Cutaneous Candidiasis. *Int. J. Pharmacol. Pharm. Sci.* **2014**, *8*, 401–409.

97. Fernandez-Campos, F.; Clares Naveros, B.; Lopez Serrano, O.; Alonso Merino, C.; Calpena Campmany, A.C. Evaluation of novel nystatin nanoemulsion for skin candidosis infections. *Mycoses* **2013**, *56*, 70–81. [CrossRef] [PubMed]

98. Offner, F.; Krcmery, V.; Boogaerts, M.; Doyen, C.; Engelhard, D.; Ribaud, P.; Cordonnier, C.; de Pauw, B.; Durrant, S.; Marie, J.P.; et al. Liposomal nystatin in patients with invasive aspergillosis refractory to or intolerant of amphotericin B. *Antimicrob. Agents Chemother.* **2004**, *48*, 4808–4812. [CrossRef]

99. El-Ridy, M.S.; Abdelbary, A.; Essam, T.; Abd El-Salam, R.M.; Aly Kassem, A.A. Niosomes as a potential drug delivery system for increasing the efficacy and safety of nystatin. *Drug Dev. Ind. Pharm.* **2011**, *37*, 1491–1508. [CrossRef]

100. Jadon, P.S.; Gajbhiye, V.; Jadon, R.S.; Gajbhiye, K.R.; Ganesh, N. Enhanced oral bioavailability of griseofulvin via niosomes. *Aaps Pharmscitech* **2009**, *10*, 1186–1192. [CrossRef]

101. Shirsand, S.B.; Keshavshetti, G.G. Formulaiton and characterization of drug loaded niosomes for antifungal activity. *Sper. J. Adv. Nov. Drug Deliv.* **2016**, *1*, 12–17.

102. Jaya raja Kumar, S.M.; Subramani, P. Antifugal Agents: New Approach for Novel Delivery Systems. *J. Pharm. Sci. Res.* **2014**, *6*, 229–235.

103. Srinivasan, A.; Lopez-Ribot, J.L.; Ramasubramanian, A.K. Overcoming antifungal resistance. *Drug Discov. Today Technol.* **2014**, *11*, 65–71. [CrossRef]

104. Pelgrift, R.Y.; Friedman, A.J. Nanotechnology as a therapeutic tool to combat microbial resistance. *Adv. Drug Deliv. Rev.* **2013**, *65*, 1803–1815. [CrossRef] [PubMed]

105. Schairer, D.O.; Chouake, J.S.; Nosanchuk, J.D.; Friedman, A.J. The potential of nitric oxide releasing therapies as antimicrobial agents. *Virulence* **2012**, *3*, 271–279. [CrossRef] [PubMed]

106. Blecher, K.; Nasir, A.; Friedman, A. The growing role of nanotechnology in combating infectious disease. *Virulence* **2011**, *2*, 395–401. [CrossRef] [PubMed]

107. Zhang, L.D.P.; Hu, C.-M.J.; Huang, C.-M. Development of Nanoparticles for Antimicrobial Drug Delivery. *Curr. Med. Chem.* **2010**, *17*, 585–594. [CrossRef]

108. Hetrick, E.M.; Shin, J.H.; Paul, H.S.; Schoenfisch, M.H. Anti-biofilm efficacy of nitric oxide-releasing silica nanoparticles. *Biomaterials* **2009**, *30*, 2782–2789. [CrossRef]

109. Hajipour, M.J.; Fromm, K.M.; Akbar Ashkarran, A.; Jimenez de Aberasturi, D.; Larramendi, I.R.d.; Rojo, T.; Serpooshan, V.; Parak, W.J.; Mahmoudi, M. Antibacterial properties of nanoparticles. *Trends Biotechnol.* **2012**, *30*, 499–511. [CrossRef]

110. Mashitah, M.D.; Chan, Y.S.; Jason, J. Antifungal nanomaterials: Syntesis, properties and applications. In *Nanobiomaterials in Antimicrobial Therapy*; William Andrew: San Diego, CA, USA, 2016; pp. 343–383.

111. Mashitah, M.D.; Chan, Y.S.; Jason, J. Antimicrobial properties of nanobiomaterials and the mechanism. In *Nanobiomaterials in Antimicrobial Therapy*; William Andrew: San Diego, CA, USA, 2016; pp. 261–312.

112. Slavin, Y.N.; Asnis, J.; Häfeli, U.O.; Bach, H. Metal nanoparticles: Understanding the mechanisms behind antibacterial activity. *J. Nanobiotechnol.* **2017**, *15*, 65. [CrossRef]

113. Qidway, A.; Pandey, A.; Kumar, R.; Shukla, S.K.; Dikshit, A. Advances in Biogenic Nanoparticles and the Mechanisms of Antimicrobial Effects. *Indian J. Pharm. Sci.* **2018**, *80*, 592–603. [CrossRef]

114. Zhang, X.F.; Liu, Z.G.; Shen, W.; Gurunathan, S. Silver Nanoparticles: Synthesis, Characterization, Properties, Applications, and Therapeutic Approaches. *Int. J. Mol. Sci.* **2016**, *17*, 1534. [CrossRef]

115. Bocate, K.P.; Reis, G.F.; de Souza, P.C.; Oliveira Junior, A.G.; Duran, N.; Nakazato, G.; Furlaneto, M.C.; de Almeida, R.S.; Panagio, L.A. Antifungal activity of silver nanoparticles and simvastatin against toxigenic species of Aspergillus. *Int. J. Food Microbiol.* **2018**, *291*, 79–86. [CrossRef]

116. Calvo, N.L.; Sreekumar, S.; Svetaz, L.A.; Lamas, M.C.; Moerschbacher, B.M.; Leonardi, D. Design and Characterization of Chitosan Nanoformulations for the Delivery of Antifungal Agents. *Int. J. Mol. Sci.* **2019**, *20*, 3686. [CrossRef] [PubMed]

117. Kucharska, M.S.; Sikora, M.; Brzoza-Malczewska, K.; Owczarek, M. Antimicrobial Properties of Chitin and Chitosan. In *Chitin and Chitosan: Properties and Applications*; John Wiley & Sons, Ltd.: West Sussex, UK, 2020.

118. Sun, Q.; Li, J.; Le, T. Zinc Oxide Nanoparticle as a Novel Class of Antifungal Agents: Current Advances and Future Perspectives. *J. Agric. Food Chem.* **2018**, *66*, 11209–11220. [CrossRef] [PubMed]

119. Singh, P.N.; Nanda, A. Antimicrobial and antifungal potential of zinc oxide nanoparticles in comparison to conventional zinc oxide particles. *J. Chem. Pharm. Res.* **2013**, *5*, 457–463.

120. Rai, M.; Yadav, A.; Bridge, P.; Gade, A. Myconanotechnology: A New and Emerging Science. In *Applied Mycology*, 14th ed.; York, C.I.N., Ed.; CABI: Wallingford, UK, 2009.

121. Ingle, A.; Rai, M.; Gade, A.; Bawaskar, M. Fusarium solani: A novel biological agent for the extracellular synthesis of silver nanoparticles. *J. Nanoparticle Res.* **2009**, *11*, 2079–2085. [CrossRef]

122. Korbekandi, H.; Ashari, Z.; Iravani, S.; Abbasi, S. Optimization of Biological Synthesis of Silver Nanoparticles using Fusarium oxysporum. *Iran. J. Pharm. Res. IJPR* **2013**, *12*, 289–298.

123. Kamil, D.; Prameeladevi, T.; Ganesh, S.; Prabhakaran, N.; Nareshkumar, R.; Thomas, S.P. Green synthesis of silver nanoparticles by entomopathogenic fungus Beauveria bassiana and their bioefficacy against mustard aphid (Lipaphis erysimi Kalt.). *Indian J. Exp. Biol.* **2017**, *55*, 555–561.

124. Huang, W.; Fang, X.; Wang, H.; Chen, F.; Duan, H.; Bi, Y.; Yu, H. Biosynthesis of AgNPs by B. maydis and its antifungal effect against Exserohilum turcicum. *IET Nanobiotechnol.* **2018**, *12*, 585–590. [CrossRef]

125. Ogar, A.; Tylko, G.; Turnau, K. Antifungal properties of silver nanoparticles against indoor mould growth. *Sci. Total Environ.* **2015**, *521–522*, 305–314. [CrossRef]

126. Hariharan, H.; Al-Harbi, N.A.; Karuppiah, P.; Rajaram, S.K. Microbial synthesis of Selenium Nanocomposite using Saccharomyces cerevisiae and its antimicrobial activity against pathogens causing nosocomial infection. *Chalcogenide Lett.* **2012**, *9*, 509–515.

127. Muthupandian Saravanan, T.A.; Letemichael, N.; Araya, G.; Arokiyaraj, S.; Vinoth, R.; Karthik, D. Extracellular Biosynthesis and Biomedical Application of Silver Nanoparticles Synthesized from Baker's Yeast. *Int. J. Res. Pharm. Biomed. Sci.* **2013**, *4*, 822–828.

128. Shanbhaga, P.P.; Janib, U. Drug delivery through nails: Present and future. *New Horiz. Transl. Med.* **2017**, *3*, 252–263.

129. Rajendra, V.B.; Baro, A.; Kumari, A.; Dhamecha, D.L.; Lahoti, S.R.; Shelke, S.D. Transungual Drug Delivery: An Overview. *J. Appl. Pharm. Sci.* **2012**, *2*, 203–209.

130. Tiwary, A.K.; Sapra, B. High failure rate of transungal drug delivery: Need for new strategies. *Ther. Deliv.* **2017**, *8*, 239–242. [CrossRef] [PubMed]

131. Beloqui, A.; Solinis, M.A.; Rodriguez-Gascon, A.; Almeida, A.J.; Preat, V. Nanostructured lipid carriers: Promising drug delivery systems for future clinics. *Nanomedicine* **2016**, *12*, 143–161. [CrossRef]

132. Puglia, C.; Blasi, P.; Ostacolo, C.; Sommella, E.; Bucolo, C.; Platania, C.B.M.; Romano, G.L.; Geraci, F.; Drago, F.; Santonocito, D.; et al. Innovative Nanoparticles Enhance N-Palmitoylethanolamide Intraocular Delivery. *Front Pharm.* **2018**, *9*, 285. [CrossRef]

133. Kumar, L.; Verma, S.; Bhardwaj, A.; Vaidya, S.; Vaidya, B. Eradication of superficial fungal infections by conventional and novel approaches: A comprehensive review. *Artif. Cells Nanomed. Biotechnol.* **2014**, *42*, 32–46. [CrossRef]

134. Soliman, G.M. Nanoparticles as safe and effective delivery systems of antifungal agents: Achievements and challenges. *Int. J. Pharm.* **2017**, *523*, 15–32. [CrossRef]

135. Mishra, G.P.; Bagui, M.; Tamboli, V.; Mitra, A.K. Recent applications of liposomes in ophthalmic drug delivery. *J. Drug Deliv.* **2011**, *2011*, 863734. [CrossRef]

136. Habib, F.S.; Fouad, E.A.; Abdel-Rhaman, M.S.; Fathalla, D. Liposomes as an ocular delivery system of fluconazole: In-vitro studies. *Acta Ophthalmol.* **2010**, *88*, 901–904. [CrossRef]

137. Fetih, G. Fluconazole-loaded niosomal gels as a topical ocular drug delivery system for corneal fungal infections. *J. Drug Deliv. Sci. Technol.* **2016**, *35*, 8–15. [CrossRef]

138. Mohanta, P.; Pandey, N.; Kapoor, D.N.; Singh, S.; Sarvi, Y.; Sharma, P. Development of surfactant-based nanocarrier system for delivery of an antifungal drug. *J. Pharm. Res.* **2017**, *11*, 1153–1158.

139. Baishakhi Bhowmik, A.R.A.S.; Gowda, D.V.; Singh, A.; Srivastava, A.; Ali, R.; Osmani, M. Recent trends and advances in fungal drug delivery. *J. Chem. Pharm. Res.* **2016**, *8*, 169–178.

140. Jensen, G.M. The care and feeding of a commercial liposomal product: Liposomal amphotericin B (AmBisome((R))). *J. Liposome Res.* **2017**, *27*, 173–179. [CrossRef]

141. Jung, S.H.; Lim, D.H.; Jung, S.H.; Lee, J.E.; Jeong, K.S.; Seong, H.; Shin, B.C. Amphotericin B-entrapping lipid nanoparticles and their in vitro and in vivo characteristics. *Eur. J. Pharm Sci.* **2009**, *37*, 313–320. [CrossRef] [PubMed]

142. Korting, H.C.; Klövekorn, W.; Klövekorn, G. Comparative Efficacy and Tolerability of Econazole Liposomal Gel 1%, Branded Econazole Conventional Cream 1% and Generic Clotrimazole Cream 1% in Tinea Pedis. *Clin. Drug Investig.* **1997**, *14*, 286–293. [CrossRef]

143. Dhawade Nakusha, K.P.; Pande, V. Emerging Trends in Topical Antifungal Therapy: A Review. *Inventi J.* **2015**, *2015*, 1–5.

144. Ascenso, A.; Raposo, S.; Batista, C.; Cardoso, P.; Mendes, T.; Praca, F.G.; Bentley, M.V.; Simoes, S. Development, characterization, and skin delivery studies of related ultradeformable vesicles: Transfersomes, ethosomes, and transethosomes. *Int. J. Nanomed.* **2015**, *10*, 5837–5851. [CrossRef]

145. Verma, S.; Utreja, P. Vesicular nanocarrier based treatment of skin fungal infections: Potential and emerging trends in nanoscale pharmacotherapy. *Asian J. Pharm. Sci.* **2018**, *14*, 117–129. [CrossRef]

146. Shaji, J.; Bajaj, R. Transethosomes: A new prospect for enhanced transdermal delivery. *Int. J. Pharm. Sci. Res.* **2018**, *9*, 2681–2685.

147. Ramasamy, T.; Khandasami, U.S.; Ruttala, H.; Shanmugam, S. Development of solid lipid nanoparticles enriched hydrogels for topical delivery of anti-fungal agent. *Macromol. Res.* **2012**, *20*, 682–692. [CrossRef]

148. Pardeike, J.; Hommoss, A.; Muller, R.H. Lipid nanoparticles (SLN, NLC) in cosmetic and pharmaceutical dermal products. *Int. J. Pharm.* **2009**, *366*, 170–184. [CrossRef] [PubMed]

149. Jain, S.; Jain, S.; Khare, P.; Gulbake, A.; Bansal, D.; Jain, S.K. Design and development of solid lipid nanoparticles for topical delivery of an anti-fungal agent. *Drug Deliv.* **2010**, *17*, 443–451. [CrossRef] [PubMed]

150. Crucho, C.I.C.; Barros, M.T. Polymeric nanoparticles: A study on the preparation variables and characterization methods. *Mater Sci. Eng. C Mater Biol. Appl.* **2017**, *80*, 771–784. [CrossRef] [PubMed]

151. Patel, E.K.; Oswal, R. Nanosponge and Micro Sponges: A Novel Drug Delivery System. *Int. J. Res. Pharm. Chem.* **2012**, *2*, 2281–2781.

152. Amaral, A.C.; Bocca, A.L.; Ribeiro, A.M.; Nunes, J.; Peixoto, D.L.; Simioni, A.R.; Primo, F.L.; Lacava, Z.G.; Bentes, R.; Titze-de-Almeida, R.; et al. Amphotericin B in poly(lactic-co-glycolic acid) (PLGA) and dimercaptosuccinic acid (DMSA) nanoparticles against paracoccidioidomycosis. *J. Antimicrob. Chemother.* **2009**, *63*, 526–533. [CrossRef]

153. Nazila Kamaly, B.Y.; Wu, J.; Farokhzad, O.C. Degradable Controlled-Release Polymers and Polymeric Nanoparticles: Mechanisms of Controlling Drug Release. *Chem. Rev.* **2016**, *116*, 2602–2663. [CrossRef]

154. Caon, T.; Porto, L.C.; Granada, A.; Tagliari, M.P.; Silva, M.A.; Simoes, C.M.; Borsali, R.; Soldi, V. Chitosan-decorated polystyrene-b-poly(acrylic acid) polymersomes as novel carriers for topical delivery of finasteride. *Eur. J. Pharm. Sci.* **2014**, *52*, 165–172. [CrossRef]

155. Ing, L.Y.; Zin, N.M.; Sarwar, A.; Katas, H. Antifungal activity of chitosan nanoparticles and correlation with their physical properties. *Int. J. Biomater.* **2012**, *2012*, 632698. [CrossRef]

156. Serrano, D.R.; Lalatsa, A.; Dea-Ayuela, M.A.; Bilbao-Ramos, P.E.; Garrett, N.L.; Moger, J.; Guarro, J.; Capilla, J.; Ballesteros, M.P.; Schätzlein, A.G.; et al. Oral Particle Uptake and Organ Targeting Drives the Activity of Amphotericin B Nanoparticles. *Mol. Pharm.* **2015**, *12*, 420–431. [CrossRef]

157. Makhmalzade, B.S.; Chavoshy, F. Polymeric micelles as cutaneous drug delivery system in normal skin and dermatological disorders. *J. Adv. Pharm. Technol. Res.* **2018**, *9*, 2–8.

158. Thambi, T.; Deepagan, V.G.; Ko, H.; Lee, D.S.; Park, J.H. Bioreducible polymersomes for intracellular dual-drug delivery. *J. Mater. Chem.* **2012**, *22*, 22028–22036. [CrossRef]

159. Lee, J.S.; Feijen, J. Polymersomes for drug delivery: Design, formation and characterization. *J. Control. Release* **2012**, *161*, 473–483. [CrossRef] [PubMed]

160. Sun, H.; Meng, F.; Cheng, R.; Deng, C.; Zhong, Z. Reduction and pH dual-bioresponsive crosslinked polymersomes for efficient intracellular delivery of proteins and potent induction of cancer cell apoptosis. *Acta Biomater.* **2014**, *10*, 2159–2168. [CrossRef] [PubMed]

161. Sherje, A.P.; Jadhav, M.; Dravyakar, B.R.; Kadam, D. Dendrimers: A versatile nanocarrier for drug delivery and targeting. *Int. J. Pharm.* **2018**, *548*, 707–720. [CrossRef] [PubMed]

162. Caminade, A.M.; Laurent, R.; Majoral, J.P. Characterization of dendrimers. *Adv. Drug Deliv. Rev.* **2005**, *57*, 2130–2146. [CrossRef] [PubMed]

163. Gupta, U.; Perumal, O. Dendrimers and Its Biomedical Applications. In *Natural and Synthetic Biomedical Polymers*; Elsevier: Amsterdam, The Netherlands, 2014; pp. 243–257.

164. Nimesh, S. Dendrimers. In *Gene therapy: Potential Applications of Nanotechnology*; Woodhead Publishing Limited: Cambridge, UK, 2013; pp. 259–285.

165. Nanjwade, B.K.; Bechra, H.M.; Derkar, G.K.; Manvi, F.V.; Nanjwade, V.K. Dendrimers: Emerging polymers for drug-delivery systems. *Eur. J. Pharm. Sci.* **2009**, *38*, 185–196. [CrossRef] [PubMed]

166. Abbasi, E.; Milani, M.; Fekri Aval, S.; Kouhi, M.; Akbarzadeh, A.; Tayefi Nasrabadi, H.; Nikasa, P.; Joo, S.W.; Hanifehpour, Y.; Nejati-Koshki, K.; et al. Silver nanoparticles: Synthesis methods, bio-applications and properties. *Crit. Rev. Microbiol.* **2016**, *42*, 173–180. [CrossRef]

167. Vallabani, N.V.S.; Singh, S. Recent advances and future prospects of iron oxide nanoparticles in biomedicine and diagnostics. *3 Biotech* **2018**, *8*, 279. [CrossRef]

168. Seabra, A.B.; Pasquoto, T.; Ferrarini, A.C.; Santos Mda, C.; Haddad, P.S.; de Lima, R. Preparation, characterization, cytotoxicity, and genotoxicity evaluations of thiolated- and s-nitrosated superparamagnetic iron oxide nanoparticles: Implications for cancer treatment. *Chem. Res. Toxicol.* **2014**, *27*, 1207–1218. [CrossRef]

169. Sankhyan, A.; Pawar, P. Recent Trends in Niosome as Vesicular Drug Delivery System. *J. Appl. Pharm. Sci.* **2012**, *2*, 20–32.

170. Boonme, P.; Kaewbanjong, J.; Amnuaikit, T.; Andreani, T.; Silva, A.M.; Souto, E.B. Microemulsion and Microemulsion-Based Gels for Topical Antifungal Therapy with Phytochemicals. *Curr. Pharm. Des.* **2016**, *22*, 4257–4263. [CrossRef] [PubMed]

171. Mehta, K.; Bhatt, D. Preparation, Optimization and in vitro Microbiological Efficacy of Antifungal Microemulsion. *Int. J. Pharm. Sci. Res.* **2011**, *2*, 2424–2429.

172. Güngör, S.; Erdal, M.S.; Aksu, B. New Formulation Strategies in Topical Antifungal Therapy. *J. Cosmet. Dermatol. Sci. Appl.* **2013**, *3*, 56–65. [CrossRef]

173. El-Hadidy, G.N.; Ibrahim, H.K.; Mohamed, M.I.; El-Milligi, M.F. Microemulsions as vehicles for topical administration of voriconazole: Formulation and in vitro evaluation. *Drug Dev. Ind. Pharm.* **2012**, *38*, 64–72. [CrossRef]

174. De Souza, M.L.; Oliveira, D.D.; Pereira, N.P.; Soares, D.M. Nanoemulsions and dermatological diseases: Contributions and therapeutic advances. *Int. J. Derm.* **2018**, *57*, 894–900. [CrossRef]

175. Zazo, H.; Colino, C.I.; Lanao, J.M. Current applications of nanoparticles in infectious diseases. *J. Control Release* **2016**, *224*, 86–102. [CrossRef]

176. Peszke, J.; Dulski, M.; Nowak, A.; Balin, K.; Zubko, M.; Sułowicz, S.; Nowak, B.; Piotrowska-Seget, Z.; Talik, E.; Wojtyniak, M.; et al. Unique properties of silver and copper silica-based nanocomposites as antimicrobial agents. *RSC Adv.* **2017**, *7*, 28092–28104. [CrossRef]

177. Botequim, D.; Maia, J.; Lino, M.M.; Lopes, L.M.; Simoes, P.N.; Ilharco, L.M.; Ferreira, L. Nanoparticles and surfaces presenting antifungal, antibacterial and antiviral properties. *Langmuir* **2012**, *28*, 7646–7656. [CrossRef]

178. Cristiana, S.; Vidal, M.; Ferreira, L.S. Antifungal Nanoparticles and Surfaces. *Biomacromolecules* **2010**, *11*, 2810–2817.

179. Camporotondi, D.E.; Foglia, M.L.; Alvarez, G.S.; Mebert, A.; Diaz, L.E.; Coradin, T. Antimicrobial properties of silica modified nanoparticles. In *Microbial Pathogens and Strategies for Combating Them: Science, Technology and Education*; Méndez-Vilas, A., Ed.; Formatex Research Center: Badajoz, Spain, 2013.

180. Derbalah, A.; Shenashen, M.; Hamza, A.; Mohamed, A.; El Safty, S. Antifungal activity of fabricated mesoporous silica nanoparticles against early blight of tomato. *Egypt. J. Basic Appl. Sci.* **2018**, *5*, 145–150. [CrossRef]

181. Newman, D.J.; Cragg, G.M. Natural Products as Sources of New Drugs from 1981 to 2014. *J. Nat. Prod.* **2016**, *79*, 629–661. [CrossRef] [PubMed]

182. Kelkawi, A.H.A.; Kajani, A.A.; Bordbar, A.K. Green synthesis of silver nanoparticles using Mentha pulegium and investigation of their antibacterial, antifungal and anticancer activity. *IET Nanobiotechnol.* **2017**, *11*, 370–376. [CrossRef]

183. El-Hossary, E.M.; Cheng, C.; Hamed, M.M.; El-Sayed Hamed, A.N.; Ohlsen, K.; Hentschel, U.; Abdelmohsen, U.R. Antifungal potential of marine natural products. *Eur. J. Med. Chem.* **2017**, *126*, 631–651. [CrossRef] [PubMed]

184. Vitali, A. Antimicrobial Peptides Derived from Marine Sponges. *Am. J. Clin. Microbiol. Antimicrob.* **2018**, *1*, 1006.

185. Ratnayake, A.S.; Bugni, T.S.; Feng, X.; Harper, M.K.; Skalicky, J.J.; Mohammed, K.A.; Andjelic, C.D.; Barrows, L.R.; Ireland, C.M. Theopapuamide, a Cyclic Depsipeptide from a Papua New Guinea Lithistid Sponge Theonella swinhoei. *J. Nat. Prod.* **2006**, *69*, 1582–1586. [CrossRef] [PubMed]

186. Alberto Plaza, G.B.; Jessica, L.K.; John, R.L.; Heather, L.B.; Carole, A.B. Celebesides A-C and Theopapuamides B-D, Depsipeptides from an Indonesian Sponge that Inhibit HIV-1 Entry. *J. Org. Chem.* **2010**, *74*, 504–512. [CrossRef]

187. Negri, M.; Salci, T.P.; Shinobu-Mesquita, C.S.; Capoci, I.R.G.; Svidzinski, T.I.E.; Kioshima, E.S. Early state research on antifungal natural products. *Molecules* **2014**, *19*, 2925–2956. [CrossRef] [PubMed]

188. Krause, J.; Tobi, G. Discovery, Development, and Regulation of Natural Products. In *Using Old Solutions to New Problems—Natural Drug Discovery in the 21st Century*; BoD–Books on Demand: Norderstedt, Germany, 2013.

189. Roemer, T.; Xu, D.; Singh, S.B.; Parish, C.A.; Harris, G.; Wang, H.; Davies, J.E.; Bills, G.F. Confronting the Challenges of Natural Product-Based Antifungal Discovery. *Chem. Biol.* **2011**, *18*, 148–164. [CrossRef]

190. Wright, G.D. Unlocking the potential of natural products in drug discovery. *Microb. Biotechnol.* **2018**, *12*, 55–57. [CrossRef]

191. Gupta, N.; Rai, D.B.; Jangid, A.K.; Kulhari, H. Use of nanotechnology in antimicrobial therapy. In *Nanotechnology*; Academic Press: Cambridge, MA, USA, 2019; pp. 143–172.

192. Pooja, D.; Kadari, A.; Kulhari, H.; Sistla, R. Lipid-based nanomedicines. In *Lipid Nanocarriers for Drug Targeting*; William Andrew: San Diego, CA, USA, 2018; pp. 509–528.

193. Koppa Raghu, P.; Bansal, K.K.; Thakor, P.; Bhavana, V.; Madan, J.; Rosenholm, J.M.; Mehra, N.K. Evolution of Nanotechnology in Delivering Drugs to Eyes, Skin and Wounds via Topical Route. *Pharmaceuticals* **2020**, *13*, 167. [CrossRef]

194. NIH. *ClinicalTrials*. Available online: https://clinicaltrials.gov/ct2/home (accessed on 10 September 2020).

 *pharmaceuticals*

Review

Advances in Antifungal Drug Development: An Up-To-Date Mini Review

Ghada Bouz * and Martin Doležal *

Faculty of Pharmacy in Hradec Králové, Charles University, 50005 Hradec Králové, Czech Republic
* Correspondence: bouzg@faf.cuni.cz (G.B.); dolezalm@faf.cuni.cz (M.D.)

Abstract: The utility of clinically available antifungals is limited by their narrow spectrum of activity, high toxicity, and emerging resistance. Antifungal drug discovery has always been a challenging area, since fungi and their human host are eukaryotes, making it difficult to identify unique targets for antifungals. Novel antifungals in clinical development include first-in-class agents, new structures for an established target, and formulation modifications to marketed antifungals, in addition to repurposed agents. Membrane interacting peptides and aromatherapy are gaining increased attention in the field. Immunotherapy is another promising treatment option, with antifungal antibodies advancing into clinical trials. Novel targets for antifungal therapy are also being discovered, allowing the design of new promising agents that may overcome the resistance issue. In this mini review, we will summarize the current status of antifungal drug pipelines in clinical stages, and the most recent advancements in preclinical antifungal drug development, with special focus on their chemistry.

Keywords: antifungals; drug discovery; drug repurposing; drug targets; invasive aspergillosis treatment

Citation: Bouz, G.; Doležal, M. Advances in Antifungal Drug Development: An Up-To-Date Mini Review. *Pharmaceuticals* **2021**, *14*, 1312. https://doi.org/10.3390/ph14121312

Academic Editors: Jong Heon Kim, Luisa W. Cheng and Kirkwood Land

Received: 2 December 2021
Accepted: 14 December 2021
Published: 16 December 2021

Publisher's Note: MDPI stays neutral with regard to jurisdictional claims in published maps and institutional affiliations.

1. Introduction

For decades, fungal infections have been difficult health conditions to treat. This fact can be attributed to the narrow spectrum and high toxicity of clinically used antifungals, long duration of treatment and the high emergence of resistance towards available agents. The seriousness of fungal infections was brought back to light during the unfortunate COVID-19 pandemic, in the form of secondary life-threatening infections in the intensive care units [1]. *Candida*, *Cryptococcus*, and *Aspergillus* are the most common causative organisms of life-threatening human fungal infections [2]. *Candida auris* is a multi-drug resistant fungus [3]. *Lomentospora prolificans* has intrinsic resistance towards all clinically used antifungals. *Aspergillus fumigatus* is becoming more resistant to treatment, making it more difficult to treat aspergillosis, with mortality rate reaching 100% in some cases. Early detection and treatment of fungal meningitis and chronic pulmonary aspergillosis can save millions of lives around the world. Fungal infections have become a silent crisis, and prompt efforts are needed before it is too late. In this review, we will highlight current state-of-the-art developments in antifungal pipeline, both in clinical and preclinical stages, with special focus on their chemistry, in order to provide the reader with a comprehensive, up-to-date source that will influence future synthetic efforts.

2. Clinically Used Antifungals

The limited number of available antifungals with there narrow safety margin contributes to the increasing morbidity and mortality of invasive fungal infections. Clinically used antifungals can be classified according to their mechanism of action, as shown in Figure 1.

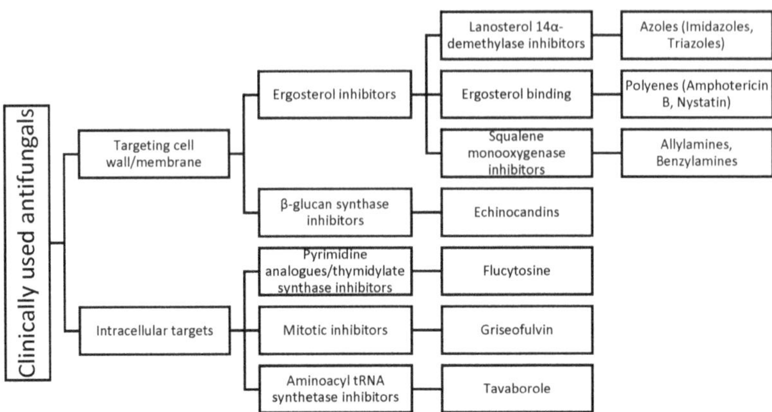

Figure 1. Clinically used antifungals grouped according to their mechanism of action with the most common agents in practice as examples.

Among all antifungals, only polyenes, flucytosine, azoles, and echinocandins are licensed for the treatment of invasive, life-threatening fungal infections (refer to Figure 2 for chemical structures) [4]. Amphotericin B is the clinically used polyene that is preserved as a second-line agent due to its high nephrotoxicity, which can lead, in some cases, to kidney failure. Several lipid-incorporated formulations of amphotericin B were developed as an attempt to reduce its nephrotoxicity; however, the high cost of such formulations limits their utility. Nystatin is a polyene macrolide antifungal that is used topically or orally to treat oropharyngeal candidiasis (for local effect, as it is not absorbed via the oral route) [5]. The major drawbacks of the pyrimidine antifungal flucytosine are the rapid development of resistance and high toxicity, both hepatological and hematological [6]. Due to their fungistatic mode of action, azoles are associated with a high rate of resistance, yet their wide safety margin contributes to their popularity. Azoles include compounds with either imidazole moiety, such as clotrimazole, or triazole moiety, such as fluconazole. Among the classical antifungals, the echinocandins were the last to be discovered in 1970s, taking them 30 years to progress from bench to bedside, to be marketed later in the year 2000 [7]. Although echinocandins have low rates of resistance, recently some fungi, especially non-albicans *candida*, started to develop resistance towards echinocandins by acquiring mutations [8]. The long stagnant phase in antifungal development ended with The Food and Drug Administration (FDA) approval of tavaborole in 2014 (refer to Figure 2 for chemical structure). Tavaborole is the first oxaborole and first tRNA synthetase-inhibitor antifungal to be approved for clinical use. It exerts its antifungal activity by inhibiting the cytosolic leucyl-transfer RNA synthetase (LeuRS), which plays a vital role in protein synthesis. It is licensed for topical use for the treatment of onychomycosis [9].

Figure 2. The chemical structures of amphotericin B (an example of a polyene); caspofungin (an example of an echinocandin); flucytosine; fluconazole (an example of an azole); and tavaborole.

3. Novel Antifungals (In Clinical Settings)

The previously mentioned limitations of available antifungals urge the need for novel agents that can overcome these problems. Novel antifungals include first-in-class agents, new structures for a known target, new derivatives/analogues in an established class of drugs, and modification to the formulations of approved agents, in addition to repurposed agents (refer to Figure 3). Obtaining first-in-class designation significantly speeds up approval process and in-return market availability. Chemistry-wise, new antifungals with new structures include cyclic peptides (hexapeptides such as rezafungin and VL-2397 and depsipeptides such as aureobasidin A), triterpenoids (ibrexafungerp), tetrazoles (VT-1129, VT-1161, VT-1598), orotomides (olorofim), siderophores (VL-2397), and arylamidines (T-2307). Each agent will be discussed separately (ordered according to Figure 3) with a focus on their chemical aspects.

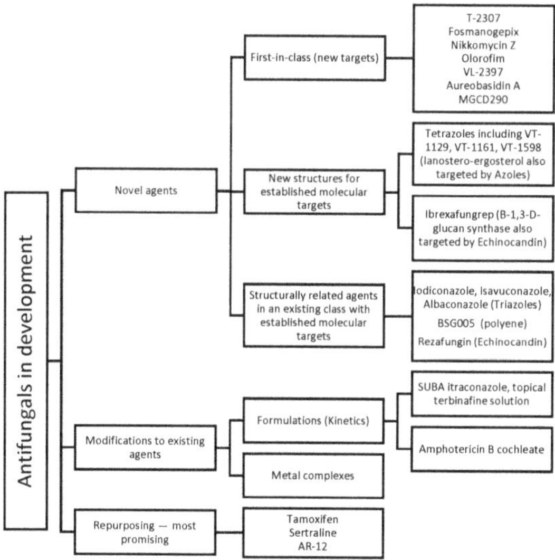

Figure 3. A summative figure showing the classification of new molecules as antifungals in clinical development.

3.1. Aryldiamidines—T-2307

T-2307 is a first-in-class antifungal that exerts its fungicidal activity by inhibiting respiratory chain complexes and thus disrupting mitochondrial membrane potential. It is selectively transported into fungal cells through a polyamine transporter [10]. Structurally, it is an aromatic diamidine that is structurally related to the antiprotozoal agent pentamidine with a characteristic plane of symmetry. Pentamidine is used to treat pneumocystis, leishmaniasis, and trypanosomiasis by a similar mechanism (refer to Figure 4 for chemical structures). Pre-clinical data show T-2307 as very potent antifungal and a perhaps superior agent to azoles and polyenes in the treatment of invasive fungal infections [11]. Further information on T-2307 activity is summarized elsewhere [12].

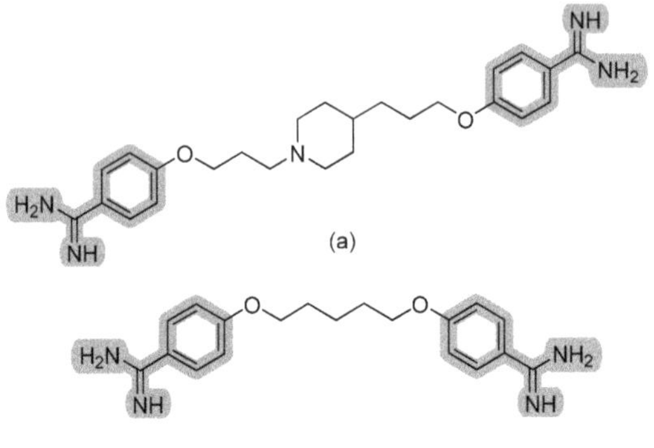

Figure 4. The chemical structures of (**a**) T-2307 and (**b**) pentamidine. The characteristic aryldiamidine moiety is highlighted in grey.

3.2. Fosmanogepix and Manogepix

Fosmanogepix is the *N*-phosphonooxymethylene prodrug of manogepix (APX001A) that is hydrolyzed by systemic phosphatases (refer to Figure 5 for chemical structures). Manogepix was first identified during lead optimization studies to improve the potency of 1-[4-butylbenzyl]isoquinoline, a hit structure found to suppress the expression of surface glycosyl phosphatidylinositol (GPI)-mannoproteins, specifically the Gwt1 protein, in *Saccharomyces cerevisiae* and *Candida albicans*, and subsequently inhibit their growth [13,14]. Therefore, manogepix is a first-in-class antifungal that inhibits the fungal Gwt1 protein. Gwt1 is an enzyme that catalyzes inositol acylation, which is an early step in the glycosylphosphatidylinositol (GPI)-anchor biosynthesis pathway [15]. In 2019, fosmanogepix obtained orphan drug designation. Further information on fosmanogepix activity is reviewed elsewhere [16].

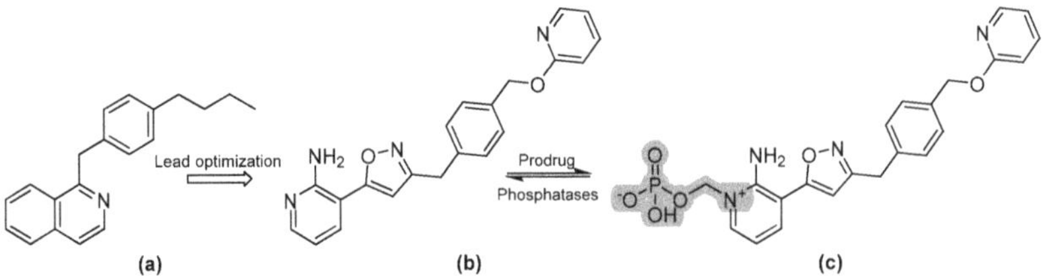

Figure 5. The chemical structures of (**a**) 1-[4-butylbenzyl]isoquinoline, (**b**) manogepix, and (**c**) fosmanogepix.

3.3. Nikkomycin Z

The discovery of nikkomycin Z goes back to the 70s. Structurally, nikkomycin Z resembles uridine diphosphate (UDP)-*N*-acetyl glucosamine, which is a precursor of chitin, and, thus, nikkomycin Z is a competitive inhibitor of chitin synthase (refer to Figure 6 for chemical structures) [17]. As a stand-alone drug, nikkomycin Z has poor in vitro fungicidal activity against Candida. However, taking into consideration that chitin is a major component of fungal cell walls, nikkomycin Z has synergistic activity with other antifungal cell-wall inhibitors, such as echinocandins [18]. Further information on nikkomycin activity is summarized elsewhere [19].

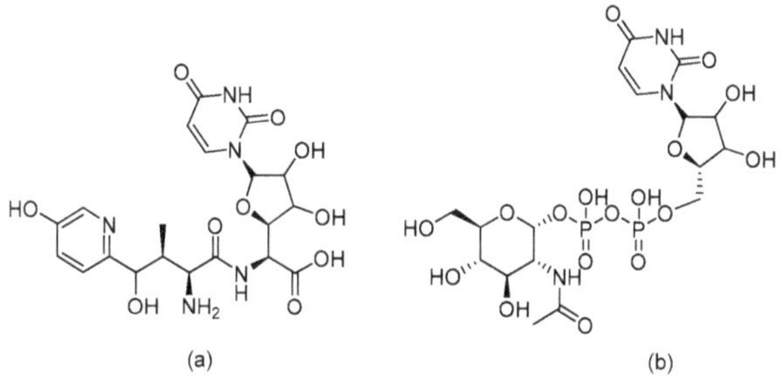

(a)　　　　　　　(b)

Figure 6. The chemical structures of (**a**) Nikkomycin Z and (**b**) Uridine Diphosphate (UDP)-*N*-acetyl glucosamine.

3.4. Orotomides—Olorofim

The orotomides are a new class of antifungals that exert their fungicidal activity via a novel mechanism of targeting dihydroorotate dehydrogenase, which is a vital enzyme in fungal pyrimidine biosynthesis. Subsequently, nucleic acid and phospholipid synthesis is disrupted [20]. One agent belonging to this new class is olorofim (F901318). Olorofim is a potent antifungal agent that has time-dependent activity against drug resistant *Aspergillus* spp. and other uncommon molds with promising results in invasive and refractory fungal infections (refer to Figure 7 for chemical structure) [21]. Olorofim is 2000-fold more selective toward fungal dihydroorotate dehydrogenase than the human enzyme homologue [20]. Having a unique mechanism of action and potent antifungal activity against hard-to-treat fungi granted olorifim orphan drug designation (ODD) by the U.S. Food and Drug Administration (FDA) in March 2020 for the treatment of invasive aspergillosis and *Lomentospora/Scedosporium* infections and later in June 2020 for the treatment of coccidioidomycosis. Similarly, The European Medicines Agency Committee for Orphan Products granted orphan drug status to olorofim for the treatment of invasive aspergillosis and rare mold infections caused by *Scedosporium* spp. Further information on olorofim's mechanism of action, pharmacokinetic profile, and clinical efficacy is summarized elsewhere [22].

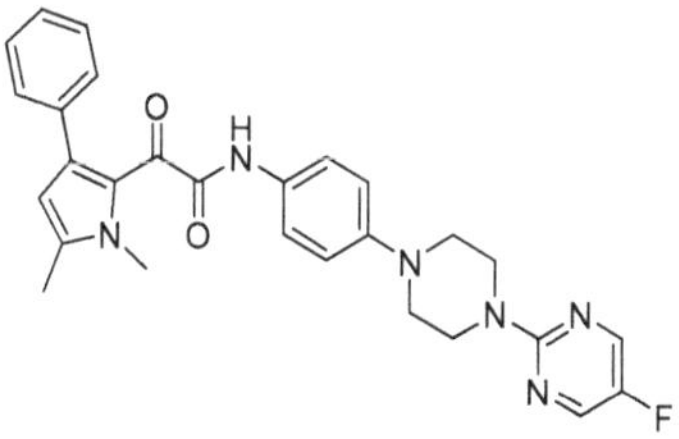

Figure 7. The chemical structures of olorofim.

3.5. Cyclic Peptides—VL-2397

VL-2397 (previously annotated as ASP2397) is a natural cyclic hexapeptide isolated from Acremonium persicinum cultures [23]. It was identified as a potential antifungal while screening various natural secondary metabolites in a silkworm infection model [23]. Structurally, VL-2397 resembles fungal ferrichrome siderophores (high-affinity, iron-chelating compounds secreted from microorganisms that serve as transporters to transport iron across cell membranes) and enters the fungal cells through specific transporters known as siderophore iron transporter 1 (Sit1) (refer to Figure 8 for chemical structures) [24]. Sit1 transporters are not present in mammalian cells, limiting possible toxicity to human cells. Once inside the cell, VL-2397 disrupts important intracellular processes through unknown mechanisms, and hence exerts its fungicidal activity [24].

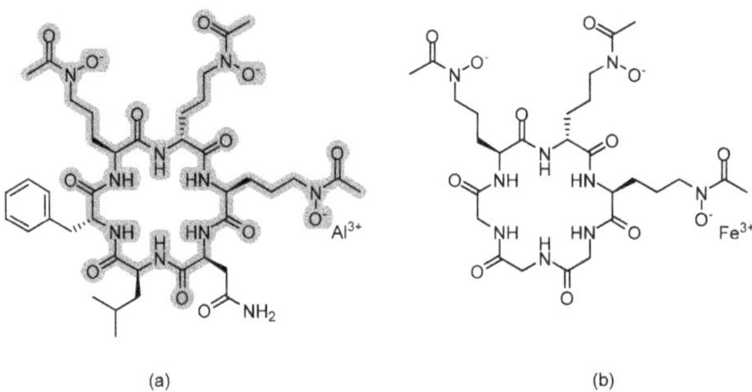

(a)   (b)

Figure 8. The chemical structures of (**a**) VL-2397 and (**b**) ferrichrome siderophore. To better visualize the structural resemblance, common fragments with ferrochrome siderophore are highlighted in grey.

3.6. Cyclic Peptides—Aureobasidin A

Aureobasidin A is a natural cyclic depsipeptide isolated from the fungus *Aureobasidium pullulans* R106 that targets the essential inositol phosphorylceramide (sphingolipid) synthase in fungi (refer to Figure 9 for chemical structure) [25]. Interestingly, this broad-spectrum antifungal also exerts significant antiprotozoal activity against the proliferative tachyzoite form of Toxoplasma [26]. Structurally, Aureobasidin A consists of eight α-amino acid units and one hydroxy acid unit. Upon acid hydrolysis, Ikai and colleagues identified these units to be 2(*R*)-hydroxy-3(*R*)-methylpentanoic acid, beta-hydroxy-*N*-methyl-L-valine, *N*-methyl-L-valine, L-proline, allo-L-isoleucine, *N*-methyl-L-phenylalanine, L-leucine, and L-phenyl-alanine [25]. Several Aureobasidin A analogues have been prepared by Kurome and colleagues. They found that analogues with 4–6 carbon chain ester derivatives at the γ-carboxyl group of Glu6 or aIle6 exhibited the best antifungal activity [27].

Figure 9. The chemical structure of Aureobasidin A.

3.7. MGCD290

MGCD290 (chemical structure is not presented in the literature) is an oral fungal Hos2 histone deacetylase (HDAC) inhibitor that also targets non-histone proteins, such as Hsp90, all of which play important roles in gene regulation [28]. Interfering with such fungal proteins suppresses fungal stress responses, which favors the use of MGCD290 as an adjuvant therapy to cell wall/membrane inhibitors [29]. Despite the in vitro synergic activity between MGCD290 and azoles/echinocandins, MGCD290 failed to show clinical importance in clinical settings, and hence its further development was suspended after a phase II clinical trial [28,30].

3.8. Tetrazoles

Tetrazoles are a new azole-like group bearing difluoromethyl-pyridines moiety, including the agents VT-1129 (quilseconazole), VT-1161 (oteseconazole), and VT-1598 (refer to Figure 10 for chemical structures). Tetrazoles are more selective to fungal CYP51 (lanosterol demethylase) rather than human enzymes, unlike triazoles, and hence have lower side effects and drug–drug interactions than triazoles [31]. VT-1161 and VT-1598 form H-bonds with the His377 of *Candida albicans* CYP51 and with the His374 of *Aspergillus fumigates* CYP51B [32]. The rationale behind the design of tetrazoles by W. J. Hoekstra et al. was to replace the metal binding group in triazoles (MBG, which is triazole) which has strong MBG/metal interaction with another MGB that has weaker MBG/metal interaction (tetrazole), as an attempt to improve selectivity. VT-1129 is a potent oral inhibitor for *Cryptococcus* species, including drug resistant strains with half-lives of six days [33]. VT-1161 is being investigated as a promising antifungal agent against both fluconazole-sensitive and fluconazole-resistant *Candida albicans* with clinical application in treating tinea pedis, onychomycosis, and vaginal candidiasis [34]. VT-1161 has half-life of 48 h [34]. VT-1598 was found to have anticandidal activity, especially against *Candida auris* [35].

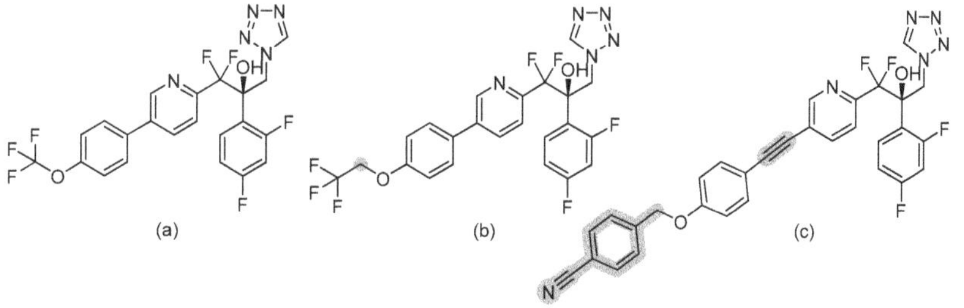

Figure 10. Chemical structures of (**a**) VT-1129, (**b**) VT-1161, and (**c**) VT-1598. Differences in chemical structures are highlighted in grey.

3.9. Triterpenoids—Ibrexafungerp

During high-throughput screening of natural products produced by endophytic fungi, enfumafungin was identified as a promising antifungal hemiacetal triterpenoid glycoside. Structural modifications to improve pharmacokinetic properties and, most importantly, oral bioavailability led to the development of the semi-synthetic derivative ibrexafungerp (previously known as SCY-078 or MK-3118) (refer to Figure 11 for chemical structure). Ibrexafungerp exerts its antifungal activity, similarly to echinocandins, via inhibiting β-glucan synthase, yet it has a distinct chemical structure. It is fungicidal against *Candida* spp. and fungistatic against *Aspergillus* spp. [36]. Echinocandins are administered intravenously due to their poor per oral stability, which necessitates switching the patient to an oral, and perhaps less potent, antifungal upon discharge. Nevertheless, there is the emerging problem of resistance that urges the need for a safer agent that can be administered at higher doses. Thus, being an orally administered agent, ibrexafungerp is superior to known echinocandins when it comes to patients' convenience.

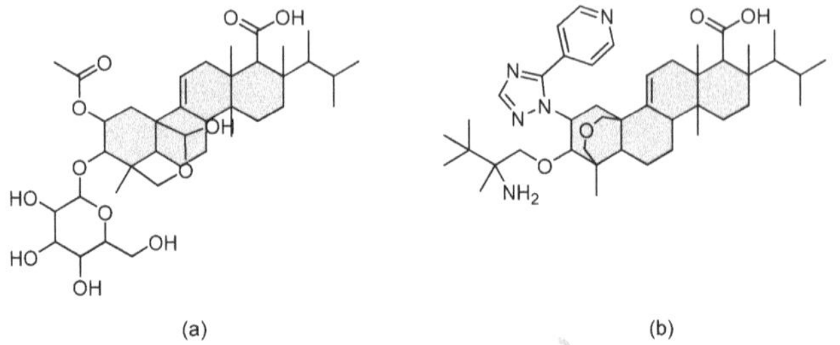

Figure 11. The chemical structures of (**a**) enfumafungin and (**b**) ibrexafungerp. The triterpenoid unit is shaded in grey.

3.10. Triazoles

Since they are the most widely used antifungal class, there were several attempts to improve the efficacy of triazoles. Isavuconazonium sulfate is a water-soluble prodrug of isavuconazole, suitable for both oral and IV administration [37]. The [*N*-(3-acetoxypropyl)-*N*-methylamino]-carboxymethyl group is linked by an ester functionality to the triazole nitrogen of isavuconazole. The prodrug is cleaved by human plasma esterases releasing isavuconazole and low levels of cleavage by-product (refer to Figure 12) [38]. Isavuconazole has long half-life, broad spectrum of activity, and good efficacy [38]. It was FDA approved in 2015 for the treatment of aspergillosis and invasive mucormycosis.

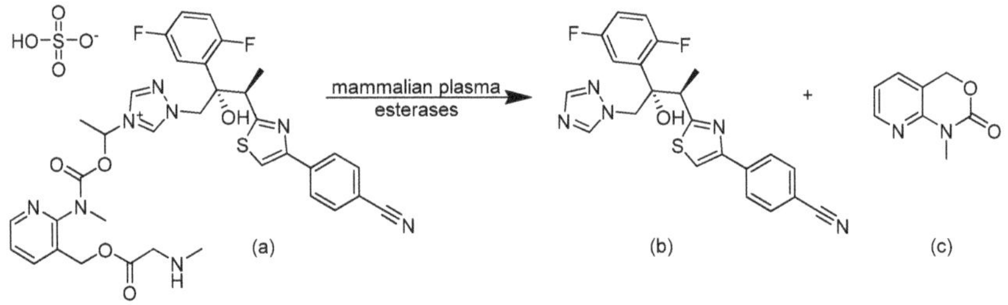

Figure 12. Hydrolysis reaction of the prodrug (**a**) isavuconazonium by mammalian plasma esterases to yield, (**b**) isavuconazole, and (**c**) by-product.

Albaconazole is another 7-chloro triazole under development for the treatment of acute candida vulvovaginitis and onychomycosis (refer to Figure 13 for chemical structure). It has improved pharmacokinetic properties that resulted in excellent oral bioavailability [39].

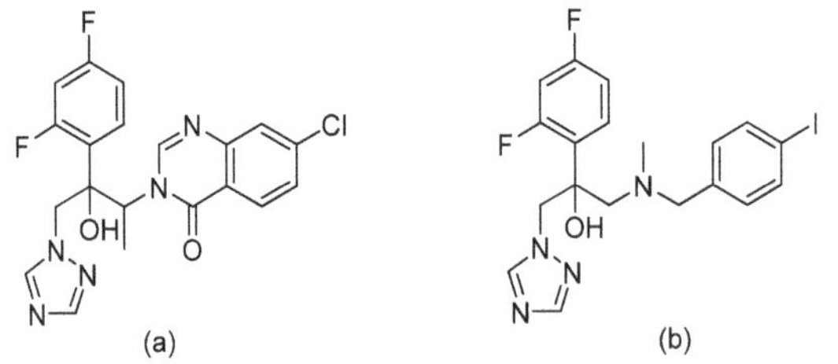

Figure 13. The chemical structures of (**a**) albaconazole and (**b**) iodiconazole.

Iodiconazole is a new topical triazole for the treatment of dermatophytosis in humans (refer to Figure 13 for chemical structure). It has a broad spectrum and potent activity against different fungal strains [40].

3.11. BSG005

BSG005 is a natural antifungal isolated from *Streptomyces noursei*. Structurally, BSG005 is an improved version of the polyene nystatin A (heptaene nystatin analogue) (refer to Figure 14 for chemical structures) [41]. It exerts fungicidal activity against a wide range of fungal strains, including azole- and echinocandin- resistant *Aspergillus* spp. and *Candida* spp., by a similar mechanism to nystatin. BSG005 does not cause nephrotoxicity, overcoming the main drawback of polyenes [41].

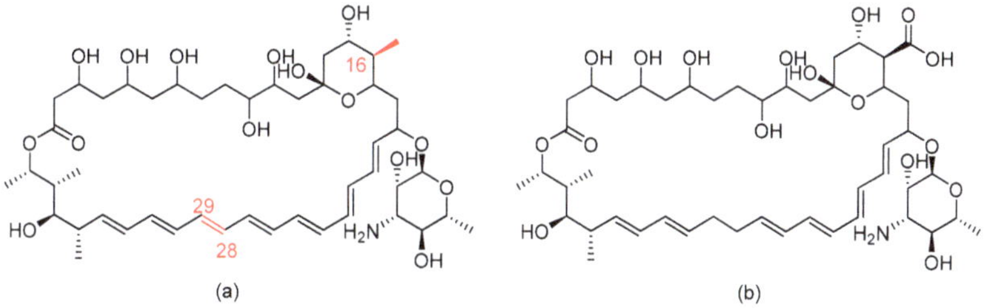

Figure 14. The chemical structures of (**a**) BSG005 and (**b**) nystatin. The structural modifications are shown in red.

3.12. Cyclic Peptides—Rezafungin

Rezafungin (previously known as CD101) is a novel echinocandin, another cyclic hexapeptide derivative. It is an analogue of the echinocandin anidulafungin obtained by replacing the hemiaminal region at the C5 ornithine position of anidulafungin with choline ether (refer to Figure 15). This chemical modification led to an increased stability towards host degradation and subsequently prolonged the half-life, allowing once-weekly dosing [42]. When incubated in human plasma for 44 h at 37 °C, rezafungin showed a stability of 79–94%, while anidulafungin had 7–15% stability [43]. The prolonged pharmacokinetic property of rezafungin prolonged tissue exposure and hence permitted the use of rezafungin in prophylaxis regimens instead of the regular azoles [44]. This additional choline ether prevented the formation of toxic intermediates, favoring rezafungin safety and allowing its administration at higher doses to prevent resistance [43].

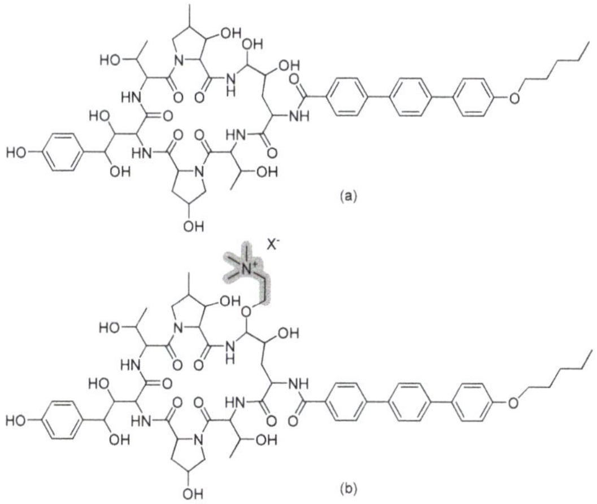

Figure 15. The chemical structures of (**a**) anidulafungin and (**b**) rezafungin. The chemical modification in the structure of rezafungin over anidulafungin is highlighted in grey.

3.13. SUBA Itraconazole

Super-bioavailability-itraconazole (SUBA itraconazole) overcame the low bioavailability issue of itraconazole by dispersing itraconazole in a pH-dependent polymer matrix as capsules. This change to the formulation increased the bioavailability of itraconazole by 173% [45]. SUBA itraconazole was granted FDA approval in 2018 for the treatment of aspergillosis, histoplasmosis, and blastomycosis in patients with contraindications to the use of amphotericin B (intolerant or refractory).

3.14. Topical Terbinafine Solution

Oral terbinafine is a gold-standard treatment for onychomycosis; however the treatment is lengthy, as it takes terbinafine a long time to concentrate in the nail plate and bed, increasing the potential for the development of systemic side effects and resistance. One solution was to prepare terbinafine as a solution for topical use (MOB-01) [46]. Pharmacokinetic studies showed that, when compared with oral terbinafine, topical terbinafine achieves higher concentrations in the nail plate ($\approx$10,003 times more) and nail bed ($\approx$403 times more) with limited or no systemic absorption. Similar examples include two new ophthalmic solutions for the treatment of candida infections; one containing hexamidine diisethionate 0.05% (keratosept) [47], and the other containing povidone-iodine 0.6% (IODIM®) [48].

3.15. Amphotericin B Cochleate

Amphotericin B cochleate is a polyene formulation of amphotericin B that is stable against gastric degradation and hence is suitable for oral administration [49]. Cochleates are made up of phosphatidylserine with phospholipid-calcium precipitates in a multilayer system that has spiral configuration. When the cochleates reach the blood stream, the spiral structure opens once the calcium concentration drop in the cochleate, releasing the encapsulated amphotericin B [49]. The possibility to administer amphotericin B via the oral route overcame infusion-related complications and offered a practical, patient-convenient, broad-spectrum antifungal treatment.

3.16. Metal Complexes and Chelates

Metal complexes may be a novel promising class of antifungals due to improved properties, especially those related to stereochemistry, redox potential, and lipophilicity [50]. Such an approach may provide an easy solution to overcome the issue of fungal resistance. Successful examples of established antifungals complexed with metal ions include organoruthenium complexes conjugated with three different azoles (namely clotrimazole, tioconazole, and miconazole) reported by Kljun et al. [51]. The resulting mono-, bis-, and tris–azoles complexes had millimolar inhibitory concentrations against *Culvularia lunata*. Another example is reported by Stevanovic et al., who showed that a fluconazole zinc (II) complex had significantly better antifungal activity against *Candida krusei* and *Candida parapsilosis* than fluconazole [52]. Other new metal coordinates, not complexed with known antifungals, are at preclinical stages [53,54]. On the other hand, Polvi and colleagues screened a library of pharmacologically active compounds that do not themselves possess antifungal activity as an attempt to identify compounds that can potentiate the efficacy of caspofungin against echinocandin-resistant *Candida albicans* strains. They identified the broad-spectrum chelator diethylenetriamine pentaacetate (DTPA) as a promising compound that synergizes with capsofungin (refer to Figure 16 for chemical structure). They justified this potentiating activity of DTPA by magnesium chelation [55]. The potential of metal-based compounds as antifungals is thoroughly reviewed elsewhere [56].

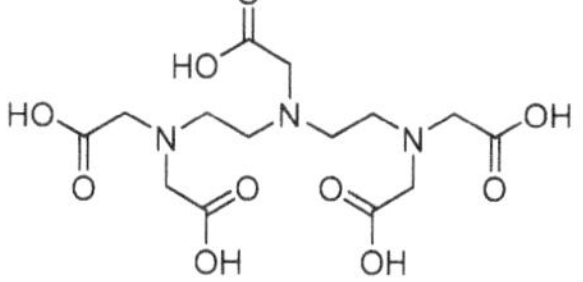

Figure 16. The chemical structure of the broad-spectrum chelator diethylenetriamine pentaacetate (DTPA).

The current status for each of the antifungals discussed in this review, including the name of companies and clinical trial registration numbers, are summarized in Table 1 below.

Table 1. The spectrum of activity, route of administration, production company, and current status of antifungals in clinical development. Clinical registration numbers were obtained from www.clinicaltrials.gov (accessed on 26 November 2021).

Agent	Spectrum of Activity	Route of Administration	Company/Sponsor	Current Status	Clinical Trial Registration Number
T-2307	Broad spectrum	IV	Toyama Chemical Ltd.	Phase I completed (as stated in [57], but no actual data available in the literature)	Not available
Fosmanogepix	*Candida* spp. (*except C. krusei*) and *Aspergillus* spp.	PO/IV	Amplyx Pharmaceuticals	Phase II recruiting	NCT04240886
Nikkomycin Z	*Candida* spp. and *Aspergillus* spp.	PO	University of Arizona	Phase I completed; however lack of funding and volunteers caused the termination of phase II studies	NCT00834184
Olorofim	*Aspergillus* spp. and uncommon molds	PO/IV	F2G Ltd.	Phase II Phase III planned (not yet recruiting)	NCT03583164 NCT05101187
VL-2397	*Aspergillus* spp. and some *Candida* spp.	IV	Vical Biotechnology	No current development plans (phase II trial terminated early, because of a business decision)	NCT03327727
Aureobasidin A	Board spectrum and proliferative tachyzoite form of toxoplasma (antiprotozoal)	PO/IV	Takara Bio Group	Preclinical	Not available
MGCD290	*Candida* spp. and *Aspergillus* spp.	PO	MethylGene, Inc.	Further development suspended after phase II clinical trial	NCT01497223
VT-1129	*Candida* spp. and *Cryptococus* spp.	PO	Viamet Pharmaceuticals Inc.	Preclinical	Not available
VT-1161	*Candida* spp., *coccidioides* spp., and *Rhizopus* spp.	PO	Mycovia Pharmaceuticals	Phase III completed	NCT03561701

Table 1. *Cont.*

Agent	Spectrum of Activity	Route of Administration	Company/Sponsor	Current Status	Clinical Trial Registration Number
VT-1598	*Candida* spp., *Aspergillus* spp., and *Cryptococus* spp.	PO	Mycovia Pharmaceuticals Clinical trial sponsored by National Institute of Allergy and Infectious Diseases (NIAID)	Phase I	NCT04208321
Ibrexafungerp	*Candida* spp. and *Aspergillus* spp.	PO/IV	Scynexis, Inc.	Phase III	NCT03059992
Isavuconazole	Broad spectrum	PO (Isavuconazonium sulfate PO/IV)	Basilea and Astellas Clinical trials sponsored by Memorial Sloan Kettering Cancer Center and M.D. Anderson Cancer Center, respectively.	Phase II trials completed FDA approved in 2015 for the treatment of aspergillosis and invasive mucormycosis	NCT03149055 NCT03019939
Albaconazole	Broad spectrum	PO	Palau Pharma Clinical trial sponsored by GSK	Phase II completed	NCT00730405
Iodiconazole	Broad spectrum	Topical	Second Military Medical University and Anhui Jiren Pharmaceutical	Phase III (as stated in [58], but no actual data available in the literature)	Not available
BSG005	Broad spectrum	IV	Biosergen AS	Phase I	NCT04921254
Rezafungin	*Candida* spp., *Aspergillus* spp., and *Pneumocystis jirovecii*	IV	Cidara Therapeutics, Inc.	Phase III	NCT03667690 NCT04368559
SUBA-itraconazole	Broad spectrum	PO	Mayne Pharma Ltd.	Phase II FDA approved in 2018 for the treatment of aspergillosis, histoplasmosis, and blastomycosis in patients contraindicated to amphotericin B	NCT03572049
Topical terbinafine solution	Onychomycosis	Topical 10% solution	Moberg Pharma	Phase III	NCT02859519
Amphotericin B cochleate	Broad spectrum	PO	Matinas BioPharma	Phase II	NCT02629419

4. New Compounds as Potential Antifungals (In Preclinical Stages)

Several research groups, both in academia and the industry, are focused on developing new, potentially active antifungals. We will briefly mention some of the most recent (published in 2021), successful synthetic efforts from academia (refer to Table 2 for general structures).

Ravu and colleagues recently reported the synthesis and in vitro antifungal evaluation of a series of phloeodictine analogues [59]. Phloeodictines are marine-derived alkaloids, and the phloeodictine-based 6-hydroxy-2,3,4,6-tetrahydropyrrolo[1,2-*a*]pyrimidinium moiety with an *n*-tetradecyl side chain at C-6 has shown antifungal activity and serves as a template for further derivatization. They identified three promising analogues (compounds 24, 36, and 48 in the original paper) with potent activity (MIC ≈ 1 μM) against *Candida neoformans* and low toxicity against mammalian Vero cells (IC_{50} > 40 μM) [59].

Choi et al. tested their in-house library where they identified a 2-amino-*N*-(2-(3,4-dichloro-[1,1-biphenyl]-4-yl)ethyl)-pentanamide hydrochloride (compound 22h in the original paper) as a promising hit with potent, fast fungicidal activity against *Candida neoformans* (MIC ≈ 2.5 μM) and *Candida albicans* (MIC ≈ 5 μM) [60]. The latter compound also showed synergistic in vitro activity with clinically available antifungals and potent in vivo efficacy in a subcutaneous infection mouse model and an ex vivo human nail infection model. The authors claim that their hit compound exerts its antifungal activity by interfering with fungal cell wall integrity.

On the other hand, Kato et al. prepared thiazoyl guanidine derivatives that inhibit fungal ergosterol biosynthesis [61]. Their hit compound *N*-(2′-(4-(methylsulfonyl)phenyl)-[4,4′-bithiazol]-2-yl)-tetrahydropyrimidin-2(1H)-imine (compound 6h in original paper) is structurally related to the antifungal abafungin and showed potent in vitro antifungal activity against *Aspergillus fumigatus* (MIC = 4.7 μM) with a favorable pharmacokinetic profile. In addition, the latter compound exhibited antifungal activity comparable to voriconazole in a murine model of *Aspergillus fumigatus* infection.

Furthermore, Li and colleagues prepared a series of carboline fungal histone deacetylase (HDAC) inhibitors in an attempt to develop a promising compound for the combinational treatment of azole-resistant candidiasis [62]. Among all synthesized compounds, 2-(4-(3-(8-chloro-1,2,3,4-tetrahydro-9H-pyrido[3,4-b]indol-9-yl)propoxy)phenyl)-*N*-hydroxyacetamide (compound D12 in original paper) showed excellent in vitro and in vivo synergistic antifungal activity with fluconazole to treat azole-resistant candidiasis.

Fungal fatty acid (FA) synthase and desaturase (responsible for introducing double bonds to yield unsaturated FAs) are essential enzymes for the growth and virulence of fungal pathogens [63]. They are structurally distinct from their mammalian homologues, making them suitable targets for antifungal development. DeJarnette et al. performed whole-cell screening using *Candida albicans* with varying levels of FA synthase or desaturase [64]. They identified four acyl hydrazides as the most promising candidates (compounds 2, 40, 41, and 48 in the original paper) with broad-spectrum activity against *Candida albicans*, *Candida auris*, and mucormycetes, including activity against azole-resistant *Candida* and low in vitro cytotoxicity in HepG2 liver cancer cell line.

Lastly, Lowes and colleagues designed and synthesized dual inhibitors targeting fungal acetohydroxy-acid synthase and NLRP3 inflammasome [65]. Their design rationale was based on the structural similarities between the two types of inhibitors. Fungal acetohydroxy-acid synthase is the common enzyme in the branched-chain amino acid-synthesis pathway [66], while NLRP3 inflammasome is a mammalian cytosolic receptor that mediates innate inflammatory responses to offending fungi by releasing mature interlukein-1 beta (IL-1β) upon activation [67]. The authors established the essential molecular scaffold required for dual activity for the first time, which shall serve as a template for future synthetic efforts. Among their prepared inhibitors, they identified 2-((1-(4-fluorophenyl)-3-oxo-3-phenylpropyl)thio)benzoate (compound 10 in original paper) as the most promising dual inhibitor, since it significantly decreased IL-1β release (IC_{50} = 2.3 ± 0.8 μM) without affecting mammalian cell viability (viability = 101.5 ± 1.4%) and exerted potent in vitro antifungal activity against *Candida albicans* (MIC = 6.4 ± 2.6 μM).

Table 2. The general structures of some of the most recent (published in 2021), successful synthetic efforts from academia.

Structural Type	General Structure *	Target	Ref.
Phloeodictine analogues	Refer to original publication for R	Fungal ergosterol biosynthesis	[59]
Biphenylethylaminoacetamides	Refer to original paper for X, R^1, R^2	Fungal cell wall integrity	[60]
Thiazoyl guanidine derivatives	Refer to original paper for R	Fungal ergosterol biosynthesis	[61]
Carboline HDAC inhibitors	Refer to original paper for R	Fungal histone deacetylase	[62]
Acyl hydrazides	Refer to original paper for R	Fungal fatty acid biosynthesis	[64]
Thiobenzoate scaffolds	Refer to original paper for R	Dual target: fungal acetohydroxyacid synthase and NLRP3 inflammasome	[65]

* Adapted with permission from refs. [59–62,64,65]. Copyright 2021 American Chemical Society.

Aside from the mentioned small molecules with intracellular targets, membrane-interacting antifungal antimicrobial peptides (AMPs) are another example of a promising class of antifungals. AMPs generally consist of 12 to 54 amino acids with a net positive charge at physiological pH [68]. They are amphipathic in nature, which enhances their interaction with target membranes [69]. Since they target fungal plasma membranes or cell walls, they have the advantage of avoiding intracellular resistance mechanisms. AMPs are selective, with multiple modes of action (for example interaction with membrane phospholipids, sphingolipids, or proteins), and low toxicity to mammalian cells. Detailed information on antifungal AMPs has recently been reviewed elsewhere [70].

In addition, several studies were conducted to explore the role of essential oils in treating fungal infections. In their recent work, Donadu et al. investigated the antifungal activity of the essential oil of the Colombian rue, *Ruta graveolens* (REO) [71]. They found that REO exerts fungicidal activity against *Candida tropicalis* and fungistatic activity against

Candida albicans by disrupting cellular membrane integrity. REO also showed synergistic activity with amphotericin B [71]. Furthermore, an oil macerate of *Helichrysum microphyllum Cambess.* subsp. *tyrrhenicum Bacch., Brullo & Giusso* showed potent inhibiting activity on candida growth, making it a promising agent to be used topically in the treatment of candidiasis [72]. Another example is the essential oil of *Austroeupatorium inulaefolium*, which showed strong species-dependent antifungal activity against *Penicillium brevicompactum* and *Fusarium oxysporum* [73].

5. Repurposing

Repurposing established drugs with possible antifungal activity and pushing them into the antifungal development pipeline saves time and resources, especially given that the pharmacokinetic and pharmacodynamic profiles are already known for such agents. The most promising candidates to be used in antifungal regimens are the selective estrogen receptor modulator tamoxifen (used for breast cancer) and the serotonin reuptake inhibitor sertraline (used for depression) (refer to Figure 17 for chemical structures). Tamoxifen showed anticryptococcal activity and may be a promising synergistic agent with fluconazole [74]. The most advanced repurposing attempt is for sertraline. An ongoing phase III trial is investigating the role of sertraline as an adjuvant therapy to the standard treatment for cryptococcal meningitis (NCT01802385). Furthermore, AR-12 is a celecoxib derivative that was assessed for safety in a phase I oncology clinical trial (refer to Figure 16 for chemical structure). However, it showed consistent antifungal activity against certain yeasts and molds, and thus it was repurposed as a promising adjuvant therapy to fluconazole in the treatment of invasive fungal infections [75].

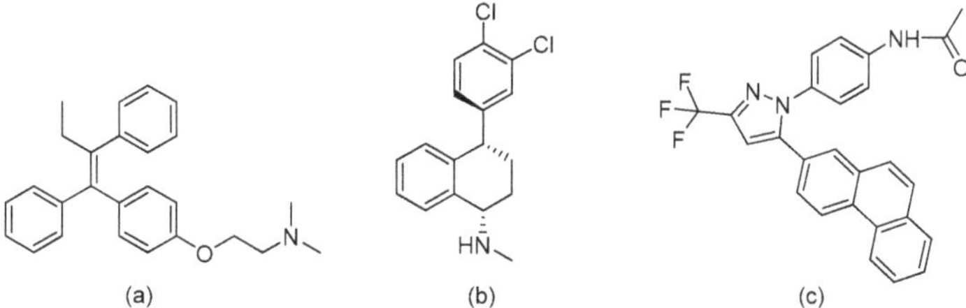

(a)   (b)   (c)

Figure 17. The chemical structures of (**a**) tamoxifen, (**b**) sertraline, and (**c**) AR-12.

6. Immunotherapy

Immunotherapy is a new promising strategy to modulate the host immune system and strengthen the innate and adaptive immune response to fight fungal infections. Immunotherapy approaches in treating fungal infections include the administration of recombinant growth factors and cytokines, granulocyte and granulocyte-macrophage colony-stimulating factors, and antibodies. Immunocompromised patients can also benefit from cell therapy, during which innate and adaptive immune cells are introduced to enhance the immune response against the offending fungi. In this review we will focus on antifungal antibodies, which are also known as antifungal passive immunotherapy. Other immunotherapy approaches are reviewed elsewhere [76].

Currently, there only two antifungal antibodies in clinical development. MAb 2G8 is a monoclonal antibody that targets laminarin (consisting mainly of β glucans). It binds to the walls of *Candida albicans* and *Cryptococcus neoformans*, inhibiting their growth and capsule formation [77]. Mambro and colleagues reported the development of a new humanized monoclonal antibody derived from MAb 2G8 that specifically targets β-1,3 glucans of pathogenic fungi, such as *Candida* spp. [78]. The new derivative showed potent in vitro antifungal activity against *Candida auris*.

Efungumab (also known as mycograb) is a single-chain variable-fragment antibody that targets heat shock protein 90 (HSP90). Efungumab was tested in combination with amphotericin B in clinical settings, where it showed a reduction in the mortality and an improvement in the survival of patients infected with *Candida*. It must be noted that efungumab was also investigated in clinical trials as an adjuvant to docetaxel in patients with breast cancer (NCT00217815).

One step beyond antifungal antibodies is radio-immunotherapy, where the antifungal antibody is linked to radioisotopes in order to specifically release fungicidal radiation in fungal cells [79]. This approach showed promising results in treating drug-resistant *Cryptococcus neoformans* infections [79].

7. New Promising Targets for Antifungal Development

Their classification as eukaryotes makes fungal infections a more challenging condition to treat in the drug discovery pipeline. It is important to discover unique targets that are present in fungi and not in humans in order to improve the selectivity and subsequently the safety profile of such agents. One interesting approach to discover new antifungals is what Novartis did by screening their chemical dark matter database for potential antifungals. Chemical dark matter libraries include molecules that have no bioactivity in human targets and thus are of increased value in screening for active molecules against other eukaryotes. During their screening campaign, Novartis have identified a novel antifungal agent and a novel fungal target pathway, which is hemebiosynthesis. The identification of molecular targets opened the window for medicinal chemists to design inhibitors following a target-based drug design approach.

Another example of a new, promising fungal pathway for antifungal drug development is a sphingolipid synthesis pathway. Sphingolipid synthesis is highly preserved among eukaryotes; however, a number of vital structures are unique to fungi, making them promising treatment targets. In fungi, sphingolipids such as inositolphosphoryl ceramide (IPC) and glucosylceramide (GlcCer) play an important role in fungal pathogenicity and fungal growth [80]. Structures that inhibit the enzymes responsible for the synthesis of the latter two sphingolipids were found to disrupt the virulence of *Candida albicans*, *Cryptococcus neoformans*, and *Aspergillus* spp. [81,82]. Promising pre-clinical molecules targeting sphingolipid biosynthesis include sphingosine *N*-acyltransferase (e.g., australifungin) [83], IPC inhibitors (e.g., the antifungal non-glycosidic macrolide galbonolide A (also known as rustmicin)) [84], GlcCer synthase inhibitors (e.g., D-threo-PDMP) [81], and GlcCer inhibitors (e.g., BHBM) [85], among others (refer to Figure 18 for chemical structures). For such agents, human toxicity is of great concern, except for IPC inhibitors, since mammals do not express IPC [80].

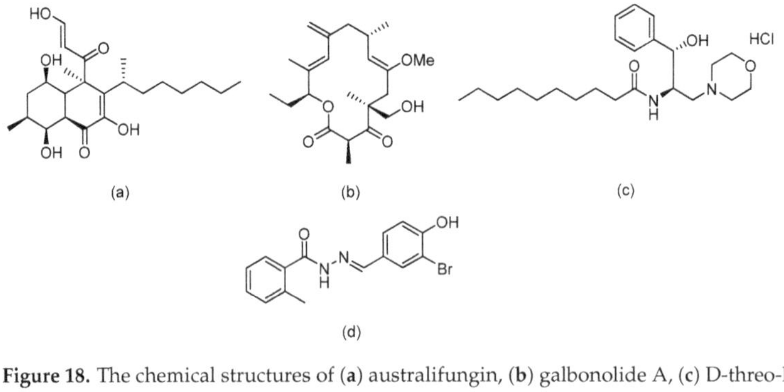

Figure 18. The chemical structures of (**a**) australifungin, (**b**) galbonolide A, (**c**) D-threo-PDMP, and (**d**) BHBM.

In addition, Krystufek and colleagues described in their latest work the great potential of aspartic proteases, especially Major aspartyl peptidase 1 (May1) secreted from *Cryptococcus neoformans*, as a potential target for antifungal drug development [86]. Inhibitors of such targets shall be of great benefit, especially in treating cryptococcosis caused by *Cryptococcus neoformans* or *Cryptococcus gattii* in immunocompromised patients with HIV, since such inhibitors are also antiretrovirals. In addition to identifying a new promising target, the authors identified promising inhibitors derived from an *N*-terminally carboxybenzylated phenylstatine scaffold (refer to Figure 18e for general structure). Other fungal targets, such as the Ras pathway, trehalose pathway, metabolic glyoxylate cycle, high-osmolarity glycerol pathway, etc., are also under investigation.

8. Conclusions and Remarks

When comparing older antifungals with the newer ones mentioned in this review, one can sense the influence of the powerful recent advances in structural biology and medicinal chemistry on phenotypic target discovery, drug discovery, and target-based drug design. It must be noted that in order to truly evaluate the efficacy of new antifungals in treating life-threatening infections, infection-specific biomarkers should be relied on rather than death as an endpoint, since such targeted patients have multiple health comorbidities. By doing so, we ensure that no potential hit is lost as a false negative. One important observation from the examples of antifungals in development mentioned earlier in text is the presence of halogen atoms, especially fluorine and chlorine, in their chemical structures. This suggests the significant importance of halogen bonding for the interactions with their corresponding targets.

Funding: This research was funded by EFSA-CDN, grant number CZ.02.1.01/0.0/0.0/16_019/0000841, co-funded by ERDF. The APC was funded by EFSA-CDN, grant number CZ.02.1.01/0.0/0.0/16_019/0000841.

Institutional Review Board Statement: Not applicable.

Informed Consent Statement: Not applicable.

Data Availability Statement: Not applicable.

Acknowledgments: We would like to thank Daria Nawrot for reviewing the manuscript.

Conflicts of Interest: Authors declare no conflicts of interest.

References

1. Chowdhary, A.; Tarai, B.; Singh, A.; Sharma, A. Multidrug-Resistant Candida auris Infections in Critically Ill Coronavirus Disease Patients, India, April–July 2020. *Emerg. Infect. Dis.* **2020**, *26*, 2694–2696. [CrossRef]
2. Boral, H.; Metin, B.; Dogen, A.; Seyedmousavi, S.; Ilkit, M. Overview of selected virulence attributes in Aspergillus fumigants, Candida albicans, Cryptococcus neoformans, Trichophyton rubrum, and Exophiala dermatitidis. *Fungal Genet. Biol.* **2018**, *111*, 92–107. [CrossRef]
3. Du, H.; Bing, J.; Hu, T.R.; Ennis, C.L.; Nobile, C.J.; Huang, G.H. Candida auris: Epidemiology, biology, antifungal resistance, and virulence. *PLoS Pathog.* **2020**, *16*, e1008921. [CrossRef] [PubMed]
4. Odds, F.C.; Brown, A.J.; Gow, N.A. Antifungal agents: Mechanisms of action. *Trends Microbiol.* **2003**, *11*, 272–279. [CrossRef]
5. Pienaar, E.D.; Young, T.; Holmes, H. Interventions for the prevention and management of oropharyngeal candidiasis associated with HIV infection in adults and children. *Cochrane Database Syst. Rev.* **2010**, *2010*, CD003940. [CrossRef] [PubMed]
6. Vermes, A.; Guchelaar, H.J.; Dankert, J. Flucytosine: A review of its pharmacology, clinical indications, pharmacokinetics, toxicity and drug interactions. *J. Antimicrob. Chemother.* **2000**, *46*, 171–179. [CrossRef]
7. Emri, T.; Majoros, L.; Toth, V.; Pocsi, I. Echinocandins: Production and applications. *Appl. Microbiol. Biotechnol.* **2013**, *97*, 3267–3284. [CrossRef]
8. Maligie, M.A.; Selitrennikoff, C.P. Cryptococcus neoformans resistance to echinocandins: (1,3)beta-glucan synthase activity is sensitive to echinocandins. *Antimicrob. Agents Chemother.* **2005**, *49*, 2851–2856. [CrossRef]
9. Ciaravino, V.; Coronado, D.; Lanphear, C.; Shaikh, I.; Ruddock, W.; Chanda, S. Tavaborole, a novel boron-containing small molecule for the topical treatment of onychomycosis, is noncarcinogenic in 2-year carcinogenicity studies. *Int. J. Toxicol.* **2014**, *33*, 419–427. [CrossRef]
10. Nishikawa, H.; Sakagami, T.; Yamada, E.; Fukuda, Y.; Hayakawa, H.; Nomura, N.; Mitsuyama, J.; Miyazaki, T.; Mukae, H.; Kohno, S. T-2307, a novel arylamidine, is transported into Candida albicans nby a high-affinity spermine and spermidine carrier regulated by Agp2. *J. Antimicrob. Chemother.* **2016**, *71*, 1845–1855. [CrossRef]
11. Yamada, E.; Nishikawa, H.; Nomura, N.; Mitsuyama, J. T-2307 Shows Efficacy in a Murine Model of Candida glabrata Infection despite In Vitro Trailing Growth Phenomenon. *Antimicrob. Agents Chemother.* **2010**, *54*, 3630–3634. [CrossRef] [PubMed]

12. Wiederhold, N.P. Review of T-2307, an Investigational Agent That Causes Collapse of Fungal Mitochondrial Membrane Potential. *J. Fungi* **2021**, *7*, 130. [CrossRef]

13. Mann, P.A.; McLellan, C.A.; Koseoglu, S.; Si, Q.; Kuzmin, E.; Flattery, A.; Harris, G.; Sher, X.; Murgolo, N.; Wang, H.; et al. Chemical Genomics-Based Antifungal Drug Discovery: Targeting Glycosylphosphatidylinositol (GPI) Precursor Biosynthesis. *ACS Infect. Dis.* **2015**, *1*, 59–72. [CrossRef] [PubMed]

14. Nakamoto, K.; Tsukada, I.; Tanaka, K.; Matsukura, M.; Haneda, T.; Inoue, S.; Murai, N.; Abe, S.; Ueda, N.; Miyazaki, M.; et al. Synthesis and evaluation of novel antifungal agents-quinoline and pyridine amide derivatives. *Bioorganic Med. Chem. Lett.* **2010**, *20*, 4624–4626. [CrossRef] [PubMed]

15. Watanabe, N.A.; Miyazaki, M.; Horii, T.; Sagane, K.; Tsukahara, K.; Hata, K. E1210, a new broad-spectrum antifungal, suppresses Candida albicans hyphal growth through inhibition of glycosylphosphatidylinositol biosynthesis. *Antimicrob. Agents Chemother.* **2011**, *56*, 960–971. [CrossRef] [PubMed]

16. Shaw, K.J.; Ibrahim, A.S. Fosmanogepix: A Review of the First-in-Class Broad Spectrum Agent for the Treatment of Invasive Fungal Infections. *J. Fungi* **2020**, *6*, 239. [CrossRef] [PubMed]

17. Hector, R.F.; Pappagianis, D. Inhibition of Chitin Synthesis in the Cell-Wall of Coccidioides-Immitis by Polyoxin-D. *J. Bacteriol.* **1983**, *154*, 488–498. [CrossRef]

18. Fortwendel, J.R.; Juvvadi, P.R.; Pinchai, N.; Perfect, B.Z.; Alspaugh, J.A.; Perfect, J.R.; Steinbach, W.J. Differential Effects of Inhibiting Chitin and 1,3-beta-D-Glucan Synthesis in Ras and Calcineurin Mutants of Aspergillus fumigatus. *Antimicrob. Agents Chemother.* **2009**, *53*, 476–482. [CrossRef]

19. Larwood, D.J. Nikkomycin Z-Ready to Meet the Promise? *J. Fungi* **2020**, *6*, 261. [CrossRef]

20. Oliver, J.D.; Sibley, G.; Beckmann, N.; Dobb, K.S.; Slater, M.J.; McEntee, L.; du Pré, S.; Livermore, J.; Bromley, M.J.; Wiederhold, N.P.; et al. F901318 represents a novel class of antifungal drug that inhibits dihydroorotate dehydrogenase. *Proc. Natl. Acad. Sci. USA* **2016**, *113*, 12809–12814. [CrossRef]

21. Negri, C.E.; Johnson, A.; McEntee, L.; Box, H.; Whalley, S.; Schwartz, J.A.; Ramos-Martín, V.; Livermore, J.; Kolamunnage-Dona, R.; Colombo, A.L.; et al. Pharmacodynamics of the Novel Antifungal Agent F901318 for Acute Sinopulmonary Aspergillosis Caused by Aspergillus flavus. *J. Infect. Dis.* **2018**, *217*, 1118–1127. [CrossRef]

22. Wiederhold, N.P. Review of the Novel Investigational Antifungal Olorofim. *J. Fungi* **2020**, *6*, 122. [CrossRef] [PubMed]

23. Nakamura, I.; Kanasaki, R.; Yoshikawa, K.; Furukawa, S.; Fujie, A.; Hamamoto, H.; Sekimizu, K. Discovery of a new antifungal agent ASP2397 using a silkworm model of Aspergillus fumigatus infection. *J. Antibiot. (Tokyo)* **2017**, *70*, 41–44. [CrossRef]

24. Nakamura, I.; Ohsumi, K.; Takeda, S.; Katsumata, K.; Matsumoto, S.; Akamatsu, S.; Mitori, H.; Nakai, T. ASP2397 Is a Novel Natural Compound That Exhibits Rapid and Potent Fungicidal Activity against Aspergillus Species through a Specific Transporter. *Antimicrob. Agents Chemother.* **2019**, *63*, e02689-18. [CrossRef]

25. Ikai, K.; Takesako, K.; Shiomi, K.; Moriguchi, M.; Umeda, Y.; Yamamoto, J.; Kato, I.; Naganawa, H. Structure of aureobasidin A. *J. Antibiot.* **1991**, *44*, 925–933. [CrossRef]

26. Alqaisi, A.; Mbekeani, A.J.; Llorens, M.B.; Elhammer, A.P.; Denny, P.W. The antifungal Aureobasidin A and an analogue are active against the protozoan parasite Toxoplasma gondii but do not inhibit sphingolipid biosynthesis—Corrigendum. *Parasitology* **2018**, *145*, 156. [CrossRef] [PubMed]

27. Kurome, T.; Inoue, T.; Takesako, K.; Kato, I.; Inami, K.; Shiba, T. Syntheses of antifungal aureobasidin A analogs with alkyl chains for structure-activity relationship. *J. Antibiot.* **1998**, *51*, 359–367. [CrossRef]

28. Pfaller, M.A.; Messer, S.A.; Georgopapadakou, N.; Martell, L.A.; Besterman, J.M.; Diekema, D.J. Activity of MGCD290, a Hos2 Histone Deacetylase Inhibitor, in Combination with Azole Antifungals against Opportunistic Fungal Pathogens. *J. Clin. Microbiol.* **2009**, *47*, 3797–3804. [CrossRef] [PubMed]

29. Cowen, L.E. The fungal Achilles' heel: Targeting Hsp90 to cripple fungal pathogens. *Curr. Opin. Microbiol.* **2013**, *16*, 377–384. [CrossRef]

30. Pfaller, M.A.; Rhomberg, P.R.; Messer, S.A.; Castanheira, M. In vitro activity of a Hos2 deacetylase inhibitor, MGCD290, in combination with echinocandins against echinocandin-resistant Candida species. *Diagn. Microbiol. Infect. Dis.* **2015**, *81*, 259–263. [CrossRef] [PubMed]

31. Hoekstra, W.J.; Garvey, E.P.; Moore, W.R.; Rafferty, S.W.; Yates, C.M.; Schotzinger, R.J. Design and optimization of highly-selective fungal CYP51 inhibitors. *Bioorg. Med. Chem. Lett.* **2014**, *24*, 3455–3458. [CrossRef]

32. Zhang, J.X.; Li, L.P.; Lv, Q.Z.; Yan, L.; Wang, Y.; Jiang, Y.Y. The Fungal CYP51s: Their Functions, Structures, Related Drug Resistance, and Inhibitors. *Front. Microbiol.* **2019**, *10*, 691. [CrossRef] [PubMed]

33. Warrilow, A.G.S.; Parker, J.E.; Price, C.L.; Nes, W.D.; Garvey, E.P.; Hoekstra, W.J.; Schotzinger, R.J.; Kelly, D.E.; Kelly, S.L. The Investigational Drug VT-1129 Is a Highly Potent Inhibitor of Cryptococcus Species CYP51 but Only Weakly Inhibits the Human Enzyme. *Antimicrob. Agents Chemother.* **2016**, *60*, 4530–4538. [CrossRef]

34. Garvey, E.P.; Hoekstra, W.J.; Schotzinger, R.J.; Sobel, J.D.; Lilly, E.A.; Fidel, P.L., Jr. Efficacy of the Clinical Agent VT-1161 against Fluconazole-Sensitive and -Resistant Candida albicans in a Murine Model of Vaginal Candidiasis. *Antimicrob. Agents Chemother.* **2015**, *59*, 5567–5573. [CrossRef]

35. Wiederhold, N.P.; Lockhart, S.R.; Najvar, L.K.; Berkow, E.L.; Jaramillo, R.; Olivo, M.; Garvey, E.P.; Yates, C.M.; Schotzinger, R.J.; Catano, G.; et al. The Fungal Cyp51-Specific Inhibitor VT-1598 Demonstrates In Vitro and In Vivo Activity against Candida auris. *Antimicrob. Agents Chemother.* **2019**, *63*, e02233-18. [CrossRef]

36. Bowman, J.C.; Hicks, P.S.; Kurtz, M.B.; Rosen, H.; Schmatz, D.M.; Liberator, P.A.; Douglas, C.M. The antifungal echinocandin caspofungin acetate kills growing cells of Aspergillus fumigatus in vitro. *Antimicrob. Agents Chemother.* **2002**, *46*, 3001–3012. [CrossRef]

37. Odds, F.C. Drug evaluation: BAL-8557—A novel broad-spectrum triazole antifungal. *Curr. Opin. Investig. Drugs* **2006**, *7*, 766–772. [PubMed]

38. Schmitt-Hoffmann, A.; Roos, B.; Heep, M.; Schleimer, M.; Weidekamm, E.; Brown, T.; Roehrle, M.; Beglinger, C. Single-ascending-dose pharmacokinetics and safety of the novel broad-spectrum antifungal triazole BAL4815 after intravenous infusions (50, 100, and 200 milligrams) and oral administrations (100, 200, and 400 milligrams) of its prodrug, BAL8557, in healthy volunteers. *Antimicrob. Agents Chemother.* **2006**, *50*, 279–285.

39. Bartroli, J.; Turmo, E.; Algueró, M.; Boncompte, E.; Vericat, M.L.; Conte, L.; Ramis, J.; Merlos, M.; García-Rafanell, J.; Forn, J. New azole antifungals. 2. Synthesis and antifungal activity of heterocyclecarboxamide derivatives of 3-amino-2-aryl-1-azolyl-2-butanol. *J. Med. Chem.* **1998**, *41*, 1855–1868. [CrossRef] [PubMed]

40. Sun, N.; Xie, Y.; Sheng, C.; Cao, Y.; Zhang, W.; Chen, H.; Fan, G. In vivo pharmacokinetics and in vitro antifungal activity of iodiconazole, a new triazole, determined by microdialysis sampling. *Int. J. Antimicrob. Agents* **2013**, *41*, 229–235. [CrossRef]

41. Brautaset, T.; Sletta, H.; Degnes, K.F.; Sekurova, O.N.; Bakke, I.; Volokhan, O.; Andreassen, T.; Ellingsen, T.E.; Zotchev, S.B. New nystatin-related antifungal polyene macrolides with altered polyol region generated via biosynthetic engineering of Streptomyces noursei. *Appl. Environ. Microbiol.* **2011**, *77*, 6636–6643. [CrossRef] [PubMed]

42. Krishnan, B.R.; James, K.D.; Polowy, K.; Bryant, B.J.; Vaidya, A.; Smith, S.; Laudeman, C.P. CD101, a novel echinocandin with exceptional stability properties and enhanced aqueous solubility. *J. Antibiot.* **2017**, *70*, 130–135. [CrossRef] [PubMed]

43. Ong, V.; James, K.D.; Smith, S.; Krishnan, B.R. Pharmacokinetics of the Novel Echinocandin CD101 in Multiple Animal Species. *Antimicrob. Agents Chemother.* **2017**, *61*, e01626-16. [CrossRef]

44. Zhao, Y.; Prideaux, B.; Nagasaki, Y.; Lee, M.H.; Chen, P.Y.; Blanc, L.; Ho, H.; Clancy, C.J.; Nguyen, M.H.; Dartois, V.; et al. Unraveling Drug Penetration of Echinocandin Antifungals at the Site of Infection in an Intra-abdominal Abscess Model. *Antimicrob. Agents Chemother.* **2017**, *61*, e01009-17. [CrossRef] [PubMed]

45. Abuhelwa, A.Y.; Foster, D.J.R.; Mudge, S.; Hayes, D.; Upton, R.N. Population Pharmacokinetic Modeling of Itraconazole and Hydroxyitraconazole for Oral SUBA-Itraconazole and Sporanox Capsule Formulations in Healthy Subjects in Fed and Fasted States. *Antimicrob. Agents Chemother.* **2015**, *59*, 5681–5696. [CrossRef]

46. Gupta, A.K.; Surprenant, M.S.; Kempers, S.E.; Pariser, D.M.; Rensfeldt, K.; Tavakkol, A. Efficacy and safety of topical terbinafine 10% solution (MOB-015) in the treatment of mild to moderate distal subungual onychomycosis: A randomized, multicenter, double-blind, vehicle-controlled phase 3 study. *J. Am. Acad. Dermatol.* **2021**, *85*, 95–104. [CrossRef]

47. Pinna, A.; Donadu, M.G.; Usai, D.; Dore, S.; Boscia, F.; Zanetti, S. In Vitro Antimicrobial Activity of a New Ophthalmic Solution Containing Hexamidine Diisethionate 0.05% (Keratosept). *Cornea* **2020**, *39*, 1415–1418. [CrossRef]

48. Pinna, A.; Donadu, M.G.; Usai, D.; Dore, S.; D'Amico-Ricci, G.; Boscia, F.; Zanetti, S. In vitro antimicrobial activity of a new ophthalmic solution containing povidone-iodine 0.6% (IODIM (R)). *Acta Ophthalmol.* **2020**, *98*, E178–E180. [CrossRef]

49. Santangelo, R.; Paderu, P.; Delmas, G.; Chen, Z.W.; Mannino, R.; Zarif, L.; Perlin, D.S. Efficacy of oral cochleate-amphotericin B in a mouse model of systemic candidiasis. *Antimicrob. Agents Chemother.* **2000**, *44*, 2356–2360. [CrossRef]

50. Gasser, G.; Metzler-Nolte, N. The potential of organometallic complexes in medicinal chemistry. *Curr. Opin. Chem. Biol.* **2012**, *16*, 84–91. [CrossRef]

51. Kljun, J.; Scott, A.J.; Rizner, T.L.; Keiser, J.; Turel, I. Synthesis and Biological Evaluation of Organoruthenium Complexes with Azole Antifungal Agents. First Crystal Structure of a Tioconazole Metal Complex. *Organometallics* **2014**, *33*, 1594–1601. [CrossRef]

52. Stevanović, N.L.; Aleksic, I.; Kljun, J.; Bogojevic, S.S.; Veselinovic, A.; Nikodinovic-Runic, J.; Turel, I.; Djuran, M.I.; Glišić, B.Đ. Copper(II) and Zinc(II) Complexes with the Clinically Used Fluconazole: Comparison of Antifungal Activity and Therapeutic Potential. *Pharmaceuticals* **2021**, *14*, 24. [CrossRef] [PubMed]

53. Andrejević, T.P.; Aleksic, I.; Počkaj, M.; Kljun, J.; Milivojevic, D.; Stevanović, N.L.; Nikodinovic-Runic, J.; Turel, I.; Djuran, M.I.; Glišić, B.Đ. Tailoring copper(II) complexes with pyridine-4,5-dicarboxylate esters for anti-Candida activity. *Dalton Trans.* **2021**, *50*, 2627–2638. [CrossRef]

54. Savic, N.D.; Vojnovic, S.; Glisic, B.Đ.; Crochet, A.; Pavic, A.; Janjic, G.V.; Pekmezovic, M.; Opsenica, I.M.; Fromm, K.M.; Nikodinovic-Runic, J.; et al. Mononuclear silver(I) complexes with 1,7-phenanthroline as potent inhibitors of Candida growth. *Eur. J. Med. Chem.* **2018**, *156*, 760–773. [CrossRef]

55. Polvi, E.J.; Averette, A.F.; Lee, S.C.; Kim, T.; Bahn, Y.; Veri, A.O.; Robbins, N.; Heitman, J.; Cowen, L.E. Metal Chelation as a Powerful Strategy to Probe Cellular Circuitry Governing Fungal Drug Resistance and Morphogenesis. *PLoS Genet.* **2016**, *12*, e1006350. [CrossRef] [PubMed]

56. Lin, Y.; Betts, H.; Keller, S.; Cariou, K.; Gasser, G. Recent developments of metal-based compounds against fungal pathogens. *Chem. Soc. Rev.* **2021**, *50*, 10346–10402. [CrossRef] [PubMed]

57. Nishikawa, H.; Fukuda, Y.; Mitsuyama, J.; Tashiro, M.; Tanaka, A.; Takazono, T.; Saijo, T.; Yamamoto, K.; Nakamura, S.; Imamura, Y.; et al. In vitro and in vivo antifungal activities of T-2307, a novel arylamidine, against Cryptococcus gattii: An emerging fungal pathogen. *J. Antimicrob. Chemother.* **2017**, *72*, 1709–1713. [CrossRef]

58. Sun, N.; Wen, J.; Lu, G.; Hong, Z.; Fan, G.; Wu, Y.; Sheng, C.; Zhang, W. An ultra-fast LC method for the determination of iodiconazole in microdialysis samples and its application in the calibration of laboratory-made linear probes. *J. Pharm. Biomed. Anal.* **2010**, *51*, 248–251. [CrossRef] [PubMed]

59. Ravu, R.R.; Jacob, M.R.; Khan, S.I.; Wang, M.; Cao, L.; Agarwal, A.K.; Clark, A.M.; Li, X. Synthesis and Antifungal Activity Evaluation of Phloeodictine Analogues. *J. Nat. Prod.* **2021**, *84*, 2129–2137. [CrossRef] [PubMed]

60. Choi, J.W.; Lee, K.; Kim, S.; Lee, Y.R.; Kim, H.J.; Seo, K.J.; Lee, M.H.; Yeon, S.K.; Jang, B.K.; Park, S.J.; et al. Optimization and Evaluation of Novel Antifungal Agents for the Treatment of Fungal Infection. *J. Med. Chem.* **2021**, *64*, 15912–15935. [CrossRef] [PubMed]

61. Kato, I.; Ukai, Y.; Kondo, N.; Nozu, K.; Kimura, C.; Hashimoto, K.; Mizusawa, E.; Maki, H.; Naito, A.; Kawai, M. Identification of Thiazoyl Guanidine Derivatives as Novel Antifungal Agents Inhibiting Ergosterol Biosynthesis for Treatment of Invasive Fungal Infections. *J. Med. Chem.* **2021**, *64*, 10482–10496.

62. Li, Z.; Tu, J.; Han, G.Y.; Liu, N.; Sheng, C.Q. Novel Carboline Fungal Histone Deacetylase (HDAC) Inhibitors for Combinational Treatment of Azole-Resistant Candidiasis. *J. Med. Chem.* **2021**, *64*, 1116–1126. [CrossRef] [PubMed]

63. Krishnamurthy, S.; Plaine, A.; Albert, J.; Prasad, T.; Prasad, R.; Ernst, J.F. Dosage-dependent functions of fatty acid desaturase Ole1p in growth and morphogenesis of Candida albicans. *Microbiology-Sgm* **2004**, *150*, 1991–2003. [CrossRef] [PubMed]

64. DeJarnette, C.; Meyer, C.J.; Jenner, A.R.; Butts, A.; Peters, T.; Cheramie, M.N.; Phelps, G.A.; Vita, N.A.; Loudon-Hossler, V.C.; Lee, R.E.; et al. Identification of Inhibitors of Fungal Fatty Acid Biosynthesis. *ACS Infect. Dis.* **2021**, *7*, 3210–3223. [CrossRef] [PubMed]

65. Lowes, D.J.; Miao, J.; Al-Waqfi, R.A.; Avad, K.A.; Hevener, K.E.; Peters, B.M. Identification of Dual-Target Compounds with Anti-fungal and Anti-NLRP3 Inflammasome Activity. *ACS Infect. Dis.* **2021**, *7*, 2522–2535. [CrossRef] [PubMed]

66. Pue, N.; Guddat, L.W. Acetohydroxyacid Synthase: A Target for Antimicrobial Drug Discovery. *Curr. Pharm. Des.* **2014**, *20*, 740–753. [CrossRef] [PubMed]

67. Camilli, G.; Griffiths, J.S.; Ho, J.; Richardson, J.P.; Naglik, J.R. Some like it hot: Candida activation of inflammasomes. *PLoS Pathog.* **2020**, *16*, e1008975. [CrossRef]

68. Teixeira, V.; Feio, M.J.; Bastos, M. Role of lipids in the interaction of antimicrobial peptides with membranes. *Prog. Lipid Res.* **2012**, *51*, 149–177. [CrossRef]

69. Gomes, B.; Augusto, M.T.; Felício, M.R.; Hollmann, A.; Franco, O.L.; Gonçalves, S.; Santos, N.C. Designing improved active peptides for therapeutic approaches against infectious diseases. *Biotechnol. Adv.* **2018**, *36*, 415–429. [CrossRef]

70. Struyfs, C.; Cammue, B.P.A.; Thevissen, K. Membrane-Interacting Antifungal Peptides. *Front. Cell Dev. Biol.* **2021**, *9*, 649875. [CrossRef]

71. Donadu, M.G.; Peralta-Ruiz, Y.; Usai, D.; Maggio, F.; Molina-Hernandez, J.B.; Rizzo, D.; Bussu, F.; Rubino, S.; Zanetti, S.; Paparella, A. Colombian Essential Oil of Ruta graveolens against Nosocomial Antifungal Resistant Candida Strains. *J. Fungi* **2021**, *7*, 383. [CrossRef] [PubMed]

72. Donadu, M.G.; Usai, D.; Marchetti, M.; Usai, M.; Mazzarello, V.; Molicotti, P.; Montesu, M.A.; Delogu, G.; Zanetti, S. Antifungal activity of oils macerates of North Sardinia plants against Candida species isolated from clinical patients with candidiasis. *Nat. Prod. Res.* **2020**, *34*, 3280–3284. [CrossRef] [PubMed]

73. Bua, A.; Usai, D.; Donadu, M.G.; Delgado Ospina, J.; Paparella, A.; Chaves-Lopez, C.; Serio, A.; Rossi, C.; Zanetti, S.; Molicotti, P. Antimicrobial activity of Austroeupatorium inulaefolium (HBK) against intracellular and extracellular organisms. *Nat. Prod. Res.* **2018**, *32*, 2869–2871. [CrossRef] [PubMed]

74. Dolan, K.; Montgomery, S.; Buchheit, B.; DiDone, L.; Wellingtone, M.; Krysan, D.J. Antifungal Activity of Tamoxifen: In Vitro and In Vivo Activities and Mechanistic Characterization. *Antimicrob. Agents Chemother.* **2009**, *53*, 3337–3346. [CrossRef] [PubMed]

75. Koselny, K.; Green, J.; DiDone, L.; Halterman, J.P.; Fothergill, A.W.; Wiederhold, N.P.; Patterson, T.F.; Cushion, M.T.; Rappelye, C.; Wellington, M.; et al. The Celecoxib Derivative AR-12 Has Broad-Spectrum Antifungal Activity In Vitro and Improves the Activity of Fluconazole in a Murine Model of Cryptococcosis. *Antimicrob. Agents Chemother.* **2016**, *60*, 7115–7127. [CrossRef]

76. Posch, W.; Wilflingseder, D.; Lass-Florl, C. Immunotherapy as an Antifungal Strategy in Immune Compromised Hosts. *Curr. Clin. Microbiol. Rep.* **2020**, *7*, 57–66. [CrossRef]

77. Rachini, A.; Pietrella, D.; Lupo, P.; Torosantucci, A.; Chiani, P.; Bromuro, C.; Proietti, C.; Bistoni, F.; Cassone, A.; Vecchiarelli, A. An anti-beta-glucan monoclonal antibody inhibits growth and capsule formation of Cryptococcus neofonnans in vitro and exerts therapeutic, anticryptococcal activity in vivo. *Infect. Immun.* **2007**, *75*, 5085–5094. [CrossRef]

78. Mambro, T.D.; Vanzolini, T.; Bruscolini, P.; Perez-Gaviro, S.; Marra, E.; Roscilli, G.; Bianchi, M.; Fraternale, A.; Schiavano, G.F.; Canonico, B.; et al. A new humanized antibody is effective against pathogenic fungi in vitro. *Sci. Rep.* **2021**, *11*, 19500. [CrossRef] [PubMed]

79. Nosanchuk, J.D.; Dadachova, E. Radioimmunotherapy of fungal diseases: The therapeutic potential of cytocidal radiation delivered by antibody targeting fungal cell surface antigens. *Front. Microbiol.* **2012**, *3*, 283. [CrossRef]

80. Heung, L.J.; Luberto, C.; Del Poeta, M. Role of sphingolipids in microbial pathogenesis. *Infect. Immun.* **2006**, *74*, 28–39. [CrossRef]

81. Levery, S.B.; Momany, M.; Lindsey, R.; Toledo, M.S.; Shayman, J.A.; Fuller, M.; Brooks, K.; Doong, R.L.; Straus, A.H.; Takahashi, H.K. Disruption of the glucosylceramide biosynthetic pathway in Aspergillus nidulans and Aspergillus fumigatus by inhibitors of UDP-Glc: Ceramide glucosyltransferase strongly affects spore germination, cell cycle, and hyphal growth (vol 525, pg 59, 2002). *Febs Lett.* **2002**, *526*, 151. [CrossRef]

82. Rittershaus, P.C.; Kechichian, T.B.; Allegood, J.C.; Merrill, A.H., Jr.; Hennig, M.; Luberto, C.; Del Poeta, M. Glucosylceramide synthase is an essential regulator of pathogenicity of Cryptococcus neoformans. *J. Clin. Investig.* **2006**, *116*, 1651–1659. [CrossRef] [PubMed]
83. Mandala, S.M.; Thornton, R.A.; Frommer, B.R.; Curotto, J.E.; Rozdilsky, W.; Kurtz, M.B.; Giacobbe, R.A.; Bills, G.F.; Cabello, M.A.; Martín, I. The Discovery of Australifungin, a Novel Inhibitor of Sphinganine N-Acyltransferase from Sporormiella-Australis—Producing Organism, Fermentation, Isolation, and Biological-Activity. *J. Antibiot.* **1995**, *48*, 349–356. [CrossRef] [PubMed]
84. Achenbach, H.; Mühlenfeld, A.; Fauth, U.; Zähner, H. The galbonolides. Novel, powerful antifungal macrolides from Streptomyces galbus ssp. eurythermus. *Ann. N. Y. Acad. Sci.* **1988**, *544*, 128–140. [CrossRef] [PubMed]
85. Mora, V.; Rellaa, A.; Farnouda, A.M.; Singha, A.; Munshia, M.; Bryana, A.; Naseema, S.; Konopkaa, J.B.; Ojimab, I.; Bullesbachc, E.; et al. Identification of a New Class of Antifungals Targeting the Synthesis of Fungal Sphingolipids. *Mbio* **2015**, *6*, e00647-15. [CrossRef] [PubMed]
86. Kryštůfek, R.; Šácha, P.; Starková, J.; Brynda, J.; Hradilek, M.; Tloušt'ová, E.; Grzymska, J.; Rut, W.; Boucher, M.J.; Drag, M.; et al. Re-emerging Aspartic Protease Targets: Examining Cryptococcus neoformans Major Aspartyl Peptidase 1 as a Target for Antifungal Drug Discovery. *J. Med. Chem.* **2021**, *64*, 6706–6719. [CrossRef] [PubMed]

pharmaceuticals

Article

Synthesis and Study of Antifungal Properties of New Cationic Beta-Glucan Derivatives

Kamil Kaminski [1,*], Magdalena Skora [2], Paweł Krzyściak [2], Sylwia Stączek [3], Agnieszka Zdybicka-Barabas [3] and Małgorzata Cytryńska [3]

[1] Faculty of Chemistry, Jagiellonian University, Gronostajowa 2 St., 30-387 Krakow, Poland
[2] Department of Infections Control and Mycology, Chair of Microbiology, Jagiellonian University Medical College, Czysta 18 St., 31-121 Krakow, Poland; magdalena.skora@uj.edu.pl (M.S.); pawel.krzysciak@uj.edu.pl (P.K.)
[3] Department of Immunobiology, Institute of Biological Sciences, Faculty of Biology and Biotechnology, Maria Curie-Sklodowska University, Akademicka 19 St., 20-033 Lublin, Poland; s.staczek@poczta.umcs.lublin.pl (S.S.); barabas@poczta.umcs.lublin.pl (A.Z.-B.); cytryna@poczta.umcs.lublin.pl (M.C.)
* Correspondence: kaminski@chemia.uj.edu.pl

Abstract: The interaction of positively charged polymers (polycations) with a biological membrane is considered to be the cause of the frequently observed toxicity of these macromolecules. If it is possible to obtain polymers with a predominantly negative effect on bacterial and fungal cells, such systems would have great potential in the treatment of infectious diseases, especially now when reports indicate the growing risk of fungal co-infections in COVID-19 patients. We describe in this article cationic derivatives of natural beta-glucan polymers obtained by reacting the polysaccharide isolated from *Saccharomyces boulardii* (SB) and *Cetraria islandica* (CI) with glycidyl trimethyl ammonium chloride (GTMAC). Two synthesis strategies were applied to optimize the product yield. Fungal diseases particularly affect low-income countries, hence the emphasis on the simplicity of the synthesis of such drugs so they can be produced without outside help. The three structures obtained showed selective anti-mycotic properties (against, i.e., *Scopulariopsis brevicaulis*, *Aspergillus brasiliensis*, and *Fusarium solani*), and their toxicity established using fibroblast 3T3-L1 cell line was negligible in a wide range of concentrations. For one of the polymers (SB derivative), using in vivo model of *Aspergillus brasiliensis* infection in *Galleria mellonella* insect model, we confirmed the promising results obtained in the preliminary study.

Keywords: antifungal; beta-glucan; polycations; *Galleria mellonella* model

Citation: Kaminski, K.; Skora, M.; Krzyściak, P.; Stączek, S.; Zdybicka-Barabas, A.; Cytryńska, M. Synthesis and Study of Antifungal Properties of New Cationic Beta-Glucan Derivatives. *Pharmaceuticals* **2021**, 14, 838. https://doi.org/10.3390/ph14090838

Academic Editors: Jong Heon Kim, Luisa W. Cheng and Kirkwood Land

Received: 20 July 2021
Accepted: 23 August 2021
Published: 24 August 2021

1. Introduction

Fungal infections are one of the most common diseases affecting people all over the world. Fungi are responsible for over a billion infections globally and cause more than 1.5 million deaths worldwide [1]. Most people during their lifetime will suffer from superficial or mucocutaneous fungal infections, which generally do not have serious health consequences but often significantly reduce the quality of life. However, millions will develop life-threatening deep infections (invasive fungal infections—IFI), frequently a multiorgan mycosis including fungaemia, that are much more difficult to diagnose and treat. Mortality rates of IFI depend on the causative agent and the presence of host predisposing factors but often exceed 50% [2]. An increase in the frequency of mycoses has been observed for years. Paradoxically it is related to the progress of medicine, which allows patients with serious underlying diseases to be kept alive, contributing directly or as a result of therapy to increased susceptibility to mycoses (e.g., HIV infection and AIDS, iatrogenic immunosuppression, invasive medical procedures, antibiotic therapy, steroid therapy, *diabetes mellitus*).

Our reality has been irrevocably changed by a global COVID-19 pandemic that has been ongoing since the beginning of 2020. In the recent scientific literature, we find reports indicating that critically ill COVID-19 patients have increased risk of serious fungal infections, such as invasive pulmonary aspergillosis, invasive candidiasis, or *Pneumocystis jirovecii* pneumonia [3,4]. The latest news indicates that COVID-19 may also predispose to mucormycosis, a rare angioinvasive disease with high morbidity and mortality caused by multi-drug resistant fungi of the order *Mucorales* [5–11]. Prophylaxis of airborne infections includes the use of protective masks. In the era of a COVID-19 pandemic, masks have become mandatory almost all over the world, especially indoors. Wearing masks that are not always handled according to the manufacturer's instructions and also frequent disinfection of the skin make us more susceptible to the development of diseases caused by various pathogens, including fungi. This applies in particular to the skin which, under the influence of alcohol-based disinfectants, loses part of its natural insulating components and natural skin microbiota that protects against the attack of pathogenic microorganisms. Similarly, moist and not often enough changed masks covering the nose and mouth are a perfect place for fungi to grow and may lead to the development of superficial or even respiratory mycoses. These facts indicate that among the social and economic problems that will remain for humanity to face after the pandemic, there may also be another epidemic—this time, one that is associated with fungal diseases.

The therapeutic options for fungal infections are very limited. The currently available antifungal drugs belong to only a few classes based on their mode of action—namely, polyenes that bind to ergosterol in the fungal cell membrane, leading to cell lysis, azoles, allylamines, and morpholine analog (amorolphine) that target ergosterol biosynthesis; echinocandins, which inhibit the synthesis of beta-glucan in fungal cell wall; analog of pyrimidine (flucytosine) that interferes with pyrimidine metabolism, RNA/DNA, and protein synthesis; and mitotic inhibitor (griseofulvine) that interferes with microtubule function. Many drugs have very limited use due to the route of administration, toxicity and side effects, and drug interactions, as well as due to bioavailability at the target site (e.g., penetration into brain tissue) [12]. As for treatment of most the common fungal infection, candidiasis, there have been several breakthroughs in recent years using prediction models for patients susceptible to such infections due to associated diseases or other immune system impairments [13,14].

The treatments associated with fungi are undoubtedly a complex issue; nevertheless, there is an undeniable need for new substances with antifungal activity. The development of new antimycotics that are cost-effective and active against a variety of fungi is not only an essential medical need but also a socio-economic necessity. In particular, access to established antifungal drugs is often limited to developed countries, while geographic and sanitary conditions make them most relevant to countries with lower incomes. The production of such drugs should be technically simple and not burdened with profit-driven restrictions imposed by big pharma, which in practice means the need to create a completely new concept. The concept of polymeric drugs fits into all this in a very good way.

Beta-glucans—a subgroup of polysaccharides that are distinguished by the type of bond connecting glucose molecules in the main polymer chain—have recently attracted particular interest in medicine [15]. There are two main postulated applications of beta-glucans related to the treatment of diabetes [16] and to use as immunomodulators [17]. It is speculated that beta-glucans may be one of the factors responsible for the lower incidence of cancer on the Asian continent, due to their presence in mushrooms, an important component of the Asian diet [18]. Beta-glucans, especially the ones that are not soluble in water, are mainly obtained from mushrooms [19]. Limited solubility in water may in the future become a problem in terms of a treatment's route of administration, so it is reasonable to study the activity of charged beta-glucan derivatives whose solubility would be much better, increasing the chance that the biological activity will be retained. Particularly interesting here may be cationization, i.e., adding a positive charge to the structure of polymers using GTMAC. This is a frequently reported method in the literature to cationize polysaccharides for medical purposes [20]. The polysaccharides that result from this reaction often have antibacterial properties [21]. Combining these two facts, i.e., positive charge and origin from fungi, and thus resulting in a spatial structure similar to the natural component of the fungal cell wall, we can assume that cationic beta-glucans will have antifungal properties.

Polymers with a positive charge that can easily interact with a negatively charged biological membrane seem to be particularly promising in this regard [22]. This feature leads polycations to exhibit biocidal activity, especially bactericidal and bacteriostatic activity, which has been documented in literature [23]. In view of this observation, it seems reasonable to hypothesize here that this property will also apply to fungi. From the point of view of medical applications, more important than showing the antimycotic property itself is to demonstrate a polycation structure that will not be toxic to mammalian cells at concentrations that are toxic to fungal cells. This is exactly what we were able to show in this work involving obtained cationic derivatives of beta-glucans of *Saccharomyces boulardii* and *Cetraria islandica*.

2. Results

2.1. Chemical Characteristics of the Obtained Materials

2.1.1. Elemental Analysis

The mass percentage of a given chemical element in the material in subsequent purification stages allows for a preliminary assessment of the effectiveness of the polysaccharides isolation from biological material. In the case of polysaccharides isolation procedures, the most undesirable group of contaminants is proteins. Elemental analysis allows determination of the amount of protein indirectly as the amount of nitrogen in the sample.

Comparing the preliminary and final stages of purification (data in Table 1) in the case of both sources, it can be concluded that the material was effectively deproteinized because the amount of nitrogen decreased. In the case of *C. islandica* (CI), where the amount of protein in the starting material was low from the beginning, it dropped to zero. In the case of *S. boulardii* (SB), containing more proteins, a negligible amount of nitrogen remained in the final material used for cationization, which may indicate a marginal quantity of proteins remaining in the sample. Cationization of polysaccharides is carried out by attaching a quaternary amine, which will cause an increase in the amount of nitrogen in the structure of the molecule. We observed this phenomenon also for our results, which confirmed the effectiveness of cationization.

Table 1. Elemental composition in the obtained materials at subsequent stages of isolation and modification of polysaccharides. The results are presented as mean of three independent measurements ± standard deviation (SD).

	The Elemental Composition of the Obtained Material (%)			
	N	C	H	S
Freeze-dried *Saccharomyces boulardii* (SB)	6.02 ± 0.03	44.44 ± 0.03	6.80 ± 0.04	0.00 ± 0.01
Dry lysed SB pellet (lSBp)	5.33 ± 0.04	42.18 ± 0.18	6.75 ± 0.01	0.00 ± 0.01
Purified beta-glucan from SB (SBB)	0.32 ± 0.01	38.60 ± 0,01	6.31 ± 0.10	0.00 ± 0.01
SBB cationized using GTMAC (SBBGTMAC)	1.44 ± 0.03	41.65 ± 0.05	7.04 ± 0.02	0.00 ± 0.01
SB raw cell wall components cationized using GTMAC (SBBGTMAC2)	6.68 ± 0.17	45.76 ± 0.24	7.60 ± 0.01	0.00 ± 0.01
Dried and well ground *Cetraria islandica* (CI)	0.22 ± 0.01	40.55 ± 0.01	6.58 ± 0.06	0.70 ± 0.02
CI after ethanol impurities extraction (EPCI)	1.45 ± 0.02	43.81 ± 0.23	6.45 ± 0.09	0.00 ± 0.01
Product of initial CI warm water extraction (IECI)	0.03 ± 0.01	36.21 ± 0.13	6.27 ± 0.12	0.00 ± 0.01
Purified beta-glucan from CI (BCI)	0.00 ± 0.01	36.16 ± 0.23	6.75 ± 0.14	0.00 ± 0.01
CI cationized using GTMAC (CIGTMAC)	2.35 ± 0.02	41.92 ± 0.11	7.53 ± 0.01	0.00 ± 0.01

2.1.2. Zeta Potential of the Systems Studied

The zeta potential value is the most compelling evidence for the effectiveness of cationization. In the case of the new polymer structures proposed in this work, as we postulated, the positive charge of the molecule is a necessary condition for obtaining biological activity. Zeta potential measurements were carried out for all obtained polycations, which are a good approximation of the charge in the case of polyelectrolytes [Table 2]. Additionally, the zeta potential of the water-soluble substrate for the synthesis of the yeast-derived polysaccharide was measured to evaluate the charge of these polymers [Table 2].

Table 2. Zeta potential values measured for crude polysaccharide and corresponding cationic derivatives. The results are presented as mean of three independent measurements ± standard deviation (SD).

	SBB	SBBGTMAC	SBBGTMAC2	CIGTMAC
Zeta potential (mV)	-5.31 ± 0.56	32.93 ± 0.35	43.77 ± 1.05	36.16 ± 0.40

2.1.3. GPC Chromatography Assessment of Molecular Weight and Its Distribution

Polysaccharides occur in nature as a collections of macromolecules of different masses; therefore, when examining these polymers, one should always think in terms of a heterogeneous system of greater or lesser molecular weight dispersion. In this work we isolated mixtures of macromolecules and then obtained cationic derivatives of their systems, which is why assessment of mass and mass distribution using GPC chromatography is necessary. Full chromatograms are included in the Supplementary Materials (Figure S3) and a summary of the results is presented in Table 3.

Table 3. The molecular weight determined from GPC measurements for crude polysaccharide and corresponding cationic derivatives.

	SBB	SBBGTMAC	SBBGTMAC2	CIGTMAC
Average molecular weight (kDa) (based on GPC)	24.95 *	29.30 **	23.92 **	53.48 **
Mass dispersion index	1.1	1.2	1.5	1.5

* PEG as mass standard; ** PVP as mass standard.

According to the literature, the molecular weight of beta-glucans may range from 0.1 to 1000 kDa [24]; therefore, it can be said that the materials we received were in the middle of this range. Regardless of the source of raw substrate and method of cationization, we obtained materials with a similar mass, which may result from the fact that the cationization and isolation were carried out under similar moderate basic conditions in all cases (alkaline treatment was necessary due to the negligible solubility of most beta-glucans in water). This suggests that we were dealing with partially alkaline transformed beta-glucans. The second important parameter studied here was mass dispersion. In the case of the unmodified yeast beta-glucan, the results show that this simple isolation procedure gave a material that was surprisingly homogeneous in terms of weight distribution. The cationization of this compound worsened these parameters to some extent, which is to be expected in any reaction involving macromolecules not completely modified (not all OH groups present in the polymer react due to steric hindrance). In the case of SBBGTMAC2 and CIGTMAC, the mass dispersion was closer to that which would be expected for systems based on additionally modified natural polymers.

2.1.4. Material Analysis Based on FT-IR Spectra

GPC gave us information about the molar masses of chemical compounds present in the tested samples, while more detailed data about the chemical composition (presence of functional groups) can be obtained from IR spectroscopy. We were dealing here with mixtures of similar compounds, so it was not possible to indicate directly complete structures, but it was possible to indicate with a high probability the chemical composition of the main components and potential impurities. In this case, these data allowed mainly for the assessment of material deproteinization at subsequent stages of purification (disappearance of bands at about 2850 cm^{-1} from $NH_3{}^+$ stretching oscillations, appearance of a broad well-defined bell-shaped band at about 3300 cm^{-1} from OH groups bound in hydrogen bonding) and confirmation of the successful cationization of polysaccharides (vibration of methyl groups in a quaternary amine at 1480 cm^{-1}; no clearly separated peak, but an increase in absorbance on the slope of the stronger band is seen for this length). The graphs of the full FT IR spectra can be found in the Supplementary Materials (Figure S1 for SBB derivatives and Figure S2 for CI derivative).

2.2. Initial Assessment of Biological Properties of New Materials

2.2.1. Antifungal Properties of the Obtained Polycations

The cationic beta-glucan derivatives were inactive against *Candida* strains. In the tested concentration range, polycations did not inhibit or weaken the growth of yeasts, and MIC values could not be identified. For filamentous fungi, the antifungal activity of the polymers was varied. The best antifungal effect was observed for *S. brevicaulis*, for which SBBGTMAC2 MIC was 62.5 mg/L, and SBBGTMAC MIC ranges were 250–1000 mg/L. For CIGTMAC, we noted a marked reduction in growth of the fungus relative to the growth control well without polycation at concentrations $\geq$ 3.9 mg/L, but complete growth inhibition did not occur. Similar results with significant reduction in fungal growth were obtained for SBBGTMAC2 against *A. brasiliensis* and *F. solani* for concentrations $\geq$ 125 mg/L and 250 mg/L, respectively. SBBGTMAC showed antifungal activity against two of the three tested *Aspergillus* species, i.e., *A. brasiliensis*, for which MIC values ranged 62.5–125 mg/L, and to a lesser extent *A. flavus*, the growth of which was diminished at concentrations $\geq$ 125–500 mg/L. The detailed results of the study of the antifungal properties of the polycations are presented in Table 4.

Table 4. MIC values (mg/L) for CIGTMAC, SBBGTMAC2, and SBBGTMAC against various fungal species. The results obtained from three replications are presented.

Name of Fungal Strain	MIC (mg/L)		
	CIGTMAC	**SBBGTMAC2**	**SBBGTMAC**
Candida albicans ATCC 90028	>1000	>1000	>1000
Candida glabrata ATCC 15454	>1000	>1000	>1000
Candida krusei ATCC 6258	>1000	>1000	>1000
Aspergillus flavus ATCC 204304	>1000	>1000	>1000 concentration $\geq$ 125–500 mg/L—diminished growth compared with control
Aspergillus fumigatus	>1000	>1000	>1000
Aspergillus brasiliensis ATCC 16404	>1000	>1000 concentration $\geq$ 125 mg/L—diminished growth compared with control	62.5–125
Trichophyton mentagrophytes ATCC 18748	>1000	>1000	>1000
Fusarium solani	>1000	>1000 concentration $\geq$ 250 mg/L—diminished growth compared with control	>1000
Scopulariopsis brevicaulis	>1000 concentration $\geq$ 3.9 mg/L—diminished growth compared with control	62.5	250–1000

2.2.2. Evaluation of the Toxicity of the Obtained Polycations In Vitro Using Fibroblast Cell Line

A chemical compound that is a good candidate for an active ingredient in antimycotic formulations cannot exhibit universal biocidal properties. The toxicity towards eukaryotic cells, which should be low for such compounds, is especially important because it will let us assess, in advance, the potential side effects. Hence, an evaluation of the cytotoxicity of the obtained polycations against fibroblasts, the most common cells present in the mammalian body, was carried out (Figure 1).

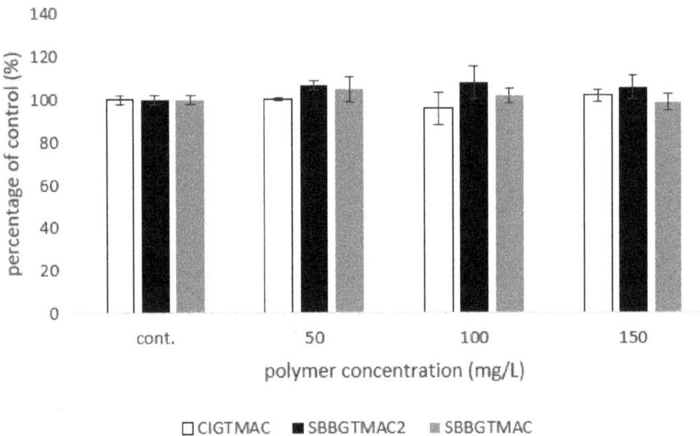

Figure 1. Preliminary toxicity studies of the obtained polycations using the 3T3-L1 fibroblast cell line. In the experiment, serum-free DMEM medium was used; exposure to the polymer lasted for 24 h.

The obtained data show that for concentrations up to 150 mg/L, we did not observe a negative effect on fibroblasts for all tested polymers (no statistically significant difference vs. control was found). We also did not observe any differences between the actions of individual cationic polysaccharides. Due to the large unavoidable differences in the culture conditions of the mammalian and fungal cells, the therapeutic index in the case of antifungal use of the polymers cannot be directly calculated here. However, using data from this and earlier subsections, it can be concluded that for concentrations safe for eukaryotic cells, we are observing a negative effect on the growth of selected strains of fungi for selected polymers (pairs *A. brasiliensis*/SBBGTMAC and SBBGTMAC2 or *S. brevicaulis*/SBBGTMAC2).

2.3. An In Vivo Model of Fungal Infection

For evaluation of the antifungal activity of the tested polymers using an in vivo infection model, only SBBGTMAC2 was chosen due to the best synthesis yield of this compound, accompanied by relatively high activity in preliminary studies (for two fungal strains). The insect model host *Galleria mellonella* combined with *A. brasiliensis* was used due to the fact that *G. mellonella* was described as a good model for testing *Aspergillus* pathogenicity and that in this case it can give us a reliable proof of concept [25–27]. These compromises in the scale of these experiments are due to the fact that these were preliminary studies, and the simplicity and efficiency aspect of the synthesis was assumed to be a priority in this work.

The *G. mellonella* larvae were injected with live *A. brasiliensis* conidia at a dose of 5×10^5 and then with SBBGTMAC2 at a concentration corresponding to an MIC value determined in vitro. The survival probability of the infected individuals treated with SBBGTMAC2 increased significantly in comparison with the control larvae treated with insect physiological saline (IPS) after infection with *A. brasiliensis* ($p = 0.00028$) (Figure 2). The survival analysis showed that the probability that larvae injected with *A. brasiliensis* and then IPS would survive 60 h was approximately 20%, whereas for the infected larvae treated with SBBGTMAC2 the probability was approximately 60%. Probability of survival of 120 h was approximately 5% and 25%, for the infected larvae treated with IPS and those treated with SBBGTMAC2, respectively. The effect of SBBGTMC2 on the insects' survival was similar to that calculated for the amphotericin B (AmB)-treated larvae. All the larvae infected with *A. brasiliensis* and then injected with IPS were dead at the end of observation, i.e., at 125 h post-treatment. In contrast, at the same time, 17% and 10% of the larvae injected with SBBGTMAC2 and AmB, respectively, were still alive (Figure 2).

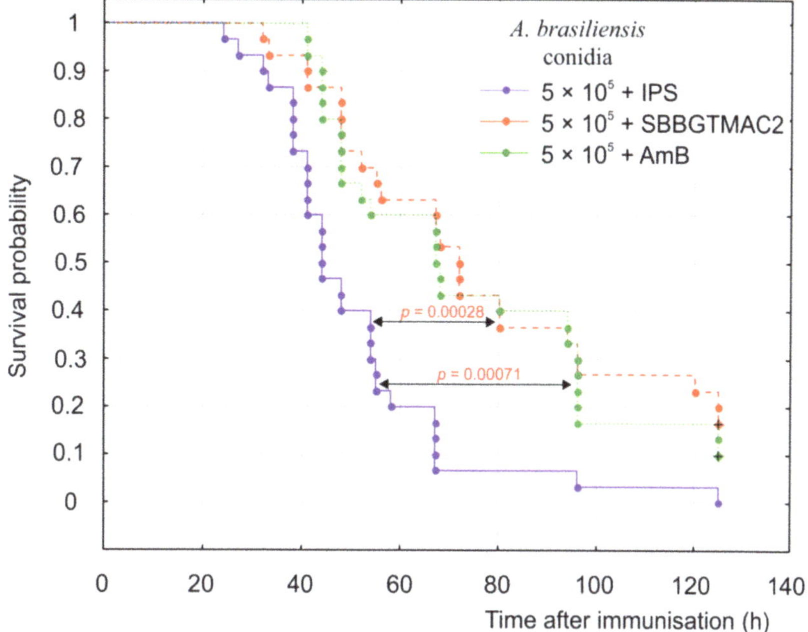

Figure 2. The effect of SBBGTMAC2 on survival of *G. mellonella* larvae infected with *A. brasiliensis*. The larvae were injected with suspension containing 5×10^5 of *A. brasiliensis* conidia and, after 2 h-incubation, with IPS (control) or SBBGTMAC2 or AmB. The larvae survival was evaluated, and survival probability was estimated using the Kaplan–Meier estimator. Survivability comparisons between control and two other groups were done using log-rank test. The effects of SBBGTMAC2 and AmB were comparable, as no statistically significant difference was found between these groups. The graph shows results from three independent experiments.

The survival experiments conducted using the insect model host *G. mellonella* indicated that the tested polycation SBBGTMAC2 exhibited in vivo antifungal activity against *A. brasiliensis* at a concentration corresponding to the MIC value determined in vitro. These results confirm the antifungal activity of SBBGTMAC2 in an in vivo model and, simultaneously, demonstrate the usefulness of this insect model for testing the antimicrobial activity of such polycationic molecules.

3. Discussion

Beta-glucans are polysaccharides found in the cell walls of certain bacteria, fungi, algae, lichens, and plants. They are used in medicine as natural preparations with antitumor properties, which decrease the levels of cholesterol and glucose in the serum and also stimulate the immune system [19,28]. The antibacterial properties of beta-glucans are also known and well described [19–31], but little is known about their direct action on fungi and possible direct antifungal activity. So far, research has focused on understanding the interactions between host cells (mainly animal) and beta-glucan and the host anti-fungal immunity induced by this compound [32].

This work describes the synthesis of three cationic beta-glucan derivatives using raw material isolated from *S. boulardii* and *C. islandica*. The purity of the materials obtained and the effectiveness of the cationization of the final products were confirmed. In the case of *S. boulardii*, two synthesis strategies were used, of which the one omitting the isolation of unmodified beta-glucan proved to be significantly more efficient (20 times higher yield SBBGTMAC2 vs. SBBGTMAC). In the case of these two derivatives, although the initial substrate for the synthesis was the same, we obtained polymers differing in their physicochemical properties—in particular, differences in their molecular weight and their dispersion (Table 3). These differences also resulted in noticeable differences in biological properties—in particular, antifungal properties expressed as MIC values (Table 4). However, these were not qualitative changes because, for the majority of tested strains, where one derivative has antifungal properties, the other also has them. Beta glucans are mixtures of macromolecules, and the method of isolation (or subsequent chemical modification) will affect the composition of the fractions that one obtains, and this is an inescapable problem with acquiring this type of natural compound.

These compounds have been investigated for their antifungal activity on selected fungal strains of yeasts and filamentous fungi. Beta-glucan derivatives showed selective in vitro antifungal activity against some molds—*S. brevicaulis*, *F. solani*, *A. brasiliensis*, and *A. flavus*—in the concentration range tested (1.95–1000 mg/L). The antimycotic properties of one of the cationic derivatives tested (SBBGTMAC2) were confirmed in *G. mellonella* larvae infected with *A. brasiliensis*. The administration of the compound resulted in a reduction in mortality, and the effect was similar to the result obtained for the known antifungal drug amphotericin B.

Candida albicans is still the most common etiological factor of candidiasis. Other Candida species cause infections at varying rates, often depending on the population (adults, children), but the rates also vary geographically. The choice of *Candida albicans* was obvious and due to epidemiological data. It was decided to test *Candida glabrata* and *Candida krusei* due to their natural resistance and reduced sensitivity to azole drugs commonly used in the prevention and treatment of candidiasis. The conducted studies were preliminary; therefore, the activity in relation to selected species of fungi was assessed. Other Candida species can be included in larger studies, but the obtained results do not indicate an antimycotic effect of tested beta-glucan derivatives on this genus. The applied methodology of antifungal activity testing of beta-glucan derivatives was not suitable for the assessment of the antimycotic effect on *Pneumocystis jirovecii*. This species does not grow in vitro in fungal culture media.

The results of our studies are promising and require research on more strains of different fungal species. Higher concentrations of beta-glucan derivatives should be investigated as well, which may prove to be fungistatic or fungicidal on more species. So far, mainly the antibacterial activity of beta-glucans has been examined. Studies of their various derivatives isolated from various organisms showed in vitro antibacterial activity at concentrations of 250 mg/L [29] or higher—1000–5000 mg/L [30]. In vivo studies also confirmed the protective effect of beta-glucan against various bacterial pathogens [33] but few clinical trials involving humans have been performed, and results remain controversial [34–36]. The studies conducted so far on glucan and fungi are sparse; however, they do indicate that this compound might have antimycotic activity, even against *Candida albicans*, which we did not show in our in vitro studies. Rice et al. demonstrated in an animal in vivo model that oral glucan administration in a dose of 1 mg/kg increased survival in mice challenged with *C. albicans* [37]. While Indian scientists found antifungal activity of β-D-glucan nanoparticles against *Pythium aphanidermatum*—a soil-borne plant pathogen [38]. It should be emphasized that the results of research on the antimicrobial activity of beta-glucans are difficult to compare with each other. This is due to the fact that these compounds are usually obtained from various organisms, subjected to various modifications, and therefore may show differences in biological activity. Our research is innovative in this field because it involves many different fungal species and also includes an in vivo model of fungal infection and beta-glucan-derivative activity. Obviously, far more studies are needed to confirm the antimycotic effects of cationic beta-glucan derivatives, but due to the growing need to search for new antifungals, such research seems even more significant and justified.

4. Materials and Methods

4.1. Chemistry

4.1.1. General Instrumentation and Materials

Saccharomyces boulardii strain CNCM I-745 (Enterol 250, Biocodex, Poland), dried well-ground *Cetraria islandica* (vegetative aboveground part)—herbal raw material (ASKOM, Wroclaw, Poland), potato dextrose broth PDB (Sigma-Aldrich, St. Louis, MO, USA), glycidyltrimethylammonium chloride (technical, $\geq$90%, Sigma-Aldrich, St. Louis, MO, USA), sodium hydroxide p.a. (POCH, Gliwice, Poland), and hydrochloric acid, acetic acid (Idali, Radom, Poland) were obtained. All dialyses were performed using a cellulose tube (cut of 3.5 kDa) produced by Carl Roth Company (Karlsruhe, Germany). Poly(2-vinylpyridine) and PEG molecular weight standards PSS were obtained from PSS Polymer Standards Service GmbH, Mainz, Germany. Evaluation of elemental composition was commissioned on a Simultaneous CHNS combustion analyzer apparatus (Micro Cube elemental analyzer, Elementar Vario). Three independent measurements were performed for each sample; the result presented is the arithmetic mean $\pm$ standard deviation (SD) of the obtained results. IR spectra were measured using an FT-IR instrument (Nicolet iS10, SMART iTX, Thermo Scientific, Waltham, MA, USA).

4.1.2. Culture Conditions for Saccharomyces Boulardii

The yeast was grown in potato dextrose broth medium (250 mg of inoculum lyophilisate per 1 L of medium) for 24 h at 25 °C with constant orbital shaking (200 rpm). An amount of 2 L of obtained yeast suspension was used for the isolation of beta-glucans.

4.1.3. Isolation of Polymers from Biological Material

Saccharomyces Boulardii

Insulation of Raw Components of the Cell Wall

The isolations were carried out using a modified procedure described in a previously published work [39]. Briefly, 2 L of yeast suspension obtained from the culture was centrifuged at 500× *g* for 2 min. The supernatant was then gently removed, the remaining pellets were pooled in one vessel, and PBS buffer was added to the obtained suspension to obtain 10 mL. The suspension was cooled on ice and then, without removing it from the ice bath, it was subjected to an immersion source of ultrasound with a power of about 10 W (Ultrasonic Processor with a titanium sonotrode tip, Sonics, Vibracell). The lysis of yeast was performed with 8 repetitions consisting of a two-stage cycle: 5 min ultrasound, 2 min of mixing on ice to remove excess heat. The non-lysed yeast was removed by centrifuging the obtained suspension twice at 500× *g* for 2 min. The finally obtained supernatant was centrifuged at 1000× *g* at 4 °C for 20 min. The pellet composed of components of the cell wall was rinsed twice with sterile water, and then the vessel containing it was immersed in boiling water for 3 min to deactivate all present enzymes.

Extraction of Beta-Glucans

Material obtained as described in the previous subsection after cooling down to room temperature was suspended in 10 mL 2% sodium hydroxide, and then the mixture was heated to 90 °C for 5 h with constant stirring. After cooling, the suspension was centrifuged at 3000× *g* for 10 min, and the obtained supernatant was neutralized with 2 M acetic acid and mixed with 3 volumes of ethanol to precipitate the raw product. The pellets were rinsed with 2 mL of ethanol and dried under vacuum without heating. Then, the crude product was dissolved in 3% acetic acid and centrifuged ($3000\times g$ for 10 min) to remove the remaining proteins. The recovered supernatant was neutralized with 2 M NaOH and dialyzed against water (water change every 12 h to fresh and sterile) for 4 days. The finished product was obtained by freeze-drying the obtained solution. The yield of the resulting product was ~2%.

Material from *Cetraria islandica*

The isolation of beta-glucan from *C. islandica* was performed in a way similar to the procedures described previously [40,41]. Briefly, 10 g of dried well-ground *C. islandica* powder was suspended in 250 mL of ethanol (99%) with constant stirring for 1 h. The preliminary extraction of impurities was repeated twice. The brown-colored ethanol was then removed by filtration, and the solid was dried under reduced pressure without heating. Then the dry material was suspended in 100 mL of distilled water and heated to 100 °C. Extractions were carried out at this temperature for 4 h with continuous vigorous mixing. After this time, the still-warm mixture was centrifuged $3000\times g$ for 10 min, and the resulting supernatant was poured into ethanol to precipitate the crude polymer (1 volume of supernatant to 2 volumes of ethanol). The resulting suspension was again centrifuged in an analogous manner to obtain a precipitate. The pellet was then dried under reduced pressure without raising the temperature. The last step of the purification consisted of redissolving the dry crude polysaccharide in water at 100 °C, hot centrifugation ($3000\times g$ for 10 min), and dialysis of the obtained supernatant against sterile distilled water. After 24 h of dialysis in the tube, polymer fibrous lumps had precipitated, which were filtered, rinsed with ethanol and dried at 40 °C under reduced pressure for 4 h. The yield of the resulting product was ~20%.

4.1.4. Synthesis of Polycations
Using Purified Beta-Glucans

For beta-glucans from both sources, the cationizations were performed in the manner used to obtain the cationic dextran described in our earlier work [20,42]. Briefly, 200 mg of beta-glucan was dissolved in 50 mL of distilled water containing an additional 200 mg of NaOH. The dissolution was carried out at 60 °C, and then 6 mL of GTMAC was added. The reaction was continued at this temperature for another 4 h, and then the mixture was cooled to room temperature and dialyzed against water. Final products were isolated from the obtained solutions by freeze-drying. The total yield of the resulting product was ~1% for SBBGTMAC and ~10% for CIGTMAC. A scheme of beta-glucan cationization reaction is presented in Figure 3.

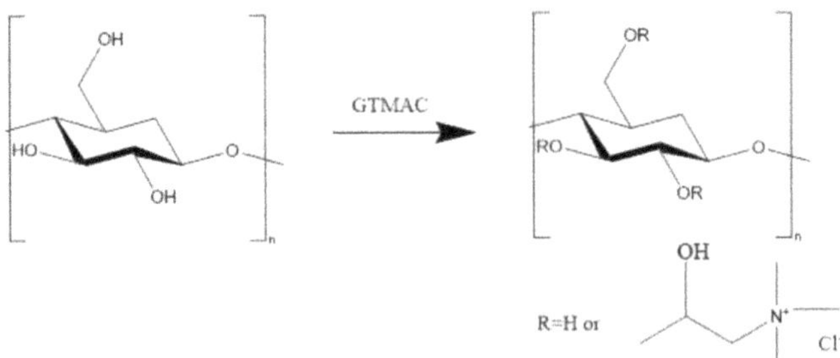

Figure 3. Scheme of beta-glucan cationization reaction used in this study.

Cationization of *Saccharomyces boulardii* Raw Cell Wall Components

Isolated as described in the previous section, raw components of the cell wall were suspended in 10 mL of sterile distilled water. After 1 day of dialysis (sterile water change every 12 h), the resulting suspension was centrifuged at $1000\times g$ for 20 min. Then pellet was washed thoroughly with ethanol with two 10 mL portions and dried under reduced pressure without heating. Next, 200 mg of obtained material was dissolved in 50 mL of distilled water containing 200 mg of NaOH. Dissolution was carried out at 60 °C, and then 6 mL of GTMAC was added. The reaction was continued at this temperature for another 4 h, and then the mixture was cooled to room temperature and dialyzed against water. Obtained raw products were isolated from the obtained solutions by freeze-drying. Then this material was suspended in a mixture of 15 mL of ethanol and 5 mL of water and shaken for 4 h at room temperature. The resulting suspension was centrifuged at $1000 \times g$ at 4 °C for 20 min, and the product was isolated from the supernatant by evaporating the solvent using a rotary evaporator. The last step of purification consisted of dissolving the dry crude product in water and centrifugation at $1000 \times g$ for 20 min, and the final product was isolated from the supernatant by freeze-drying. The total yield of the resulting product SBBGTMAC2 was ~20%.

4.1.5. Gel Permeation Chromatography (GPC) and Zeta Potential Measurements

The purpose of GPC measurements was to assess the average molecular weight and dispersion of the molecular weight for the obtained polymers [20,43] and thus to find out how heterogeneous was the beta-glucan mixture that was obtained [20,43]. Measurement was possible only in the case of the beta-glucans from yeast since the material in the obtained form dissolved in water. This polymer was composed of uncharged macromolecules (confirmed experimentally, details later in this chapter), and this is why the PEG standards were used to assess molecular mass (eluent composed of 0.1 M NaCl in water). In the case of the material derived from *C. islandica*, we obtained a polymer that was insoluble in water at a temperature below 80 °C, which made GPS measurements impossible in classical systems. In the case of products after cationization, they were all well soluble in aqueous solutions, so GPC measurements were performed. Due to the positive charge of these polymers (confirmed by measurements of the zeta potential), poly(2-vinylpyridine) standards were used for molecular mass determination (eluent consisting of 0.3 M Na_2SO_4 aqueous solution containing 0.5 M acetic acid). In all cases, the column PolySep™-SEC GFC-P Linear, LC Column 300 × 7.8 mm (Phenomenex, Torrance, CA, USA) was used; the flow rate, injection volume, and polymer concentration were, respectively, 0.8 mL/min, 100 µL, and 5 g/L. The measurement of the zeta potential was performed using a Zetasizer Nano ZS (Malvern Panalytical, Malvern, UK) apparatus and polymer solutions in demineralized water (pH 6.0, conductivity 0.5 µS) with a concentration of 5 g/L [20,44]. Three independent measurements were made for each sample; the result presented is the arithmetic mean ± standard deviation (SD) of the obtained results.

4.2. Preliminary Assessment of Cytotoxicity on Mammalian Cells

Embryo mouse fibroblast 3T3-L1 (ATCC CL-173) (Manassas, USA) was used to assess the toxicity of polycations. The medium for this cell line was Dulbecco's modified Eagle's medium (DMEM, Sigma-Aldrich, St. Louis, MO, USA), supplemented with fetal bovine serum (Sigma-Aldrich, St. Louis, MO, USA), for a final concentration of 10% (v/v) and 1% (v/v) penicillin–streptomycin solution (Sigma-Aldrich, St. Louis, MO, USA). Cultures were incubated at 37 °C in atmosphere containing 5% of carbon dioxide (CO_2). Cultures were seeded 6×10^4 cells per well in 24-well plates and grown for 24 h. After that, the medium was changed to serum-free, and cells were treated with polycations solution (in serum-free media) for the next 24 h to assess cytotoxicity using neutral red uptake assay [45]. Three independent measurements were made for each concentration; the result presented is the arithmetic mean ± standard deviation (SD) of the obtained results. Additionally, the lack of statistical differences in cell viability was confirmed using the Mann–Whitney U test.

4.3. Assessment of Antifungal Properties

4.3.1. General Instrumentation and Materials

Strains for antifungal activity testing were *Candida albicans* ATCC 90028, *C. glabrata*, ATCC 15454, *C. krusei* ATCC 6258, *Aspergillus flavus* ATCC 204304, *A. brasiliensis* ATCC

16404, *A. fumigatus* clinical strain (Department of Infections Control and Mycology, Jagiellonian University Medical College culture collection), *Trichophyton mentagrophytes* ATCC 18748, *Fusarium solani* (Department of Infections Control and Mycology, Jagiellonian University Medical College culture collection), *Scopulariopsis brevicaulis* (Department of Infections Control and Mycology, Jagiellonian University Medical College culture collection). *Aspergillus fumigatus* and *Scopulariopsis brevicaulis* were identified using phenotypic methods based on fungal morphology [46]. *Fusarium solani* identification was established with matrix-assisted laser desorption ionization time-of-flight mass spectrometry (MALDI-TOF MS) [47]. Phenotypic identification was not confirmed by genetic methods.

We obtained RPMI-1640 medium with L-glutamine, without sodium bicarbonate (Sigma-Aldrich, St. Louis, USA); glucose (Chempur, Piekary Slaskie, Poland); 3-(N-morpholino) propanesulfonic acid (MOPS) (Glentham Life, Corsham, UK); flat-bottom polypropylene 96-well microdilution plates (VWR, Radnor, PA, USA); Sabouraud glucose agar with chloramphenicol: neopeptone (Difco Laboratories Inc., Franklin Lakes, NJ, USA); glucose (Chempur, Piekary Slaskie, Poland); agar (Biocorp, Waszawa Poland); chloramphenicol (Farm-Impex, Gliwice, Poland); Czapek yeast extract agar: $ZnSO_4 \times 7H_2O$ (POCH, Gliwice, Poland), $CuSO_4 \times 7H_2O$ (ACROS, Belgium); $MgSO_4 \times 7H_2O$ (POCH, Gliwice, Poland); KCl (Chempur, Piekary Slaskie, Poland); $NaNO_3$ (POCH, Poland); saccharose (Chempur, Piekary Slaskie, Poland); K_2HPO_4 (Chempur, Piekary Slaskie, Poland); agar (Biocorp, Warszawa, Poland); and yeast extract (Oxoid, Pratteln, Switzerland).

Apparatuses included a densitometer (Biosan, Poland), incubator (POL-EKO, Poland), and vortex mixer (Labnet, Poland).

4.3.2. Antifungal Activity Testing

The assessment of antimycotic activity of beta-glucan derivatives was performed using our own microdilution method in a liquid culture medium based on European Committee on Antimicrobial Susceptibility Testing methodology for antifungal susceptibility testing for yeasts and molds (EUCAST DEFINITIVE DOCUMENT E.DEF: 7.3.2. Method for the determination of broth dilution minimum inhibitory concentrations of antifungal agents for yeasts; 9.3.2. Method for the determination of broth dilution minimum inhibitory concentrations of antifungal agents for conidia forming moulds) and the Hancock Lab [48] procedure for cationic antimicrobial peptides.

Preparation of Stock Solutions and Working Solutions

Stock solutions were prepared in sterile distilled water to obtain concentrations of 10 g/L. The working solutions were prepared as a 2-fold dilution series of stock solutions in sterile distilled water.

Preparation of Microdilution Plates

Wells 1 to 10 of each column of the flat-bottom polypropylene 96-well microdilution plates were filled with 20 µL of the corresponding concentration of CIGTMAC, SBBGTMAC, and SBBGTMAC2, while each well of columns 11 and 12 was filled with 20 µL of sterile distilled water.

Inocula Preparation

The inocula were prepared from fresh cultures (24 h for yeasts, 2- to 7-day-old for filamentous fungi) on solid agar media—Sabouraud glucose agar with chloramphenicol for *Candida* species and *Trichopyton mentagropytes* and Czapek yeast extract agar for molds. Yeasts inocula were prepared by suspending a few representative colonies in sterile distilled water. Filamentous fungi colonies were covered with approximately 5 mL of sterile water, then the conidia were rubbed with a sterile cotton swab and transferred to a sterile tube. The suspensions were homogenized with a gyratory vortex mixer, and the cell density was adjusted to 0.5 McFarland. The inocula were used for testing within 30 min of preparation.

Preparation of Working Suspensions

The working suspensions were prepared by dilution of the primary inocula in RPMI-1640 medium with L-glutamine, without sodium bicarbonate, with 2% glucose, buffered to pH 7 with MOPS (0.165 mol/L). Next, 20-fold dilutions were prepared, which gave the following inocula densities: 0.5–2.5×10^5 CFU/mL for yeasts and 1–2.5×10^5 CFU/mL for filamentous fungi.

Inoculation of Microdilution Plates

The microdilution plates were inoculated with 180 µL of the working suspensions, except sterility control wells, which contained 180 µL of microbial-free RPMI-1640 with L-glutamine, without sodium bicarbonate, with 2% glucose, buffered to pH 7 with MOPS (0.165 mol/L). The range of final concentrations of beta-glucan derivatives on microplates after the addition of the fungal suspension was 1.95–1000 mg/L.

Incubation of Microdilution Plates

The microdilution plates were incubated without agitation at 37 °C (yeasts) or 27 °C (molds, dermatophyte) in ambient air for 24–48 h.

Reading Results

The antimicrobial activity was estimated visually. Minimal inhibitory concentration (MIC) values of tested compounds were defined as no visible growth of fungi by eye.

4.4. Galleria Mellonella Larvae Survival Experiments

The last instar larvae (250–300 mg weight) of the greater wax moth *Galleria mellonella* (Lepidoptera: Pyralidae) from a continuous laboratory culture were used. They were maintained at 30 °C in the dark and reared on honeybee nest debris. The larvae were injected with 3 µL of insect physiological saline (IPS; 0.1 M Tris–HCl pH 6.9, 150 mM NaCl, 5 mM KCl) containing *Aspergillus brasiliensis* conidia (5×10^5 per larva). After 2 h incubation at 37 °C, the larvae were randomly divided into three groups (10 larvae per group) and injected with 3 µL of IPS (control) or 3 µL of SBBGTMAC2 solution in IPS (80 µg per larva—the final concentration in larval hemolymph corresponded to an MIC value 1000 mg/L) or 3 µL of AmB solution (2 µg per larva in 30% DMSO; final concentration of DMSO in larval hemolymph—approximately 1.25%). The larvae were incubated at 37 °C, and survival was checked until 125 h after treatment. The entire experimental layout was repeated at three independent occasions (in total 90 larvae were used). The larvae survival was evaluated, and survival probability was estimated using the Kaplan–Meier estimator [49,50]. Survivability comparisons between control and the two other groups were done using the log-rank test. The graph shows results from three independent experiments.

5. Conclusions

Fungi and fungal-related diseases are a significant medical and socio-economic problem. The possibilities of combating fungi are currently very limited, and the search for new, effective, and inexpensive methods of their elimination is a very current need. Fungi are a heterogeneous group of organisms, which makes it difficult to find a compound showing a broad antifungal spectrum that in addition has low toxicity to humans, animals, and plants. Our research on antifungal activity of cationic beta-glucan derivatives indicates that these substances exhibit selective antifungal activity against certain species. No antifungal activity was found for unicellular yeasts of the genus *Candida*, while antimycotic activity against some filamentous fungi, i.e., *Scopulariopsis brevicaulis*, *Aspergillus brasiliensis*, and *Fusarium solani*, was shown. *S. brevicaulis* and *F. solani* are multi-resistant species, showing high MIC values for most of the antifungals currently available. The obtained promising preliminary results for these fungi encourage research to be conducted on a greater number of strains of these species and to extend the research to other species of filamentous fungi resistant to currently used drugs. Our research indicates that these compounds may not necessarily completely inhibit the growth of fungi, but the compounds can significantly limit it, so they could be successfully used as agents preventing the growth of fungi. In addition, the known immune-stimulating effect may potentiate the antifungal activity.

Supplementary Materials: The following are available online at https://www.mdpi.com/article/10.3390/ph14090838/s1, Figure S1: IR spectra of various stages of SBB purification (left) and products of beta-glucan modification using GTMAC (right), Figure S2: IR spectra of various stages of CI purification (left) and products of beta-glucan modification using GTMAC (right), Figure S3: GPC/SEC chromatogram of a beta-glucan isolated from *Saccharomyces boulardii* (left) and the obtained cationic beta-glucans (right).

Author Contributions: Conceptualization, K.K.; methodology K.K., M.S., S.S. and A.Z.-B.; software, S.S.; validation, S.S., A.Z.-B. and M.C.; formal analysis, S.S., K.K., M.S. and P.K.; resources, S.S. and M.C.; data curation, M.S. and S.S.; writing—original draft preparation, K.K. and M.S.; writing—review and editing, M.C.; visualization, S.S. and A.Z.-B.; supervision, M.C.; project administration, M.C.; funding acquisition, M.S. and S.S. All authors have read and agreed to the published version of the manuscript.

Funding: This research was funded by the Centre for Technology Transfer CITTRU, Jagiellonian University, grant number N41/DBS/000591. The in vivo experiments using *G. mellonella* model were financed by the research project for young scientists (MN/2021/2) from the Institute of Biological Sciences (UMCS, Lublin, Poland).

Institutional Review Board Statement: Not applicable.

Informed Consent Statement: Not applicable.

Data Availability Statement: Not applicable.

Acknowledgments: The authors thank Monika Koziej, (Department of Immunobiology, Institute of Biological Sciences, UMCS, Lublin, Poland) for conducting the *G. mellonella* laboratory culture and for preparing the insects for the experiments.

Conflicts of Interest: The authors declare no conflict of interest.

References

1. Bongomin, F.; Gago, S.; Oladele, R.O.; Denning, D.W. Global and multi-national prevalence of fungal diseases—Estimate precision. *J. Fungi* **2017**, *3*, 57. [CrossRef]
2. Brown, G.D.; Gow, N.A.R.; Levitz, S.M.; Netea, M.G.; White, T.C. Hidden killers: Human fungal infections. *Sci. Transl. Med.* **2012**, *4*, 165rv13. [CrossRef]
3. Song, G.; Liang, G.; Liu, W. Fungal co-infections associated with global COVID-19 pandemic: A clinical and diagnostic perspective from China. *Mycopathologia* **2020**, *185*, 599–606. [CrossRef] [PubMed]
4. Pemán, J.; Ruiz-Gaitán, A.; García-Vidal, C.; Salavert, M.; Ramírez, P.; Puchades, F.; García-Hita, M.; Alastruey-Izquierdo, A.; Quindós, G. Fungal co-infection in COVID-19 patients: Should we be concerned? *Rev. Iberoam. Micol.* **2020**, *37*, 41–46. [CrossRef] [PubMed]
5. Sarkar, S.; Gokhale, T.; Choudhury, S.S.; Deb, A.K. COVID-19 and orbital mucormycosis. *Indian J. Ophthalmol.* **2021**, *69*, 1002–1004. [PubMed]
6. Revannavar, S.M.; Supriya, P.S.; Samaga, L.; Vineeth, K.V. COVID-19 triggering mucormycosis in a susceptible patient: A new phenomenon in the developing world? *BMJ Case Rep.* **2021**, *14*. [CrossRef]
7. Mehta, S.; Pandey, A. Rhino-orbital mucormycosis associated with COVID-19. *Cureus* **2020**, *12*, e10726. [CrossRef]
8. Mekonnen, Z.K.; Ashraf, D.C.; Jankowski, T.; Grob, S.R.; Vagefi, M.R.; Kersten, R.C.; Simko, J.P.; Winn, B.J. Acute invasive rhino-orbital mucormycosis in a patient with COVID-19-associated acute respiratory distress syndrome. *Ophthalmic Plast. Reconstr. Surg.* **2021**, *37*, 40–80. [CrossRef]
9. Werthman-Ehrenreich, A. Mucormycosis with orbital compartment syndrome in a patient with COVID-19. *Am. J. Emerg. Med.* **2021**, *42*, 264.e5–264.e8. [CrossRef]
10. Waizel-Haiat, S.; Guerrero-Paz, J.A.; Sanchez-Hurtado, L.; Calleja-Alarcon, S.; Romero-Gutierrez, L. A case of fatal rhino-orbital mucormycosis associated with new onset diabetic ketoacidosis and COVID-19. *Cureus* **2021**, *13*, e13163.
11. Pasero, D.; Sanna, S.; Liperi, C.; Piredda, D.; Branca, G.P.; Casadio, L.; Simeo, R.; Buselli, A.; Rizzo, D.; Bussu, F.; et al. A challenging complication following SARS-CoV-2 infection: A case of pulmonary mucormycosis. *Infection* **2020**. [CrossRef]
12. Campoy, S.; Adrio, J.L. Antifungals. *Biochem. Pharmacol.* **2017**, *133*, 86–96. [CrossRef] [PubMed]

13. Yoo, J.; Kim, S.-H.; Hur, S.; Ha, J.; Huh, K.; Cha, W.C. Candidemia Risk Prediction (CanDETEC) Model for Patients with Malignancy: Model Development and Validation in a Single-Center Retrospective Study. *JMIR Med. Inform.* **2021**, *9*, e24651. [CrossRef]
14. Rauseo, A.M.; Aljorayid, A.; Olsen, M.A.; Larson, L.; Lipsey, K.L.; Powderly, W.G.; Spec, A. Clinical predictive models of invasive Candida infection: A systematic literature review. *Med. Mycol.* **2021**, myab043. [CrossRef] [PubMed]
15. Vetvicka, V.; Vannucci, L.; Sima, P.; Richter, J. Beta glucan: Supplement or drug? From Laboratory to Clinical Trials. *Molecules* **2019**, *24*, 1251. [CrossRef] [PubMed]
16. Frid, A.; Tura, A.; Pacini, G.; Ridderstrale, M. Effect of oral pre-meal administration of betaglucans on glycaemic control and variability in subjects with type 1 diabetes. *Nutrients* **2017**, *9*, 1004. [CrossRef] [PubMed]
17. van Steenwijk, H.P.; Bast, A.; de Boer, A. Immunomodulating effects of fungal beta-glucans: From traditional use to medicine. *Nutrients* **2021**, *13*, 1333. [CrossRef]
18. Moore, M.A. Diverse influences of dietary factors on cancer in Asia. *Asian Pac. J. Cancer Prev.* **2009**, *10*, 981–986.
19. Camilli, G.; Tabouret, G.; Quintin, J. The Complexity of fungal β-glucan in health and disease: Effects on the mononuclear phagocyte system. *Front. Immunol.* **2018**, *9*. [CrossRef] [PubMed]
20. Kaminski, K.; Plonka, M.; Ciejka, J.; Szczubiałka, K.; Nowakowska, M.; Lorkowska, B.; Korbut, R.; Lach, R. Cationic derivatives of dextran and hydroxypropylcellulose as novel potential heparin antagonists. *J. Med. Chem.* **2011**, *54*, 6586–6596. [CrossRef] [PubMed]
21. Kim, Y.H.; Choi, H.-M.; Yoon, J.H. Synthesis of a quaternary ammonium derivative of chitosan and its application to a cotton antimicrobial finish. *Text. Res. J.* **1998**, *68*, 428–434. [CrossRef]
22. Chen, A.; Peng, H.; Blakey, I.; Whittaker, A.K. Biocidal Polymers: A Mechanistic Overview. *Polym. Rev.* **2017**, *57*, 276–310. [CrossRef]
23. Ikeda, T.; Hirayama, H.; Yamaguchi, H.; Tazuke, S.; Watanabe, M. Polycationic biocides with pendant active groups: Molecular weight dependence of antibacterial activity. *Antimicrob. Agents Chemother.* **1986**, *30*, 132–136. [CrossRef] [PubMed]
24. Ramandeep, K.R.; Sharma, M.; Ji, D.; Xu, M.; Agyei, D. Structural features, modification, and functionalities of beta-glucan. *Fibers* **2020**, *8*, 1. [CrossRef]
25. Maurer, E.; Browne, N.; Surlis, C.; Jukic, E.; Moser, P.; Kavanagh, K.; Lass-Florl, C.; Binder, U. *Galleria mellonella* as a host model to study *Aspergillus terreus* virulence and amphotericin B resistance. *Virulence* **2015**, *6*, 591–598. [CrossRef] [PubMed]
26. Slater, J.L.; Gregson, L.; Denning, D.W.; Warn, P.A. Pathogenicity of *Aspergillus fumigatus* mutants assessed in *Galleria mellonella* matches that in mice. *Med. Mycol.* **2011**, *49*, 107–113. [CrossRef]
27. Durieux, M.-F.; Melloul, É.; Jemel, S.; Roisin, L.; Dardé, M.-L.; Guillot, J.; Dannaoui, É.; Botterel, F. *Galleria mellonella* as a screening tool to study virulence factors of *Aspergillus fumigatus*. *Virulence* **2021**, *12*, 818–834. [CrossRef] [PubMed]
28. Rahar, S.; Swami, G.; Nagpal, N.; Nagpal, M.A.; Singh, G.S. Preparation, characterization, and biological properties of β-glucans. *J. Adv. Pharm. Technol. Res.* **2011**, *2*, 94–103. [CrossRef]
29. Chamidah, A.; Hardoko, H.; Prihanto, A.A. Antibacterial activities of β-glucan (laminaran) against gram-negative and gram-positive bacteria. In *AIP Conference Proceedings, Proceedings of the 2nd International Conference on Composite Materials and Material Engineering (ICCMME 2017), Chengdu, China, 17–19 February 2017*; AIP Publishing: Melville, NY, USA, 2017; Volume 1844. [CrossRef]
30. Wan-Mohtar, W.A.; Young, L.; Abbott, G.M.; Clements, C.; Harvey, L.M.; McNeil, B. Antimicrobial properties and cytotoxicity of sulfated (1,3)-β-D-glucan from the mycelium of the mushroom *Ganoderma lucidum*. *J. Microbiol. Biotechnol.* **2016**, *26*, 999–1010. [CrossRef] [PubMed]
31. Hetland, G.; Ohno, N.; Aaberge, I.S.; Lovik, M. Protective effect of beta-glucan against systemic Streptococcus pneumoniae infection in mice. *Immunol. Med. Microbiol.* **2000**, *27*, 111–116.
32. Desamero, M.J.M.; Chung, S.H.; Kakuta, S. Insights on the functional role of beta-glucans in fungal immunity using receptor-deficient mouse models. *Int. J. Mol. Sci.* **2021**, *22*, 4778. [CrossRef] [PubMed]
33. Vetvick, V.; Vetvickova, J. Anti-infectious and antitumor activities of β-glucans. *Anticancer Res.* **2020**, *40*, 3139–3145. [CrossRef]
34. Babineau, T.J.; Hackford, A.; Kenler, A.; Bistrian, B.; Forse, R.A.; Fairchild, P.G.; Heard, S.; Keroack, M.; Caushaj, P.; Benotti, P. A phase II multicenter, double-blind, randomized, placebo-controlled study of three dosages of an immunomodulator (PGG-glucan) in high-risk surgical patients. *Arch. Surg.* **1994**, *129*, 1204–1210. [CrossRef]
35. Babineau, T.J.; Marcello, P.; Swails, W.; Kenler, A.; Bistrian, B.; Forse, R.A. Randomized phase I/II trial of a macrophage-specific immunomodulator (PGG-glucan) in high-risk surgical patients. *Ann. Surg.* **1994**, *220*, 601–609. [CrossRef] [PubMed]
36. Dellinger, E.P.; Babineau, T.J.; Bleicher, P.; Kaiser, A.B.; Seibert, G.B.; Postier, R.G.; Vogel, S.B.; Norman, J.; Kaufman, D.; Galandiuk, S.; et al. Effect of PGG-glucan on the rate of serious postoperative infection or death observed after high-risk gastrointestinal operations. *Arch. Surg.* **1999**, *134*, 977–983. [CrossRef]
37. Rice, P.J.; Adams, E.L.; Ozment-Skelton, T.; Gonzalez, A.J.; Goldman, M.P.; Lockhart, B.E.; Barker, L.A.; Breuel, K.F.; Deponti, W.K.; Kalbfleisch, J.H.; et al. Oral delivery and gastrointestinal absorption of soluble glucans stimulate increased resistance to infectious challenge. *J. Pharmacol. Exp. Ther.* **2005**, *314*, 1079–1086. [CrossRef] [PubMed]
38. Anusuya, S.; Sathiyabama, M. Preparation of β-D-glucan nanoparticles and its antifungal activity. *Int. J. Biol. Macromol.* **2014**, *70*, 440–443. [CrossRef] [PubMed]

39. Shokri, H.; Asadi, F.; Khosravi, A.L. Isolation of β-glucan from the cell wall of *Saccharomyces cerevisiae*. *Nat. Prod. Res.* **2008**, *22*, 414–421. [CrossRef]

40. Surayot, U.; Yelithao, K.; Tabarsa, M.; Lee, D.H.; Palanisamy, S.; Prabhu, N.M.; Lee, J.; You, S. Structural characterization of a polysaccharide from Certaria islandica and assessment of immunostimulatory activity. *Process Biochem.* **2019**, *83*, 214–221. [CrossRef]

41. Ingolfsdottir, K.; Jurcic, K.; Fischer, B.; Wagner, H. Immunologically active polysaccharide from *Cetraria islandica*. *Planta Med.* **1994**, *60*, 527–531. [CrossRef]

42. Kaminski, K.; Stalińska, K.; Niziołek, A.; Wróbel, M.; Nowakowska, M.; Kaczor-Kaminska, M. Cell proliferation induced by modified cationic dextran. *Bio-Algorithms Med-Syst.* **2018**, *14*. [CrossRef]

43. Williams, T. Gel permeation chromatography: A review. *J. Mater. Sci.* **1970**, *5*, 811–820. [CrossRef]

44. Rasmussen, M.K.; Pedersen, J.N.; Marie, R. Size and surface charge characterization of nanoparticles with a salt gradient. *Nat. Commun.* **2020**, *11*, 2337. [CrossRef] [PubMed]

45. Janik-Hazuka, M.; Kaminski, K.; Kaczor-Kamińska, M.; Szafraniec-Szczesny, J.; Kmak, A.; Kassassir, H.; Watala, C.; Wróbel, M.; Zapotoczny, S. Hyaluronic Acid-Based Nanocapsules as Efficient Delivery Systems of Garlic Oil Active Components with Anticancer Activity. *Nanomaterials* **2021**, *11*, 1354. [CrossRef]

46. de Hoog, G.S.; Guarro, J.; Gene, J.; Figueras, M.J. *Atlas of Clinical Fungi*, 2nd ed.; Centraalbureau voor Schimmelcultures (CBS): Utrecht, The Netherlands, 2000.

47. Marinach-Patricea, C.; Lethuillier, A.; Brossas, J.-Y.M.; Gené, J.; Symoens, F.; Datry, A.; Guarro, J.; Mazier, D.; Hennequin, C. Use of mass spectrometry to identify clinical Fusarium isolates. *Clin. Microbiol. Infect.* **2009**, *15*, 634–642. [CrossRef]

48. Modified MIC Method for Cationic Antimicrobial Peptides. Available online: http://cmdr.ubc.ca/bobh/method/modified-mic-method-for-cationic-antimicrobial-peptides (accessed on 5 August 2021).

49. Kaplan-Meier, A.; Kaplan, E.L.; Meier, P. Nonparametric estimation from incomplete observations. *J. Am. Stat. Assoc.* **1958**, *53*, 457–481. [CrossRef]

50. Rasmussen, L.; Pratt, N.; Hansen, M.R.; Hallas, J.; Pottegård, A. Using the "proportion of patients covered" and the Kaplan-Meier survival analysis to describe treatment persistence. *Pharmacoepidemiol. Drug Saf.* **2018**, *27*, 867–871. [CrossRef]

pharmaceuticals

Article

Echinocandin Drugs Induce Differential Effects in Cytokinesis Progression and Cell Integrity

Natalia Yagüe, Laura Gómez-Delgado, M. Ángeles Curto, Vanessa S. D. Carvalho, M. Belén Moreno, Pilar Pérez, Juan Carlos Ribas and Juan Carlos G. Cortés *

Instituto de Biología Funcional y Genómica, Consejo Superior de Investigaciones Científicas (CSIC) and Universidad de Salamanca, 37007 Salamanca, Spain; natalia98@usal.es (N.Y.); laugomdel@usal.es (L.G.-D.); emecur@usal.es (M.Á.C.); vanessa.sdc@usal.es (V.S.D.C.); mbms@usal.es (M.B.M.); piper@usal.es (P.P.); ribas@usal.es (J.C.R.)
* Correspondence: cortes@usal.es; Tel.: +34-(92)-3294880

Abstract: Fission yeast contains three essential β(1,3)-D-glucan synthases (GSs), Bgs1, Bgs3, and Bgs4, with non-overlapping roles in cell integrity and morphogenesis. Only the $bgs4^+$ mutants *pbr1-8* and *pbr1-6* exhibit resistance to GS inhibitors, even in the presence of the wild-type (WT) sequences of $bgs1^+$ and $bgs3^+$. Thus, Bgs1 and Bgs3 functions seem to be unaffected by those GS inhibitors. To learn more about echinocandins' mechanism of action and resistance, cytokinesis progression and cell death were examined by time-lapse fluorescence microscopy in WT and *pbr1-8* cells at the start of treatment with sublethal and lethal concentrations of anidulafungin, caspofungin, and micafungin. In WT, sublethal concentrations of the three drugs caused abundant cell death that was either suppressed (anidulafungin and micafungin) or greatly reduced (caspofungin) in *pbr1-8* cells. Interestingly, the lethal concentrations induced differential phenotypes depending on the echinocandin used. Anidulafungin and caspofungin were mostly fungistatic, heavily impairing cytokinesis progression in both WT and *pbr1-8*. As with sublethal concentrations, lethal concentrations of micafungin were primarily fungicidal in WT cells, causing cell lysis without impairing cytokinesis. The lytic phenotype was suppressed again in *pbr1-8* cells. Our results suggest that micafungin always exerts its fungicidal effect by solely inhibiting Bgs4. In contrast, lethal concentrations of anidulafungin and caspofungin cause an early cytokinesis arrest, probably by the combined inhibition of several GSs.

Keywords: fungi; invasive fungal infections; fission yeast; cell wall; β(1,3)-D-glucan synthase; antifungal drugs; echinocandin drugs; echinocandin resistance; Fks resistance hot spots; cytokinesis; septation; cell separation; cell integrity; cell lysis

Citation: Yagüe, N.; Gómez-Delgado, L.; Curto, M.Á.; Carvalho, V.S.D.; Moreno, M.B.; Pérez, P.; Ribas, J.C.; Cortés, J.C.G. Echinocandin Drugs Induce Differential Effects in Cytokinesis Progression and Cell Integrity. *Pharmaceuticals* **2021**, 14, 1332. https://doi.org/10.3390/ph14121332

Academic Editors: Jong Heon Kim, Luisa W. Cheng and Kirkwood Land

Received: 23 November 2021
Accepted: 14 December 2021
Published: 20 December 2021

Publisher's Note: MDPI stays neutral with regard to jurisdictional claims in published maps and institutional affiliations.

1. Introduction

Owing to the increasing population of immunocompromised patients, life-threatening systemic fungal infections have become a major risk for public health. A weakened immune system because of illnesses, such as cancer, diabetes, or HIV/AIDS syndrome, together with the practice of treatments or medicines, such as chemotherapy, organ transplantation, or corticoids, might predispose the afflicted to secondary fungal infections [1–3]. Thus, it has been estimated that over 1.6 million people die each year because of severe fungal infections [1,4]. Only four classes of antifungal drugs (echinocandins, polyenes, pyrimidine analogues, and triazoles) are used for treating systemic fungal infections [5–7]. Thus, the problem of systemic fungal infections is also worsened by the emergence of fungi resistant to one or several classes of the available antifungals [8,9].

The cell wall is a structure external to the plasma membrane present in all fungal cells. Its integrity is crucial for fungal survival, making the boundary between the cytoplasm and the external milieu and maintaining the internal turgor pressure [10,11]. Because the cell wall is not present in the infected animal hosts, molecules that inhibit its synthesis are appealing as potential antifungal drugs [12,13]. Structural microfibrils of the β(1,3)-D-glucan polysaccharide are the most abundant components in the cell wall framework [14–16]. The fission yeast *Schizosaccharomyces pombe* is an attractive model for exploring the role of the β(1,3)-D-glucan in cell wall synthesis, morphogenesis, and cell integrity. The *S. pombe* cell wall has no detectable amounts of chitin [16–18]; instead, it contains three different and essential β-glucans: a major branched β(1,3)-D-glucan that is the main responsible for the cell wall structure and integrity; a minor linear β(1,3)-D-glucan, which is responsible for the primary septum structure; and a minor branched β(1,6)-D-glucan [19,20].

In all the studied fungi, the catalytic subunit of the β(1,3)-D-glucan synthase (GS), the glycosyltransferase in charge of the β(1,3)-D-glucan synthesis, is constituted by the family of integral membrane proteins Bgs/Fks [19,21,22]. *S. pombe* contains four essential GS catalytic subunits: Bgs1 synthesizes the linear β(1,3)-D-glucan of the primary septum [19] and is required for septum ingression [19,23–25]; Bgs2 is essential for spore wall maturation [26,27]; Bgs3 is essential, although its function remains unknown [28]; and Bgs4 is the only subunit that has been shown to form part of the GS enzyme. It synthesizes the branched β(1,3)-D-glucan and is responsible for most GS activity [19,29]. The branched β(1,3)-D-glucan produced by Bgs4 is vital for maintaining cell shape and septum formation and completion. Differently from Bgs1 and Bgs3, the function of Bgs4 is essential for cell integrity, and thus, Bgs4 depletion causes cell lysis and cytoplasm leakage in the growing poles and mainly in the medial region at the onset of septum degradation and cell separation [30].

Treatment with inhibitors of the GS blocks the incorporation of novel β(1,3)-D-glucan to the wall surrounding the sites of active growth and causes osmotic fragility and cell lysis in exponentially growing cells. Thus, treated fungal cells die by the focalized rupture of the underneath plasma membrane and the subsequent cytoplasm leakage [31–34]. Three families of antifungals, lipopeptides (echinocandins), acidic terpenoids (enfumafungin), or glycolipids (papulacandins), have been described to alter the integrity of the cell wall by targeting the GS [10,13,35]. From all of them, only the echinocandins anidulafungin, caspofungin, and micafungin have been authorized as drugs for treating systemic fungal infections (Table 1) [5,6,13,35]. Thus, because of their fungicidal effect in yeasts, echinocandins are proposed as the forefront treatment for *Candida albicans* infections [36,37].

Resistance to GS inhibitors is well conserved in fungi and is generally associated with point mutations located in highly conserved short regions (hot spots) of the Bgs/Fks proteins [29,38–40]. In *S. pombe*, *pbr1-8* and *pbr1-6* are the only isolated mutants exhibiting resistance to specific GS inhibitors [29]. These two mutants are exclusively due to point mutations in the *bgs4⁺* sequence. A single mutation within *bgs4⁺* confers resistance to specific GS inhibitors in the presence of the wild-type (WT) sequences of *bgs1⁺* and *bgs3⁺*, suggesting that the function of the two encoding synthases, Bgs1 and Bgs3, is not inhibited by the available GS inhibitors [29]. The mode of action of echinocandins inhibits the synthesis of the β(1,3)-D-glucan through a noncompetitive inhibition of the GS activity [41–43]. However, the integral membrane GS catalytic subunit has not been homogenously purified, and, thus, neither the biochemistry and structure of the GS complex, nor the molecular mechanism of echinocandin inhibition of the GS activity, have been entirely elucidated [22,44–46].

The absence of cell wall chitin and the presence of three essential GS catalytic subunits exhibiting both differential antifungal susceptibility and non-redundant roles in β(1,3)-D-glucan synthesis render *S. pombe* an ideal tool for studying echinocandins mechanism of action and resistance [19,29]. Bgs proteins are essential for different aspects of cytokinesis and cell integrity. Thus, we have compared the effect of several echinocandins in these cellular processes. For this purpose, *S. pombe* WT and *pbr1-8* strains were imaged through time-lapse fluorescence microscopy in the presence of the drugs. In trying to minimize the effects of cell wall compensatory mechanisms that might alter the echinocandins-derived phenotypes [47,48], time-lapses were performed at very early times of treatment (up to 3 h) with both sublethal (non-inhibitory) and lethal (inhibitory) concentrations of anidulafungin, caspofungin, and micafungin. Our study shows that, depending on the strain and the used concentration, the echinocandin drugs differentially affected the cytokinesis progression and cell integrity. Sublethal and lethal concentrations of micafungin selectively inhibited Bgs4, suggesting that Bgs1 and Bgs3 are not susceptible to this antifungal. Similarly, sublethal concentrations of anidulafungin and caspofungin primarily affected Bgs4, confirming that Bgs4 is the GS subunit most susceptible to echinocandins. Remarkably, lethal concentrations of anidulafungin and caspofungin severely affected cytokinesis progression, showing that besides their effect on Bgs4 activity, they also likely affect the function of Bgs1 and/or Bgs3.

Table 1. Chemical structure of the echinocandins derivatives approved for therapeutic use (table adapted from [10]).

Chemical Structure	Compound (Commercial Name and Proprietary Company)
	Anidulafungin (Eraxis® or Ecalta®, Pfizer)
	Caspofungin (Cancidas®, Merck, Sharp and Dhome, MSD)
	Micafungin (Mycamine®, Astellas Pharma)

2. Results and Discussion

2.1. Susceptibilities of WT and pbr1-8 Strains to Echinocandins

The susceptibility of yeasts to echinocandins depends on the number of inoculated cells. Thus, greater cell densities require higher drug concentrations to inhibit organism growth [41]. Because the imaging and preparation of yeast cells for time-lapse fluorescence microscopy requires cultures containing a higher density of cells than cultures used for visualizing the presence or absence of growth in susceptibility assays, we compared the susceptibilities to echinocandins of the WT and *pbr1-8* strains in cultures inoculated either with a typical lower cell density (5×10^5 cells/mL) or with a higher cell density (5×10^6 cells/mL) that will later be used in time-lapse fluorescence microscopy experiments (Table 2). The minimal inhibitory concentration (MIC) for the WT strain varied between 1 and 10 µg/mL at lower cell densities (Table 2, left column). As expected, the MIC for the WT increased to 10–20 µg/mL at higher cell densities (Table 2, right column). Thus, based on the obtained MICs with a higher number of cells, sublethal (non-inhibitory) and lethal (inhibitory) concentrations for the WT of 2 and 20 µg/mL, respectively, were chosen for performing the comparative time-lapse experiments (see next section). The *pbr1-8* strain exhibited a complete resistance to micafungin that was independent of the density of cells in the culture (a MIC of more than 80 µg/mL). In contrast, this strain presented some susceptibility to caspofungin and anidulafungin, with the MICs varying from 2–10 (lower cell density, Table 2, left column) to 20–40 µg/mL (higher cell density, Table 2, right column). Finally, micro-cultures starting at a higher cell density were used to examine the cell morphology by phase-contrast microscopy after 24 h of growth in the presence of increasing concentrations of the echinocandins. The three echinocandins led to the WT cells becoming aggregated, rounded, and swollen. This phenotype started to be observed in cells treated with sublethal concentrations of 1 µg/mL for caspofungin and 4 µg/mL for both anidulafungin and micafungin (Figure 1, WT). None of the applied concentrations of the micafungin was sufficient to induce the phenotype of swollen and rounded cells in the *pbr1-8* resistant strain; however, lethal concentrations of anidulafungin and caspofungin were able to induce it. (Figure 1, *pbr1-8*). These results show that micafungin selectively inhibits the function of Bgs4, whereas Bgs1 and Bgs3 are not affected by this antifungal. The same results were previously described for the other two classes of GS inhibitors, the glycopeptide papulacandin B and the acidic terpenoid enfumafungin that also selectively inhibited Bgs4 [29].

Table 2. Susceptibilities of *S. pombe* wild-type and *pbr1-8* strains to echinocandins [1].

Strain	Lower Cell Density [2]	Higher Cell Density [3]
	Anidulafungin MIC (µg/mL)	
WT	4	10
pbr1-8	10	20
	Caspofungin MIC (µg/mL)	
WT	1	10
pbr1-8	2	40
	Micafungin MIC (µg/mL)	
WT	10	20
pbr1-8	>80	>80

[1] The MIC was defined as the concentration of echinocandin at which growth of the corresponding *S. pombe* strain was completely inhibited after 24 h at 28 °C in YES medium. [2] Cells were diluted to a cell density of 5×10^5 cells/mL and grown in microcultures of YES liquid medium with increasing concentrations (0, 1, 2, 4, 10, 20, 40, and 80 µg/mL) of echinocandins. [3] Cells were diluted to a cell density of 5×10^6 cells/mL and grown in microcultures of YES liquid medium with increasing concentrations (0, 1, 2, 4, 10, 20, 40, and 80 µg/mL) of echinocandins.

2.2. Differential Effects of the Echinocandins in Cytokinesis Progression

Cytokinesis is the last event of the cell cycle, where a cleavage furrow partitions the cell giving rise to two independent cells. In fungi, the ingression of the cleavage furrow is tightly coupled to the synthesis of a specialized cross wall named division septum [19,25,30,49]. In contrast to the filamentous fungi, yeast cytokinesis also requires the separation of daughter cells through the enzymatic degradation of the septum. To avoid cell lysis during cell separation, strict coordination between degradation and synthesis of the cell wall surrounding the septum is needed [30,50]. Thus, defects in the synthesis of the main cell wall glucans frequently lead to excessive wall degradation and cell lysis at the onset of cell separation [30]. Here, time-lapse fluorescence microscopy was used to examine the dynamics of septation and cell separation in the WT and *pbr1-8* strains during the first 3 h of echinocandin treatment. To follow the cytokinesis progression, cells were stained with very low doses of calcofluor white (CW), a fluorochrome that specifically and with high affinity binds to the linear β(1,3)-D-glucan of the primary septum, thus allowing the monitoring of cytokinesis since the very early steps of primary septum synthesis [49]. As described above, cells of the two strains were treated with both sublethal (2 µg/mL) and lethal (20 µg/mL) concentrations of the three echinocandins. As a control for the experiments, we first compared the cytokinesis progression in the WT and *pbr1-8* cells growing in the absence of echinocandins (control, Figure 2), observing that septum progression (double arrow depicted in green) and cell separation onset (double arrow depicted in orange) were similar in both strains. Table 3 compiles the data from the performed time-lapse microscopy experiments, showing the elapsed minutes from the beginning to the end of septum formation (septation) and from the end of septum formation to the start of cell separation (cell separation) in the two strains (WT and *pbr1-8*) growing in the absence (Figure 2, control) or the presence of sublethal and lethal concentrations of the three echinocandins (Figures 3–5). Depending on the strain and concentration of the echinocandin, different effects in the cytokinesis were induced: (1) cytokinesis was normal, with the elapsed times of septation and separation being similar to those of control cells (Table 3, boxes depicted in green); (2) cytokinesis was not blocked, but cell death occurred at the onset of cell separation (Table 3, boxes depicted in red); and (3) cytokinesis was blocked or delayed in the septum progression, and interrupted (or greatly delayed) the cell separation onset, and, thus, cell death did not occur (Table 3, boxes depicted in other colors). Thus, we analyzed how each echinocandin affects the cytokinesis progression (see below in the following subsection) and cell integrity (see below in the last section).

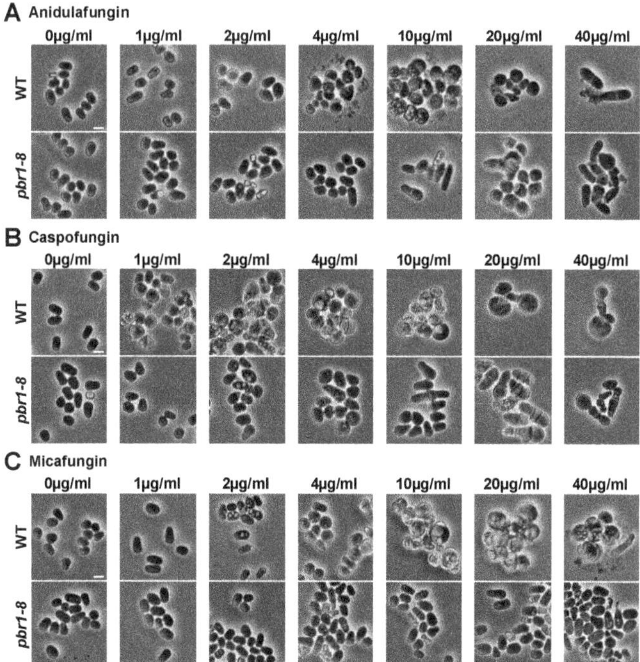

Figure 1. Morphology of WT and *pbr1-8* cells after 24 h of growth in the presence of increasing concentrations of anidulafungin (**A**), caspofungin (**B**), and micafungin (**C**) drugs. Early logarithmic-phase cells of the WT and *pbr1-8* strains growing in YES liquid medium at 28 °C were diluted to a high cell density of 5×10^6 in micro-cultures of YES liquid medium containing either DMSO (0.8%, control) or increasing concentrations (1, 2, 4, 10, 20, and 40 µg/mL) of the drugs, grown with shaking for 24 h and imaged by phase-contrast microscopy. The data of this figure are developed in Table 2. Scale bars, 10 µm.

Table 3. Cytokinesis phenotypes and elapsed times for septation and cell separation periods in *S. pombe* cells growing in culture chamber slides containing sublethal (2 µg/mL) or lethal (20 µg/mL) concentrations of the three echinocandins.

Drug (Concentration)	Cytokinesis Phenotype	Septation [1]		Cell Separation [2]	
		WT	*pbr1-8*	WT	*pbr1-8*
Control (+DMSO)	Normal cytokinesis	22.4 ± 2.4 $n = 30$	20.2 ± 1.6 $n = 33$	14.5 ± 2.4 $n = 27$	16.5 ± 3.4 $n = 33$
Anidulafungin (2 µg/mL)	Cell death [3]	45.9 ± 10.5 $n = 25$	$n = 0$	36.3 ± 6.5 $n = 25$	$n = 0$
	Normal cytokinesis	$n = 0$	22.6 ± 2.3 $n = 60$	$n = 0$	21.4 ± 4.6 $n = 55$
	Slow septation Blocked separation [4]	48.0 ± 0.0 $n = 1$	$n = 0$	≥84.0 ± 0.0 $n = 1$	$n = 0$
Anidulafungin (20 µg/mL)	Blocked cytokinesis [5]	≥155.1 ± 24.9 $n = 23$	≥160.8 ± 15.4 $n = 31$	≥145.9 ± 36.1 $n = 7$	≥172.0 ± 0.0 $n = 1$
Caspofungin (2 µg/mL)	Cell death [3]	31.0 ± 4.3 $n = 20$	32.0 ± 9.0 $n = 8$	32.4 ± 6.2 $n = 21$	26.0 ± 9.2 $n = 10$
	Normal cytokinesis	$n = 0$	24.3 ± 5.2 $n = 12$	$n = 0$	22.8 ± 4.9 $n = 17$
	Blocked separation	20.0 ± 2.9 $n = 14$	25.1 ± 4.3 $n = 11$	≥80.0 ± 26.2 $n = 14$	≥81.5 ± 36.5 $n = 11$
Caspofungin (20 µg/mL)	Extremely slow septation Blocked separation [4]	≥121.6 ± 26.5 $n = 24$	$n = 0$	≥ 75.3 ± 28.7 $n = 7$	≥105.1 ± 23.1 $n = 29$
	Slow septation	$n = 0$	51.4 ± 8.4 $n = 25$	$n = 0$	$n = 0$
Micafungin (2 µg/mL)	Cell death [3]	22.8 ± 1.8 $n = 23$	$n = 0$	26.1 ± 6.8 $n = 31$	$n = 0$
	Normal cytokinesis	$n = 0$	20.1 ± 1.3 $n = 36$	$n = 0$	17.3 ± 2.7 $n = 33$
	Blocked separation	20.7 ± 1.8 $n = 6$	$n = 0$	≥52.0 ± 21.3 $n = 6$	$n = 0$
Micafungin (20 µg/mL)	Cell death [3]	27.6 ± 2.0 $n = 11$	$n = 0$	33.7 ± 4.3 $n = 12$	$n = 0$
	Normal cytokinesis	$n = 0$	19.8 ± 1.2 $n = 46$	$n = 0$	25.9 ± 5.0 $n = 39$

Values are minutes ± SD. [1] Elapsed time between septation onset and septation completion. [2] Elapsed time between septation completion and cell separation onset. [3] Cell death was at the onset of cell separation. [4] Separation onset is blocked or largely delayed. [5] Septation and cell separation are both blocked. n = number of cells that exhibit the corresponding phenotype within 3 h of imaging. Green boxes: normal cytokinesis; red boxes: cytokinesis with cell death at the onset of cell separation; and boxes in other colors: cytokinesis is either blocked or delayed.

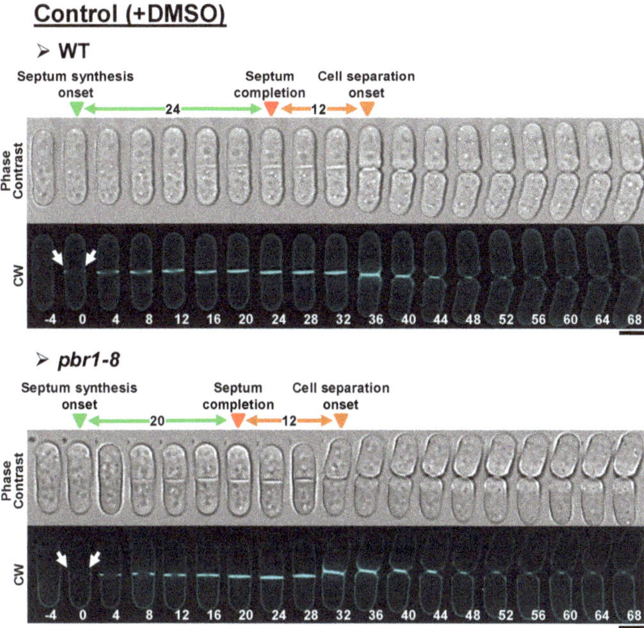

Figure 2. Normal cytokinesis (septum synthesis and cell separation) in WT and *pbr1-8* cells growing in the absence of the three echinocandin drugs. Early logarithmic-phase cells of the indicated strains growing in YES liquid medium at 28 °C were diluted to a cell density of 5×10^6 in liquid YES medium containing both calcofluor white (CW, 1.25 µg/mL) and DMSO (0.8%), and then imaged by time-lapse fluorescence microscopy (1 medial z slice, 4 min elapsed time) for 3 h, as described in the Materials and Methods section. The data of this figure are developed in Tables 3–5. White arrow: first CW-stained septum synthesis. Arrowheads: blue, septum synthesis onset (time 0 for the elapsed time until septum synthesis completion); red, septum completion onset (time 0 for the elapsed time until cell separation onset); orange, cell separation onset. Scale bars, 5 µm.

2.2.1. Anidulafungin

Treatment of the WT strain with sublethal concentrations of anidulafungin for 3 h mainly induced an increase in the elapsed times of both septum formation and the onset of cell separation, without blocking cytokinesis progression (Table 3, Figure 3A, and Figure S1A). In contrast, the elapsed times for septation and cell separation onset in the *pbr1-8* mutant cells treated with sublethal concentrations of the drug were similar to those observed in the non-treated control cells (Table 3, Figure 3A, and Figure S1A). Interestingly, in both strains, lethal concentrations of anidulafungin arrested septum formation and consequently impeded the cell-separation onset (Table 3, Figure 3B, and Figure S1B). These results show that anidulafungin induces defects in the synthesis of β(1,3)-D-glucan that affect the progression of septation. These defects are also proportional to the antifungal concentration: low concentrations slow down septation, whereas higher concentrations completely block it. Thus, sublethal concentrations of the drug primarily induced two phenotypes in the WT strain; slower cytokinesis progression and cell death at the start of cell separation, which were similar to those described for Bgs4 depletion [30]. Additionally, the fact that both phenotypes were absent in the *pbr1-8* strain (Figure 3A and Figure S1A) indicates that lower concentrations of anidulafungin cause the sole inhibition of Bgs4. In contrast, the septation arrest observed in the presence of lethal concentrations of the drug was not corrected in the *pbr1-8* strain, indicating that these concentrations of the drug lead to the combined inhibition of Bgs4 together with Bgs1 and/or Bgs3 (Figure 3B). In agreement, septum ingression is similarly impaired when the function of Bgs1 is depleted or reduced in some mutant alleles [19,23–25].

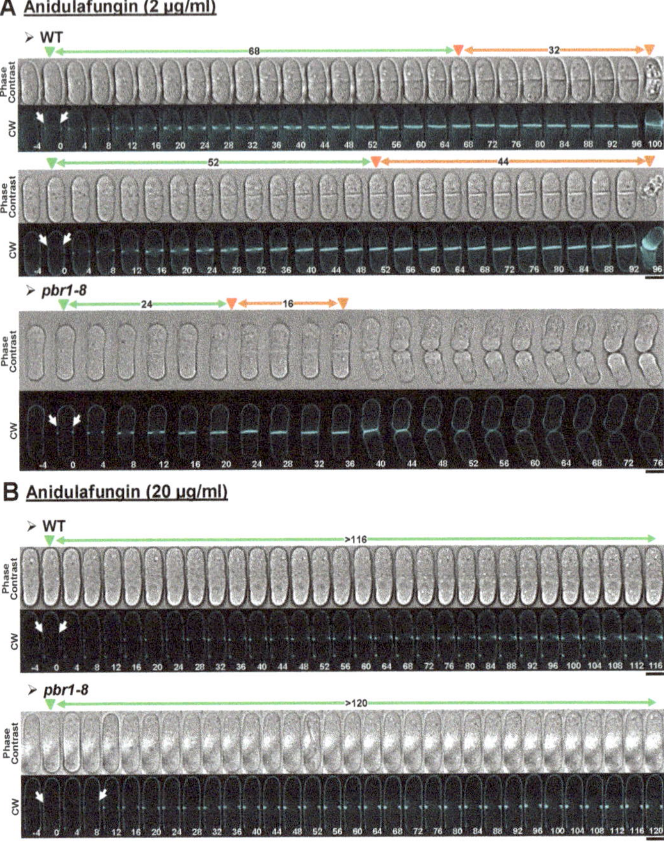

Figure 3. Cytokinesis phenotypes in WT and *pbr1-8* cells growing in the presence of sublethal and lethal concentrations of anidulafungin. The indicated strains were grown and imaged as in Figure 2 in the presence of either sublethal ((**A**), 2 µg/mL) or lethal ((**B**), 20 µg/mL) concentrations of the drug. The data of this figure are developed in Tables 3–5. Arrows and arrowheads are as in Figure 2. Scale bars, 5 µm.

2.2.2. Caspofungin

Next, cytokinesis progression in the presence of caspofungin was examined. Sublethal concentrations of the drug resulted in a slight slowdown of the cytokinesis progression, with many WT cells dying at the onset of cell separation (Table 3, Figure 4A, WT, Type 1, and Figure S2A). After approximately 1.5 h of growth in the presence of the antifungal, many WT cells started to exhibit a progressive swelling of the growing pole, giving rise to cells with a bulb or drumstick appearance. Interestingly, due to a blockage of the onset of cell separation, most of these drumstick-like cells were still alive by the end of the 3 h treatment (Table 3, Figure 4A, WT, Type 2, and Figure S2A). The same drumstick-like cells with blocked separation were also observed in the presence of sublethal concentrations of both anidulafungin and micafungin (Table 3, Figures S1A and S3A, yellow arrows), as previously described for the echinocandin lipopeptide aculeacin A [32]. The *pbr1-8* mutation partially suppressed the phenotype of cell death at the start of cell separation, with many cells at the earliest treatment times exhibiting normal cytokinesis (Table 3, data not shown). However, most of the resistant cells that entered cytokinesis at later treatment times acquired the same drumstick appearance of the WT strain (Table 3, Figure 4A, and Figure S2A). Although lethal concentrations of caspofungin did not block the cytokinesis progression like anidulafungin (Figure 3B), the septum completion was extremely slowed down, and the cytokinesis was still incomplete by the end of the 3 h treatment (Table 3 and Figure 4A). The *pbr1-8* mutation partially suppressed the septum progression defects of cells exposed to higher doses of caspofungin ($\geq$121.6 $\pm$ 26.5 min in the WT compared to 51.4 $\pm$ 8.4 min in the resistant strain), although, because of the blocked cell separation, cytokinesis progression was still very delayed and incomplete by the end of the 3 h treatment (Table 3 and Figure 4B). The fact that the *pbr1-8* mutation did not completely suppress the phenotypes observed in the WT strain suggests that caspofungin also acted against Bgs4 and Bgs1 and/or Bgs3, like anidulafungin. In agreement and similarly to anidulafungin, lethal concentrations of caspofungin led to an extremely slow cytokinesis development.

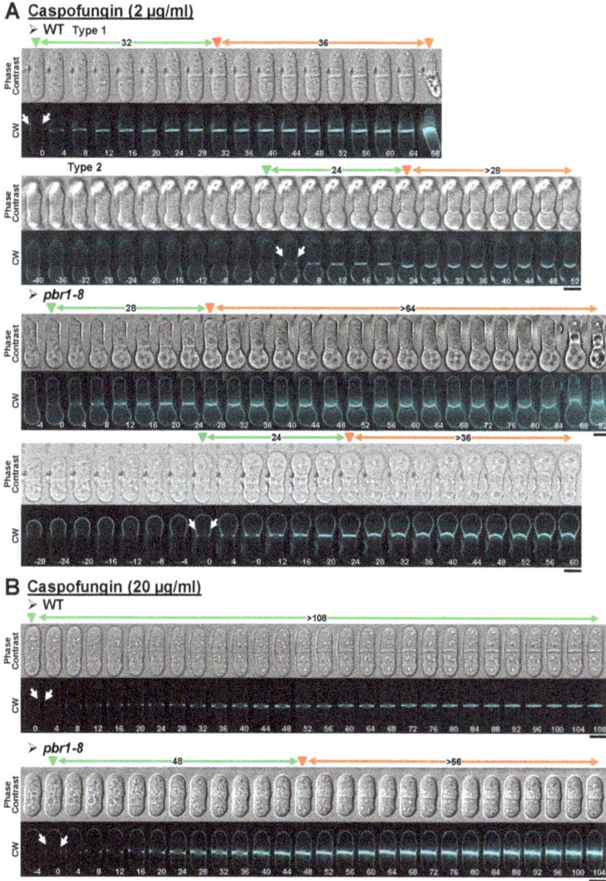

Figure 4. Cytokinesis phenotypes in WT and *pbr1-8* cells growing in the presence of sublethal and lethal concentrations of caspofungin. The indicated strains were grown and imaged as in Figure 2 in the presence of sublethal ((**A**), 2 µg/mL) or lethal ((**B**), 20 µg/mL) concentrations of the drug. The data of this figure are developed in Tables 3–5. Arrows and arrowheads are as in Figure 2. Scale bars, 5 µm.

2.2.3. Micafungin

In contrast to the other two echinocandins, the studies performed with micafungin showed that this drug solely inhibits the activity of Bgs4 at both sublethal and lethal concentrations (Table 3 and Figure 5). Thus, lower micafungin concentrations did not significatively affect cytokinesis progression, inducing a phenotype of cell death at the onset of cell separation (Table 3, Figure 5A and Figure S3A). At later treatment times, a few drumstick-like cells with blocked separation were also observed (Table 3 and Figure S3A, yellow arrows). Interestingly, both phenotypes were entirely suppressed in the resistant *pbr1-8* strain (Table 3, Figure 5A and Figure S3A). Similarly, lethal concentrations of the drug only caused a slight slowdown of the cytokinesis progression, with all the cells in cytokinesis dying at the cell separation onset. In these conditions, no drumstick-like cells were observed within the 3 h treatment (Table 3, Figure 5B and Figure S3B), suggesting that those drumstick-like cells may be the germ for the survivors observed after 24 h of treatment with sublethal concentrations of the three drugs. Again, non-cell death was observed in the treated cells of the *pbr1-8* strain, which also exhibited a normal cytokinesis progression (Table 3, Figure 5B and Figure S3B). It is noteworthy to mention that the cytokinesis phenotypes observed in the WT strain with lethal concentrations of micafungin were the opposite of those with anidulafungin and caspofungin. Additionally, while the *pbr1-8* mutation suppressed all the phenotypes caused by both sublethal and lethal concentrations of the micafungin, the mutation did not totally suppress the phenotypes caused by the other two echinocandins. All these results show that both sublethal and lethal concentrations of micafungin selectively inhibit Bgs4. In contrast, the other two echinocandins, Bgs1 and Bgs3, seem to not be affected by micafungin. This observation also might suggest that these two enzymes are not the targets for micafungin.

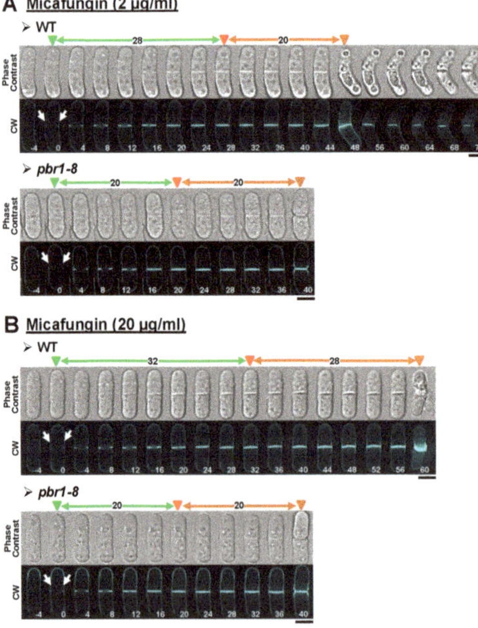

Figure 5. Cytokinesis phenotypes in WT and *pbr1-8* cells growing in the presence of sublethal and lethal concentrations of micafungin. The indicated strains were grown and imaged as in Figure 2 in the presence of sublethal ((**A**), 2 µg/mL) or lethal ((**B**), 20 µg/mL) concentrations of the drug. The data of this figure are developed in Tables 3–5. Arrows and arrowheads are as in Figure 2. Scale bars, 5 µm.

Currently, it is unknown why Bgs4 is susceptible to all the examined GS inhibitors, while Bgs1 and Bgs3 are susceptible to some inhibitors (anidulafungin and caspofungin) but not susceptible to others (micafungin, papulacandin B, and enfumafungin). Although Bgs1, Bgs3, and Bgs4 show high degrees of conservation (approximately 55% identity), they might still acquire different tertiary structures, making Bgs1 and Bgs3 less accessible to some echinocandins and more accessible to others. Another explanation might be due to changes in the lipid composition of the microdomains containing the GS subunits. Lipid composition impacts the structural and functional properties of the cell membranes, thus influencing the function of integral membrane proteins, such as the GS enzymes. Besides this, specific lipids may distribute asymmetrically between leaflets, originating different nanoscale domains along the plasma membrane [51]. Echinocandins are cyclic hexapeptides with a lipid side chain (Table 1) that is essential for their antibiotic activity and toxicity [52]. The lipid side chain is thought to interact with the outer leaflet of the plasma membrane containing the GS enzymes [45]. Thus, the differential susceptibility of the Bgs proteins to the echinocandins might be caused by differences in the organization and/or composition of the Bgs-containing lipid microenvironment. Consequently, it has been proposed that changes in the membrane lipids' composition might modulate the interaction of echinocandins with the plasma membrane depending on the structure of the lipid side chain [53,54]. Alternatively, caspofungin and anidulafungin might affect the function of Bgs1 and/or Bgs3 through a secondary mechanism. Caspofungin and anidulafungin treatments induce a reduction in the levels of Bgs1 in the cleavage furrow (our unpublished results). In the budding yeast, the echinocandin B inhibits the localization of Spa2 [55], a cell polarity protein that regulates the localization of Chs2, which is responsible for the chitin synthesis of the primary septum [56]. Similarly, *S. pombe* Spa2 interacts with the F-BAR protein Cdc15 that is essential for the localization of Bgs1 to the division site [19,56].

2.3. Differential Effects of the Echinocandins in Cell Integrity

Next, the phenotypes of cell death, as well as the percentages corresponding to each phenotype, were analyzed in the time-lapse microscopy experiments described in the previous section. Depending on the effect of echinocandins in cytokinesis progression, differential levels of cell death were observed. Sublethal concentrations of any of the three antifungals did not block cytokinesis (Table 3), inducing death in 80–90% of the cells in the examined fields (Table 4). The cell death occurred either at the start of cell separation (Figures 3A, 4A and 5A) or during polarized growth in interphase (Figures S1A–S3A, and S4). Interestingly, these high percentages of initial death were not enough for arresting the cell growth (Table 2), probably because of the emergence of drumstick-like survivors at the later times of the 3 h treatment (Figures S1A–S3A, yellow arrows). Something similar is observed at longer times of Bgs4 depletion, where a small population of drumstick-like cells is also able to survive ([30], and our unpublished results). As expected, lethal concentrations also induced differential effects in cell integrity (Table 4). Both anidulafungin and caspofungin were primarily fungistatic and caused low percentages of cell death (22–27%) because they blocked the cell separation. In contrast, micafungin did not block cell separation, mainly being a fungicide and inducing high percentages of cell death (97%) (Table 4).

As expected, the *pbr1-8* resistant mutation entirely suppressed the cell death caused by both sublethal and lethal concentrations of micafungin, reinforcing the finding that micafungin only inhibits Bgs4 activity (Tables 4 and 5). Similarly, the *pbr1-8* mutation either suppressed or reduced the death caused by sublethal and lethal concentrations of anidulafungin, respectively (Tables 4 and 5). In contrast, the *pbr1-8* mutation partially suppressed the death observed at lower concentrations (from 80% to 22%) but was largely ineffective with higher concentrations of caspofungin (22% in both WT and *pbr1-8*). Again, these results reinforce that caspofungin and anidulafungin inhibit several GS subunits to different degrees.

Table 4. Percentages and types of cell death observed in the time-lapses of WT and *pbr1-8* cells growing in the presence of sublethal (2 µg/mL) or lethal (20 µg/mL) concentrations of the three echinocandin drugs.

Drug (Concentration)	Strain	Number of Cells	Cell Death [1]	Cell Lysis [2]	Non-Cell Lysis [3]
Control	**WT**	*n = 30*	0 [4] (0) [5]	0 [4] (0) [5]	0 [4] (0) [5]
(+DMSO)	*pbr1-8*	*n = 33*	0 (0)	0 (0)	0 (0)
Anidulafungin	**WT**	*n = 79*	89.9 (100)	65.2 (72.5)	24.7 (27.5)
(2 µg/mL)	*pbr1-8*	*n = 61*	0 (0)	0 (0)	0 (0)
Anidulafungin	**WT**	*n = 60*	21.7 (100)	13.3 (61.5)	8.4 (38.5)
(20 µg/mL)	*pbr1-8*	*n = 45*	8.9 (100)	2.2 (25)	6.7 (75)
Caspofungin	**WT**	*n = 96*	82.3 (100)	58.3 (70.9)	24 (29.1)
(2 µg/mL)	*pbr1-8*	*n = 85*	22.4 (100)	5.9 (26.3)	16.5 (73.7)
Caspofungin	**WT**	*n = 33*	27.3 (100)	12.1 (44.4)	15.2 (55.6)
(20 µg/mL)	*pbr1-8*	*n = 64*	22.7 (100)	0 (0)	22.7 (100)
Micafungin	**WT**	*n = 99*	80.3 (100)	13.6 (17)	69.7 (83)
(2 µg/mL)	*pbr1-8*	*n = 36*	0 (0)	0 (0)	0 (0)
Micafungin	**WT**	*n = 61*	97.5 (100)	97.5 (100)	0 (0)
(20 µg/mL)	*pbr1-8*	*n = 46*	0 (0)	0 (0)	0 (0)

[1] Cell death is the sum of the percentages of cell lysis and non-cell lysis. [2] Cells shrunk and died with cell breakage and cytoplasm leakage. [3] Cells shrunk and died without cell breakage and cytoplasm leakage. [4] Value is the percentage of dead cells with respect to the total number of visualized cells (*n*). [5] Value is the percentage of cell lysis or non-cell lysis with respect to the total number of dead cells.

The three echinocandins induced two death phenotypes: death with cytoplasm leakage (Figures S1–S3, lysis, red arrows) and without cytoplasm leakage (Figures S1–S3, non-lysis, white arrows). Sublethal concentrations of anidulafungin and caspofungin mainly induced a lytic phenotype that mostly occurred at the onset of cell separation and was almost entirely suppressed in the *pbr1-8* strain (Tables 4 and 5). In contrast, the cell death without an apparent leakage of the cytoplasm was still observed in the *pbr1-8* cells treated with caspofungin and was residual in the case of anidulafungin (Tables 4 and 5). Intriguingly, the results of micafungin show a phenotypic difference in cell death that depends on the used concentration (Figure 5 and Figure S3). Sublethal concentrations of this antifungal induced, in the WT strain, a very high percentage of death without lysis (70%), while, with lethal concentrations, the totality of cell death involved the release of the cytoplasmic content (Tables 3 and 5). Both death phenotypes were completely suppressed in the *pbr1-8* strain, indicating that both micafungin effects are due to the Bgs4 function. The phenotype of cell lysis is clearly the result of a localized rupture of the plasma membrane located under the weakened and/or reduced cell wall because of the reduction in the Bgs4-derived branched $\beta(1,3)$-D-glucan [19,30]. In contrast, death without lysis does not involve release of cytoplasmic content and might be due to a general alteration of plasma membrane permeability. The same phenotype of death without lysis was also observed in a few cells during Bgs4 depletion (our unpublished results) or when the function of the regulatory subunit of the GS complex, GTPase Rho1, is affected in the mutant allele *rho1-596* [57]. Endoplasmic reticulum (ER) stress in the budding yeast induces the permeabilization of the vacuolar membrane, causing cell acidification and death [58]. Interestingly, it has been described in several yeast species that cell wall stress activates an ER stress-like response that requires the entry of calcium and the function of the phosphatase calcineurin [57]. Thus, the cell damage caused by micafungin-inhibition of Bgs4 might activate a calcineurin-dependent ER stress-like response in fission yeast. In agreement, many mutants resistant to the calcineurin inhibitor FK506 exhibit hypersensitivity to micafungin [59]

Table 5. Percentages of cell lysis either at interphase or at cell separation onset observed in the time-lapses of WT and *pbr1-8* cells growing in the presence of sublethal (2 µg/mL) or lethal (20 µg/mL) concentrations of the three echinocandin drugs.

Drug (Concentration)	Strain	Number of Cells	Total Cell Lysis [1]	Lysis at Separation [1]	Lysis at Interphase [1]
Control	**WT**	*n = 30*	0 [2] (0) [3]	0 [2] (0) [3]	0 [2] (0) [3]
(+DMSO)	*pbr1-8*	*n = 33*	0 (0)	0 (0)	0 (0)
Anidulafungin	**WT**	*n = 79*	65.2 (72.5)	45.6 (50.7)	19.6 (21.8)
(2 µg/mL)	*pbr1-8*	*n = 61*	0 (0)	0 (0)	0 (0)
Anidulafungin	**WT**	*n = 60*	13.3 (61.5)	10.0 (46.1)	3.3 (15.4)
(20 µg/mL)	*pbr1-8*	*n = 45*	2.2 (25)	0 (0)	2.2 (25)
Caspofungin	**WT**	*n = 96*	58.3 (70.9)	40.6 (49.4)	17.7 (21.5)
(2 µg/mL)	*pbr1-8*	*n = 85*	5.9 (26.3)	2.4 (10.5)	3.5 (15.8)
Caspofungin	**WT**	*n = 33*	12.1 (44.4)	12.1 (44.4)	0 (0)
(20 µg/mL)	*pbr1-8*	*n = 64*	0 (0)	0 (0)	0 (0)
Micafungin	**WT**	*n = 99*	13.6 (17)	3.0 (3.8)	10.6 (13.2)
(2 µg/mL)	*pbr1-8*	*n = 36*	0 (0)	0 (0)	0 (0)
Micafungin	**WT**	*n = 61*	97.5 (100)	44.3 (45.4)	53.3 (54.6)
(20 µg/mL)	*pbr1-8*	*n = 46*	0 (0)	0 (0)	0 (0)

Cell lysis is cell death accompanied by cell breakage and cytoplasm leakage. [1] Total cell lysis is the sum of the percentages of both cell lysis at interphase and at cell separation onset. [2] Value is the percentage of lysed cells with respect to the total number of cells (*n*). [3] Value is the percentage of lysed or non-lysed cells with respect to the total number of dead cells.

3. Materials and Methods

3.1. Strains and Culture Conditions

The *S. pombe* strains examined in this study were isogenic to the WT strain h⁻ 972. The *pbr1-8* mutant was isolated by ethyl methane sulfonate mutagenesis and selection in the presence of 20 µg/mL papulacandin B [29]. The standard rich yeast growth medium (YES from "Yeast Extract with Supplements") has been described previously [60]. Cell growth was monitored by measuring the A600 of early log-phase cell cultures in a Smart-Spec 3000 spectrophotometer (Bio-Rad, Hercules, CA, USA; A600 0.1 = 10^6 cells/mL).

3.2. Antifungal Drugs and Susceptibility Assays

The echinocandins used in this study were generous gifts from Pfizer, New York, NY, USA (anidulafungin), Merck Sharp and Dohme, Kenilworth, NJ, USA (caspofungin), and Astellas Pharma, Chūō, Tokyo, Japan (micafungin). The echinocandins were kept at −20 °C in stock solution (10 mg/mL in DMSO) and assayed at the final concentrations specified in the text, tables, and figures. For micro-culture assays of a large number of samples, log-phase cultures grown in YES medium were diluted to a cell density of 5×10^5 cells/mL (lower cell density) or 5×10^6 cells/mL (higher cell density) in YES medium containing increasing concentrations of echinocandins (1, 2, 4, 10, 20, 40, and 80 µg/mL) or an equal volume of solvent (0.8% DMSO), which was the control cell culture. The cell cultures were grown in an orbital roller at 28 °C, and turbidity was examined after 24 h of growth. The MIC was the minimal concentration of echinocandin that induced a complete inhibition of the cell growth after 24 h of growth. The values were calculated from at least three independent experiments.

3.3. Microscopy Techniques and Data Analysis

Images of cells after 24 h of growth in the presence of the echinocandins (Figure 1) were directly obtained from the micro-cultures used for the susceptibility assays with a Nikon Eclipse 50i microscope, a Nikon Plan FLUOR 20 ×/0.45 objective, a Nikon Ds-Fi1 digital camera, and a Nikon Digital Sight DS-L2 control unit. Time-lapse imaging was performed as previously described [49]. A volume of 0.3 mL of logarithmic-phase cells grown in YES medium at 28 °C was collected at the cell density of 5×10^6 cells/mL. The cells were gently centrifuged (1 min at 1000g), and the cell pellet was suspended in the same YES medium containing calcofluor white (CW) at a very low final concentration of 1.25 µg/mL together with 0.8% DMSO as control or the corresponding echinocandin at a final concentration of 2 µg/mL (sublethal, non-inhibitory) or 20 µg/mL (lethal, inhibitory), and placed in a well from a µ-Slide 8 well (80,821-Uncoated; Ibidi, Gräfelfing, Germany) previously covered with 5 µL of 1 mg/mL soybean lectin (L1395; Sigma-Aldrich, Burlington, MA, USA). All the time-lapse experiments were performed at 28 °C by acquiring epifluorescence and phase contrast cell images in single planes every 4 min and 1×1 binning on an inverted microscope (Olympus IX71) equipped with a PlanApo $100 \times/1.40$ IX70 objective and a Personal DeltaVision system (Applied Precision LLC., Issaquah, WA, USA). Images were obtained using CoolSnap HQ2 monochrome camera (Photometrics, Tucson, AZ, USA) and softWoRx 5.5.0 release 6 imaging software (Applied Precision LLC.). Subsequently, CW time-lapse images were restored and corrected by 3D Deconvolution (conservative ratio, 10 iterations, and medium noise filtering) through soft-WoRx imaging software. Finally, images were processed with Image J (National Institutes of Health, Bethesda, MD, USA) and Adobe Photoshop software. All the time-lapse videos were repeated at least twice, and the data were calculated from at least two independent experiments.

4. Conclusions

Here, we have compared the cytokinesis and cell integrity processes in cells treated with sublethal and lethal concentrations of three echinocandin drugs: anidulafungin, caspofungin, and micafungin. We found that micafungin only inhibits activity due to Bgs4. Thus, and similarly to Bgs4 depletion, both sublethal and lethal concentrations of this drug primarily caused a phenotype of cell death in the WT that was entirely suppressed by the $bgs4^+$ mutation in *pbr1-8* cells. Our results indicate that all three echinocandins appear to act via Bgs4. In comparison, from the examination of septation and cell separation, it is evident that anidulafungin appears to affect, in some degree, the functions of Bgs1 and/or Bgs3, while caspofungin affects to a lower extent, and micafungin does not affect at all. To our knowledge, this is the first study analyzing, in detail, the progression and dynamics of cytokinesis in the presence of the echinocandin drugs. This experimental approximation has helped us identify that echinocandins induce differential effects in cytokinesis and cell integrity. Globally, these results show that caspofungin and anidulafungin are the most effective echinocandins against the resistant strain *pbr1-8*, probably because they combinedly affect the function of Bgs4 and other GS catalytic subunits.

Supplementary Materials: The following are available online at https://www.mdpi.com/article/10.3390/ph14121332/s1, Figure S1: Representative time-lapse field of WT and *pbr1-8* cells growing in the presence of sublethal and lethal concentrations of anidulafungin, Figure S2: Representative time-lapse field of WT and *pbr1-8* cells growing in the presence of sublethal and lethal concentrations of caspofungin, Figure S3: Representative time-lapse fields of WT and *pbr1-8* cells growing in the presence of sublethal and lethal concentrations of micafungin, Figure S4: Cell lysis and cytoplasm leakage during interphase in WT cells growing in the presence of the echinocandin drugs.

Author Contributions: Conceptualization, J.C.G.C., P.P. and J.C.R.; methodology, N.Y., L.G.-D., M.Á.C., V.S.D.C., M.B.M., P.P., J.C.R. and J.C.G.C.; formal analysis, N.Y., L.G.-D., M.Á.C., V.S.D.C., M.B.M., P.P., J.C.R. and J.C.G.C.; investigation, N.Y.; resources, J.C.R.; writing—original draft preparation, J.C.G.C. and J.C.R.; writing—review and editing, J.C.G.C. and J.C.R.; supervision, J.C.G.C., P.P. and J.C.R.; project administration, P.P. and J.C.R.; funding acquisition, P.P. and J.C.R. All authors have read and agreed to the published version of the manuscript.

Funding: This research was funded by grants PGC2018-098924-B-I00 (the Spanish Ministry of Science and Innovation, MICINN, Spain; the European Regional Developmental Fund, FEDER, EU), and CSI150P20 and "Escalera de Excelencia" CLU-2017-03 (Regional Government of Castile and Leon, JCYL, Spain; and the European Regional Developmental Fund, FEDER, EU).

Institutional Review Board Statement: Not applicable.

Informed Consent Statement: Not applicable.

Data Availability Statement: Data is contained within the article and Supplementary Material.

Acknowledgments: Vanessa S. D. Carvalho and Laura Gómez-Delgado recognize financial support received through contracts from the Regional Government of Castile and Leon (JCYL, Spain) and the European Regional Developmental Fund (FEDER, EU), and the Spanish Ministry of Science and Innovation (MICINN, Spain), respectively. The authors are grateful to the companies Astellas Pharma, Pfizer, and Merck Sharp and Dohme for the generous gifts of micafungin, caspofungin, and anidulafungin, respectively.

Conflicts of Interest: The authors declare no conflict of interest.

References

1. Bongomin, F.; Gago, S.; Oladele, R.O.; Denning, D.W. Global and multi-national prevalence of fungal diseases-estimate precision. *J. Fungi* **2017**, *3*, 57. [CrossRef]
2. Rudramurthy, S.M.; Hoenigl, M.; Meis, J.F.; Cornely, O.A.; Muthu, V.; Gangneux, J.P.; Perfect, J.; Chakrabarti, A.; ECMM; ISHAM. ECMM/ISHAM recommendations for clinical management of COVID-19 associated mucormycosis in low- and middle-income countries. *Mycoses* **2021**, *64*, 1028–1037. [CrossRef]
3. Limper, A.H.; Adenis, A.; Le, T.; Harrison, T.S. Fungal infections in HIV/AIDS. *Lancet Infect. Dis.* **2017**, *17*, e334–e343. [CrossRef]
4. LIFE. Leading International Fungal Education: The Burden of Fungal Disease 2017. Available online: http://www.life-worldwide.org/media-centre/article/the-burden-of-fungal-disease-new-evidence-to-show-the-scale-of-the-problem (accessed on 3 November 2021).
5. Perfect, J.R. The antifungal pipeline: A reality check. *Nat. Rev. Drug Discov.* **2017**, *16*, 603–616. [CrossRef]
6. Ostrosky-Zeichner, L.; Casadevall, A.; Galgiani, J.N.; Odds, F.C.; Rex, J.H. An insight into the antifungal pipeline: Selected new molecules and beyond. *Nat. Rev. Drug Discov.* **2010**, *9*, 719–727. [CrossRef]
7. Denning, D.W.; Bromley, M.J. Infectious Disease. How to bolster the antifungal pipeline. *Science* **2015**, *347*, 1414–1416. [CrossRef]
8. Arastehfar, A.; Gabaldón, T.; Garcia-Rubio, R.; Jenks, J.D.; Hoenigl, M.; Salzer, H.J.F.; Ilkit, M.; Lass-Florl, C.; Perlin, D.S. Drug-resistant fungi: An emerging challenge threatening our limited antifungal armamentarium. *Antibiotics* **2020**, *9*, 877. [CrossRef]
9. Ksiezopolska, E.; Schikora-Tamarit, M.A.; Beyer, R.; Nunez-Rodriguez, J.C.; Schuller, C.; Gabaldón, T. Narrow mutational signatures drive acquisition of multidrug resistance in the fungal pathogen Candida glabrata. *Curr. Biol.* **2021**. [CrossRef]
10. Cortés, J.C.G.; Curto, M.A.; Carvalho, V.S.D.; Pérez, P.; Ribas, J.C. The fungal cell wall as a target for the development of new antifungal therapies. *Biotechnol. Adv.* **2019**, *37*, 107352. [CrossRef]
11. Hopke, A.; Brown, A.J.P.; Hall, R.A.; Wheeler, R.T. Dynamic Fungal Cell Wall Architecture in Stress Adaptation and Immune Evasion. *Trends Microbiol.* **2018**, *26*, 284–295. [CrossRef]
12. Georgopapadakou, N.H.; Tkacz, J.S. The fungal cell wall as a drug target. *Trends Microbiol.* **1995**, *3*, 98–104. [CrossRef]
13. Curto, M.A.; Butassi, E.; Ribas, J.C.; Svetaz, L.A.; Cortés, J.C.G. Natural products targeting the synthesis of β(1,3)-D-glucan and chitin of the fungal cell wall. Existing drugs and recent findings. *Phytomedicine* **2021**, *88*, 153556. [CrossRef]
14. Cabib, E.; Arroyo, J. How carbohydrates sculpt cells: Chemical control of morphogenesis in the yeast cell wall. *Nat. Rev. Microbiol.* **2013**, *11*, 648–655. [CrossRef]
15. Lesage, G.; Bussey, H. Cell wall assembly in *Saccharomyces cerevisiae*. *Microbiol. Mol. Biol. Rev.* **2006**, *70*, 317–343. [CrossRef]
16. Carvalho, V.S.D.; Gómez-Delgado, L.; Curto, M.A.; Moreno, M.B.; Pérez, P.; Ribas, J.C.; Cortés, J.C.G. Analysis and application of a suite of recombinant endo-β(1,3)-D-glucanases for studying fungal cell walls. *Microb. Cell Fact.* **2021**, *20*, 126. [CrossRef]
17. Horiseberger, M.; Rosset, J. Localization of α-galactomannan on the surface of *Schizosaccharomyces pombe* cells by scanning electron microscopy. *Arch. Microbiol.* **1977**, *112*, 123–126. [CrossRef]
18. Kreger, D.R. Observations on cell walls of yeasts and some other fungi by x-ray diffraction and solubility tests. *Biochim. Biophys. Acta* **1954**, *13*, 1–9. [CrossRef]
19. Cortés, J.C.G.; Ramos, M.; Osumi, M.; Pérez, P.; Ribas, J.C. The Cell Biology of Fission Yeast Septation. *Microbiol. Mol. Biol. Rev.* **2016**, *80*, 779–791. [CrossRef]
20. Humbel, B.M.; Konomi, M.; Takagi, T.; Kamasawa, N.; Ishijima, S.A.; Osumi, M. In situ localization of β-glucans in the cell wall of *Schizosaccharomyces pombe*. *Yeast* **2001**, *18*, 433–444. [CrossRef]
21. Liu, J.; Balasubramanian, M.K. 1,3-β-Glucan synthase: A useful target for antifungal drugs. *Curr. Drug Targets-Infect. Disord.* **2001**, *1*, 159–169. [CrossRef]
22. Latge, J.P. The cell wall: A carbohydrate armour for the fungal cell. *Mol. Microbiol.* **2007**, *66*, 279–290. [CrossRef]
23. Le Goff, X.; Woollard, A.; Simanis, V. Analysis of the *cps1* gene provides evidence for a septation checkpoint in *Schizosaccharomyces pombe*. *Mol. Gen. Genet.* **1999**, *262*, 163–172. [CrossRef]
24. Liu, J.; Wang, H.; Balasubramanian, M.K. A checkpoint that monitors cytokinesis in *Schizosaccharomyces pombe*. *J. Cell Sci.* **2000**, *113*, 1223–1230. [CrossRef]

25. Ramos, M.; Cortés, J.C.G.; Sato, M.; Rincón, S.A.; Moreno, M.B.; Clemente-Ramos, J.A.; Osumi, M.; Pérez, P.; Ribas, J.C. Two *S. pombe* septation phases differ in ingression rate, septum structure, and response to F-actin loss. *J. Cell Biol.* **2019**, *218*, 4171–4194. [CrossRef]
26. Liu, J.; Tang, X.; Wang, H.; Balasubramanian, M. Bgs2p, a 1,3-β-glucan synthase subunit, is essential for maturation of ascospore wall in *Schizosaccharomyces pombe*. *FEBS Lett.* **2000**, *478*, 105–108. [CrossRef]
27. Roncero, C.; Sánchez, Y. Cell separation and the maintenance of cell integrity during cytokinesis in yeast: The assembly of a septum. *Yeast* **2010**, *27*, 521–530. [CrossRef]
28. Martín, V.; García, B.; Carnero, E.; Durán, A.; Sánchez, Y. Bgs3p, a putative 1,3-β-glucan synthase subunit, is required for cell wall assembly in *Schizosaccharomyces pombe*. *Eukaryot. Cell* **2003**, *2*, 159–169. [CrossRef]
29. Martins, I.M.; Cortés, J.C.G.; Muñoz, J.; Moreno, M.B.; Ramos, M.; Clemente-Ramos, J.A.; Durán, A.; Ribas, J.C. Differential activities of three families of specific β(1,3)glucan synthase inhibitors in wild-type and resistant strains of fission yeast. *J. Biol. Chem.* **2011**, *286*, 3484–3496. [CrossRef]
30. Muñoz, J.; Cortés, J.C.G.; Sipiczki, M.; Ramos, M.; Clemente-Ramos, J.A.; Moreno, M.B.; Martins, I.M.; Pérez, P.; Ribas, J.C. Extracellular cell wall β(1,3)glucan is required to couple septation to actomyosin ring contraction. *J. Cell Biol.* **2013**, *203*, 265–282. [CrossRef]
31. Johnson, B.F. Lysis of yeast cell walls induced by 2-deoxyglucose at their sites of glucan synthesis. *J. Bacteriol.* **1968**, *95*, 1169–1172. [CrossRef]
32. Miyata, M.; Kitamura, J.; Miyata, H. Lysis of growing fissin-yeast cells induced by aculeacin A, a new antifungal antibiotic. *Arch. Microbiol.* **1980**, *127*, 11–16. [CrossRef]
33. Cassone, A.; Mason, R.E.; Kerridge, D. Lysis of growing yeast-form cells of *Candida albicans* by echinocandin: A cytological study. *Sabouraudia* **1981**, *19*, 97–110. [CrossRef]
34. Yamaguchi, H.; Hiratani, T.; Baba, M.; Osumi, M. Effect of aculeacin A, a wall-active antibiotic, on synthesis of the yeast cell wall. *Microbiol. Immunol.* **1985**, *29*, 609–623. [CrossRef]
35. Vicente, M.F.; Basilio, A.; Cabello, A.; Peláez, F. Microbial natural products as a source of antifungals. *Clin. Microbiol. Infect.* **2003**, *9*, 15–32. [CrossRef]
36. Suwunnakorn, S.; Wakabayashi, H.; Kordalewska, M.; Perlin, D.S.; Rustchenko, E. FKS2 and FKS3 Genes of Opportunistic Human Pathogen *Candida albicans* Influence Echinocandin Susceptibility. *Antimicrob. Agents Chemother.* **2018**, *62*, e02299-17. [CrossRef]
37. Pappas, P.G.; Kauffman, C.A.; Andes, D.R.; Clancy, C.J.; Marr, K.A.; Ostrosky-Zeichner, L.; Reboli, A.C.; Schuster, M.G.; Vázquez, J.A.; Walsh, T.J.; et al. Clinical Practice Guideline for the Management of Candidiasis: 2016 Update by the Infectious Diseases Society of America. *Clin. Infect. Dis.* **2016**, *62*, e1–e50. [CrossRef]
38. Perlin, D.S. Echinocandin Resistance in *Candida*. *Clin. Infect. Dis.* **2015**, *61* (Suppl. 6), S612–S617. [CrossRef]
39. Pfaller, M.A.; Messer, S.A.; Jones, R.N.; Castanheira, M. Antifungal susceptibilities of *Candida*, *Cryptococcus neoformans* and *Aspergillus fumigatus* from the Asia and Western Pacific region: Data from the SENTRY antifungal surveillance program (2010–2012). *J. Antibiot.* **2015**, *68*, 556–561. [CrossRef]
40. Johnson, M.E.; Katiyar, S.K.; Edlind, T.D. New Fks hot spot for acquired echinocandin resistance in Saccharomyces cerevisiae and its contribution to intrinsic resistance of Scedosporium species. *Antimicrob. Agents Chemother.* **2011**, *55*, 3774–3781. [CrossRef]
41. Sawistowska-Schroder, E.T.; Kerridge, D.; Perry, H. Echinocandin inhibition of 1,3-β-D-glucan synthase from *Candida albicans*. *FEBS Lett.* **1984**, *173*, 134–138. [CrossRef]
42. Douglas, C.M.; Marrinan, J.A.; Li, W.; Kurtz, M.B. A *Saccharomyces cerevisiae* mutant with echinocandin-resistant 1,3-β–D-glucan synthase. *J. Bacteriol.* **1994**, *176*, 5686–5696. [CrossRef]
43. Taft, C.S.; Stark, T.; Selitrennikoff, C.P. Cilofungin (LY121019) inhibits *Candida albicans* (1-3)-β-D-glucan synthase activity. *Antimicrob. Agents Chemother.* **1988**, *32*, 1901–1903. [CrossRef]
44. Perlin, D.S. Resistance to echinocandin-class antifungal drugs. *Drug Resist. Updates* **2007**, *10*, 121–130. [CrossRef]
45. Johnson, M.E.; Edlind, T.D. Topological and mutational analysis of *Saccharomyces cerevisiae* Fks1. *Eukaryot. Cell* **2012**, *11*, 952–960. [CrossRef]
46. Jiménez-Ortigosa, C.; Jiang, J.; Chen, M.; Kuang, X.; Healey, K.R.; Castellano, P.; Boparai, N.; Ludtke, S.J.; Perlin, D.S.; Dai, W. Preliminary structural elucidation of β-(1,3)-glucan synthase from *Candida glabrata* using cryo-electron tomography. *J. Fungi* **2021**, *7*, 120. [CrossRef]
47. García, R.; Itto-Nakama, K.; Rodríguez-Pena, J.M.; Chen, X.; Sanz, A.B.; de Lorenzo, A.; Pavon-Verges, M.; Kubo, K.; Ohnuki, S.; Nombela, C.; et al. Poacic acid, a β-1,3-glucan-binding antifungal agent, inhibits cell-wall remodeling and activates transcriptional responses regulated by the cell-wall integrity and high-osmolarity glycerol pathways in yeast. *FASEB J.* **2021**, *35*, e21778. [CrossRef]
48. Roncero, C.; Celador, R.; Sánchez, N.; García, P.; Sánchez, Y. The role of the cell integrity pathway in septum assembly in yeast. *J. Fungi* **2021**, *7*, 729. [CrossRef]
49. Cortés, J.C.G.; Ramos, M.; Konomi, M.; Barragán, I.; Moreno, M.B.; Alcaide-Gavilán, M.; Moreno, S.; Osumi, M.; Pérez, P.; Ribas, J.C. Specific detection of fission yeast primary septum reveals septum and cleavage furrow ingression during early anaphase independent of mitosis completion. *PLoS Genet.* **2018**, *14*, e1007388. [CrossRef]
50. Sipiczki, M. Splitting of the fission yeast septum. *FEMS Yeast Res.* **2007**, *7*, 761–770. [CrossRef]

51. Makarova, M.; Peter, M.; Balogh, G.; Glatz, A.; MacRae, J.I.; Lopez Mora, N.; Booth, P.; Makeyev, E.; Vigh, L.; Oliferenko, S. Delineating the rules for structural adaptation of membrane-associated proteins to evolutionary changes in membrane lipidome. *Curr. Biol.* **2020**, *30*, 367–380.e368. [CrossRef]

52. Boeck, L.D.; Fukuda, D.S.; Abbott, B.J.; Debono, M. Deacylation of echinocandin B by *Actinoplanes utahensis*. *J. Antibiot.* **1989**, *42*, 382–388. [CrossRef]

53. Healey, K.R.; Katiyar, S.K.; Raj, S.; Edlind, T.D. CRS-MIS in *Candida glabrata*: Sphingolipids modulate echinocandin-Fks interaction. *Mol. Microbiol.* **2012**, *86*, 303–313. [CrossRef]

54. Satish, S.; Jiménez-Ortigosa, C.; Zhao, Y.; Lee, M.H.; Dolgov, E.; Kruger, T.; Park, S.; Denning, D.W.; Kniemeyer, O.; Brakhage, A.A.; et al. Stress-induced changes in the lipid microenvironment of β-(1,3)-D-glucan synthase cause clinically important echinocandin resistance in *Aspergillus fumigatus*. *mBio* **2019**, *10*, e00779-19. [CrossRef]

55. Okada, H.; Ohnuki, S.; Roncero, C.; Konopka, J.B.; Ohya, Y. Distinct roles of cell wall biogenesis in yeast morphogenesis as revealed by multivariate analysis of high-dimensional morphometric data. *Mol. Biol. Cell* **2014**, *25*, 222–233. [CrossRef]

56. Foltman, M.; Filali-Mouncef, Y.; Crespo, D.; Sánchez-Díaz, A. Cell polarity protein Spa2 coordinates Chs2 incorporation at the division site in budding yeast. *PLoS Genet.* **2018**, *14*, e1007299. [CrossRef]

57. Viana, R.A.; Pinar, M.; Soto, T.; Coll, P.M.; Cansado, J.; Pérez, P. Negative functional interaction between cell integrity MAPK pathway and Rho1 GTPase in fission yeast. *Genetics* **2013**, *195*, 421–432. [CrossRef]

58. Kim, H.; Kim, A.; Cunningham, K.W. Vacuolar H^+-ATPase (V-ATPase) promotes vacuolar membrane permeabilization and nonapoptotic death in stressed yeast. *J. Biol. Chem.* **2012**, *287*, 19029–19039. [CrossRef]

59. Ma, Y.; Jiang, W.; Liu, Q.; Ryuko, S.; Kuno, T. Genome-wide screening for genes associated with FK506 sensitivity in fission yeast. *PLoS ONE* **2011**, *6*, e23422. [CrossRef]

60. Alfa, C.; Fantes, P.; Hyams, J.; McLeod, M.; Warbrick, E. (Eds.) *Experiments with Fission Yeast: A Laboratory Course Manual*; Cold Spring Harbor Laboratory Press: Cold Spring Harbor, NY, USA, 1993; p. 186.

 *pharmaceuticals*

Review

Clotrimazole for Vulvovaginal Candidosis: More Than 45 Years of Clinical Experience

Werner Mendling [1], Maged Atef El Shazly [2] and Lei Zhang [2,*]

[1] German Center for Infections in Obstetrics and Gynaecology, Heusnerstrasse 40, D-42283 Wuppertal, Germany; w.mendling@t-online.de
[2] Bayer Consumer Care AG, Peter Merian-Strasse 84, CH-4002 Basel, Switzerland; maged.el-shazly@bayer.com
* Correspondence: lei.zhang2@bayer.com; Tel.: +41-58-272-7497; Fax: +41-58-272-7902

Received: 1 September 2020; Accepted: 23 September 2020; Published: 25 September 2020

Abstract: Vulvovaginal candidosis is a common disease, and various treatment strategies have emerged over the last few decades. Clotrimazole belongs to the drugs of choice for the treatment of vulvovaginal candidosis. Although available for almost 50 years, systematic reviews on the usefulness of topical clotrimazole across disease severity and populations affected are scarce. Thus, we conducted a systematic literature search in the PubMed and Embase databases to summarize the effectiveness and safety of topical clotrimazole in the treatment of uncomplicated (acute) and complicated vulvovaginal candidosis. In total, 37 randomized controlled studies in women suffering from vaginal yeast infections qualified for inclusion in our review. In women with uncomplicated vulvovaginal candidosis, single intravaginal doses of clotrimazole 500 mg vaginal tablets provided high cure rates and were as effective as oral azoles. A single dose of clotrimazole 500 mg was equipotent to multiple doses of lower dose strengths. Prolonged treatment regimens proved to be effective in severe and recurrent cases as well as in symptomatic pregnant women. It is therefore expected that in the general population, clotrimazole will continue to be widely used in the field of vaginal health in the upcoming years; more so as clotrimazole resistance in vaginal candidosis is rare.

Keywords: canesten; clotrimazole; vulvovaginal; vaginitis; mycosis; candidosis; yeast infection; candida; candida albicans; vaginal health

---

1. Introduction

Clotrimazole is an imidazole antimycotic agent that was discovered in the 1960s. It has a specific chemical structure consisting of four aromatic rings, out of which one represents an imidazole ring [1]. Clotrimazole has a broad antimicrobial activity against *Candida albicans* and other fungal species. Like other azole-type antifungal drugs, the antimycotic properties are mediated by an interaction with ergosterol synthesis (via inhibition of the fungal cytochrome 14α-demethylase enzyme) eventually resulting in increased fungal cell wall leakiness with disruption of the structure and function of the cell wall [2]. Topical clotrimazole is widely used for the treatment of tinea pedis (athlete's foot), cutaneous mycoses, and oropharyngeal candidosis [1,3]. It also belongs to the drugs of choice for the topical treatment of vulvovaginal candidosis and *Candida* balanitis [4,5].

Clotrimazole was first registered as Canesten® in Germany more than 45 years ago (in 1973) [6]. The initial formulation for local treatment of vulvovaginal candidosis was the vaginal tablet [7] followed by internal vaginal cream, external cream and soft ovule (soft capsule). Additional formulations are marketed under other trade names. Drug combinations (e.g., clotrimazole plus fluconazole) are also available nowadays. Clotrimazole monopreparations for the management of vulvovaginal candidosis are available over the counter in most countries and cover a dose range from 100 to 500 mg (solid systems). Comparable local clotrimazole exposure can be achieved by administration of semi-solid systems (e.g., creams containing clotrimazole 1%, 2% or 10%) to the vagina and vulva [1]. While many preparations are available as generics, Canesten® is still the market leader in this field [8].

Clotrimazole has a poor oral bioavailability. When administered intravaginally, approximately 3% of the dose is systemically available [9]. The latter explains the favorable systemic tolerability of clotrimazole following vaginal application.

Approximately 70–75% of childbearing aged women experience symptomatic vulvovaginal candidosis at least once during their life and 40–50% will suffer from repeated episodes during their lifetime. About 5–8% of adult women may experience recurrent vulvovaginal candidosis (i.e., ≥4 episodes per year) [10,11]. *Candida albicans* is the most common pathogen, but other non-albicans *Candida* species may also be causative microorganisms, particularly in association with disease recurrence [8,11]. The treatment of vulvovaginal candidosis depends on whether the infection is uncomplicated or complicated. Complicated cases comprise recurrent, severe and non-albicans vulvovaginal candidosis as well as vulvovaginal yeast infections during pregnancy and in subjects with immunocompromised conditions [4]. According to current treatment guidelines, topical clotrimazole may play a role in the treatment of both uncomplicated and complicated cases [4,12–14]. Despite the fact that in certain patient subpopulations resistance to clotrimazole has been reported [1,15], clotrimazole resistance in vaginal candidosis is rare and susceptibility testing is usually not recommended [12,16]. It is therefore expected that in the general population, clotrimazole will continue to be widely used in the field of female intimate health in the upcoming years.

This article presents the results of a systematic literature search on the effectiveness and safety of topical clotrimazole when used for the treatment of uncomplicated (acute) and complicated vulvovaginal candidosis. In addition, the scientific evidence for its topical use in men with *Candida* balanitis will also be explored.

2. Methods

A systematic literature search was performed in the PubMed and Embase databases. The following various combinations of terms were queried: clotrimazole, Canesten, vulvovaginal, vaginal, vaginitis, mycosis, candidiasis, candidosis, yeast infection, balanitis, study. Moreover, references quoted in identified publications were tracked. Articles for this review were restricted to prospective randomized controlled clinical trials in which mycological cure was part of the efficacy endpoints. Active-controlled trials in uncomplicated vulvovaginal candidosis must have included an arm studying an oral imidazole/triazole antifungal agent to judge the relative effectiveness of intra-vaginal clotrimazole in comparison with oral treatment. This approach was in accordance with a previous systematic review on anti-fungal agents for uncomplicated vulvovaginal candidosis [17]. There were no date or size restrictions. In principle, only English written articles were considered. However, since the early studies were frequently conducted in Germany, German written papers were also included in this review.

3. Results

In total, the search identified 273 publications. Of these, 37 articles were included to assess the effectiveness and safety of topical clotrimazole in the treatment of uncomplicated (acute) and complicated vulvovaginal candidosis. Two additional studies conducted in men with *Candida* balanitis also qualified for inclusion. Most common reasons for article rejection were missing randomization or control group, comparator other than oral imidazole/triazole antifungal agents, use of clotrimazole combination products, non-reporting of mycological cure rate, and inclusion of largely mixed populations (e.g., subjects with uncomplicated and complicated disease or pregnant and non-pregnant women); or the publication was out of scope for this review (e.g., experimental or mechanistic studies).

3.1. Clotrimazole in Uncomplicated (Acute) Vulvovaginal Candidosis

Approximately 80–90% of symptomatic vulvovaginal candidosis cases are uncomplicated in nature [4]. Among them, about 75–90% are caused by overgrowth of *Candida albicans* [14,18]. The symptoms are of mild to moderate severity and occur in otherwise healthy, non-pregnant women with a frequency of less than four episodes per year [10]. The condition is uncommon in postmenopausal women without hormonal replacement therapy and in premenarchal girls because in these populations, the vagina is infrequently colonized by *Candida* species [12,14,19,20]. In many women of childbearing age, vaginal *Candida* colonization is asymptomatic and needs no treatment [14]. Therefore, the diagnosis of vulvovaginal candidosis necessitates both the presence of symptoms (e.g., itching, burning, inflammation, or dysuria) and proof of *Candida* in the vagina/vulvar region [11].

In our systematic review, 27 articles were included that studied intravaginal clotrimazole in adolescents (usually ≥16 years) and adults with uncomplicated (acute) symptomatic vulvovaginal candidosis. Clotrimazole was compared with placebo in three trials and versus an oral imidazole/triazole antifungal agent in 15 studies (Table 1). Ten trials compared different formulations or treatment regimens of clotrimazole (Table 2).

Generally, single doses of clotrimazole 500 mg vaginal tablet provided mycological cure rates of 70–95% at 1 or 2 weeks following treatment. In studies reporting longer-term cure rates (i.e., after approximately 4 weeks), mycological cure rates ranged roughly between 60% and 90%. Once-daily doses of clotrimazole 200 and 100 mg vaginal tablets for 3 and 6–7 days, respectively, provided similar results. The single use of clotrimazole 10% internal cream resulted in a 1-week mycological cure rate of 85–91%. Combined treatment of clotrimazole 500 mg vaginal tablet and clotrimazole 1% external cream led to mycological cure in 80–95% of subjects within 2 weeks after treatment, whereas the combined application of clotrimazole 10% internal cream and 2% external cream yielded a 2-week cure rate of 74% in one study [21].

Topical imidazole was more effective than placebo in the treatment of uncomplicated (acute) symptomatic vulvovaginal candidosis. In most head-to-head comparisons, intravaginal clotrimazole showed a comparable effectiveness to oral imidazole/triazole antifungal agents. Specifically, in five trials, single-dose therapies with topical clotrimazole and oral fluconazole were compared. There, clotrimazole 500 mg vaginal tablet provided mycological cure rates of 75–95% at 1 or 2 weeks after therapy compared to 76–87% with oral fluconazole 150 mg. When intravaginal clotrimazole formulations or treatments were compared versus each other, it turned out that a single dose of clotrimazole 500 mg was as effective as multiple doses of lower dose strengths (e.g., 200 mg for three days). Similarly, the single use of clotrimazole 10% internal cream or clotrimazole 500 mg ovule (soft capsule) yielded mycological cure rates that were comparable to those following a single dose of clotrimazole 500 mg vaginal tablet. In cases of vulvar involvement, the combined use of intravaginal clotrimazole and external clotrimazole cream provided more favorable treatment results than intravaginal clotrimazole alone [22]. In all studies, topical clotrimazole was well tolerated.

Previous systematic reviews or meta-analyses reported similar findings. Watson et al. [17] included 17 studies in a meta-analysis to compare the effectiveness of oral and intravaginal therapies for uncomplicated vulvovaginal candidosis. There were no statistically significant differences between oral and intravaginal treatments in terms of mycological and clinical cure rates. A subsequent Cochrane review on the same subject came to identical conclusions [23]. Two systematic reviews, including more than 20 studies, inferred that intravaginal and oral imidazoles appear to be equally effective in treating uncomplicated vulvovaginal candidosis, and that single-dose regimens may provide similar results as multiple-dose regimens [24,25].

Table 1. Placebo- and active controlled studies investigating intravaginal clotrimazole in women with uncomplicated (acute) vulvovaginal candidosis.

Reference and Population	Design	Fungal Verification Method	*Candida* Species	Drug and Formulation	Regimen	Outcomes	Adverse Events
[26] Adolescents ≥16 y and adults	r, db, pc	Culture and microscopy	*C. albicans*	CLO vaginal tablet or placebo	CLO: 500 mg, single dose, $n = 10$ Placebo: $n = 13$	Mycological cure rate at D 5–31 after Rx (CLO vs. placebo): 90% vs. 0%, $p = 0.0001$ Clinical cure rate at D 5–31 after Rx (CLO vs. placebo): 90% vs. 0%, $p = 0.0001$	None occurred
[27] Adolescents ≥16 y and adults	r, db, pc, mc	Culture	*C. albicans* (65% of subjects)	CLO vaginal tablet or placebo	CLO: 500 mg, single dose, $n = 55$ Placebo: $n = 40$	Mycological cure rate at W 1 after Rx (CLO vs. placebo): 62% vs. 25%, $p < 0.001$	nr
[28] Adults	r, db *, pc, ac	Culture and microscopy	*C. albicans* (98% of subjects)	CLO vaginal tablet or oral itraconazole capsule or oral placebo capsule	CLO: 200 mg/d for 3 d, $n = 20$ Itraconazole: 200 mg/d for 3 d, $n = 48$ Placebo for 3 d, $n = 22$	Mycological cure rate at W 1 after Rx (CLO vs. itraconazole vs. placebo): 95% vs. 73% vs. 32%, $p < 0.005$ for CLO vs. placebo Mycological cure rate at W 4 (CLO vs. itraconazole vs. placebo): 83% vs. 89% vs. 57%, $p < 0.05$ for CLO vs. placebo Clinical cure rate at W 1 after Rx (CLO vs. itraconazole vs. placebo): 65% vs. 73% vs. 45%, $p < 0.05$ for CLO vs. placebo Clinical cure rate at W 4 (CLO vs. itraconazole vs. placebo): 61% vs. 77% vs. 57%, ns for CLO vs. placebo, ns for CLO vs. itraconazole	CLO: 1 subject with AEs Itraconazole: 17 subjects with AEs Placebo: 9 subjects with AEs
[29] Adults	r, o, ac, mc	Culture and microscopy	*C. albicans* (93% of subjects)	CLO vaginal tablet or oral fluconazole capsule	CLO: 200 mg qd for 3 d, $n = 181$ Fluconazole: 150 mg, single dose, $n = 188$	Mycological cure rate at D 5–16 after Rx (CLO vs. fluconazole): 81% vs. 85%, ns Mycological cure rate at D 27–62 (CLO vs. fluconazole): 62% vs. 72%, ns	CLO: 9 subjects with mild AEs Fluconazole: 8 subjects with mild AEs

Table 1. *Cont.*

Reference and Population	Design	Fungal Verification Method	*Candida* Species	Drug and Formulation	Regimen	Outcomes	Adverse Events
[30] Adolescents ≥15 y and adults	r, o, ac	Culture and microscopy	*C. albicans*	CLO vaginal tablet or oral fluconazole capsule	CLO: 500 mg, single dose, $n = 20$ Fluconazole: 150 mg, single dose, $n = 23$	Mycological cure rate at D 8 after Rx (CLO vs. fluconazole): 75% vs. 87%, $p = 0.05$ Mycological cure rate at D 32 (CLO vs. fluconazole): 60% vs. 87%, $p = 0.05$	CLO: 2 subjects with mild AEs Fluconazole: 1 subject with mild AEs
[31] Adults	r, o, ac, mc	Culture	nr	CLO vaginal tablet or oral fluconazole capsule	CLO: 200 mg qd for 3 d, $n = 95$ Fluconazole: 50 mg qd for 3 d, $n = 90$	Mycological cure rate at D 7–10 after Rx (CLO vs. fluconazole): 93% vs. 89%, nc Mycological cure rate at D 30–35 (CLO vs. fluconazole): 86% vs. 77%, nc Clinical cure rate at D 7–10 after Rx (CLO vs. fluconazole): 88% vs. 84%, nc Clinical cure rate at D 30–35 (CLO vs. fluconazole): 83% vs. 79%, nc Mycological cure rate at D 6–8 after Rx (CLO vs. fluconazole): 84% vs. 86%, ns	CLO: 5 subjects with mild/moderate AEs Fluconazole: 12 subjects with mild/moderate AEs One subject on fluconazole discontinued due to diarrhea
[32] Adolescents ≥16 y and adults	r, o, ac, mc	Culture and microscopy	*C. albicans* (92% of subjects)	CLO vaginal tablet + CLO external 1% cream or oral fluconazole capsule	CLO: 500 mg, single dose + cream on vulva bid, $n = 92$ Fluconazole: 150 mg, single dose, $n = 93$	Mycological cure rate at D 28–32 (CLO vs. fluconazole): 85% vs. 89%, ns Clinical cure rate at D 6–8 after Rx (CLO vs. fluconazole): 80% vs. 87%, ns Clinical cure rate at D 28–32 (CLO vs. fluconazole): 84% vs. 92%, ns	CLO: 3 subjects with AEs Fluconazole: 2 subjects with AEs
[33] Adults	r, o, ac, mc	Culture	nr	CLO vaginal tablet or oral fluconazole capsule	CLO: 500 mg, single dose Fluconazole: 150 mg, single dose Total $n = 537$	Mycological cure rate at D 7 after Rx (CLO vs. fluconazole): 76% vs. 82%, ns Mycological cure rate at D 28 (CLO vs. fluconazole): 72% vs. 75%, ns	CLO: 110 AEs Fluconazole: 112 AEs Both drugs well accepted

Table 1. *Cont.*

Reference and Population	Design	Fungal Verification Method	*Candida* Species	Drug and Formulation	Regimen	Outcomes	Adverse Events
[34]	r, o, ac	Culture and microscopy	nr	CLO vaginal tablet or oral fluconazole	CLO: 100 mg qd for 6 d, $n = 50$ Fluconazole 1: 50 mg qd for 6 d, $n = 90$ Fluconazole 2: 150 mg, single dose, $n = 50$	Mycological cure rate at D 5–15 after Rx (CLO vs. fluconazole 1 vs. fluconazole 2): 72% vs. 88% vs. 76%, nc Mycological cure rate at D 30–60 (CLO vs. fluconazole 1 vs. fluconazole 2): 60% vs. 80% vs. 70%, nc Clinical cure rate at D 5–15 after Rx (CLO vs. fluconazole 1 vs. fluconazole 2): 72% vs. 92% vs. 80%, ns Clinical cure rate at D 30–60 (CLO vs. fluconazole 1 vs. fluconazole 2): 58% vs. 88% vs. 76%, ns	None occurred
[35] Adults	r, sb, ac, mc	Culture	*C. albicans* (85–88% of subjects)	CLO vaginal tablet or oral fluconazole capsule	CLO: 100 mg qd for 7 d, $n = 214$ Fluconazole: 150 mg, single dose, $n = 218$	Mycological cure rate at D 14 after start of Rx (CLO vs. fluconazole): 72% vs. 77%, ns Mycological cure rate at D 35 (CLO vs. fluconazole): 57% vs. 63%, ns Clinical cure rate at D 14 after start of Rx (CLO vs. fluconazole): 67% vs. 73%, ns Clinical cure rate at D 35 (CLO vs. fluconazole): 75% vs. 75%, ns	CLO: 37 subjects with mild/moderate AEs Fluconazole: 59 subjects with mild/moderate AEs
[36] Adults	r, sb, ac	Culture	nr	CLO vaginal tablet or oral fluconazole capsule	CLO: 100 mg bid for 3 d, $n = 50$ Fluconazole: 150 mg, single dose, $n = 53$	Mycological cure rate at W 1 after Rx (CLO vs. fluconazole): 80% vs. 79%, ns Mycological cure rate at W 4 (CLO vs. fluconazole): 66% vs. 60%, ns	CLO: 11 subjects had mild vaginal burning sensations Fluconazole: 4 subjects with mild nausea or dizziness

Table 1. *Cont.*

Reference and Population	Design	Fungal Verification Method	*Candida* Species	Drug and Formulation	Regimen	Outcomes	Adverse Events
[37] Adolescents ≥16 y and adults	r, sb, ac	Culture and microscopy	nr	CLO vaginal tablet + CLO external 1% cream or oral fluconazole or oral itraconazole	CLO: 500 mg, single dose, $n = 82$ Fluconazole: 150 mg, single dose, $n = 72$ Itraconazole: 200 mg bid for 1 d, $n = 75$	Mycological cure rate at D 7–10 (CLO vs. fluconazole vs. itraconazole): 95% vs. 83% vs. 96%, $p = 0.033$ for CLO vs. fluconazole Clinical cure rate at D 7–10 (CLO vs. fluconazole vs. itraconazole): 80% vs. 62% vs. 80%, $p = 0.027$ for CLO vs. fluconazole	nr
[38] Adults	r, o, ac	Culture and microscopy	nr	CLO vaginal tablet or oral fluconazole or oral itraconazole	CLO: 100 mg qd for 6 d, $n = 50$ Fluconazole: 150 mg, single dose, $n = 50$ Itraconazole: 200 mg qd for 3 d, $n = 50$	Mycological cure rate at D 5–15 after Rx (CLO vs. fluconazole vs. itraconazole): 72% vs. 76% vs. 80%, nc Mycological cure rate at D 30–60 (CLO vs. fluconazole vs. itraconazole): 60% vs. 70% vs. 74%, nc Clinical cure rate at D 5–15 after Rx (CLO vs. fluconazole vs. itraconazole): 72% vs. 80% vs. 84%, nc Clinical cure rate at D 30–60 (CLO vs. fluconazole vs. itraconazole): 58% vs. 76% vs. 78%, nc	None occurred
[21] nr	r, sb, ac, mc	Culture and microscopy	*C. albicans* (95% of subjects)	CLO vaginal tablet + CLO external 1% cream (VT) or CLO 10% vaginal cream + CLO external 2% cream (VC) or oral fluconazole capsule	CLO VT: 500 mg, single dose, $n = 226$ CLO VC: 1 tube, single dose, $n = 226$ Fluconazole: 150 mg, single dose, $n = 227$	Mycological cure rate at D 14 (CLO VT vs. CLO VC vs. fluconazole): 80% vs. 74% vs. 76%, CLO VT and CLO VC non-inferior to fluconazole Clinical cure rate at D 14 (CLO VT vs. CLO VC vs. fluconazole): 66% vs. 61% vs. 59%, CLO VT and CLO VC non-inferior to fluconazole	CLO VT: 26 subjects with mild AEs CLO VC: 32 subjects with mild AEs Fluconazole: 29 subjects with mild AEs

Table 1. *Cont.*

Reference and Population	Design	Fungal Verification Method	*Candida* Species	Drug and Formulation	Regimen	Outcomes	Adverse Events
[39] Adolescents >15 y and adults	r, o, ac	Microscopy	nr	CLO vaginal tablet or oral fluconazole	CLO: 200 mg qd for 6 d, n = 70 Fluconazole: 150 mg, single dose, n = 72	Mycological cure rate at D 7 after Rx (CLO vs. fluconazole): 70% vs. 83%, p = 0.01 Clinical cure rate at D 7 after Rx (CLO vs. fluconazole): 59% vs. 74%, p = 0.001	CLO: pelvic pain Fluconazole: headache No difference in AE frequencies
[40] Adolescents ≥16 y and adults	r, sb, ac, mc	Culture	nr	CLO vaginal tablet or oral ketoconazole	CLO: 100 mg qd for 6 d, n = 29 Ketoconazole: 200 mg bid for 5 d, n = 34	Mycological cure rate at W 2 (CLO vs. ketoconazole): 86% vs. 83%, nc	CLO: 1 subject with AEs Ketoconazole: 5 subjects AEs
[41] Adolescents ≥16 y and adults	r, o, ac	Culture and microscopy	*C. albicans* (96% of subjects)	CLO vaginal ovule or oral ketoconazole or CLO + lactic acid in vaginal ovule	CLO: 500 mg, single dose, n = 25 Ketoconazole: 200 mg bid for 5 d, n = 25 CLO + lactic acid: 500 mg, single dose, n = 25	Mycological cure rate at W 3 after Rx (CLO vs. ketoconazole vs. CLO + lactic acid): 84% vs. 76% vs. 92%, nc	nr
[42] Adults	r, sb, ac, mc	Culture	nr	CLO vaginal tablet or oral itraconazole capsule	CLO: 500 mg, single dose, n = 105 Itraconazole: 200 mg bid for 1 d, n = 109	Mycological cure rate at W 1 after Rx (CLO vs. itraconazole): 72% vs. 74%, ns Mycological cure rate at W 6 (CLO vs. itraconazole): 50% vs. 51%, ns	nr

ac: active-controlled; AEs: adverse events; bid: twice-daily; CLO: clotrimazole; d or D: day; db: double-blind; M: month; mc: multicenter; n: number of subjects; nc: not compared; nr: not reported; ns: not statistically significant; o: open; pc: placebo-controlled; qd: once-daily; r: randomized; Rx: treatment; sb: single-blind; W: week; y: year. * Comparison with clotrimazole was not blinded.

Table 2. Randomized controlled studies comparing different treatment regimens or formulations of intravaginal clotrimazole in women with uncomplicated (acute) vulvovaginal candidosis.

Reference and Population	Design	Fungal Verification Method	*Candida* Species	Formulations	Regimen	Outcomes	Adverse Events
[43] Adults	r, db, ac	Culture	*C. albicans*	CLO vaginal tablet 500 mg or CLO vaginal tablet 200 mg	CLO 1: 500 mg, single dose, n = 35 CLO 2: 200 mg qd for 3 d, n = 37	Mycological cure rate at D 7 after Rx (CLO 1 vs. CLO 2): 94% vs. 89%, ns Clinical cure rate at W 4 after Rx (CLO 1 vs. CLO 2): 86% vs. 92%, ns	None occurred

Table 2. *Cont.*

Reference and Population	Design	Fungal Verification Method	*Candida* Species	Formulations	Regimen	Outcomes	Adverse Events
[44] Adults	r, db, ac, mc	Culture and microscopy	nr	CLO vaginal tablet 500 mg or CLO vaginal tablet 100 mg	CLO 1: 500 mg, single dose, $n = 53$ CLO 2: 2 × 100 mg qd for 3 d, $n = 50$	Mycological + clinical cure rate at D 5–10 after Rx (CLO 1 vs. CLO 2): 90% vs. 89%, ns Mycological + clinical cure rate at ≥D 27 (CLO 1 vs. CLO 2): 75% vs. 72%, ns	CLO 2: 1 subject had moderate edema of vulva
[45] Adults	r, db, ac	Culture and microscopy	nr	CLO vaginal tablet 500 mg or CLO vaginal tablet 100 mg	CLO 1: 500 mg, single dose, $n = 18$ CLO 2: 2 × 100 mg qd for 3 d, $n = 18$	Mycological + clinical cure rate at M 1 after Rx (CLO 1 vs. CLO 2): 89% vs. 83%, ns	CLO 2: 1 subject had moderate edema of vulva
[46] Adults	r, db, ac	Culture and microscopy	*C. albicans*	CLO vaginal tablet 500 mg or CLO vaginal tablet 100 mg	CLO 1: 500 mg, single dose, $n = 14$ CLO 2: 2 × 100 mg qd for 3 d, $n = 13$	Mycological + clinical cure rate at M 1 after Rx (CLO 1 vs. CLO 2): 86% vs. 85%, ns	None occurred
[47] Adults	r, db, ac	Culture and microscopy	nr	CLO vaginal tablet 500 mg or CLO vaginal tablet 100 mg	CLO 1: 500 mg, single dose, $n = 20$ CLO 2: 100 mg qd for 6 d, $n = 20$	Mycological cure rate at D 3 after Rx (CLO 1 vs. CLO 2): 85% vs. 95%, nc Mycological cure rate at W 4 after Rx (CLO 1 vs. CLO 2): 95% vs. 80%, nc	1 subject reported vaginal burning
[21] nr	r, sb, ac, mc	Culture and microscopy	*C. albicans* (95% of subjects)	CLO vaginal tablet 500 mg + CLO external 1% cream (VT) or CLO 10% vaginal cream + CLO external 2% cream (VC) or oral fluconazole capsule	CLO VT: 500 mg, single dose, $n = 226$ CLO VC: 1 tube, single dose, $n = 226$ Fluconazole: 150 mg, single dose, $n = 227$	Mycological cure rate at D 14 (CLO VT vs. CLO VC vs. fluconazole): 80% vs. 74% vs. 76%, CLO VT and CLO VC non-inferior to fluconazole Clinical cure rate at D 14 (CLO VT vs. CLO VC vs. fluconazole): 66% vs. 61% vs. 59%, CLO VT and CLO VC non-inferior to fluconazole	CLO VT: 26 subjects with mild AEs CLO VC: 32 subjects with mild AEs Fluconazole: 29 subjects with mild AEs
[48] Adolescents ≥16 y and adults	r, o, ac	Culture	*C. albicans*	CLO 10% cream or CLO 2% cream	CLO 10%: 5 g, single dose, $n = 55$ CLO 2%: 5 g, bid for 3 d, $n = 55$	Mycological cure rate at D 7 after start of Rx (CLO 10% vs. CLO 2%): 91% vs. 96%, nc Mycological cure rate at D 35 (CLO 10% vs. CLO 2%): 84% vs. 81%, nc	None occurred

Table 2. *Cont.*

Reference and Population	Design	Fungal Verification Method	*Candida* Species	Formulations	Regimen	Outcomes	Adverse Events
[49] Adolescents ≥14 y and adults	r, o, ac	Culture and microscopy	*C. albicans* (96% of subjects)	CLO 10% cream or CLO vaginal tablet 500 mg (VT)	CLO 10%: 5 g, single dose, $n = 81$ * CLO VT: 500 mg, single dose, $n = 82$ #	Mycological cure rate at W 1 after Rx (CLO 10% vs. CLO VT): 85% vs. 87%, ns Mycological cure rate at W 4 after Rx (CLO 10% vs. CLO VT): 73% vs. 80%, nc Clinical cure rate at W 1 after Rx (CLO 10% vs. CLO VT): 94% vs. 91%, ns Clinical cure rate at W 4 after Rx (CLO 10% vs. CLO VT): 85% vs. 94%, nc	CLO 10%: 3% of subjects had mild AEs
[22] Adults	r, db, pc, mc	Culture and microscopy	nr	CLO vaginal suppository 200 mg + CLO external 2% cream (T1) or CLO vaginal suppository 200 mg + placebo external cream (T2)	CLO 1 (T1): 200 mg, qd for 3 d, $n = 79$ CLO 2 (T2): 200 mg qd for 3 d, $n = 79$ Cream was applied to the vulva	Mycological cure rate (vagina) at D 6 after Rx (CLO 1 vs. CLO 2): 92% vs. 90%, ns Mycological cure rate (vulva) at D 6 after Rx (CLO 1 vs. CLO 2): 73% vs. 52%, $p = 0.005$ Mycological cure rate (vagina) at W 4 (CLO 1 vs. CLO 2): 23/25 (92%) vs. 21/23 (91%), ns Mycological cure rate (vulva) at W 4 (CLO 1 vs. CLO 2): 16/23 (70%) vs. 13/21 (62%), ns	CLO 1: 1 subject reported vaginal burning CLO 2: 1 subject reported skin peeling With CLO 1, significantly less local itching and extravaginal redness
[50] Adolescents ≥14 y and adults	r, sb, ac, mc	Culture and microscopy	nr	CLO ovule 500 mg (O) or CLO vaginal tablet 500 mg (VT)	CLO 1 (O): 500 mg, single dose, $n = 237$ CLO 2 (VT): 500 mg, single dose, $n = 228$	Mycological cure rate at W 2 after Rx (CLO 1 vs. CLO 2): 81% vs. 78%, ns Mycological cure rate at W 6–8 (CLO 1 vs. CLO 2): 78% vs. 81%, ns Clinical cure rate at W 2 after Rx (CLO 1 vs. CLO 2): 88% vs. 84%, ns Clinical cure rate at W 6–8 (CLO 1 vs. CLO 2): 93% vs. 93%, ns	CLO 1: 1 subject reported a mild drug-related AE (vulvovaginal discomfort)

ac: active-controlled; AEs: adverse events; bid: twice-daily; CLO: clotrimazole; d or D: day; db: double-blind; mc: multicenter; *n*: number of subjects; nc: not compared; nr: not reported; ns: not statistically significant; o: open; pc: placebo-controlled; qd: once-daily; r: randomized; Rx: treatment; sb: single-blind; W: week. * Seven women were pregnant; # One woman was pregnant.

3.2. Clotrimazole in Complicated Vulvovaginal Candidosis

3.2.1. Vaginal Yeast Infection during Pregnancy

During pregnancy, up to 50% of women experience vulvovaginal candidosis [51]. In addition, in pregnant women, recurrent vaginal yeast infections and insufficient therapeutic responses are more frequently recorded than in non-pregnant women [10,52]. An increased content of glycogen and lower pH value in the vagina as well as an intensified binding of yeast cells to the vaginal mucosa have been made accountable for this observation [10,51]. During pregnancy, the therapy of symptomatic vaginal yeast infections should be intense, but restricted to topical preparations; oral antifungals should be avoided [10,14]. Specifically, topical azoles can be used at all stages of pregnancy because there is no or only minimal systemic exposure following intravaginal administration [53]. The FDA assigned topical clotrimazole to pregnancy category B. The other topical imidazoles and triazoles have been assigned to category C. In fact, several clinical trials confirmed the safety of clotrimazole in pregnancy; no association was observed between vaginal application of clotrimazole and congenital abnormalities [54].

Our systematic literature search identified five articles which qualified for inclusion (Table 3). In pregnant women with symptomatic vulvovaginal candidosis, clotrimazole 100 mg (vaginal tablet), administered for approximately 1 week, provided high mycological cure rates (78–88% at 1 or 2–4 weeks after therapy). Similar results were observed following 1-week application of clotrimazole 1% internal cream. Intravaginal clotrimazole was significantly more effective than placebo treatment [55]. In the same and other studies, the prophylactic use of clotrimazole during pregnancy significantly lowered the frequency of *Candida* presence on the neonatal skin [55–59]. Moreover, mycological cure rates were markedly higher with multiple-dose clotrimazole compared with nystatin, while there was no difference to terconazole. There was a trend towards reduced effectiveness of clotrimazole when taken as a single intravaginal 500 mg dose. In all five studies, topical clotrimazole was well tolerated.

Two additional randomized controlled trials studied the potential of clotrimazole in the prevention of preterm birth when administered to pregnant women with asymptomatic vaginal candidosis. Kiss et al. [60] evaluated whether a screening strategy in pregnant women presenting for their prenatal visits early in the second trimester lowers the rate of preterm delivery. Subjects with pathological vaginal flora were randomized to an intervention group or control group (no treatment). Women who received intravaginal clotrimazole 100 mg once-daily for 6 days due to asymptomatic vulvovaginal candidosis showed a significantly lower spontaneous preterm birth rate than untreated women (8/294 (2.7%) vs. 22/292 (7.5%); risk ratio (RR): 0.36; 95% confidence interval (CI): 0.16, 0.80) [61]. In a smaller study conducted by Roberts and colleagues [62], 98 pregnant women at <20 weeks of gestation were randomized to intravaginal clotrimazole 100 mg once-daily for 6 days or no treatment. There was a reduction in preterm birth rate in women treated with clotrimazole (4.0% vs. 6.3%), but the difference did not reach statistical significance. A meta-analysis of both trials concluded that treatment of asymptomatic candidosis may reduce the risk of preterm birth, but further prospective, sufficiently powered studies are required [61].

Our findings are in agreement with a previous systematic review on the topical treatment of symptomatic vulvovaginal candidosis in pregnancy. A Cochrane review from 2001 [63] concluded that topical imidazoles are more effective than nystatin and should be used during pregnancy. In addition, short-term treatments with topical imidazole drugs were found to be less effective than 7-day administrations, and topical treatments beyond 7 days provided no additional benefit.

Table 3. Randomized controlled studies investigating intravaginal clotrimazole in pregnant women with symptomatic vulvovaginal candidosis.

Reference	Design	Fungal Verification Method	*Candida* Species	Drug and Formulation	Regimen	Outcomes	Adverse Events
[55]	r, db, pc	Culture	*C. albicans*	CLO vaginal tablets 100 mg + CLO external cream or placebo	CLO: 100 mg + cream qd for 6 d, $n = 50$ Placebo: $n = 50$	Mycological cure rate at W 2–4 after Rx (CLO vs. placebo): 88% vs. 42%, $p < 0.05$	nr
[64]	r, sb, ac	Culture and microscopy	*C. albicans* (90% of subjects)	CLO vaginal tablet 100 mg or nystatin vaginal tablet 100,000 IU	CLO: 100 mg qd for 6 d, $n = 33$ Nystatin: 200,000 IU qd for 6 d, $n = 29$	Mycological cure rate at W 1 after Rx (CLO vs. nystatin): 78% vs. 41%, $p < 0.01$ Mycological cure rate at W 5 after Rx (CLO vs. nystatin): 21/23 (91%) vs. 7/23 (32%), $p < 0.005$	None occurred
[65]	r, o, ac	Culture and microscopy	nr	CLO vaginal tablet 100 mg or nystatin vaginal tablet 100,000 IU	CLO: 100 mg qd for 11 d, $n = 21$ Nystatin: 200,000 IU qd for 11 d, $n = 19$	Mycological cure rate at D 3 after Rx (CLO vs. nystatin): 76% vs. 63%, ns Mycological cure rate at W 6 after Rx (CLO vs. nystatin): 5/7 (71%) vs. 1/7 (14%), ns	None occurred
[66]	r, db, ac	Culture and microscopy	*C. albicans* (99% of subjects)	CLO 1% cream or terconazole 0.4% cream	CLO: 5 g qd for 7 d, $n = 19$ Terconazole: 5 g qd for 7 d, $n = 19$	Mycological + clinical cure rate at W 1 after Rx (CLO vs. terconazole): 95% vs. 95%, ns Mycological + clinical cure rate at W 4 after Rx (CLO vs. terconazole): 84% vs. 79%, ns	Terconazole: 1 subject reported mild vaginal burning
[67]	r, db, ac	Culture and microscopy	nr	CLO vaginal tablet 500 mg or CLO vaginal tablet 100 mg	CLO 1: 500 mg, single dose, $n = 48$ * CLO 2: 2×100 mg qd for 3 d, $n = 53$ [#]	Mycological cure rate at $\geq$W 4 after Rx (CLO 1 vs. CLO 2): 67% vs. 75%, ns	CLO 1: 1 subject reported vulvar lesion CLO 2: 2 subjects with AEs (rash, vulvar lesion)

ac: active-controlled; AEs: adverse events; CLO: clotrimazole; d or D: day; db: double-blind; IU: international unit; *n*: number of subjects; nr: not reported; ns: not statistically significant; o: open; pc: placebo-controlled; qd: once-daily; r: randomized; Rx: treatment; sb: single-blind; W: week. * Four women were not pregnant; [#] Five women were not pregnant.

3.2.2. Recurrent Vaginal Yeast Infection

Recurrent symptomatic vaginal yeast infections may have a severe adverse impact on the quality of life. Defined as ≥4 culture-proven episodes per year, it is a long-term condition causing significant morbidity in women [68]. The highest prevalence (9%) has been observed in young women aged 25–34 years [8]. The main fungal species causing recurrent vulvovaginal candidosis is still *Candida albicans*. In fact, in about 85–95% of cases, azole-sensitive *Candida albicans* can be identified as the responsible pathogen. The latter implies that peculiarities of the host (e.g., genetic factors that facilitate vaginal colonization and persistence) must contribute to the development of disease recurrence [68]. Optimized treatment regimens are required to combat this debilitating and complex disease. Usually, an induction course followed by maintenance or intermittent treatment for at least 6 months is pursued [14].

We found four randomized controlled studies investigating intravaginal clotrimazole in non-pregnant women with recurrent symptomatic vulvovaginal candidosis (Table 4). Clotrimazole induction treatment for 1–2 weeks (e.g., clotrimazole 100 mg once-daily) resulted in short-term mycological and clinical cure rates of >80%. Intermittent once-monthly prophylactic administration of clotrimazole 500 mg (solid system) for 6 months prevented clinical disease recurrence in up to 70% of women, but did not prevent vaginal recolonization with *Candida albicans* in the majority of cases. More subjects on intermittent intravaginal clotrimazole remained asymptomatic than subjects receiving intermittent oral itraconazole or placebo. In one head-to-head study, clotrimazole and oral ketoconazole induction treatments were equally effective at achieving high short-term cure rates, but without immediate initiation of maintenance/intermittent therapy, longer-term recurrence rates were high in both groups [69]. Except for occasional vulvovaginal burning, topical clotrimazole was well tolerated in all four studies. Adverse events were significantly more common with oral itraconazole and oral ketoconazole than with intravaginal clotrimazole.

Our findings summarized in Table 4 must be put into perspective with the fact that higher cure rates have been reported with oral fluconazole. In a randomized controlled study in 387 women with recurrent symptomatic vulvovaginal candidosis, oral fluconazole 150 mg was administered on days 0, 3, 6, followed by once-weekly intakes for 6 months. The proportion of women who remained clinically cured until 6 months was 91% in the fluconazole group and 36% in the placebo group ($p < 0.001$). During the maintenance phase, 2.9% of patients in the fluconazole group and 1.2% in the placebo group reported at least one adverse event that led to discontinuation of the study medication [70].

Table 4. Randomized controlled studies investigating intravaginal clotrimazole in women with recurrent symptomatic vulvovaginal candidosis.

Reference and Population	Design	Fungal Verification Method	*Candida* Species	Drug and Formulation	Regimen	Outcomes	Adverse Events
[71] Adults	r, db, pc	Culture and microscopy	*C. albicans* (86% of subjects)	CLO vaginal suppository or placebo	Phase 1 (Rx) (all subjects) CLO: 500 mg qw for 2 w, $n = 42$ Phase 2 (Pro) * CLO: 500 mg qm for 6 m, $n = 15$ Placebo: $n = 12$	Phase 1 Mycological cure rate at W 2 after Rx: 83% Clinical cure rate at W 2 after Rx: 90% Phase 2 Mycologically still cured at M 6 (CLO vs. placebo): ~20% # vs. ~5% #, ns Clinically still cured at M 6 (CLO vs. placebo): 47% vs. 33%, ns	None occurred
[72] Adults	r, db, pc	Culture and microscopy	*C. albicans*	CLO vaginal tablet or placebo	Phase 1 (Rx) (all subjects) CLO: 500 mg, single dose Phase 2 (Pro) $ CLO: 500 mg qm for 6 m, $n = 33$ Placebo: $n = 29$	Phase 1 Mycological and clinical cure rates: nr Phase 2 Mycologically still cured at M 6 (CLO vs. placebo): 30% vs. 14%, ns Clinically still cured at M 6 (CLO vs. placebo): 70% vs. 21%, $p < 0.001$	None occurred
[73] Adults	r, o, ac	Culture and microscopy	*C. albicans*	CLO vaginal ovules or oral itraconazole	CLO: 200 mg qd for 5 d, then 200 mg biw for 6 m, $n = 22$ Itraconazole: 100 mg bid for 5 d, then 200 mg biw for 6 m, $n = 22$	Mycological + clinical cure rate at M 6 after Rx (CLO vs. itraconazole): 17/17 (100%) vs. 14/21 (67%), $p = 0.02$	Itraconazole: 32% of subjects had mild AEs
[69] Adults	r, o, mc	Culture and microscopy	nr	CLO vaginal suppository or oral ketoconazole	CLO: 100 mg qd for 7 d, $n = 77$ Ketoconazole: 400 mg qd for 14 d, $n = 74$ No maintenance treatment in both groups	Mycological cure rate at W 1 after Rx (CLO vs. ketoconazole): 82% vs. 80%, ns Clinical cure rate at W 1 after Rx (CLO vs. ketoconazole): 82% vs. 86%, ns Clinical cure rate at M 2 after Rx (CLO vs. ketoconazole): 37% vs. 48%, ns	AEs occurred significantly more frequently with ketoconazole CLO: 3.1% had mild vulvovaginal burning

ac: active-controlled; AEs: adverse events; bid: twice-daily; biw: twice-weekly; CLO: clotrimazole; d or D: day; db: double-blind; mc: multicenter; m or M: month; *n*: number of subjects; nr: not reported; ns: not statistically significant; o: open; pc: placebo-controlled; Pro: prophylaxis; qd: once-daily; qw: once-weekly; qm: once-monthly; r: randomized; Rx: treatment; w or W: week. * Only asymptomatic women, culture status was ignored; # Estimated from graphical display; $ Only asymptomatic women with negative microscopy and culture.

3.2.3. Severe Vaginal Yeast Infection

Severe vulvovaginal candidosis is characterized by extensive vulvar erythema, swelling, excoriation, itching and fissure formation. Standard treatments as used in uncomplicated vulvovaginal candidosis are insufficient in women with severe disease; intensified treatment regimens comprising repeated doses are required [4,10,14].

We identified one article qualifying for inclusion in our review. Zhou et al. [74] conducted a prospective, open, randomized (1:1) study in 240 women with severe vulvovaginal candidosis to assess whether two 500 mg doses of clotrimazole vaginal tablet are as effective as two 150 mg doses of oral fluconazole. In each group, the two doses were administered 3 days apart. Study participants were to be not pregnant and at least 18 years old. In 90% of cases, *Candida albicans* was the causative pathogen. The mycological cure rates at days 7–14 after therapy were 78% and 74% ($p = 0.147$) in the clotrimazole group and the fluconazole group, respectively. The corresponding clinical cure rate was 89% in both groups. At days 30–35, the mycological cure rates amounted to 54% and 56% ($p = 0.813$) in the clotrimazole group and the fluconazole group, respectively; the clinical cure rates at that time point were 72% and 78% ($p = 0.298$), respectively. Systemic adverse events were more common with oral fluconazole (e.g., headache) than with intravaginal clotrimazole, while local adverse events (e.g., mild burning) occurred more frequently in the clotrimazole group. It was concluded that the two tested treatment regimens are equally effective, with a safety benefit for clotrimazole.

3.2.4. Vaginal Yeast Infection in Immunocompromised Host

Vulvovaginal candidosis is of particular concern in immunocompromised subjects such as women with poorly controlled diabetes mellitus, HIV-infection, or on immunosuppressive drugs. Usually, these populations require prolonged conventional antifungal therapy for 7–14 days, including intravaginal azole therapy [4,11,14].

We found no qualifying study that investigated the therapeutic effectiveness of intravaginal clotrimazole in this setting. However, Williams and colleagues performed a randomized, double-blind, placebo-controlled study on the potential of intravaginal clotrimazole to prevent vaginal candidosis in adult women with HIV [75]. In the clotrimazole arm, study participants applied the drug (capsules containing clotrimazole powder 100 mg) once a week. At 6-month intervals, vaginal samples were collected and examined. The risk of experiencing an episode of vulvovaginal candidosis during clotrimazole use was significantly reduced compared to placebo (risk ratio: 0.4; 95% CI: 0.2, 0.9].

3.2.5. Non-Albicans Vaginal Yeast Infection

In about 10% of cases, non-*albicans Candida* species are responsible for acute symptomatic vulvovaginal candidosis [14]. Among them, *Candida glabrata* is the most frequently identified strain [68]. Non-*albicans Candida* species as causative pathogens of vulvovaginal candidosis are an emerging threat. Risk factors include uncontrolled type 2 diabetes and advanced age as well as intake of glycosuria-inducing agents to manage type 2 diabetic patients [68]. The optimal treatment of non-*albicans* vaginal yeast infections has not been established [4]. Non-*albicans Candida* infections are unlikely to respond to standard treatments. In fact, azoles are poorly effective in the treatment of *Candida glabrata*-caused vulvovaginal candidosis [10]. Boag and co-workers [76] investigated the effect of oral fluconazole and intravaginal clotrimazole on the vaginal microbial flora. The vaginal flora was unaltered after both therapies. In women suffering from vaginitis due to *Candida glabrata* or *Candida krusei*, the yeasts persisted longer and revealed a poorer treatment response to either treatment.

Our literature search did not identify qualifying clotrimazole-studies in this setting.

3.3. Clotrimazole in Men with Candida Balanitis

Balanitis affects approximately 3–11% of men at least once during lifetime. Fungal microorganisms—especially *Candida albicans*—are most frequently responsible for this itching inflammation of the glans penis [5]. *Candida* balanitis is not considered a sexually transmitted disease, but in approximately 20% of male partners of women with recurrent vulvovaginal candidosis, *Candida* species can be isolated on their penises [10]. In one study, 43% of men with *Candida* balanitis had partners with *Candida* vaginitis [77]. In another study, 107 male partners of women with acute vaginal candidosis were examined, and 45% exhibited symptoms of balanitis [48]. Those men who develop *Candida* balanitis benefit from topical antifungal treatment [4].

In a randomized (1:1), open-label study, Stary et al. [77] compared the effectiveness and safety of a single oral dose of fluconazole 150 mg with clotrimazole 1% external cream applied topically twice-daily for 7 days in 157 adult men with balanitis. In those study participants with a positive baseline culture for *Candida albicans*, 49/63 (78%) subjects in the fluconazole group and 53/64 (83%) subjects in the clotrimazole group were mycologically cured at days 8–11 after initiation of therapy ($p > 0.05$). At the same time-point, 92% and 91% ($p > 0.05$), respectively, were clinically cured or had improved. At the one-month follow-up visit, 26/36 (72%) subjects in the fluconazole group and 25/33 (76%) subjects in the clotrimazole group were mycologically cured. The median time to relief of symptoms (erythema) was 6 days during fluconazole treatment and 7 days with clotrimazole. Both treatments were well tolerated. It was concluded that both treatments are equipotent in the treatment of *Candida* balanitis.

Maw et al. [78] conducted an open, comparative study of bifonazole and clotrimazole in adults with candidal balanoposthitis (i.e., both the glans and the foreskin were affected). Study participants were randomized to receive either bifonazole 1% cream once-daily for 6 days or clotrimazole 1% cream twice-daily for 6 days. On day 7 after treatment initiation, 21/27 (78%) subjects in the bifonazole group and 20/26 (77%) subjects in the clotrimazole group were mycologically cured. The difference was not statistically significant. No adverse events were reported in the clotrimazole group while one subject in the bifonazole group reported increased erythema following initial application.

In an uncontrolled study in 138 men with *Candida* balanitis, clotrimazole 1% external cream applied topically twice-daily for 7 days yielded mycological cure rates of 90% at day 7 after the start of therapy. Treatment was acceptable even to subjects who continued to engage in sexual activity during treatment [79].

4. Summary

More than 45 years ago, the first topical clotrimazole formulation (Canesten® vaginal tablet) was registered. Today, it is available in different dosages and formats allowing treatments according to individual needs and preferences. Various studies and its therapeutic use over decades confirmed the antimycotic activity of clotrimazole in the treatment of vulvovaginal candidosis. Despite its use over many years, clotrimazole resistance in vaginal candidosis is rare, with the caveat that drug resistance has emerged particularly in immunocompromised patients [1,12]. Resistance has been associated with the overexpression of efflux pump genes. Changes in the clotrimazole target, the fungal cytochrome 14α-demethylase enzyme, may also play a role in some cases [1]. We conducted a systematic literature search on the effectiveness and local tolerability of topical clotrimazole when used for the treatment of uncomplicated and complicated vulvovaginal candidosis, and when used topically by men with *Candida* balanitis. In total, 39 randomized controlled studies qualified for inclusion in our review.

In women with uncomplicated (acute) vulvovaginal candidosis, single topical doses of a clotrimazole 500 mg vaginal tablet provided mycological cure rates of up to 95%. The single use of clotrimazole 10% internal cream or clotrimazole 500 mg ovule resulted in similar results. This is in accordance with the observation that fungicidal concentrations of clotrimazole could be determined in vaginal secretions up to three days after insertion of one vaginal 500 mg tablet [80]. In head-to-head comparative studies, single-dose regimens of intravaginal clotrimazole and oral antifungals were equally effective. In cases when the candidosis extended to the vulvar region, the combined use of intravaginal clotrimazole and external clotrimazole cream provided added benefit (e.g., less local itching). In fact, two observational studies including over 5800 women with vulvovaginal mycosis showed that more than 75% of physicians prefer to treat their patients with a combination of clotrimazole to be applied intravaginally (vaginal tablet or cream) and clotrimazole cream to be applied externally to the vulva and surrounding areas [19,81].

Topical clotrimazole has also proved its effectiveness in the therapy of complicated vulvovaginal candidosis. In pregnant women with symptomatic vaginal yeast infection, one-week treatments with clotrimazole 100 mg or 1% internal cream were associated with high mycological cure rates. Furthermore, the prophylactic use of clotrimazole during pregnancy significantly lowered the risk for *Candida* infections of newborns and preterm birth.

In recurrent symptomatic vulvovaginal candidosis, clotrimazole induction treatment for 1–2 weeks resulted in short-term mycological cure rates of >80%, and topical treatments applied intermittently (monthly) prevented disease recurrence and breakthrough vaginitis in up to 70%. However, higher clinical cure rates were observed with once-weekly oral fluconazole. In one head-to-head comparative study in women with severe vulvovaginal candidosis, two 500 mg doses of clotrimazole vaginal tablet, administered 3 days apart, were as effective as two 150 mg doses of oral fluconazole, whereas intravaginal clotrimazole was better tolerated. No high-quality studies were found on the usefulness of topical clotrimazole in the therapy of non-*albicans* vaginal yeast infections and in the treatment of vulvovaginal candidosis in immunocompromised hosts.

Results from two studies provided convincing scientific evidence that clotrimazole 1% external cream, applied topically twice-daily for one week, is associated with high mycological cure rates in men with *Candida* balanitis. In fact, the latter regimen was as effective as a single oral dose of fluconazole 150 mg.

In all studies, topical clotrimazole was well tolerated; mild local adverse events (e.g., burning) were reported occasionally.

Based on our findings, principal clotrimazole treatment regimens can be inferred. Those are summarized in Table 5.

The results of our literature search are also mirrored in several treatment guidelines. In the guideline issued by the U.S. Centers for Disease Control and Prevention (CDC), it reads: "Short-course topical formulations (i.e., single dose and regimens of 1–3 days) effectively treat uncomplicated vulvovaginal candidosis. The topically applied azole drugs are more effective than nystatin." [4]. In various other treatment guidelines, short course intravaginal clotrimazole (i.e., vaginal tablet 500 mg as a single dose or 200 mg once-daily for three days) belongs to the recommended regimens for treating uncomplicated vulvovaginal candidosis [12,14,74]. In the World Health Organization (WHO) guideline on the management of vaginal discharge, it reads: "The Guidelines Group recommends that the current best treatment for candida in pregnant women are topical azole preparations. Topical azoles can be used at any stage of pregnancy for treatment of symptomatic candidosis" [14]. This is in agreement with other guidelines [4,12]. In terms of recurrent vulvovaginal candidosis, the most frequently recommended regimen consists of an induction course and maintenance therapy with fluconazole (weekly for six months) [13]. However, in cases when this regimen is not feasible, intermittent topical treatments are also suitable options [4,14]. The American College of Obstetricians and Gynecologists suggests a maintenance therapy with topical clotrimazole 500 mg once-weekly if affected women are "unable or unwilling" to take oral fluconazole [13]. Specific induction therapies can be found in the Canadian Guidelines. Among the recommended regimens, there is oral fluconazole 150 mg, once every 72 h for three doses, and a topical azole for 10–14 days. Maintenance therapy may consist of oral fluconazole 150 mg once-weekly or clotrimazole 500 mg intravaginally once-monthly [82]. According to the CDC, severe vulvovaginal candidosis should be treated with a topical azole for 7–14 days or with two oral doses of fluconazole 150 mg (administered three days apart) [4]. Of note, two doses of clotrimazole vaginal tablet 500 mg (administered three days apart) were shown to be as effective as two doses of oral fluconazole 150 mg, suggesting that both regimens represent suitable treatment options in subjects with severe vulvovaginal candidosis [74].

Table 5. Overview of suggested clotrimazole treatment regimens.

Uncomplicated VVC	Complicated VVC		Other Population
≥16 Years of Age	VVC in Pregnancy	Recurrent VVC	Severe VVC	Balanitis
Clotrimazole 500 mg vaginal tablet or vaginal soft ovule as single dose for 1 day or Clotrimazole 10% vaginal cream as single dose for 1 day	Clotrimazole 100 mg vaginal tablet once-daily for 7 days	Induction: Clotrimazole 100 mg vaginal tablet once-daily for 1–2 weeks Maintenance: Clotrimazole 500 mg vaginal tablet or vaginal soft ovule once-monthly for 6 months	Clotrimazole 500 mg vaginal tablet or vaginal soft ovule in two doses 3 days apart	Clotrimazole 1% cream twice-daily for 7 days

VVC: vulvovaginal candidosis.

For the treatment of balanitis/balanoposthitis, the WHO recommended topical application of clotrimazole twice-daily for 7 days or nystatin [83]. In fact, imidazoles such as topical clotrimazole 1% twice-daily belong to the drugs of choice in the treatment of *Candida* balanitis [5].

5. Conclusions

Almost 50 years ago, the first topical clotrimazole formulation was registered for the local treatment of vulvovaginal candidosis. Since then, different formulations have been developed and tested in a variety of studies. Cumulative evidence from numerous randomized controlled trials provided convincing scientific evidence that topical clotrimazole is effective and safe in the therapy of uncomplicated (acute) and certain types of complicated vaginal yeast infections, as well as *Candida* balanitis. The drug still belongs to the first-line treatments in these settings as specified in several treatment guidelines. It is therefore expected that clotrimazole will continue to be widely used in the field of vaginal health in future, more so as clotrimazole resistance in vaginal candidosis occurs rarely, despite its global use over many years.

Author Contributions: Writing, review and editing: W.M., M.A.E.S. and L.Z.; W.M. provided substantial contribution to the conception of this work and the interpretation of study data. All authors have read and agreed to the published version of the manuscript.

Funding: Writing assistance was provided by Edgar A. Mueller, 3P Consulting and has been funded by Bayer Consumer Care AG, Basel, Switzerland.

Conflicts of Interest: W.M. has received honoraria for oral presentations at scientific congresses or meetings of experts. M.A.E.S. and L.Z. are employees of Bayer Consumer Care AG, Basel, Switzerland.

References

1. Crowley, P.D.; Gallagher, H.C. Clotrimazole as a pharmaceutical: Past, present and future. *J. Appl. Microbiol.* **2014**, *117*, 611–617. [CrossRef] [PubMed]
2. Hitchcock, C.A.; Dickinson, K.; Brown, S.B.; Evans, E.G.; Adams, D.J. Interaction of azole antifungal antibiotics with cytochrome P-450-dependent 14 alpha-sterol demethylase purified from Candida albicans. *Biochem. J.* **1990**, *266*, 475–480. [CrossRef] [PubMed]
3. Taudorf, E.H.; Jemec, G.B.E.; Hay, R.J.; Saunte, D.M.L. Cutaneous candidiasis—An evidence-based review of topical and systemic treatments to inform clinical practice. *J. Eur. Acad. Dermatol. Venereol.* **2019**, *33*, 1863–1873. [CrossRef]
4. Workowski, K.A.; Bolan, G.A. Centers for Disease Control and Prevention. Sexually transmitted diseases treatment guidelines, 2015. *MMWR Recomm. Rep.* **2015**, *64*, 1–137.
5. Wray, A.A.; Velasquez, J.; Khetarpal, S. Balanitis. In *StatPearls Treasure Island (FL)*; StatPearls Publishing: Treasure Island, FL, USA, 2020; pp. 1–8.
6. Mendling, W.; Stock, I.; Becker, N. Therapie von Vulvovaginalmykosen im Wandel der Zeit (Teil 2): Behandlungsprinzipien in der Antimykotikatherapie. *Gyne* **2014**, *35*, 26–31.
7. Sawyer, P.R.; Brogden, R.N.; Pinder, R.M.; Speight, T.M.; Avery, G.S. Clotrimazole: A review of its antifungal activity and therapeutic efficacy. *Drugs* **1975**, *9*, 424–447. [CrossRef]
8. Denning, D.W.; Kneale, M.; Sobel, J.D.; Rautemaa-Richardson, R. Global burden of recurrent vulvovaginal candidiasis: A systematic review. *Lancet Infect. Dis.* **2018**, *18*, e339–e347. [CrossRef]
9. Benziger, D.P.; Edelson, J. Absorption from the vagina. *Drug Metab. Rev.* **1983**, *14*, 137–168. [CrossRef]
10. Sobel, J.D. Vulvovaginal candidosis. *Lancet* **2007**, *369*, 1961–1971. [CrossRef]
11. Jeanmonod, R.; Jeanmonod, D. Vaginal candidiasis (vulvovaginal candidiasis). In *StatPearls Treasure Island (FL)*; StatPearls Publishing: Treasure Island, FL, USA, 2020; pp. 1–4.
12. Mendling, W.; Brasch, J.; Cornely, O.A.; Effendy, I.; Friese, K.; Ginter-Hanselmayer, G.; Hof, H.; Mayser, P.; Mylonas, I.; Ruhnke, M.; et al. Guideline: Vulvovaginal candidosis (AWMF 015/072), S2k (excluding chronic mucocutaneous candidosis). *Mycoses* **2015**, *58*, 1–15. [CrossRef]
13. Matheson, A.; Mazza, D. Recurrent vulvovaginal candidiasis: A review of guideline recommendations. *Aust. N. Z. J. Obstet. Gynaecol.* **2017**, *57*, 139–145. [CrossRef] [PubMed]

14. Sherrard, J.; Wilson, J.; Donders, G.; Mendling, W.; Jensen, J.S. 2018 European (IUSTI/WHO) International Union against sexually transmitted infections (IUSTI) World Health Organisation (WHO) guideline on the management of vaginal discharge. *Int. J. STD AIDS* **2018**, *29*, 1258–1272. [CrossRef] [PubMed]

15. White, T.C.; Holleman, S.; Dy, F.; Mirels, L.F.; Stevens, D.A. Resistance mechanisms in clinical isolates of Candida albicans. *Antimicrob. Agents Chemother.* **2002**, *46*, 1704–1713. [CrossRef] [PubMed]

16. Mukasa, K.J.; Herbert, I.; Daniel, A.; Sserunkuma, K.L.; Joel, B.; Frederick, B. Antifungal susceptibility patterns of vulvovaginal *candida* species among women attending antenatal clinic at Mbarara regional referral hospital, South Western Uganda. *Br. Microbiol. Res. J.* **2015**, *5*, 322–331. [CrossRef] [PubMed]

17. Watson, M.C.; Grimshaw, J.M.; Bond, C.M.; Mollison, J.; Ludbrook, A. Oral versus intra-vaginal imidazole and triazole anti-fungal agents for the treatment of uncomplicated vulvovaginal candidiasis (thrush): A systematic review. *BJOG* **2002**, *109*, 85–95. [CrossRef]

18. Mendling, W.; Stock, I.; Becker, N. Therapie von Vulvovaginalmykosen im Wandel der Zeit (Teil 1): Vom Knoblauchextrakt zur topischen Eintagstherapie mit Clotrimazol. *Gyne* **2014**, *35*, 34–40.

19. Becker, N.; Gessner, U. Canesten® in vaginal mycosis: Therapeutic experience with 3784 patients. *Matern. Child Health* **1996**, *1*, 2–6.

20. Fischer, G.; Bradford, J. Vulvovaginal candidiasis in postmenopausal women: The role of hormone replacement therapy. *J. Low. Genit. Tract. Dis.* **2011**, *15*, 263–267. [CrossRef]

21. Mendling, W.; Krauss, C.; Fladung, B. A clinical multicenter study comparing efficacy and tolerability of topical combination therapy with clotrimazole (Canesten®, two formats) with oral single dose fluconazole (Diflucan®) in vulvovaginal mycoses. *Mycoses* **2004**, *47*, 136–142. [CrossRef]

22. Mendling, W.; Schlegelmilch, R. Three-day combination treatment for vulvovaginal candidosis with 200 mg clotrimazol vaginal suppositories and clotrimazol cream for the vulva is significantly better than treatment with vaginal suppositories alone—An earlier, multi-centre, placebo-controlled double blind study. *Geburtshilfe Frauenheilkd.* **2014**, *74*, 355–360. [CrossRef]

23. Nurbhai, M.; Grimshaw, J.; Watson, M.; Bond, C.; Mollison, J.; Ludbrook, A. Oral versus intra-vaginal imidazole and triazole anti-fungal treatment of uncomplicated vulvovaginal candidiasis (thrush). *Cochrane Database Syst. Rev.* **2007**, *4*, CD002845. [CrossRef] [PubMed]

24. Spence, D. Candidiasis (vulvovaginal). *BMJ Clin. Evid.* **2010**, *0815*, 1–39.

25. Martin Lopez, J.E. Candidiasis (vulvovaginal). *BMJ Clin. Evid.* **2015**, *0815*, 1–23.

26. Hughes, D.; Kriedman, T. Treatment of vulvovaginal candidiasis with a 500-mg vaginal tablet of clotrimazole. *Clin. Ther.* **1984**, *6*, 662–668. [PubMed]

27. Bro, F. Single-dose 500-mg clotrimazole vaginal tablets compared with placebo in the treatment of Candida vaginitis. *J. Fam. Pract.* **1990**, *31*, 148–152. [PubMed]

28. Stein, G.E.; Mummaw, N. Placebo-controlled trial of itraconazole for treatment of acute vaginal candidiasis. *Antimicrob. Agents Chemother.* **1993**, *37*, 89–92. [CrossRef]

29. Anonymous. A comparison of single-dose oral fluconazole with 3-day intravaginal clotrimazole in the treatment of vaginal candidiasis. Report of an international multicentre trial. *Br. J. Obstet. Gynaecol.* **1989**, *96*, 226–232. [CrossRef]

30. Adetoro, O.O. Comparative trial of a single oral dose of fluconazole (150 mg) and a single intravaginal tablet of clotrimazole (500 mg) in the treatment of vaginal candidiasis. *Curr. Ther. Res.* **1990**, *48*, 275–281.

31. Stein, G.E.; Christensen, S.; Mummaw, N. Comparative study of fluconazole and clotrimazole in the treatment of vulvovaginal candidiasis. *DICP* **1991**, *25*, 582–585. [CrossRef]

32. Petersen, E.E. Vergleich einer oralen Einmalgabe Fluconazol mit einer 1-tägigen intravaginalen Clotrimazol-Kombinationstherapie bei der Behandlung der vaginalen Candidose. *Frauenarzt* **1994**, *35*, 219–223.

33. van Heusden, A.M.; Merkus, H.M.; Euser, R.; Verhoeff, A. A randomized, comparative study of a single oral dose of fluconazole versus a single topical dose of clotrimazole in the treatment of vaginal candidosis among general practitioners and gynaecologists. *Eur. J. Obstet. Gynecol. Reprod. Biol.* **1994**, *55*, 123–127. [CrossRef]

34. Mikamo, H.; Izumi, K.; Ito, K.; Tamaya, T. Comparative study of the effectiveness of oral fluconazole and intravaginal clotrimazole in the treatment of vaginal candidiasis. *Infect. Dis. Obstet. Gynecol.* **1995**, *3*, 7–11. [CrossRef] [PubMed]

35. Sobel, J.D.; Brooker, D.; Stein, G.E.; Thomason, J.L.; Wermeling, D.P.; Bradley, B.; Weinstein, L. Single oral dose fluconazole compared with conventional clotrimazole topical therapy of *Candida* vaginitis. *Am. J. Obstet. Gynecol.* **1995**, *172*, 1263–1268. [CrossRef]

36. O-Prasertsawat, P.; Bourlert, A. Comparative study of fluconazole and clotrimazole for the treatment of vulvovaginal candidiasis. *Sex. Transm. Dis.* **1995**, *22*, 228–230. [CrossRef]

37. Woolley, P.D.; Higgins, S.P. Comparison of clotrimazole, fluconazole and itraconazole in vaginal candidiasis. *Br. J. Clin. Pract.* **1995**, *49*, 65–66.

38. Mikamo, H.; Kawazoe, K.; Sato, Y.; Hayasaki, Y.; Tamaya, T. Comparative study on the effectiveness of antifungal agents in different regimens against vaginal candidiasis. *Chemotherapy* **1998**, *44*, 364–368. [CrossRef]

39. Sekhavat, L.; Tabatabaii, A.; Tezerjani, F.Z. Oral fluconazole 150 mg single dose versus intra-vaginal clotrimazole treatment of acute vulvovaginal candidiasis. *J. Infect. Public Health* **2011**, *4*, 195–199. [CrossRef]

40. Miller, P.I.; Humphries, M.; Grassick, K. A single-blind comparison of oral and intravaginal treatments in acute and recurrent vaginal candidosis in general practice. *Pharmatherapeutica* **1984**, *3*, 582–587.

41. Sanchez Carazo, J.L.; Gimeno Carpio, E.; Arnedo Grifol, A.L.; Vilata Corell, J.J.; Nogueira Coito, J.M. Comparison of three imidazolic regimens in the treatment of vaginal candidosis. *Eur. J. Sex. Transm. Dis.* **1986**, *3*, 223–225.

42. Tobin, J.M.; Loo, P.; Granger, S.E. Treatment of vaginal candidosis: A comparative study of the efficacy and acceptability of itraconazole and clotrimazole. *Genitourin. Med.* **1992**, *68*, 36–38. [CrossRef]

43. Milsom, I.; Forssman, L. Treatment of vaginal candidosis with a single 500-mg clotrimazole pessary. *Br. J. Vener. Dis.* **1982**, *58*, 124–126. [CrossRef] [PubMed]

44. Fleury, F.; Hughes, D.; Floyd, R. Therapeutic results obtained in vaginal mycoses after single-dose treatment with 500 mg clotrimazole vaginal tablets. *Am. J. Obstet. Gynecol.* **1985**, *152*, 968–970. [CrossRef]

45. Floyd, R., Jr.; Hodgson, C. One-day treatment of vulvovaginal candidiasis with a 500-mg clotrimazole vaginal tablet compared with a three-day regimen of two 100-mg vaginal tablets daily. *Clin. Ther.* **1986**, *8*, 181–186. [PubMed]

46. Hughes, D.; Kriedman, T.; Hodgson, C. Treatment of vulvovaginal candidiasis with a single 500-mg clotrimazole vaginal tablet compared with two 100-mg tablets daily for three days. *Curr. Ther. Res.* **1986**, *39*, 773–777.

47. Westphal, J. Vulvovaginale Candidamykosen—Behandlung mit Clotrimazol. Vergleich einer Einmaldosis-Behandlung mit der Sechs-Tage-Therapie. *Fortschr. Med.* **1988**, *106*, 445–448. [PubMed]

48. Cohen, L. Treatment of vaginal candidosis using clotrimazole vaginal cream: Single dose versus 3-day therapy. *Curr. Med. Res. Opin.* **1985**, *9*, 520–523. [CrossRef]

49. Breuker, G.; Jurczok, F.; Lenaerts, M.; Weinhold, E.; Krause, U. Ein-Dosis-Therapie von Vaginalmykosen mit Clotrimazol-Vaginalcreme 10%. *Mykosen* **1986**, *29*, 427–436. [CrossRef]

50. Anonymous. An Investigator-Blinded, Active-Controlled Phase 3 Study to Prove the Non-Inferior Efficacy of a Clotrimazole Ovule (500 mg) Versus a Clotrimazole Vaginal Tablet (500 mg) in Vaginal Candidiasis. Clinical Study Synopsis, Dated 27 November 2009. Available online: http://trialfinder.bayerscheringpharma. de/html/pdf/13071_Study_Synopsis_CTP.pdf (accessed on 30 June 2020).

51. Müllegger, R.R.; Häring, N.S.; Glatz, M. Skin infections in pregnancy. *Clin. Dermatol.* **2016**, *34*, 368–377. [CrossRef]

52. Aguin, T.J.; Sobel, J.D. Vulvovaginal candidiasis in pregnancy. *Curr. Infect. Dis. Rep.* **2015**, *17*, 462. [CrossRef]

53. Pilmis, B.; Jullien, V.; Sobel, J.; Lecuit, M.; Lortholary, O.; Charlier, C. Antifungal drugs during pregnancy: An updated review. *J. Antimicrob. Chemother.* **2015**, *70*, 14–22. [CrossRef]

54. Moudgal, V.V.; Sobel, J.D. Antifungal drugs in pregnancy: A review. *Expert Opin. Drug Saf.* **2003**, *2*, 475–483. [CrossRef] [PubMed]

55. Ruiz-Velasco, V.; Rosas-Arceo, J. Prophylactic clotrimazole treatment to prevent mycoses contamination of the newborn. *Int. J. Gynaecol. Obstet.* **1978**, *16*, 70–71. [CrossRef] [PubMed]

56. Schwarze, R.; Blaschke-Hellmessen, R.; Hinkel, G.K.; Hoffmann, H.; Weigl, J. Untersuchungen zur Soorprophylaxe Neugeborener. I. Mitteilung: Wirksamkeit einer Fungicin-(Nystatin-) Prophylaxe bei gesunden Neugeborenen. *Kinderäztl. Prax.* **1976**, *7*, 305–314.

57. Schnell, J.D. *Vaginalmykose und Perinatale Pilzinfektion*, 1st ed.; Karger, S., Ed.; Karger Publishers: Basel, Switzerland, 1982.

58. Schnell, J.D. "Soorprophylaxe" in der Schwangerschaft. *Frauenarzt* **1986**, *5*, 19–26.

59. Blaschke-Hellmessen, R. Subpartale Übertragung von Candida und ihre Konsequenzen. *Mycoses* **1998**, *41* (Suppl. 2), 31–36. [CrossRef]

60. Kiss, H.; Petricevic, L.; Husslein, P. Prospective randomised controlled trial of an infection screening programme to reduce the rate of preterm delivery. *BMJ* **2004**, *329*, 371. [CrossRef]

61. Roberts, C.L.; Algert, C.S.; Rickard, K.L.; Morris, J.M. Treatment of vaginal candidiasis for the prevention of preterm birth: A systematic review and meta-analysis. *Syst. Rev.* **2015**, *4*, 31. [CrossRef]

62. Roberts, C.L.; Rickard, K.; Kotsiou, G.; Morris, J.M. Treatment of asymptomatic vaginal candidiasis in pregnancy to prevent preterm birth: An open-label pilot randomized controlled trial. *BMC Pregnancy Childbirth* **2011**, *11*, 18. [CrossRef]

63. Young, G.L.; Jewell, D. Topical treatment for vaginal candidiasis (thrush) in pregnancy. *Cochrane Database Syst. Rev.* **2001**, *4*, CD000225. [CrossRef]

64. Tan, C.G.; Good, C.S.; Milne, L.J.; Loudon, J.D. A comparative trial of six day therapy with clotrimazole and nystatin in pregnant patients with vaginal candidiasis. *Postgrad. Med. J.* **1974**, *50* (Suppl. 1), 102–105.

65. Dunster, G.D. Vaginal candidiasis in pregnancy—A trial of clotrimazole. *Postgrad. Med. J.* **1974**, *50* (Suppl. 1), 86–88.

66. Del Palacio-Hernanz, A.; Sanz-Sanz, F.; Rodriquez-Noriega, A. Double-blind investigation of R-42470 (terconazole cream 0.4%) and clotrimazole (cream 1%) for the topical treatment of mycotic vaginitis. *Chemioterapia* **1984**, *3*, 192–195. [PubMed]

67. Lebherz, T.; Guess, E.; Wolfson, N. Efficacy of single- versus multiple-dose clotrimazole therapy in the management of vulvovaginal candidiasis. *Am. J. Obstet. Gynecol.* **1985**, *152*, 965–968. [CrossRef]

68. Sobel, J.D. Recurrent vulvovaginal candidiasis. *Am. J. Obstet. Gynecol.* **2016**, *214*, 15–21. [CrossRef] [PubMed]

69. Sobel, J.D.; Schmitt, C.; Stein, G.; Mummaw, N.; Christensen, S.; Meriwether, C. Initial management of recurrent vulvovaginal candidiasis with oral ketoconazole and topical clotrimazole. *J. Reprod. Med.* **1994**, *39*, 517–520.

70. Sobel, J.D.; Wiesenfeld, H.C.; Martens, M.; Danna, P.; Hooton, T.M.; Rompalo, A.; Sperling, M.; Livengood, C., 3rd; Horowitz, B.; Von Thron, J.; et al. Maintenance fluconazole therapy for recurrent vulvovaginal candidiasis. *N. Engl. J. Med.* **2004**, *351*, 876–883. [CrossRef]

71. Sobel, J.D.; Schmitt, C.; Meriwether, C. Clotrimazole treatment of recurrent and chronic candida vulvovaginitis. *Obstet. Gynecol.* **1989**, *73*, 330–334. [CrossRef]

72. Roth, A.C.; Milsom, I.; Forssman, L.; Wåhlén, P. Intermittent prophylactic treatment of recurrent vaginal candidiasis by postmenstrual application of a 500 mg clotrimazole vaginal tablet. *Genitourin. Med.* **1990**, *66*, 357–360. [CrossRef]

73. Fong, I.W. The value of chronic suppressive therapy with itraconazole versus clotrimazole in women with recurrent vaginal candidiasis. *Genitourin. Med.* **1992**, *68*, 374–377. [CrossRef]

74. Zhou, X.; Li, T.; Fan, S.; Zhu, Y.; Liu, X.; Guo, X.; Liang, Y. The efficacy and safety of clotrimazole vaginal tablet vs. oral fluconazole in treating severe vulvovaginal candidiasis. *Mycoses* **2016**, *59*, 419–428. [CrossRef]

75. Williams, A.B.; Yu, C.; Tashima, K.; Burgess, J.; Danvers, K. Evaluation of two self-care treatments for prevention of vaginal candidiasis in women with HIV. *J. Assoc. Nurses AIDS Care* **2001**, *12*, 51–57. [CrossRef]

76. Boag, F.C.; Houang, E.T.; Westrom, R.; McCormack, S.M.; Lawrence, A.G. Comparison of vaginal flora after treatment with a clotrimazole 500 mg vaginal pessary or a fluconazole 150 mg capsule for vaginal candidosis. *Genitourin. Med.* **1991**, *67*, 232–234. [CrossRef]

77. Stary, A.; Soeltz-Szoets, J.; Ziegler, C.; Kinghorn, G.R.; Roy, R.B. Comparison of the efficacy and safety of oral fluconazole and topical clotrimazole in patients with candida balanitis. *Genitourin. Med.* **1996**, *72*, 98–102. [CrossRef] [PubMed]

78. Maw, R.D.; Horner, T.; Evans, J. A comparative trial of bifonazole 1% cream and clotrimazole 1% cream in the treatment of candidal balanoposthitis. *Mykosen* **1987**, *30*, 229–232. [CrossRef] [PubMed]

79. Waugh, M.A.; Evans, E.G.; Nayyar, K.C.; Fong, R. Clotrimazole (Canesten) in the treatment of candidal balanitis in men. With incidental observations on diabetic candidal balanoposthitis. *Br. J. Vener. Dis.* **1978**, *54*, 184–186. [CrossRef]

80. Ritter, W.; Patzschke, K.; Krause, U.; Stettendorf, S. Pharmacokinetic fundamentals of vaginal treatment with clotrimazole. *Chemotherapy* **1982**, *28* (Suppl. 1), 37–42. [CrossRef]

81. Mendling, W.; Becker, N. Clotrimazol bei Vaginalmykosen—Vergleichende Untersuchung der 1- und 3-Tages-Therapie. *Gyn* **2001**, *6*, 97–100.

82. Canadian Guidelines on Sexually Transmitted Infections. Section 4–9: Management and Treatment of Specific Syndromes—Vaginal Discharge. Available online: https://www.canada.ca/en/public-health/services/infectious-diseases/sexual-health-sexually-transmitted-infections/canadian-guidelines/sexually-transmitted-infections/canadian-guidelines-sexually-transmitted-infections-26.html (accessed on 15 July 2020).
83. World Health Organization. Guidelines for the Management of Sexually Transmitted Infections. Available online: https://www.who.int/hiv/topics/vct/sw_toolkit/guidelines_management_sti.pdf?ua=1 (accessed on 15 July 2020).

pharmaceuticals

Article

The Repurposing of Acetylsalicylic Acid as a Photosensitiser to Inactivate the Growth of Cryptococcal Cells

Adepemi O. Ogundeji [1], Nozethu Mjokane [1], Olufemi S. Folorunso [1], Carolina H. Pohl [1], Martin M. Nyaga [2] and Olihile M. Sebolai [1,*]

[1] Department of Microbiology and Biochemistry, University of the Free State, Bloemfontein 9301, South Africa; ogundejiao@ufs.ac.za (A.O.O.); nmjokane@gmail.com (N.M.); foxyphemmzy@gmail.com (O.S.F.); pohlch@ufs.ac.za (C.H.P.)

[2] Next Generation Sequencing Unit and Division of Virology, University of the Free State, Bloemfontein 9301, South Africa; nyagamm@ufs.ac.za

* Correspondence: sebolaiom@ufs.ac.za; Tel.: +27-51-401-2004

Abstract: Photodynamic treatment (PDT) is often successful when used against aerobic microbes, given their natural susceptibility to oxidative damage. To this end, the current study aimed to explore the photodynamic action of acetylsalicylic acid (ASA; aspirin, which is commonly used to treat non-infectious ailments), when administered to respiring cryptococcal cells. The treatment of cryptococcal cells, i.e., exposure to 0.5 or 1 mM of ASA in the presence of ultraviolet light (UVL) for 10 min, resulted in a significant ($p < 0.05$) reduction in the growth of tested cells when compared to non-treated (non-Rx) cells, i.e., no ASA and no UVL. The treated cells were also characterised by diseased mitochondria, which is crucial for the survival of respiring cells, as observed by a significant ($p < 0.05$) loss of mitochondrial membrane potential ($\Delta\Psi M$) and significant ($p < 0.05$) accumulation of reactive oxygen species (ROS) when compared to non-Rx cells. Moreover, the photolytic products of acetylsalicylic acid altered the ultrastructural appearance of treated cells as well as limited the expression levels of the capsular-associated gene, *CAP64*, when compared to non-Rx cells. The results of the study highlight the potential use of ASA as a photosensitiser that is effective for controlling the growth of cryptococcal cells. Potentially, this treatment can also be used as an adjuvant, to complement and support the usage of current anti-microbial agents.

Keywords: acetylsalicylic acid (ASA, aspirin); capsule; *CAP64*; *Cryptococcus*; membrane potential ($\Delta\Psi M$); photodynamic treatment; photosensitiser; ultrastructure

Citation: Ogundeji, A.O.; Mjokane, N.; Folorunso, O.S.; Pohl, C.H.; Nyaga, M.M.; Sebolai, O.M. The Repurposing of Acetylsalicylic Acid as a Photosensitiser to Inactivate the Growth of Cryptococcal Cells. *Pharmaceuticals* **2021**, *14*, 404. https://doi.org/10.3390/ph14050404

Academic Editor: Daniela De Vita

Received: 10 March 2021
Accepted: 14 April 2021
Published: 23 April 2021

Publisher's Note: MDPI stays neutral with regard to jurisdictional claims in published maps and institutional affiliations.

1. Introduction

Cryptococcosis caused by the fungus *Cryptococcus* (C.) *neoformans*, is one of the most common opportunistic infections with a significant mortality rate among AIDS patients, more especially in resource-limited countries such as in sub-Saharan Africa [1–7]. In these vulnerable subjects, i.e., with a depleted cell-mediated immune response, fungal cells can disseminate to the brain. There, the cells impair the ability of the brain to reabsorb the cerebrospinal fluid, leading to a build-up within the skull as well as causing meningitis [8,9]. Such a patient can then present with a debilitating headache, coma, or even die. It is reported that 223,100 cases of cryptococcal meningitis occur annually, of which 73% are diagnosed in sub-Saharan Africa [2–4,10]. Furthermore, the mortality rate of this disease in sub-Saharan Africa is 75% [2–4,10].

This fungus can also cause cutaneous infections, which are not as prevalent as cryptococcal lung or central nervous system infections [11,12]. The manifestation of cryptococcal skin infections, i.e., draining sinuses, acneiform lesions, among others, is typically an indication of systemic infection and not direct inoculation [11,12]. In the case of the former, cryptococcal cells may be transported via a haematogenous route or invade circulating macrophages and, in a manner akin to the Trojan horse, reach the skin [8,9]. These skin infections are reported to occur in 10–15% of patients with an invasive cryptococcal infection [11].

To control cryptococcal infections, three drugs, namely, fluconazole, amphotericin b and flucytosine, have been recommended for management purposes [13,14]. However, in South Africa, only fluconazole and amphotericin b are routinely used in public health institutions [7,15–17]. To compound this, the clinical application of these two drugs is often limited. For example, fluconazole is usually used as a first-line treatment. Although this drug has been shown to easily cross the blood–brain barrier, it is, however, also reported to have relatively poor fungal clearance, even when administered at high doses [5,18]. On the other hand, amphotericin b is a better choice due to its effectiveness but has huge toxicity, and its administration is usually intravenous [18,19]. Due to these shortcomings, these medicines are thus often associated with clinical failure [20–22].

Given the medical importance of cryptococcal infections, several scholars have looked at other strategies, such as the repurposing of non-traditional antifungals in combating this deadly pathogen [21,23–25]. Thus, the current study aims to repurpose acetylsalicylic acid (ASA; aspirin) as a photosensitiser. ASA is an old drug that is recommended to treat pain and inflammation [26,27]. It is, therefore, not surprising that ASA is on the World Health Organization's List of Essential Medicines [28]. Importantly, the wholesale cost of ASA in the developing world is estimated to be USD 0.002 to USD 0.025, as of 2014 [29].

A photosensitiser is a compound which, when excited by an appropriate light source, could generate harmful radicals [30]. ASA, like other aromatic carboxylic acids, has intense absorption bands in the UV spectral range [31]. However, the photochemical properties of molecules such as salicylates (SAs) and their derivatives are said to be poorly understood [31].

SA is the active compound of ASA, and it is responsible for its biological properties. Through using optical spectroscopy, fluorescence spectroscopy, and nanosecond laser flash photolysis, Pozdnyakov et al. [31] showed that excitation of SA could yield a salicylate anion triplet state (HSA^-), $HSA\cdot$ radical and hydrated electron (e_{eq}^-), wherein the last two species are possibly generated due to two-photon processes [32,33]. These radicals can then react with molecular oxygen to generate reactive oxygen species (ROS) typifying a Type I reaction (Figure 1).

The ability of ASA to generate ROS was also shown in a study by Ogundeji et al. (2016) [25]. In the study, these authors showed that ASA engaged a signalling programme that involved the participation of a high osmolarity glycerol (HOG) pathway. The activation of this pathway suggested that ASA-induced ROS-based oxidative stress was sufficient to kill cryptococcal cells [25]. In their study, Norman et al. [34] also showed that SA could uncouple the electron transport chain. In turn, the latter impairs the oxygen (as a final electron acceptor) from receiving a full complement of electrons, thus leading to the production of ROS [34].

In the current study, ASA was used at concentrations (0.5 and 1 mM) that are within the recommended threshold [35]. Based on the physiology of cryptococcal cells, i.e., they are non-fermentative [36], it was hypothesised that this respiring organism would be susceptible to stress induced by harmful radicals, as previously shown after treatment with ASA [25].

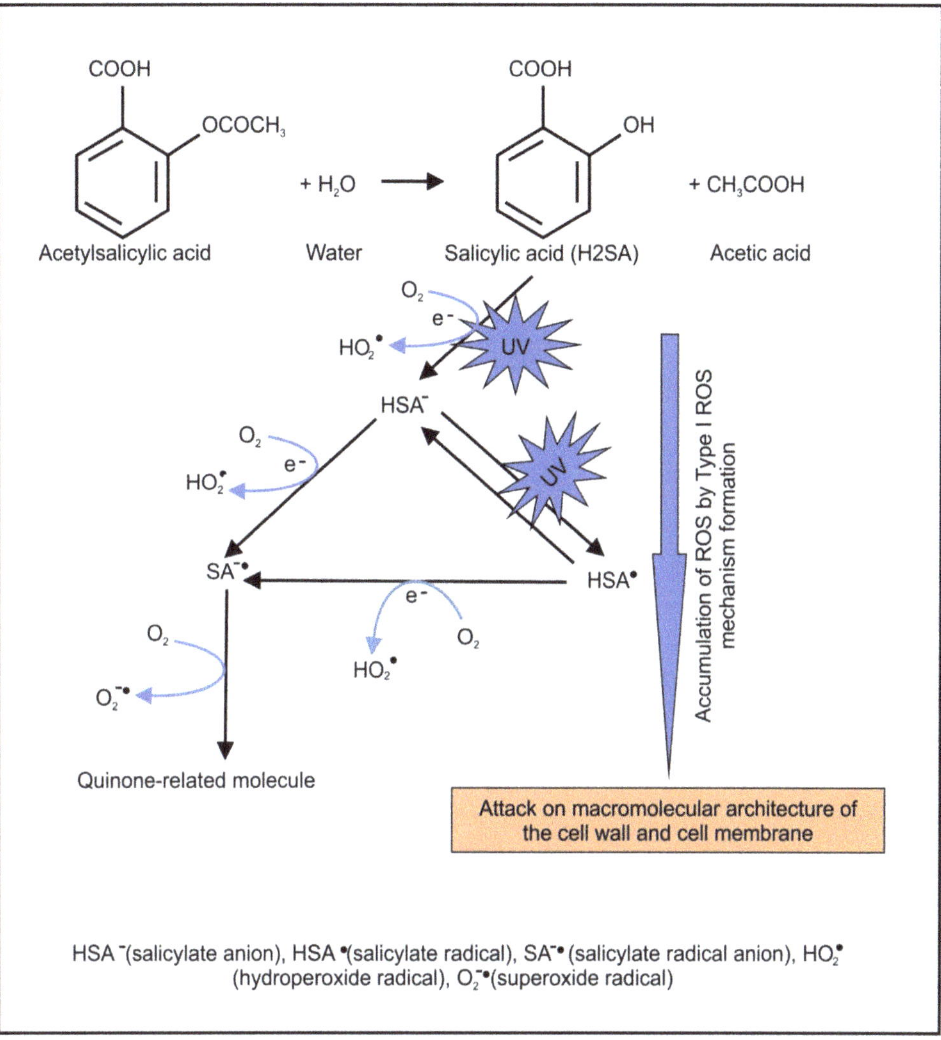

Figure 1. Proposed mechanism of UV-induced photosensitizer and accumulation of ROS via Type I mechanism.

2. Results

2.1. Cryptococcal Cells Are Susceptible to the Photodynamic Action of ASA

Figure 2 summarises the effect of PDT with ASA on the survival of the tested cryptococcal cells. As expected, the negative controls, i.e., 1 mM of ASA alone (with no cells), and 1 mM of ASA and 10 min UVL (with no cells), and no CFUs were observed on the agar plates after incubation (data not shown). When used alone, neither ASA nor UVL was able to yield a reduction in CFU counts that were above 10% when compared to non-treated (non-Rx) cells counts, i.e., no ASA and no UVL. However, when used in combination, a compounded effect was observed. Moreover, there was a significant ($p < 0.05$) reduction (equal to and above 97%) in CFU counts when PDT was applied (0.5 mM and 10 min UVL or 1 mM and 10 min UVL), when compared to non-Rx cells' counts. A similar response profile was obtained when the following cryptococcal strains were exposed to similar conditions *viz. C. neoformans* H99, *C. gattii* LMPE 052 and *C. gattii* R265 (Figure S1).

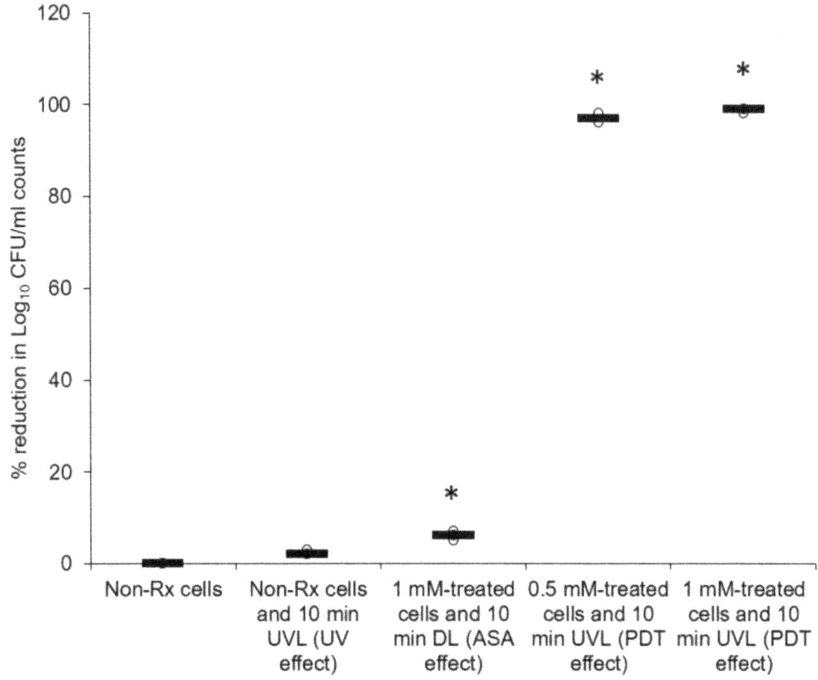

Figure 2. Effect of PDT with ASA on the survival of *C. neoformans* LMPE 046. Non-Rx, non-treated cells; UVL, ultraviolet light; DL, dark light; CFU, colony-forming unit. Data points were obtained from three biological replicates for each defined experimental condition. * Significantly different from non-Rx cells at $p < 0.05$.

2.2. Light-Activated ASA Kills Cryptococcal Cells via the Production of Harmful Radical Species That Target Cell Walls

Given that the photolytic products of ASA may be harmful, it was sought to determine if these products may also target the cytoplasmic membrane and ultrastructural features of cryptococcal cells. It was, therefore, not surprising when studying the effects of PDT on the integrity of the cell membrane (using the PI exclusion assay) to note that all treated cells (0.5 mM or 1 mM) following exposure to UVL (10 min) had significantly ($p < 0.05$) accumulated the PI stain in the cytoplasm, while the corresponding non-RX cells did not accumulate the stain (Figure 3). The above findings suggested that the membranes of the treated cells had lost their selective permeability.

The captured micrographs (representing cells obtained from the different experimental conditions) were collated into Figure 4, to aid in making deductions regarding the effect of PDT. The figure showed that the harmful ASA photolytic products may have targeted the cell walls of cells that were exposed to the combined effect of ASA and UVL. This assertion is based on comparing the morphological differences of cells between the different experimental conditions. Furthermore, *C. neoformans* LMPE 046's non-Rx cells appeared to be whole and covered with web-like extracellular matrixes (possibly the capsule) on their cell wall surfaces. Exposure of cells to ASA treatment and separately to UVL treatment did not effect a change in the appearance of these cells when compared to non-Rx cells. However, the exposure of cells to PDT led to an ultrastructural change, because cells were observed to have collapsed cell walls, and some appeared to have less of the web-like extracellular matrixes.

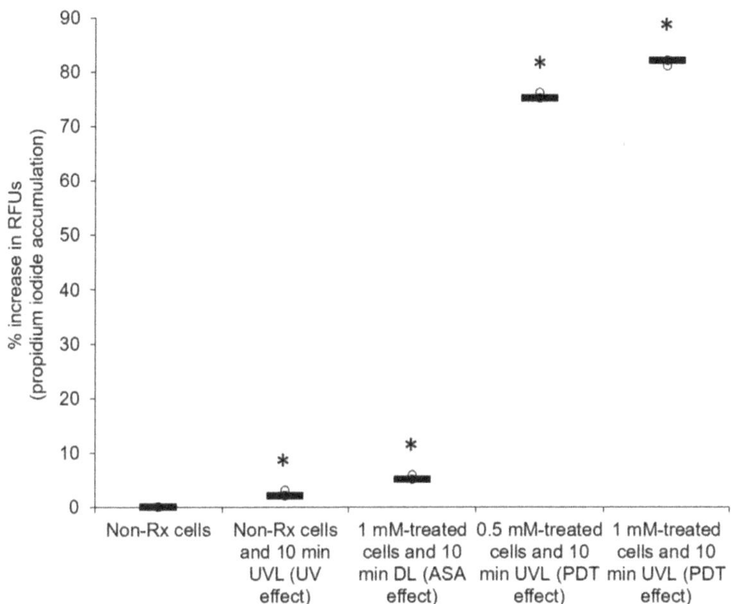

Figure 3. PDT action of ASA showing that cells of *C. neoformans* LMPE 046 lost their selective permeability. The background signal of negative controls was used for normalisation. There was a normality of data distribution. Non-Rx, non-treated cells; UVL, ultraviolet light; DL, dark light; RFUs, relative fluorescence units. Data points were obtained from three biological replicates for each defined experimental condition. * Significantly different from non-Rx cells at $p < 0.05$.

Figure 4. SEM images showing the morphological differences of cryptococcal cells' ultrastructure after being handled in designed experimental groups. The arrow points at cells which exemplify the effects of PDT, i.e., a cell that appears wrinkled. The images were taken after studying *C. neoformans* LMPE 046 cells.

The morphological changes may be incidental due to sample preparation; assessment of the impact of PDT on the expression of a capsular gene responsible for capsule formation was also sought. The effects of the different experimental conditions on the expression of the *CAP64* gene were summarised in Figure 5. When comparing the non-Rx cell data, it was noted that PDT had an effect on the capsule of *C. neoformans* at a genomic level; consequently, the expression of *CAP64* was significantly reduced ($p < 0.05$), as shown in Figure 5. Low expression of a Cap gene, particularly *CAP64* in this case, may result in an acapsular phenotype, as previously documented [37,38]. On the other hand, there was no significant difference ($p > 0.05$) in the expression of the *CAP64* gene, when the non-Rx cell data were compared to the ASA effect and separately to UVL effect. The implication of the above is that cryptococcal cells exposed to PDT with ASA would be more vulnerable to the action of antimicrobial drugs. It is documented elsewhere that the capsule can impair the transport and uptake of antifungals such as amphotericin B into the cell cytoplasm [39]. Moreover, in the context of an immune response, lack of or a compromised capsule implies that cells would be easily targeted by host immune molecules and cells [40]. It is important to note that our work is preliminary to draw concrete conclusions on the effects of PDT treatment on the cryptococcal capsule. To this end, supporting evidence related to the downregulation of additional cell structures or cell wall genes is required.

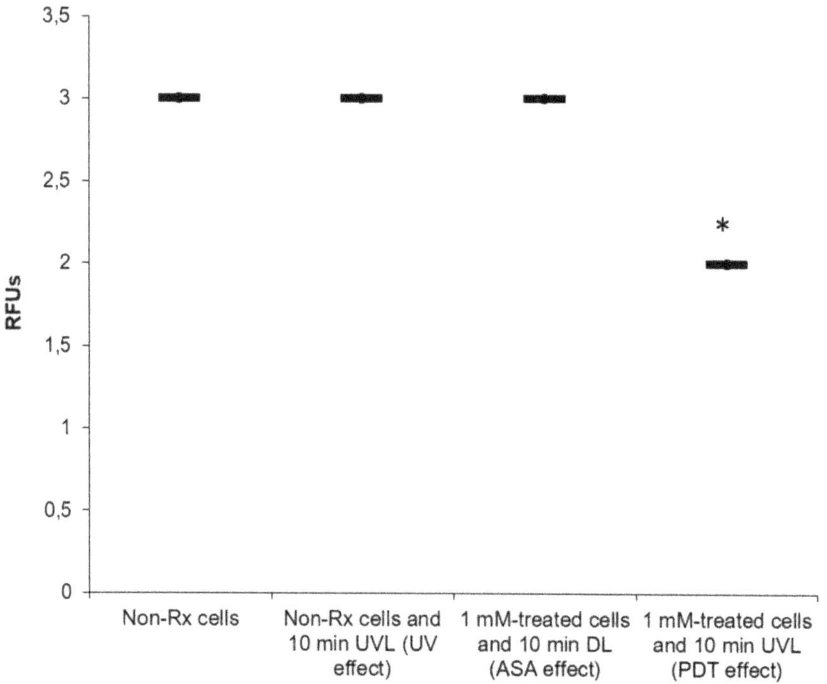

Figure 5. The effect of different experimental conditions on the expression of the *CAP64* gene in *C. neoformans* LMPE 046. Non-Rx, non-treated cells; UVL, ultraviolet light; DL, dark light; RFUs, relative fluorescence units. Data points were obtained from three biological replicates for each defined experimental condition. * Significantly different from non-Rx cells at $p < 0.05$.

2.3. The Photodynamic Action of ASA Impairs Cryptococcal Mitochondrial Function

Oxygen is an essential element for obligate, respiring microbes [41]. It is required for their survival because it is used in oxidative phosphorylation, a process that leads to cellular energy production [42]. Therefore, any impairment to the above process that results in oxygen not receiving a full complement of electrons to reduce it leads to the production of harmful oxygen radicals [43]. Herein, we show that the photosensitising effect of ASA seems to disrupt the normal functioning of mitochondria (Figure 6). When comparing the membrane potential ($\Delta\Psi$M) of all the tested *C. neoformans* strains (treated with 0.5 mM or 1 mM of ASA in the presence of UVL for 10 min) to that of their corresponding non-Rx cells, it was noted that the effect of ASA was enhanced in the presence of UVL; these cells (exposed to the combined effect of ASA and UVL) had significantly ($p < 0.05$) lost their membrane potential ($\Delta\Psi$M). ROS accumulation was also measured in all treated cells (0.5 mM or 1 mM in the presence of UVL (10 min)) as well in their corresponding non-Rx cells (Figure 7). As expected, the combined effect of ASA and UVL resulted in cells significantly accumulating more than 80% ($p < 0.05$) of ROS when compared to their respective non-Rx cells. Therefore, this turn of events in the form of a collapsing mitochondrion starves cells of potential energy to support important cellular processes such as growth [44].

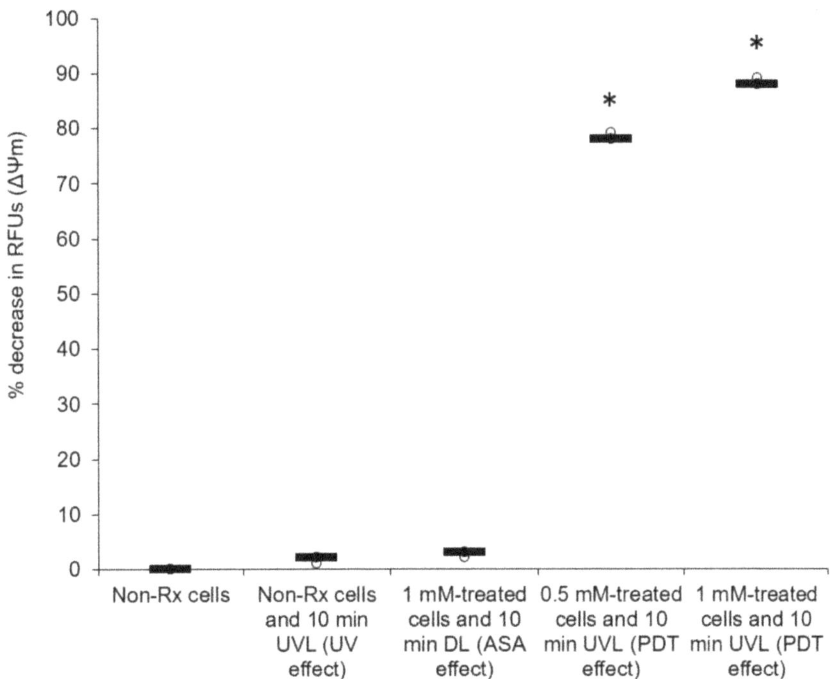

Figure 6. PDT action of ASA impairs the energising of mitochondrial membranes of *C. neoformans* LMPE 046 cells. The background signal of negative controls was used for normalisation. There was a normality of data distribution. Non-Rx, non-treated cells; UVL, ultraviolet light; DL, dark light; RFUs, relative fluorescence units. Data points were obtained from three biological replicates for each defined experimental condition. * Significantly different from non-Rx cells at $p < 0.05$.

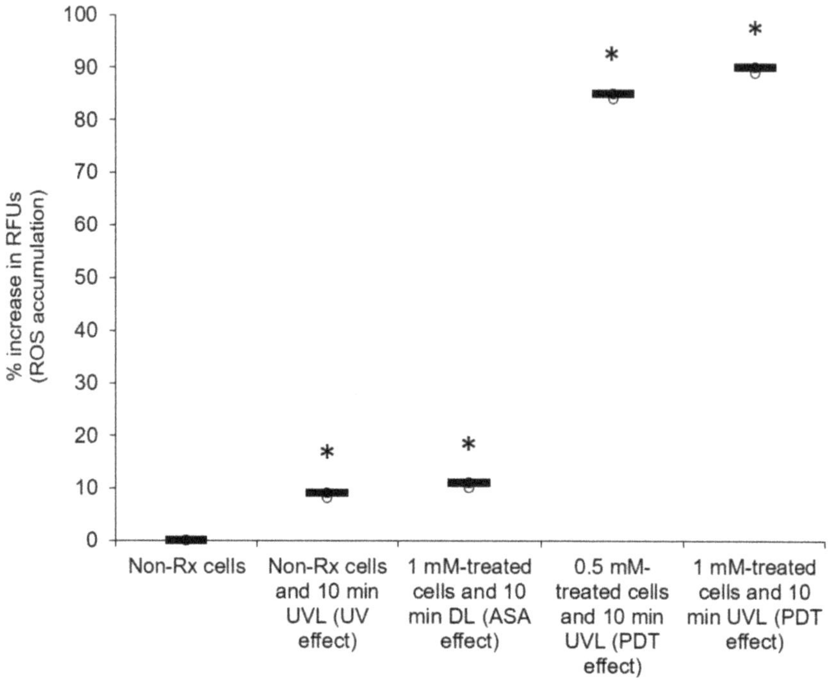

Figure 7. PDT action of ASA impairs the mitochondrial electron transport chain leading to excessive ROS production in *C. neoformans* LMPE 046. The background signal of negative controls was used for normalisation. There was a normality of data distribution. Non-Rx, non-treated cells; UVL, ultraviolet light; DL, dark light; RFUs, relative fluorescence units. Data points were obtained from three biological replicates for each defined experimental condition. * Significantly different from non-Rx cells at $p < 0.05$.

2.4. The PDT Action of ASA Does Not Adversely Affect the Health of Macrophages

Crucial to the pathogenesis of *C. neoformans* is the ability of cells to manipulate macrophages in a Trojan horse-like manner for dissemination purposes [45]. Thus, for ASA to be considered as an ideal photosensitiser, it should also not be detrimental to the host cells. Here, it was determined that PDT with ASA did not kill more than 10% of the macrophage population (0.5 mM-treated cells and 10 min UVL exposure or 1 mM-treated cells and 10 min UVL exposure) when compared to the non-Rx macrophage population (i.e., no ASA and no UVL) as seen in Figure 8. Moreover, after a 10 min exposure period, only 5% of the population was killed using 0.5 mM, while 1 mM killed 8%. Importantly, in all these cases, the application of ASA did not effect a lethal dosage of 50 (LD_{50}), wherein 50% of the tested macrophage population would have been killed.

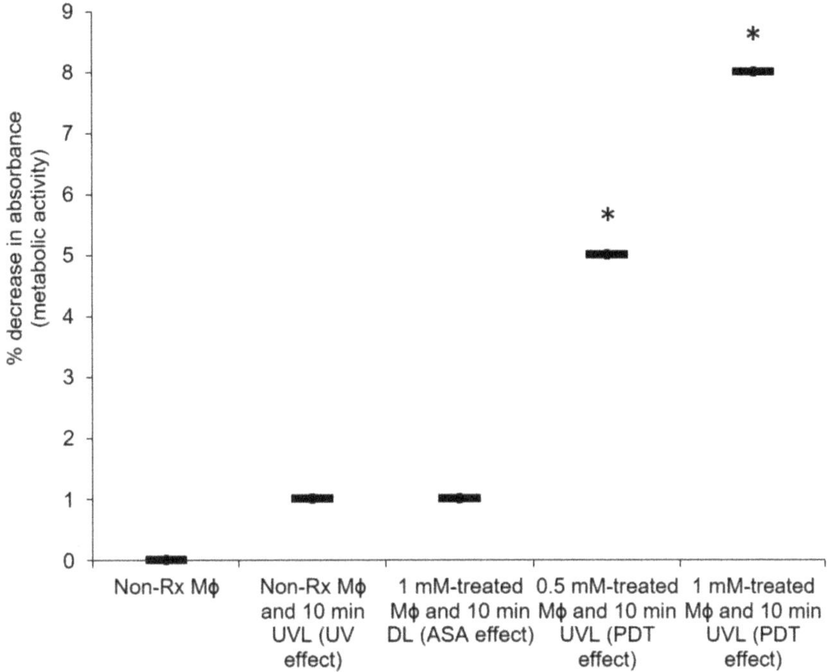

Figure 8. ASA is a suitable photosensitiser to apply against macrophages. The background signal of negative controls was used for normalisation. There was a normality of data distribution. Non-Rx, non-treated cells; UVL, ultraviolet light; DL, dark light; and Mφ, macrophages. Data points were obtained from three biological replicates for each defined experimental condition. * Significantly different from non-Rx cells at $p < 0.05$.

3. Discussion

Antimicrobial resistance has become a major contributor to mortality, and this has necessitated the quest to find alternative ways of controlling infections. One avenue that was considered in this study was the susceptibility of the fungal pathogen *C. neoformans* to PDT with ASA. We showed that when either ASA or UVL is used alone, they are unable to kill a significant number of cryptococcal cells. This speaks to the principle of PDT, wherein a photosensitiser is only effective when activated by a specific light.

The idea that *Cryptococcus* would be susceptible to PDT is not surprising, based on the theory by Kock and co-workers that organisms with a strict aerobic metabolism (such as *Cryptococcus*) are more susceptible to death induced by oxidative damage as a result of impairment to mitochondrial respiration than those that can also produce energy through an alternative anaerobic glycolytic fermentative pathway in which mitochondria are not involved [46]. The above theory was shown to be true when tested on additional strains belonging to the *C. neoformans* species complex (Figure S1).

While this technology has mainly been used in the treatment of cancers, there is a body of work that has shown a successful application against mycotic agents. To illustrate this point, in 2010, Mang and co-workers demonstrated the fungicidal effect of PDT against *Candida* species [47]. From their observations, 25 μg/mL of porfimer sodium killed more than 90% of *Candida* strains following illumination [45]. This is important because these fungi are often problematic to manage in AIDS patients, because they are resistant to fluconazole and amphotericin B [48,49]. There may be hope on the horizon; a new study by Nagy et al. [50] recently demonstrated the in vitro antimicrobial success of using 1-amino-5-isocyanonaphthalene [ICAN] against fungal pathogens [50]. Based on the aromatic fluorophore nature of ICAN, it would be prudent to also assess its photosensitising quality. PDT has also been shown to be effective against some *Cryptococcus* species. In their study, Fuchs et al. [51] demonstrated the susceptibility of *C. neoformans* towards a polycationic conjugate of polyethyleneimine and chlorin (as the photosensitiser). In 2011, Soares et al. [52] also reported the efficacy of PDT against a number of *C. gattii* isolates with different susceptibility profiles, suggesting that PDT could be an alternative strategy to inhibit cryptococcal cells.

Importantly, we further showed that harmful radical species possibly killed cryptococcal cells. Based on our results, it is conceivable that the generated radicals can then target the cell wall of the cryptococcal cells, as shown in Figures 5 and 6, and in summary, the mode of action is depicted in Figure 9. It important to note that in a biological system (infected host organism), it is possible that SA's ROS-driven photodynamic effect may be counteracted by SA's powerful hydroxyl radical-scavenging capacity [53]. To address the latter, animal studies should be considered.

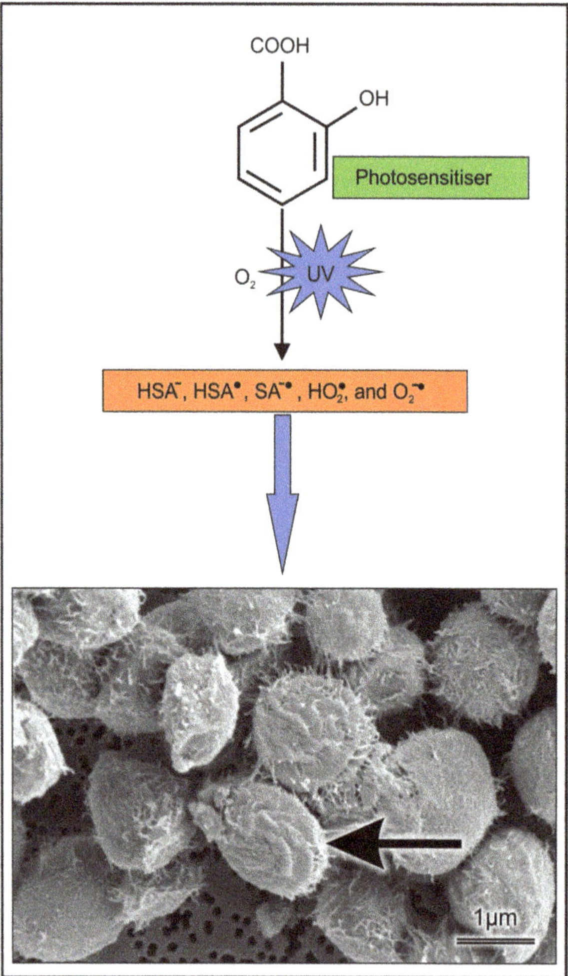

Figure 9. Deleterious effect of accumulated radicals impairs the ultrastructure and integrity of *C. neoformans* LMPE 046 cell walls.

4. Materials and Methods

4.1. Materials

Yeast extract, malt extract, peptone, glucose, agar (Merck, Johannesburg, South Africa), phosphate buffer solution (PBS) (Sigma-Aldrich, Johannesburg, South Africa), RPMI-1640 medium, foetal bovine serum (Biochrom, Berlin, Germany), penicillin, streptomycin (Sigma-Aldrich, St. Louis, MO, USA), L-glutamine (Sigma-Aldrich, Johannesburg, South Africa), trypan blue stain, sterile disposable 96-well flat-bottom microtitre plate (Greiner Bio-One, Frickenhausen, Germany), acetylsalicylic acid ($C_9H_8O_4$, MW 180.158 g/mol) (Sigma-Aldrich, Johannesburg, South Africa), absolute ethanol (Merck, South Africa), propidium iodide (PI) (Life Technologies, Carlsbad, CA, USA), sodium-phosphate-buffered 3% glutardialdehyde (Merck, Johannesburg, South Africa), sodium-phosphate-buffered 3% osmium tetroxide (Merck, Johannesburg, South Africa), Beadbug microtube homogeniser (Lasec, Johannesburg, South Africa), 1.0% (*w/v*) agarose gel, Tris-acetate-EDTA (TAE) buffer, ethidium bromide (Merck, Johannesburg, South Africa), DNAase (Qiagen, Hilden, Germany), dark reader transilluminator (Clare Chemical Research, Dolores, CO, USA), 5,5′,6,6′-tetrachloro-1,1′,3,3′-tetraethylbenzimidazolylcarbocyanine iodide (JC-1) (Life Technologies, Carlsbad, CA, USA), 2′,7-dichlorofluorescein diacetate (DCFHDA) (Sigma-Aldrich, Johannesburg, South Africa), 2,3-bis (2-methoxy-4-nitro-5-sulfophenyl)-5-[(phenylamino)carbonyl]-2H-tetrazolium hydroxide (XTT; Sigma-Aldrich, Johannesburg, South Africa), 1 mM of menadione (Sigma-Aldrich, Johannesburg, South Africa), germicidal ultraviolet light (UVL) lamp (ESCO, Johannesburg, South Africa), Airstream®Class II Biological Safety Cabinet (ESCO, Johannesburg, South Africa), SEM coating system (Bio-Rad Microscience Division, Johannesburg, South Africa), Shimadzu Superscan SSX 550 SEM (Shimadzu, Tokyo, Japan) and Fluoroskan Ascent FL microplate reader (Thermo Scientific, Waltham, MA, USA).

4.2. Cells, Cultivation and Standardisation

The fungal strain *C. neoformans* LMPE 046 was used in this study. The strain was obtained from a patient at the Universitas Academic Hospital, Bloemfontein, South Africa. The strain was grown on yeast–malt-extract (YM) agar (3 mg/mL yeast extract, 3 mg/mL malt extract, 5 mg/mL peptone, 10 mg/mL glucose, 16 mg/mL agar) at 37 °C for 48 h. Before use, 10 mL of phosphate buffer solution (PBS) was used to prepare standardised inocula with a final concentration of between 0.5×10^5 and 2.5×10^5 CFU/mL, according to the protocol of the European Committee on Antimicrobial Susceptibility Testing [54].

The RAW 264.7 macrophage cell line (originally obtained from ATCC) was cultivated in RPMI-1640 medium that was supplemented with 10% foetal bovine serum, 20 U/mL penicillin, 20 mg/mL streptomycin and 2 mM L-glutamine at 37 °C and 5% CO_2 until 80% confluence was achieved. The viability of the cells was determined to be 90% using a trypan blue stain. This stain is excluded by viable cells and accumulates inside dead cells. Next, macrophages were standardised using a haemocytometer to reach a final cell concentration of 1×10^6 cells/mL in fresh RPMI-1640 media. A 100 μL suspension of macrophages was seeded into wells of a sterile, disposable 96-well flat-bottom microtitre plate and left overnight in a humidified 5% CO_2 incubator at 37 °C.

4.3. UV Radiation

Photosensitiser: Acetylsalicylic acid ($C_9H_8O_4$, MW 180.158 g/mol) was obtained as a standard powder from Sigma-Aldrich. ASA stock solution was prepared in absolute ethanol to yield a stock solution of 10 mM. The drug was further diluted in RPMI-1640 media in order to reach final concentrations of 0.5 mM and 1 mM. The final amount of ethanol in RPMI-1640 media never exceeded 1%. The UV/Vis absorption of ASA was determined to be 250 nm (data not shown).

Light source: A germicidal ultraviolet light (UVL) lamp that was fitted inside an Airstream®Class II Biological Safety Cabinet was used as the light source. The lamp is reported to have a nominal power of 30 watts, and thus provided a germicidal UV intensity (irradiance or radiation output) of approximately 125 μW/cm^2 one metre from the lamp [55]. In the current study, the distance between the lamp and the position of the experimental microtiter plates was 20 cm.

4.4. Preparation of Cells for Experimental Assays

A 100 μL suspension of the standardised cryptococcal inoculum (0.5×10^5 and 2.5×10^5 CFU/mL) was added to sterile 96-well flat-bottom microtitre plate wells. Thereafter, 100 μL of the photosensitiser at twice the desired final concentrations was added. A number of experimental conditions were set up at room temperature and these were: (1) non-Rx cells (no ASA and incubated in dark light (DL) for 10 min): (2) non-Rx cells that were exposed to ultraviolet light (UVL) for 10 min (UV effect); (3) cells treated with 1 mM of ASA (incubated in DL for 10 min; ASA effect); (4) cells treated with 0.5 mM of ASA and exposed to UVL 10 min (PDT effect); and (5) cells treated with 1 mM of ASA and exposed to UVL 10 min (PDT effect). In addition, negative controls were included, and where appropriate, the background readings they produced were subtracted for normalisation. These controls were: 1 mM ASA (with no cells) and 1 mM ASA with 10 min UVL (with no cells).

4.5. Survival Assay of Cryptococcal Cells

Following the handling of cryptococcal cells as per the above-mentioned experimental conditions, they were plated out. Specifically, the contents of wells were aspirated and transferred to corresponding 1.5 mL plastic tubes. A 1:10 serial dilution was made using sterile, distilled water. A 25 μL suspension of the diluted sample was dispensed to the centre of a corresponding YM agar plate. The suspension was then spread to create a lawn. The plates were then incubated for 48 h at 37 °C. At the end of the incubation period, colonies were counted to determine the survival rate of cells [56].

4.6. The Effects of PDT with ASA on the Cell Membrane and Cell Wall

It was expected that the ASA photolytic products would target cellular structures; therefore, determination of the effect of the resultant radicals on membrane integrity was then sought. To this end, the integrity of the cell membrane was assessed by measuring the amount of propidium iodide (PI) accumulated through damaged membranes. For the PI assay, the cells obtained from the following conditions: (1) non-Rx cells; (2) non-Rx cells and 10 min UVL (UV effect); (3) 1 mM-treated cells and 10 min DL (ASA effect); (4) 0.5 mM-treated cells and 10 min UVL (PDT effect); and (5) 1 mM-treated cells and 10 min UVL (PDT effect), were used in this experiment. The cells were then washed twice with PBS, and 99 μL of cells (from each experimental condition) were reacted with 1 μL of PI (1 mg/mL) in a black 96-well flat-bottom microtiter plate. The plate was immediately incubated in the dark for 30 min at 37 °C. The fluorescence was measured at 485 nm excitation and the corresponding emission at 538 nm using the Fluoroskan Ascent FL microplate reader [20].

4.6.1. Effect of PDT on Cellular Outer Ultrastructure

Cells used for scanning electron microscopy (SEM) were obtained from: (1) non-Rx cells (no ASA and incubated in dark light (DL) for 10 min); (2) cells treated with 1 mM of ASA (incubated in DL for 10 min (ASA effect)); (3) non-Rx cells that were exposed to UVL for 10 min (UV effect); and (4) cells treated with 1 mM of ASA and exposed to UVL for 10 min (PDT effect).

All experimental cells were prepared for SEM according to the method of van Wyk and Wingfield [57]. Sodium-phosphate-buffered 3% glutardialdehyde and sodium-phosphate-buffered 3% osmium tetroxide were used to fix the cells before they were dehydrated in a graded series of ethanol solution. Next, the cells were critically point-dried, mounted, and coated with gold using an SEM coating system. Examination of cells was performed using a Shimadzu Superscan SSX 550 SEM. A number of images were taken from different positions on the stub. Furthermore, 100 randomly selected cells were considered, and their cell diameters were measured [57].

4.6.2. Effect of PDT on the Expression of CAP64 Gene

The pellets obtained from the cells washed with PBS and number of cells were determined ($5 \times 10^5 - 1 \times 10^7$ CFU/mL). The cells were vortexed and centrifuged at $500 \times g$ for 5 min. Next, cells were aspirated, and the supernatant was discarded. A 700 µL GLT lysis buffer was added to the cells, which were mixed with beads. The cells were disrupted at 4 °C for 1 h using a Beadbug microtube homogeniser. The lysate was carefully removed from the glass beads and the total RNA was extracted following the manufacturer's protocol. The nanodrop quantification to ensure the purity of the extracted total RNA was performed at wavelengths of A230, A260 and A280 nm, respectively. The RNA product was subjected to gel electrophoresis using 1.0% (*w/v*) agarose gel with 1X Tris-acetate-EDTA (TAE) buffer at 100 V for 35 min. The gel was then stained with 50 µL of ethidium bromide, and 100 bp DNA marker was used. The gel was visualised under a dark reader transilluminator. The associated DNA from the RNA preparations was removed by DNase. Next, the RNA samples were dissolved in RNase-free water. The real-time PCR reactions were performed in a total reaction of 25 µL containing 12.5 µL 2 × Rotor-Gene SYBR Green PCR Master Mix, 1.6 µL template DNA, 2.5 µL each of forward and reverse primers, 5.65 of RNase-free water, and 0.25 µL of Rotor-Gene RT Mix. The primers used for real-time PCR are listed in Table 1 and were designed for DNA sequences of actin and CAP genes. The RT-qPCR results of the *CAP64* gene were normalised to actin by carrying out the PCR under the same running conditions with an equal concentration of total RNA. Prior to PCR, reverse transcription was carried out. Reactions were incubated at 55 °C for 10 min. After reverse transcription, PCR was carried out and required an initial incubation step at 95 °C for 5 min. Two-step cycling was performed with 50 cycles. Each step comprised 2 steps: 95 °C for 5 s (denaturation step), and 60 °C for 10 s (annealing/extension step). The amplification was performed using a Rotor-Gene cycler. Melting curve analysis was performed after PCR completion to check the specificity of the reaction [58].

Table 1. Primer sequences.

Primer	Accession No.	Sequence (5′–3′)	Band Size
Actin	U10867	F-TGTACAATGGTATTGCCGACC R-CTGGTCCCTCAATCGTCCAC	200 bp
CAP64	L40026	F-GCCACGCCCACATTGACT R-ACTCTTCCTCGATCAATGTC	200 bp

4.7. Effect of PDT with ASA on Cryptococcal Mitochondrial Membrane Potential (ΔΨM) and ROS Accumulation

Following the preparation of all cells (as mentioned in Section 4.4.), they were aspirated and dispensed into wells of a sterile black 96-well flat-bottom microtiter plate. The cells were then stained with 10 µL of the dye 5,5′,6,6′-tetrachloro-1,1′,3,3′-tetraethylbenzimidazolylcarbocyanine iodide (JC-1) dye, according to the manufacturer's instructions. The plate was then incubated at 37 °C for 15 min. A Fluoroskan Ascent FL microplate reader was used to measure J aggregates (healthy cells) at excitation/emission wavelengths of 540/570 nm, while the monomeric forms (unhealthy cells) were measured at excitation/emission wavelengths of 485/535 nm [20].

In a separate experiment, ROS accumulation was also measured in a sterile black 96-well flat-bottom microtiter plate. A 90 µL suspension of the cells (treated as mentioned in Section 2.1.) was stained with 10 µL of the fluorescent 2′,7-dichlorofluorescein diacetate (DCFHDA; 1 mg/mL for 30 min in the dark at room temperature. Fluorescence was measured at excitation/emission wavelengths of 485/535 nm using a Fluoroskan Ascent FL microplate reader [20].

4.8. Effect of PDT Using ASA on the Health of Macrophages

The seeded macrophages (representing the standardised 1×10^6 cells/mL) that were suspended in 100 µL of the media in a microtiter plate) were grouped into the following experimental conditions: (1) non-Rx cells; (2) non-Rx cells and UVL (UV effect); (3) 1 mM-treated cells and DL (ASA effect); (4) 0.5 mM-treated cells and UVL (PDT effect); and (5) 1 mM-treated cells and UVL (PDT effect). The plates were kept at room temperature and exposed to light (UV or dark) for 10 min. Following this, the media containing macrophages were aspirated and the cells were washed three times with PBS to remove excess ASA. To measure the metabolic activity of these cells, 54 µL of 2,3-bis (2-methoxy-4-nitro-5-sulfophenyl)-5-[(phenylamino)carbonyl]-2H-tetrazolium hydroxide (XTT; Sigma-Aldrich, South Africa) with 1 mM of menadione (Sigma-Aldrich, South Africa) were reacted with the cells. After three hours of incubation in the dark in a 5% CO_2 incubator, which allowed the initiation of the tetrazolium reaction, the optical density (OD) was measured at 492 nm using a Biochrom spectrophotometer [59].

4.9. Statistical Analysis

For each study, three independent experiments were performed. No technical repeats were included for each independent experiment. Microsoft Excel was used to calculate mean values and the standard deviation of the means. The same programme was used to calculate the paired t-test to determine the statistical significance of non-Rx cells' data, i.e., no ASA and incubated in dark light (DL) for 10 min, and the respective experimental condition being compared. For proper interpretation of the data, box plots were used, as recommended by Weissgerber et al., 2015 [60].

5. Conclusions

This old technology of light therapy has provided us with some evidence for combating cryptococcal cells under controlled laboratory conditions. It will be interesting to see if similar results can be obtained in a diseased animal with a cryptococcal skin infection; additionally, to determine if this treatment would be suitable for controlling facultative organisms, which can switch to anaerobic glycolytic fermentative pathways. Given the effect of PDT with ASA on the health of macrophages (which may be manipulated to disseminate cryptococcal cells), it would be prudent to now assess the impact of PDT with ASA on macrophage phagocytosis and to elucidate the molecular changes that may be enhanced to resolve internalised cells.

The application of this form of treatment against internal cryptococcal infections may, at the moment, be limited by the lack of appropriate technology to deliver it to affected organs. However, hand-held devices can be used to resolve cutaneous cryptococcal infections. Care should also be taken to avoid continuous photoreactions, given the harmful effects of UV light on the skin, particularly when subjects have concluded PDT administration and are exposed to the sun's radiation. To this end, it is equally important to show the benefit of this treatment when administered to skin epithelial cells. Such a study could also include a molecule that could modify how skin epithelial cells receive UV radiation. Additionally, there are already some topical creams that have salicylates as an ingredient available on the market [27,49]. The latter demonstrates that there is scope to consider administering ASA as a photosensitiser to the skin.

Supplementary Materials: The following is available online at https://www.mdpi.com/article/10.3390/ph14050404/s1, **Figure S1:** Effect of PDT with ASA on the survival of *C. neoformans* H99 (**A**), *C. gattii* LMPE 053 (**B**) and *C. gattii* R265 (**C**). Non-Rx, non-treated cells; UVL, ultraviolet light; DL, dark light; CFU, colony-forming unit.

Author Contributions: Conceptualization, A.O.O. and O.M.S.; methodology, A.O.O., N.M., O.S.F. and O.M.S.; formal analysis, A.O.O., N.M., O.S.F., M.M.N., C.H.P. and O.M.S.; investigation, N.M.; resources, O.M.S.; writing—original draft preparation, N.M.; writing—review and editing, A.O.O., N.M., O.S.F., M.M.N., C.H.P. and O.M.S.; supervision, A.O.O., O.S.F., C.H.P. and O.M.S.; project administration, O.M.S.; funding acquisition, O.M.S. All authors have read and agreed to the published version of the manuscript.

Funding: This research was funded by the NATIONAL REASERCH FOUNDATION OF SOUTH AFRICA, grant number UID 114321", and the University of the Free State.

Institutional Review Board Statement: Not applicable.

Informed Consent Statement: Not applicable.

Data Availability Statement: Raw data is available from the corresponding author upon request.

Acknowledgments: Parts of this manuscript are contained in the MSc dissertation of N.M. The authors are grateful for the services and assistance offered by the following colleagues: P.W.J. van Wyk and H. Globler, for SEM work.

Conflicts of Interest: The authors declare no conflict of interest. The funders had no role in the design of the study; in the collection, analyses, or interpretation of data; in the writing of the manuscript, or in the decision to publish the results.

References

1. Maziarz, E.K.; Perfect, J.R. Cryptococcosis. *Infect. Dis. Clin. N. Am.* **2016**, *30*, 179–206. [CrossRef] [PubMed]
2. Oladele, R.O.; Bongomin, F.; Gago, S.D.; Denning, W. HIV-associated cryptococcal disease in resource-limited settings: A case for "prevention is better than cure"? *J. Fungi* **2017**, *3*, e67. [CrossRef]
3. Rajasingham, R.; Smith, R.M.; Park, B.J.; Jarvis, J.N.; Govender, N.P.; Chiller, T.M.; Denning, D.W.; Loyse, A.; Boulware, D.R. Global burden of disease of HIV-associated cryptococcal meningitis: An updated analysis. *Lancet Infect. Dis.* **2017**, *17*, 873–881. [CrossRef]
4. Ford, N.; Shubber, Z.; Jarvis, J.N.; Chiller, T.; Greene, G.; Migone, C.; Vitoria, M.; Doherty, M.; Meintjes, G. CD4 cell count threshold for cryptococcal antigen screening of HIV-infected individuals: A systematic review and meta-analysis. *Clin. Infect. Dis.* **2018**, *66*, S152–S159. [CrossRef]
5. Bongomin, F.; Atikoro, L. "Recurrence of cryptococcal meningitis and the hidden role of patient education and social support," case reports in Neurological Medicine. *Hindawi* **2018**, 1–4. [CrossRef]
6. Merry, M.; Boulware, D.R. Cryptococcal meningitis treatment strategies affected by the explosive cost of Flucytosine in the United States: A cost-effectiveness analysis. *Clin. Infect. Dis.* **2016**, *62*, 1564–1568. [CrossRef] [PubMed]
7. Lofgren, S.; Abassi, M.; Rhein, J.; Boulware, D.R. Recent advances in AIDS-related cryptococcal meningitis treatment with an emphasis on resource limited settings. *Expert Rev. Anti. Infect. Ther.* **2017**, *15*, 331–340. [CrossRef]
8. Neuville, S.; Dromer, F.; Morin, O.; Dupont, B.; Ronin, O.; Lortholary, O. French cryptococcosis study group, primary cutaneous cryptococcosis: A distinct clinical entity. *Clin. Infect. Dis.* **2003**, *36*, 337–347. [CrossRef]
9. Leopold, W.C.M.; Hole, C.R.; Wozniak, K.L.; Wormley, J.F.L. *Cryptococcus* and phagocytes: Complex interactions that influence disease outcome. *Front. Microbiol.* **2016**, *7*, 105. [CrossRef]
10. Nyazika, T.K.; Tatuene, J.K.; Kenfak-Foguena, A.; Verweij, P.E.; Meis, J.F.; Robertson, V.J.; Hagen, F. Epidemiology and aetiologies of cryptococcal meningitis in Africa, 1950–2017: Protocol for a systematic review. *BMJ Open.* **2018**, *8*, e020654. [CrossRef] [PubMed]
11. Srivastava, G.N.; Tilak, R.; Yadav, J.; Bansal, M. Cutaneous *Cryptococcus*: Marker for disseminated infection. *BMJ Case Rep.* **2015**, *2015*, bcr2015210898. [CrossRef] [PubMed]
12. Wang, J.; Bartelt, L.; Yu, D.; Joshi, A.; Weinbaum, B.; Pierson, T.; Patrizio, M.; Warren, C.A.; Hughes, M.A.; Donowitz, G. Primary cutaneous cryptococcosis treated with debridement and fluconazole monotherapy in an immunosuppressed patient: A case report and review of the literature. *Case Rep. Infect. Dis.* **2015**, *2015*, 131356. [CrossRef] [PubMed]
13. Saag, M.S.; Graybill, R.J.; Larsen, R.A.; Pappas, P.G.; Perfect, J.R.; Powderly, W.G.; Sobel, J.D.; Dismukes, W.E. Practice guidelines for the management of cryptococcal disease. Infectious Diseases Society of America. *Clin. Infect. Dis.* **2000**, *30*, 710–718. [CrossRef]
14. Perfect, J.R.; Dismukes, W.E.; Dromer, F.; Goldman, D.L.; Graybill, J.R.; Hamill, R.J.; Harrison, T.S.; Larsen, R.A.; Lortholary, O.; Nguyen, M.H.; et al. Clinical practice guidelines for the management of cryptococcal disease: 2010 update by the Infectious Diseases Society of America. *Clin. Infect. Dis.* **2010**, *50*, 291–322. [CrossRef]
15. Ross, A.J.; Ndayishimiye, E. A review of the management and outcome of patients admitted with cryptococcal meningitis at a regional hospital in KwaZulu-Natal province. *S. Afr. Fam. Pract.* **2019**, *61*, 159–164. [CrossRef]
16. Govender, N.P.; Dlamini, S. Management of HIV-associated cryptococcal disease in South Africa. *S. Afr. Med. J.* **2014**, *104*, 869. [CrossRef]
17. Govender, N.P.; Meintjes, G.; Bicanic, T.; Dawood, H.; Harrison, T.S.; Jarvis, J.N.; Karstaedt, A.S.; Maartens, G.; McCarthy, K.M.; Rabie, H.; et al. Guideline for the prevention, diagnosis and management of cryptococcal meningitis among HIV-infected persons: 2013 update. *S. Afr. J. HIV Med.* **2013**, *14*, 76–86. [CrossRef]
18. Abassi, M.; Boulware, D.R.; Rhein, J. Cryptococcal meningitis: Diagnosis and management update. *Curr. Trop. Med. Rep.* **2015**, *2*, 90–99. [CrossRef] [PubMed]

19. Tenforde, M.W.; Wake, R.; Leeme, T.; Jarvis, J.N. HIV-Associated Cryptococcal meningitis: Bridging the gap between developed and resource-limited settings. *Curr. Clin. Microbiol Rep.* **2016**, *3*, 92–102. [CrossRef]
20. Ogundeji, A.O.; Pohl, C.H.; Sebolai, O.M. The repurposing of anti-psychotic drugs, Quetiapine and Olanzapine, as anti-*Cryptococcus* drugs. *Front. Microbiol.* **2017**, *8*, 815. [CrossRef]
21. Morawski, B.; Boulware, D.; Nalintya, E.; Kiragga, A.; Kazooza, F.; Rajasingham, R.; Benjamin, J.; Park, B.J.; Manabe, Y.C.; Kaplan, J.E.; et al. Pre-ART cryptococcal antigen titer associated with preemptive fluconazole failure. In Proceedings of the Conference on Retroviruses and Opportunistic Infections (CROI), Boston, MA, USA, 22–25 February 2016; Volume 24, abstract 159.
22. Miró-Canturri, A.; Ayerbe-Algaba, R.; Smani, Y. Drug repurposing for the treatment of bacterial and fungal infections. *Front. Microbiol.* **2019**, *10*, 41. [CrossRef]
23. Truong, M.; Monahan, L.G.; Carter, D.A.; Charles, I.G. Repurposing drugs to fast-track therapeutic agents for the treatment of cryptococcosis. *PeerJ* **2018**, *6*, e4761. [CrossRef]
24. Sebolai, O.M.; Ogundeji, A.O. New antifungal discovery from existing chemical compound collections. In *Antifungals: From Genomics to Resistance and the Development of Novel Agents*; Coste, A.T., Vandeputte, P., Eds.; Caister Academic Press: Norfolk, UK, 2015; pp. 143–158. [CrossRef]
25. Ogundeji, A.O.; Pohl, C.H.; Sebolai, O.M. Repurposing of aspirin and ibuprofen as candidate anti-*Cryptococcus* drugs. *Antimicrob. Agents Chemother.* **2016**, *60*, 4799–4808. [CrossRef] [PubMed]
26. Yeomans, N.D. Aspirin: Old drug, new uses and challenges. *J. Gastroenterol. Hepatol.* **2011**, *26*, 426–431. [CrossRef] [PubMed]
27. Bubna, A.K. Aspirin in dermatology: Revisited. *Indian Dermatol. Online J.* **2015**, *6*, 428–435. [CrossRef]
28. World Health Organization. *Model List of Essential Medicines*; 21st List; World Health Organization: Geneva, Switzerland, 2019.
29. Management Sciences for Health. *International Drug Price Indicator Guide*, 2014th ed.; Management Sciences for Health: Medford, MA, USA, 2015.
30. Stapleton, M.; Rhodes, L. Photosensitizers for photodynamic therapy of cutaneous disease. *J. Dermatol. Treat.* **2003**, *14*, 107–112. [CrossRef] [PubMed]
31. Pozdnyakov, I.P.; Sosedova, Y.A.; Plyusnin, V.F.; Grivin, V.P.; Bazhin, N.M. Photochemistry of salicylate anion in aqueous solution. *Russ. Chem. Bull.* **2007**, *5*, 1318–1324. [CrossRef]
32. Pozdnyakov, I.P.; Plyusnin, V.F.; Grivin, V.P.; Vorobyev, D.Y.; Kruppa, A.I.; Lemmetyinen, H. Photochemistry of sulfosalicylic acid in aqueous solutions. *J. Photochem. Photobiol. A Chem.* **2004**, *162*, 153–162. [CrossRef]
33. Pozdnyakov, I.P.; Plyusnin, V.F.; Grivin, V.P.; Vorobyev, D.Y.; Bazhin, N.M.; Vauthey, E. Photochemistry of Fe(III) and sulfosalicylic acid aqueous solutions. *J. Photochem. Photobiol. A Chem.* **2006**, *182*, 75–81. [CrossRef]
34. Norman, C.; Howell, K.A.; Millar, A.H.; Whelan, J.M.; Day, D.A. Salicylic acid is an uncoupler and inhibitor of mitochondrial electron transport. *Plant. Physiol.* **2004**, *134*, 492–501. [CrossRef] [PubMed]
35. Levy, G. Pharmacokinetics of aspirin in man. *J. Investig. Dermatol.* **1976**, *67*, 667–668. [CrossRef]
36. Kwon-Chung, K.J. Filobasidiella Kwon-Chung (1975). In *The Yeast, a Taxonomic Study*, 5th ed.; Kurtzman, C.P., Fell, J.W., Boekhout, T., Eds.; Elsevier Sciences BV: Amsterdam, The Netherlands, 2011; pp. 1443–1456.
37. Chang, Y.C.; Penoyer, L.A.; Kwon-Chung, K.J. The second capsule gene of *Cryptococcus neoformans*, *CAP64*, is essential for virulence. *Infect. Immun.* **1996**, *64*, 1977–1983. [CrossRef]
38. Okabayashi, K.; Hasegawa, A.; Watanabe, T. Microreview: Capsule-associated genes of *Cryptococcus neoformans*. *Mycopathologia* **2007**, *163*, 1–8. [CrossRef]
39. Casadevall, A.; Coelho, C.; Cordero, R.J.B.; Dragotakes, Q.; Jung, E.; Vij, R. The capsule of *Cryptococcus neoformans*. *Virulence* **2019**, *10*, 822–831. [CrossRef]
40. Voelz, K.; May, R.C. Cryptococcal interactions with the host immune system. *Eukaryot. Cell* **2010**, *9*, 835–846. [CrossRef]
41. Aslam, S.; Lan, X.R.; Zhang, B.W.; Chen, Z.L.; Wang, L.; Niu, D.K. Aerobic prokaryotes do not have higher GC contents than anaerobic prokaryotes, but obligate aerobic prokaryotes have. *BMC Evol. Biol.* **2019**, *19*, 35. [CrossRef]
42. Ingavale, S.S.; Chang, Y.C.; Lee, H.; McClelland, C.M.; Leong, M.L.; Kwon-Chung, K.J. Importance of mitochondria in survival of *Cryptococcus neoformans* under low oxygen conditions and tolerance to cobalt chloride. *PLoS Pathog.* **2008**, *4*, e1000155. [CrossRef] [PubMed]
43. Sharma, P.; Jha, A.B.; Dubey, R.S.; Pessarakli, M. Reactive oxygen species, oxidative damage, and antioxidative defense mechanism in plants under stressful conditions. *J. Bot.* **2012**, 1–26. [CrossRef]
44. Houstek, J.; Pícková, A.; Vojtísková, A.; Mrácek, T.; Pecina, P.; Jesina, P. Mitochondrial diseases and genetic defects of ATP synthase. *Biochim. Biophys. Acta* **2006**, *1757*, 1400–1405. [CrossRef] [PubMed]
45. Ma, H.; Croudace, J.E.; Lammas, D.A.; May, R.C. Expulsion of live pathogenic yeast by macrophages. *Curr. Biol.* **2006**, *16*, 2156–2160. [CrossRef] [PubMed]
46. Kock, J.L.F.; Sebolai, O.M.; Pohl, C.H.; van Wyk, P.W.J.; Lodolo, E.J. Oxylipin studies expose aspirin as antifungal. *FEMS Yeast Res.* **2007**, *7*, 1207–1217. [CrossRef]
47. Mang, T.S.; Mikulski, L.; Hall, R.E. Photodynamic inactivation of normal and antifungal resistant *Candida* species. *Photodiagnosis Photodyn. Ther.* **2010**, *7*, 98–105. [CrossRef]
48. Eliopoulos, G.M.; Perea, S.; Patterson, T.F. Antifungal resistance in pathogenic fungi. *Clin. Infect. Dis.* **2002**, *35*, 1073–1080. [CrossRef]

49. Katiraee, F.; Teifoori, F.; Soltani, M. Emergence of azole-resistant *Candida* species in AIDS patients with oropharyngeal candidiasis in Iran. *Curr. Med. Mycol.* **2015**, *1*, 11–16. [CrossRef] [PubMed]

50. Nagy, M.; Szemán-Nagy, G.; Kiss, A.; Nagy, Z.L.; Tálas, L.; Rácz, D.; Majoros, L.; Tóth, Z.; Szigeti, Z.M.; Pócsi, I.; et al. Antifungal activity of an original amino-isocyanonaphthalene (ICAN) compound family: Promising broad spectrum antifungals. *Molecules* **2020**, *25*, 903. [CrossRef]

51. Fuchs, B.B.; Tegos, G.P.; Hamblin, M.R.; Mylonakis, E. Susceptibility of *Cryptococcus neoformans* to photodynamic inactivation is associated with cell wall integrity. *Antimicrob. Agents Chemother.* **2007**, *51*, 2929–2936. [CrossRef] [PubMed]

52. Soares, B.M.; Alves, O.A.; Ferreira, M.V.; Amorim, J.C.; Sousa, G.R.; Silveira, L.B.; Prates, R.A.; Avila, T.V.; Baltazar Lde, M.; de Souza Dda, G.; et al. *Cryptococcus gattii*: In vitro susceptibility to photodynamic inactivation. *Photochem. Photobiol.* **2011**, *87*, 357–364. [CrossRef]

53. Von Sonntag, C. *Free-Radical-Induced DNA Damage and Its Repair, a Chemical Perspective*; Springer-Verlag: Berlin/Heidelberg, Germany, 2006. [CrossRef]

54. Arendrup, M.C.; Guinea, J.; Cuenca-Estrella, M.; Meletiadis, J.; Mouton, J.W.; Lagrou, K.; Howard, S.J. *The Subcommittee on Antifungal Susceptibility Testing (AFST) of the ESCMID European Committee for Antimicrobial Susceptibility Testing (EUCAST). EUCAST Definitive Document E. Def 7.3: Method for the Determination of Broth Dilution Minimum Inhibitory Concentrations of Antifungal Agents for Yeasts*; EUCAST: Copenhagen, Denmark, 2015. Available online: http://www.eucast.org/fileadmin/src/media/PDFs/EUCAST_files/AFST/Files/EUCAST_E_Def_7_3_Yeast_testing_definitive.pdf (accessed on 1 December 2015).

55. Harrington, B.; Valigosky, M. Monitoring ultraviolet lamps in biological safety cabinets with cultures of standard bacterial strains on TSA blood agar. *Lab. Med.* **2007**, *38*, 165–168. [CrossRef]

56. Madu, U.L.; Ogundeji, A.O.; Pohl, C.H.; Albertyn, J.; Sebolai, O.M. Elucidation of the role of 3-hydroxy fatty acids in *Cryptococcus*-amoeba interactions. *Front. Microbiol.* **2017**, *8*, 765. [CrossRef]

57. Van Wyk, P.W.J.; Wingfield, M.J. Ascospore ultrastructure and development in *Ophiostoma cucullatum*. *Mycologia* **1991**, *83*, 698–707. [CrossRef]

58. Rotor-Gene SYBR Green Handbook. Rotor-Gene SYBR Green RT-PCR Kit and Additional Protocols, US. 2014. Available online: http://www.qiagen.com (accessed on 1 January 2014).

59. Madu, U.L.; Ogundeji, A.O.; Mochochoko, B.M.; Pohl, C.H. Cryptococcal 3-hydroxy fatty acids protect cells against amoebal phagocytosis. *Front. Microbiol.* **2015**, *6*, 1–12. [CrossRef] [PubMed]

60. Weissgerber, T.L.; Milic, N.M.; Winham, S.J.; Garovic, V.D. Beyond bar and line graphs: Time for a new data presentation paradigm. *PLoS Biol.* **2015**, *13*, e1002128. [CrossRef] [PubMed]

pharmaceuticals

MDPI

Review

Retinoids in Fungal Infections: From Bench to Bedside

Terenzio Cosio [1], Roberta Gaziano [2], Guendalina Zuccari [3], Gaetana Costanza [4], Sandro Grelli [4], Paolo Di Francesco [2], Luca Bianchi [1] and Elena Campione [1,*]

1   Dermatology Unit, Department of Systems Medicine, University of Rome Tor Vergata, Via Montpellier 1, 00133 Rome, Italy; terenziocosio@gmail.com (T.C.); luca.bianchi@uniroma2.it (L.B.)
2   Department of Experimental Medicine, University of Rome Tor Vergata, 00133 Rome, Italy; roberta.gaziano@uniroma2.it (R.G.); difra@uniroma2.it (P.D.F.)
3   Department of Pharmacy, University of Genoa, Viale Cembrano, 16148 Genoa, Italy; zuccari@difar.unige.it
4   Department of Experimental Medicine, University of Rome Tor Vergata, Via Montpellier 1, 00133 Rome, Italy; costanza@uniroma2.it (G.C.); grelli@med.uniroma2.it (S.G.)
*   Correspondence: elena.campione@uniroma2.it

Abstract: Retinoids—a class of chemical compounds derived from vitamin A or chemically related to it—are used especially in dermatology, oncohematology and infectious diseases. It has been shown that retinoids—from their first generation—exert a potent antimicrobial activity against a wide range of pathogens, including bacteria, fungi and viruses. In this review, we summarize current evidence on retinoids' efficacy as antifungal agents. Studies were identified by searching electronic databases (MEDLINE, EMBASE, PubMed, Cochrane, Trials.gov) and reference lists of respective articles from 1946 to today. Only articles published in the English language were included. A total of thirty-nine articles were found according to the criteria. In this regard, to date, In vitro and In vivo studies have demonstrated the efficacy of retinoids against a broad-spectrum of human opportunistic fungal pathogens, including yeast fungi that normally colonize the skin and mucosal surfaces of humans such as *Candida* spp., *Rhodotorula mucilaginosa* and *Malassezia furfur*, as well as environmental moulds such as *Aspergillus* spp., *Fonsecae monofora* and many species of dermatophytes associated with fungal infections both in humans and animals. Notwithstanding a lack of double-blind clinical trials, the efficacy, tolerability and safety profile of retinoids have been demonstrated against localized and systemic fungal infections.

Keywords: retinoids; *Candida* spp.; *Aspergillus* spp.; mycosis; onychomycosis; nanoparticles; *Malassezia* spp.; dermatophytes; microbiology; mycology; all-trans retinoic acid

Citation: Cosio, T.; Gaziano, R.; Zuccari, G.; Costanza, G.; Grelli, S.; Di Francesco, P.; Bianchi, L.; Campione, E. Retinoids in Fungal Infections: From Bench to Bedside. *Pharmaceuticals* **2021**, *14*, 962. https://doi.org/10.3390/ph14100962

Academic Editor: Fu-Gen Wu

Received: 26 August 2021
Accepted: 22 September 2021
Published: 24 September 2021

Publisher's Note: MDPI stays neutral with regard to jurisdictional claims in published maps and institutional affiliations.

1. Introduction

In the last 20 years, the incidence of invasive fungal infections (IFI) has increased significantly [1]. From a global perspective, *Candida* spp. and *Aspergillus* spp. represent the most common opportunistic fungal pathogens associated with systemic infections, especially in severely immunocompromised individuals. However, the above are not the only infections, as there are many new cases of fungal diseases linked to environmental opportunistic fungi, which were previously considered non-pathogenic for humans. Among these, *Cryptococcus neoformans* is one of the main causes of morbidity and mortality due to systemic mycosis associated with acquired immunodeficiency syndrome (AIDS), with an estimated 600,000 deaths per year. There are also a wide range of pathogens belonging to the order of Mucorales (e.g., *Rhizopus* spp.), such as ialoifomycetes (*Fusarium* and *Scedosporium* spp.) or pheoifomycetes (*Alternaria* spp. and *Cladophialophora bantiana*), which can be encountered [2]. Mycoses caused by these organisms, especially rhino-cerebral forms, are often difficult to treat and require specialist advice [3]. The high incidence of invasive and non-invasive fungal infections in recent years is due to the increased prevalence of immunocompromised subjects, especially for iatrogenic causes, such as oncological chemotherapies, corticosteroid-based therapies to cure autoimmune diseases, and subjects suffering from AIDS in countries with limited access to treatments [4]. Thanks to improved surgical management and the introduction of selective immunosuppressive drugs, there are also subjects undergoing solid organ or bone marrow transplantation, who are at high risk for the development of fungal infections in the neutropenic phase. Other predisposing conditions are represented by the increased use of intravenous devices, prolonged hospitalization in intensive care units and the administration of antibacterial therapies, which alter the normal human microbiota [4]. The Global Action Fund for Fungal Infections (GAFFI) has calculated that around the world, at least 150 patients die every hour from IFI. This means a total of 1,350,000 deaths a year. It is true to say that some neutropenic cancer patients must undergo antifungal therapy in one-third of cases when they do not respond to broad-spectrum antibiotic therapies, because of their predisposition to IFI [5]. The development of drug-resistant strains and the high toxicity of traditional antifungal drugs, especially in the case of prolonged and extensive use, have increasingly strengthened the need to find new and more effective therapeutic strategies. In particular, the search for new antifungal agents aims at identifying molecules with greater selectivity, less cytotoxicity and less chance of developing drug resistance phenomenon in fungi. There is, therefore, a need to find new effective therapies in the field of fungal infections with a safer profile for the patient and easy to access and manage clinically. Historically, vitamin A was considered as an anti-microbial agent, but the mechanisms associated with its anti-infective properties remain mostly hypothetical and are probably related to its pleiotropic effects on the immune response [6]. All-trans retinoic acid (ATRA), also known as tretinoin, is vitamin A's more potent naturally occurring derivative. It has been reported that the administration of vitamin A or its metabolite, ATRA, decreases the incidence and severity of infectious diseases, although their modulation of the immune function may also vary widely, depending on the type of infection and the immune responses involved [7,8]. The term "retinoid" concerns both natural and synthetic analogues of vitamin A. The first generation of retinoids comprises natural derivatives (retinol, tretinoin, isotretinoin, and alitretinoin), obtained by modifying the polar end group of vitamin A [9]. The second generation refers to more lipophilic monoaromatic compounds such as etretinate and acitretin, in which a benzene ring replaces the cyclohexene ring. Finally, the third generation is made up by rigid polyaromatic molecules, resulting from the cyclization of the unsaturated side chain (adapalene, bexarotene, and tazarotene). The fixed structure reduces broad interactions, thus increasing drug selectivity. Trifarotene is a recently synthetized fourth-generation retinoid, highly specific for skin RAR-γ receptors [10]. A milestone in the possible application of retinoids as antifungal agents arose after an observation of evidence-based medicine (EBM), gone unnoticed by Girmenia et al. [11] but elucidated by Campione et al. [12]. Girmenia et al. [11] analyzed the incidence and type of infections complicating the clinical course of 89 consecutive acute promyelocytic leukemia (APL) patients receiving the all-trans retinoic acid plus idarubicin (AIDA) protocol. A total of 179 febrile episodes were registered during induction and consolidation, most of them due to coagulase-negative staphylococci and viridans group streptococci, while fungal infections were only occasionally observed. This clinical observation paved the way for the possible use of retinoid in the prevention and treatment of mycosis [11]. Campione et al. [12] reported the In vitro fungistatic effect of tazarotene 0.1% against *Candida albicans* and *Candida glabrata* and the In vivo effect in the treatment of onychomycosis caused by *Trichophyton* spp. [12]. For this observation, they hypothesized if also ATRA, the active form of vitamin A, could be effective in preventing and treating fungal infections, highlighting the results obtained from Girmenia et al. [11]. Moreover, Campione et al. [13] demonstrated the efficacy of ATRA In vitro and In vivo against *Aspergillus niger*, thus suggesting the use of retinoids in mycosis [13]. From these results, retinoids have been investigated as a possible agent in the treatment of mycoses.

The aim of our review is to highlight the current application of retinoids as antimycotic agents due to their direct or indirect antifungal activity, by means of the immunoadjuvant properties of these compounds, opening up new scenarios for the future use of retinoids in the treatment of localized and disseminated mycoses.

2. Methods and Study Design

2.1. Search Strategy

We performed a comprehensive search in the following databases from 1946 to July 2021: Cochrane Central Register of Controlled Trials; MEDLINE; Embase; US National Institutes of Health Ongoing Trials Register; NIHR Clinical Research Network Portfolio Database; and the World Health Organization International Clinical Trials Registry Platform. We studied reference lists and published systematic review articles. We used the term "retinoid" with the following keywords, separately and in combination: "Fungal Infection", "fungal biofilm", "dermatophytoma", *"Entomophthoromycota"*, *"Basidiobolus"*, *"Conidiobolus"*, *"Ascomycota"*, *"Ajellomycetaceae"*, *"Paracoccidioides"*, *"Lacazia"*, *"Coccidioides"*, *"Blastomyces"*, *"Histoplasma"*, *"Sporothrix"*, *"Talaromyces"*, *"Trichophyton"*, *"Microsporum"*, *"Epidermophyton"*, *"Rhizopus"*, *"Mucor"*, *"Malassezia"*, *"Micrococcus"* *"Cladophialphora"*, *"Ramichloridium"*, *"Exophiala"*, *"Curvularia"*, *"Alternaria"*, *"Fusarium"*, *"Aspergillus"*, *"Candida"*, *"Fonsecaea"*, *"Rhodotorula"*. Only English language articles were included in the searches. Forward citation searching of the reference lists of the original studies and review articles was also conducted.

2.2. Inclusion Criteria

To investigate the direct and indirect effect of retinoids against fungal pathogens, if the study included the retinoids with other drugs, only the retinoid frame was analyzed. All human studies were included, with no restrictions on age, sex, ethnicity or type of study. Case reports and case series were included if they described the use of retinoids in diseases that were not included in reviews or trials.

2.3. Exclusion Criteria

The target intervention excluded the analyses of other pathologies not due to fungal pathogens, and non-English language articles.

3. Results

Three hundred and ninety-seven articles or trials regarding retinoids and fungal infections were identified by this quantitative research. Three hundred and fifty-six were excluded after the application of exclusion criteria. Among the forty-one articles or trials eligible for evaluation, two were excluded after abstract or full text reading. Thirty-nine articles or trials were evaluated in this review (Figure S1). The results of our research are summarized in Table 1.

3.1. Effectiveness of Retinoids against Opportunistic Fungi That Colonize the Skin and Mucosae in Humans

3.1.1. Candida

Infections due to *Candida* spp. are the major causes of morbidity and mortality among hospitalized patients and are associated with a wide variety of clinical manifestations, ranging from superficial and mucosal infections to life-threatening disseminated bloodstream candidemia [14]. Global estimates suggest that invasive candidiasis occurs in more than a quarter of a million patients every year, with incidence rates for candidemia of 2–14 per 100,000 inhabitants in population-based studies [15]. Antifungal drug resistance, including multidrug resistance (MDR) in *Candida* spp., has become increasingly important in the management of invasive fungal infections. These infections are related to high morbidity and mortality rates and can be linked to healthcare-associated transmission. In this view, the finding of novel drugs able to act against MDR *Candida* spp. should be a priority. From 2007, novel retinoid derivatives containing a benzimidazole moiety were synthesized and tested for their antimicrobial activity. Their antimicrobial properties against *Staphylococcus aureus*, including methicillin-resistant *Staphylococcus aureus* strains (MRSA), *Escherichia coli*, *Pseudomonas aeruginosa*, *Enterococcus faecalis*, *Candida krusei* and *Candida albicans* were evaluated. While some of them exhibited moderate activity against *S. aureus*, including MRSA strains, *E. faecalis*, *C. krusei and C. albicans*, none of the compounds showed any activity against *E. coli* and *P. aeruginosa* [16]. For *Candida* spp., *Candida krusei* (ATTC 6258) and *Candida albicans* (ATCC 10231), minimum inhibitory concentration (MIC) values of 50 µg/mL were obtained against the fungus for the targeted compounds (Figure 1), comparable with fluconazole for *C. krusei* [16]. Moreover, translational research is increasing our knowledge in fungistatic retinoids' effects, paving the way for their future use in the treatment of mycosis. As previously reported, Campione et al. reported the fungistatic effect of tazarotene 0.1% In vitro against *Candida albicans* and *Candida glabrata* [12]. Scardina et al. [17] investigated the efficacy of isotretinoin for the treatment of nystatin-resistant oral candidiasis. They evaluated six patients with clinical and mycological diagnosis of Candida stomatitis, treated with 0.18% isotretinoin solution applied twice daily, previously treated with nystatin for 30 days. After one month of retinoid therapy, five patients were negative for *Candida* spp., and only one patient with suspected sicca syndrome was found to have oral candidiasis 15 days after the last administration of isotretinoin. None of the patients had any complaints about the medication. These findings prove that 0.18% isotretinoin, applied twice a day for one month, can suppress nystatin-resistant *Candida* oral infection [17]. To date, the current results about In vitro and In vivo application of retinoid against *Candida* spp. deserve future attention as molecules to control candida infection.

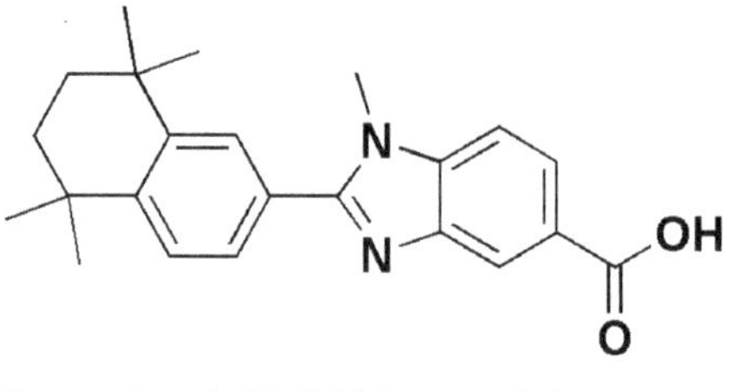

Figure 1. 1-methyl-2-(5,5,8,8-tetramethyl-5,6,7,8-tetrahydro-naphthalen-2-yl)-1H-benzoimidazole-5-carboxylic acid, a retinoid derivative showing MIC value of 50 µg/mL against Candida krusei and Candida albicans. This sketch was drawn in ChemDraw® Ultra 7.0. Perkin Elmer Italia S.p.A.

3.1.2. Rhodotorula Mucilaginosa

Human fungal infections by *Rhodotorula* spp., a genus of unicellular environmental pigmented yeasts, have increased in the last decades [18]. In China, they are among the main causes of invasive fungal infections by non-*Candida* yeasts [19] and are considered an emerging pathogen. Martini et al. [20] reported a case of 38-year-old man who had whitish nail changes on all fingers as the only symptom. The condition had developed within a few days and led to dystrophy of the proximal part of the nail plates. Microscopic examination of his nail scrapings demonstrated budding hyphae and the patient—who is a teacher—reported frequent use of a wet sponge. Antifungal therapy was started with oral itraconazole pulse therapy (400 mg day for 1 week), combined with the topical application of 5% amorolfine nail lacquer. Subsequent cultures and molecular typing methods identified *Rhodotorula mucilaginosa* [20]. This environmental yeast was repeatedly isolated despite treatment with itraconazole. As no improvement was achieved and testing of the biological activity of the fungus revealed only marginal keratolytic activity, it was considered as a colonizer of a destroyed nail matrix. Finally, a biopsy of the nail bed confirmed the diagnosis of nail psoriasis, which rapidly responded to treatment with oral acitretin (10 mg/day) and topical calcipotriol/betamethasone cream, leading to a rapid clinical improvement after 8 weeks. Fungal growth in the destroyed nails masked the underlying disease and might have triggered the psoriatic nail reaction [20]. Moreover, Jarros et al. [21] isolated *R. mucilaginosa* from a patient with chronic renal disease (CKD). This opportunistic fungus may represent a high risk of serious infection; thus, a correct identification of the yeast is the main means for an efficient treatment [21]. Additionally, as reported in the literature, *R. mucilaginosa* tends to co-infect the nails together with other fungal pathogens, such as *Candida* or *Trichophyton*, paving the way to understanding whether it can infect nails on its own or needs other fungi to develop and promote a crosstalk in a dermatophytoma nail [22,23]. Ge et al. [22] also reported a case of onychomycosis with the concomitant presence of *Rhodotorula mucilaginosa* and *Candida parapsilosis*, while Idris et al. [23] documented a mixed infection of toenails caused by *Trichosporon asahii* and *Rhodotorula mucilaginosa* [22,23]. These cases underline how environmental yeasts deserve our attention as pathogens in clinical practice and strive us to search for new molecules to prevent fungal infections in susceptible individuals.

3.1.3. Malassezia

Malassezia spp. are lipid-dependent basidiomycetous yeasts that inhabit the skin and mucosa of humans and other warm-blooded animals, representing a major component of the skin microbiome. They occur as skin commensals, but are also associated with various skin disorders such as tinea versicolor and blood-stream infections [24]. Handojo et al. [25] were the first to propose the clinical use of retinoids in the treatment of *Malassezia*. They carried out a clinical trial to evaluate the cure rate, the incidence of relapse and the tolerance of retinoic acid 0.05% cream applied topically to 50 patients suffering from tinea versicolor. In this trial, two types of topical retinoic acid were used: retinoic acid 0.05% in vanishing cream and retinoic acid 0.05% in equal parts of propylene glycol and 95% alcohol. The patients entered in this trial randomly and were divided into two groups: 25 patients received the 0.05% cream, applied twice daily, and 25 patients received the 0.05% lotion, applied twice daily. Four patients failed to complete the treatment and were thus excluded from evaluation. The number of patients followed up until the end of the treatment period can be specified as follows: (i) 23 patients (92%) in the group receiving the 0.05% cream; (ii) 23 patients (92%) in the group receiving the 0.05% lotion. The treatment period needed to obtain a favorable result was relatively short, namely: 2 weeks in 33 patients (73.33%) and 3 weeks in 45 patients (97.83%) with a mean of 2.27 weeks. The results of this study show that in vivo, retinoic acid has an anti-fungal action, as proved in this trial by the absence of the causative agent on microscopic examination and by the low incidence of relapse. Another important feature of retinoic acid is its deep penetrative capacity, which makes it possible to reach the deeply seated *Malassezia furfur* and destroy it. As far as we know, this deep penetrative capacity of retinoic acid is not found in any other anti-fungal preparation [25]. As regards the above-mentioned criteria for a successful treatment, there was no significant difference between the lotion and cream groups. It was pointed out that patients suffering from tinea versicolor are predisposed to dermatophyte infections, and that cleanliness of the skin is a simple but essential way of preventing contamination with *Malassezia furfur* [25]. Moreover, Shi et al. [26] studied the effect of adapalene gel versus 2% ketoconazole cream in pityriasis versicolor. A total of 100 patients were enrolled in the study, 80 of which were randomized into two arms: one receiving adapalene gel and the other 2% ketoconazole. Among the 80 patients, 67 fulfilled the study criteria and entered evaluation of the curative effect according to the investigator's requirements. Of the 67 patients, 35 were included in the adapalene group and 32 in the ketoconazole group. One group was treated with 2% ketoconazole cream topically twice daily for 2 weeks, whereas adapalene gel was used for the other group in a similar fashion. No significant differences in efficacy between the two groups were observed: the adapalene gel group presented clinical resolution in 31/40 (77.5%) patients and a negative mycological test in 30/40 (75%) patients after 4 weeks, whilst the ketoconazole cream 2% treated group presented clinical resolution and a negative test in 28/40 (70%) and 28/40 (70%) patients, respectively. No major side effects were noted in either group. In conclusion, the study underlines the clinical efficacy, security and tolerability of adapalene in pityriasis versicolor [26]. Adapalene can strip the abnormal keratinocytes and normalize the dysfunction of keratinization in keratinocytes and epidermal turnover time in the lesions. Moreover, adapalene can decrease the sebum secretion of sebaceous glands in the skin, and thus topical adapalene gel promotes an environment inhospitable to *Malassezia* yeasts—a disadvantage for the propagation of the yeasts—and leads to the elimination of numerous spores and hyphae together with the abnormal cell layer of keratinocytes. On the other hand, adapalene also has anti-inflammatory activity [27]. Based on its immunomodulatory actions, topical adapalene gel was discovered to effectively reduce inflammation in the lesions of pityriasis versicolor, thus improving the symptoms of the disease [27]. Therefore, although adapalene is not a special antifungal drug, it might counteract the fungal growth by affecting the environment required by *Malassezia* yeasts for their living [26]. Yazıcı et al. [28] reported a case of pityriasis versicolor in a 33-year-old man with lesions recurring for 15 years. Medical history revealed the use of many topical and systemic antifungal agents as well as topical retinoic acids with partial benefits. Only oral isotretinoin therapy at a dose of 20 mg/day (0.4 mg/kg/day) has proven to be effective with clinical resolution of pityriasis [28]. Nowadays, In vivo data support the use of retinoids in the treatment of *Malassezia* spp. human infection. Further In vitro studies are needed to elucidate the direct and indirect effects of retinoids.

3.2. Effectiveness of Retinoids against Environmental Human Pathogenic Filamentous Fungi

3.2.1. *Aspergillus* spp.

Aspergillus is a natural ubiquitous saprophyte found in air, soil and organic matter. Humans normally inhale the spore form of the fungus [29]. *Aspergillus* species cause a wide spectrum of diseases in humans [30]. Depending on the underlying immune status of the host, *Aspergillus* diseases can be roughly classified into three groups with distinct pathogenetic mechanisms, clinical manifestations, and overlapping features [31]. Moreover, aspergillus infections could arise as clinical syndromes in patients with different immune statuses, including severe asthma with fungal sensitization (SAFS), allergic bronchial pulmonary aspergillosis (ABPA), chronic pulmonary aspergillosis (CPA), invasive pulmonary aspergillosis (IPA) and invasive bronchial aspergillosis (IBA) [31]. Furthermore. *Aspergillus* infections do not affect only the respiratory system, but could also involve different apparatuses, including skin appendages. As previously reported by Girmenia et al. [11], a lower infection rate due to fungal pathogens was observed in neutropenic patients with acute promyelocytic leukemia and treated with the AIDA regimen. Apart from the reported results, there is one noteworthy fact: patients suffering from promyelocytic leukemia treated with ATRA have significantly lower incidences of fungal infections caused by *Candida* spp. and *Aspergillu* spp. [11]. Their data were elucidated and underlined by Campione et al. [12], showing a strong In vitro fungistatic effect exerted by ATRA against *Candida albicans* and *Aspergillus fumigatus*, through the inhibition of both germination and hyphal growth of *Candida* yeasts and *Aspergillus* conidia [12]. The authors also demonstrated that ATRA at sub-optimal doses is able to synergize In vitro with the conventional antifungal amphotericin B and posaconazole to counteract the germination of *Aspergillus* conidia [13]. This synergistic effect is of great importance, as it allows one to reduce the dosage of antifungal drugs, limiting exposure to the toxic effects related to their prolonged and massive use. ATRA has also proven to exert a protective effect against aspergillosis in vivo. In an experimental animal model of IPA, ATRA, administered as prophylaxis, significantly increased the survival rate of the animals, compared to the untreated control animals (60% vs. 20% after 12 days of infection). Interestingly, the antifungal efficacy of ATRA was completely comparable to that obtained with posaconazole, one of the antifungals commonly used in the treatment of IPA. Molecular docking studies suggested that the mechanism underlying the antifungal property of ATRA might be due to its ability to interfere with HSP90 activity by competing with the ATP binding-site of the HSP90 protein. HSP90 is an ATP-dependent chaperonin that plays a crucial role in fungal virulence and pathogenicity. It has been shown that ATRA down-regulates the expression of the HSP90-related gene as well as the mRNA expression levels of *AbaA*, *CrzA* and *WetA* genes involved in the conidial germination process. Intriguingly, ATRA enhances In vitro the phagocytosis of *Aspergillus* conidia by macrophages [13]. It could be hypothesized that the protective effect In vivo of this retinoid might be due not only to a direct fungistatic effect but also to the immunoadjuvant properties of the molecule due to its ability to enhance the innate immune response that play a key role in the clearance of fungal infection. ATRA could, therefore, represent a pleiotropic molecule with a great potential either in monotherapy or in a combination with conventional antimycotic drugs for the development of novel antifungal therapeutic strategies aimed at countering the increasing spread of the drug resistance phenomenon in fungal pathogens. As previously reported, *Aspergillus* infection could also affect skin appendages. Onychomycosis (OM) is a chronic fungal infection of the nail caused by dermatophytes, yeasts, and non-dermatophytes. El-Salam et al. [32] evaluated the efficacy of tazarotene 0.1% gel alone or combined with tioconazole nail paint in the treatment of onychomycosis due to *Aspergillus* spp., including *A, candidius, A. flavus, A. niger, A. nidulans* and *A. terreus* [32]. Forty patients presenting with onychomycosis underwent full history taking, as well as clinical and nail examination, including a clinical and dermoscopic assessment of severity using the Onychomycosis Severity Index (OSI), KOH test, and fungal culture. A significant treatment response was found in those patients treated with tazarotene 0.1% gel administered in combination with tioconazole, compared to tazarotene 0.1% gel alone (decrease in OSI, dermoscopic features, and mycological clearance). Tazarotene alone has been shown to have an antifungal activity especially against *Aspergillus niger* and *A. flavus*, whereas, when combined with antifungals such as tioconazole, it seems to be more effective, suggesting its potential use as adjuvant for standard systemic or topical antifungal treatments of onychomycosis [32]. Based on the immunomodulatory activity of retinoids and given that the nail unit is an immunological niche, a balance between pro-inflammatory and anti-inflammatory cytokines is paramount for the correct management of onychomycosis and other related inflammatory diseases such as psoriasis [33]. *Aspergillus* spp. are also involved in fungal keratitis, a major cause of corneal ulcers, resulting in significant visual impairment and blindness. Zhao et al. [34] investigated the effect of fenretinide in *Aspergillus fumigatus* keratitis In vivo in a mouse model as well as In vitro in a THP-1-derived macrophage cell line infected with *A. fumigatus*. In both experimental models, the pretreatment with fenretinide contributed to protecting corneal transparency during *A. fumigatus* keratitis in the early phase of the infection by reducing neutrophil recruitment, decreasing myeloperoxidase (MPO) levels and increasing apoptosis. Compared with controls, fenretinide also impaired proinflammatory cytokine interleukin 1 beta (IL-1β) production in response to *A. fumigatus* exposure through the blockade of the lectin-type oxidized LDL receptor 1 (LOX-1) and c-Jun N-terminal kinase (JNK) pathway [34]. All retinoids evaluated in *Aspergillus* spp. infections have demonstrated both clinical and In vitro efficacy, setting the stage for further investigation on the use of retinoids in clinical practice.

3.2.2. *Fonsecaea* spp.

Chromoblastomycosis is a subcutaneous fungal infection caused by dematiaceous fungi that belong to the order *Chaetothyriales* and family *Herpotrichiellaceae*, such as as *Fonsecaea pedrosoi, Phialophora verrucosa*, and *Cladophialophora carrionii* [35]. This infection is prevalent in tropical and subtropical areas and has been designated as a neglected tropical disease according to the World Health Organization (WHO) [36]. Chromoblastomycosis infection is difficult to treat, and there are limited therapeutic options, making the characterization of new drugs or approaches to treat this infection urgent. Belda et al. [35] reported two cases of extensive chromoblastomycosis lesions due to Fonsecaea spp., treated with a combination of itraconazole (200 mg/day), acitretin (50 mg/kg), and topical imiquimod, 5 days a week for 4 months. In the fourth month of treatment, both patients showed improvement in the verrucous plates, suggesting that acitretin combined with drugs already used in chromoblastomycosis therapy can decrease the duration of treatment, thus improving the patient's quality of life [37]. Bao et al. [38] reported a case of chromoblastomycosis caused by *Fonsecaea monophora* in a 60-year-old male carpenter with a 40-year history of psoriasis, from Shandong in northern China. A therapy based on 400 mg/day of itraconazole for 5 weeks did not resolve the infection. However, oral itraconazole (200 mg/day), combined with acitretin (20 mg/day), was more effective, and the lesions resolved completely after 1 month of treatment [38]. The rational use of acitretin in chromoblastomycosis is based on its keratoplastic activity and modulation of hyperkeratosis found in this pathology. However, no data regarding the direct action of acitretin against *Fonsecaea* spp. have been reported. From these clinical studies, it would appear that the combined therapy with itraconazole plus acitretin led to the regression of the chromoblastomycosis caused by *F. monophora*, suggesting a potential synergistic effect between the two compounds. Further In vitro and In vivo studies are necessary to understand their interactive effect when used in association.

3.2.3. Dermatophytes

Dermatophytes are a group of fungi able to invade keratinized tissues such as skin, hair and nails, causing infections in humans and animals, collectively named dermatophytosis. These fungi are classified into three genera, including *Trichophyton, Epidermophyton* and *Microsporum*. The genus *Trichophyton* is characterized morphologically by the development of both smooth-walled macro- and microconidia. Clinical manifestations of *Trichophyton* include "tinea", also called athlete's foot, ringworm, jock itch, and similar infections of the nail, beard, skin and scalp [39]. Within the wide range of *Trichophyton* clinical presentations, onychomycosis is the most prevalent nail disease, and it is mainly caused by two dermatophyte species, *Trichophyton rubrum* and *Trichophyton interdigitale*, with a frequency in the range of 80% and 20%, respectively [40]. Campione et al. [14] evaluated 15 patients presented with distal and lateral subungual onychomycosis and treated with topical tazarotene 0.1% gel, once per day for 12 weeks. Ten of these patients presented *T. rubrum*, two *T. mentagrophytes*, one presented *T. tonsurans and* one *Epidermophyton floccosum*. Six patients (40%) reached a mycological cure on target nail samples already after 4 weeks of treatment, while sampled material collected after 12 weeks were negative for infections in all patients. Complete clinical healing and negative cultures were reached in all patients at week 12, with a significant improvement in all clinical parameters of the infected nails [14]. Moreover, Campione et al. [14] evaluated the fungistatic effect of tazarotene 0.1% In vitro against *Trichophyton verrucosum* [14]. This agent also displayed In vitro a dose-dependent inhibitory activity, suggesting its direct antifungal effect on the fungus *Trycophyton verrucosum* [14]. Tazarotene is a synthetic third-generation retinoid derived from vitamin A, which has proved to be beneficial in modulating keratinocyte proliferation and in reducing inflammation [14]. In particular, tazarotene effect is mediated by the up-regulation of the intracellular retinoic acid-binding protein-II expression [41]. This protein transactivates nuclear retinoic acid receptors (RARs), in particular RARβ and RARγ, by binding to specific DNA sequences on their promoter gene regions, thus leading to a reduced proliferation of normal and neoplastic cells and favoring cellular differentiation and apoptosis [41]. The anti-inflammatory activity of tazarotene, due to its immunomodulatory properties, may inhibit the fungal keratinolytic proteases, thus contributing to its efficacy. Indeed, tazarotene is able to impair the release of inflammatory cytokines such as IL-6 and interferon gamma (IFN-γ) from mononuclear leukocytes in vitro, as well as the nitric oxide synthesis in a dose-dependent manner [42]. A limitation of our preliminary results is the absence of a randomized placebo control group [14]. Additionally, Handojo et al. [23] showed that the growth of dermatophytes (*Trichophyton, Epidermophyton* and *Microsporum*) on Sabouraud glucose agar medium (pH adjusted to 5.4–5.6) containing ATRA at a concentration of 0.015% was totally inhibited, as well as the growth of contaminants [23]. On the other hand, the fungal growth was only partially inhibited by a concentration of 0.01% [23]. Despite tazarotene being considered a third generation retinoid, the translational research demonstrated the efficacy of ATRA in dermatophytes infection, both for *Aspergillus* spp. and *Candida* spp. The efficacy of ATRA was demonstrated by Gaziano et al. [43]. Among the numerous components of *Cardiospermum halicacabum L. (C. halicacabum)*, an herbaceous climber belonging to the Sapindaceae family, ATRA can exert a clear dose-dependent fungistatic activity In vitro against *Trichophyton rubrum* [43]. These results are in line with Gaziano et al.'s previous analyses showing that ATRA inhibited In vitro the germination of *Candida albicans* and *Aspergillus fumigatus* in a dose-dependent manner [30]. Furthermore, the heat shock protein 90 (Hsp90) chaperone of microbial pathogens could be a possible therapeutic target for *C. halicacabum*. The potential use of chaperones as molecular drug targets could be considered unattractive due to the similarity of the molecular structure between human and microbial chaperones. However, they exhibit different dependencies on chaperone-dependent pathways and could therefore display differences in the sensitivity of their inhibition [30]. In conclusion, Hsp90 may not be the only possible target for *C. halicacabum*, since multiple plant bioactive compounds may interact with different molecular targets both in a single and multiple intracellular pathways [43]. Both In vitro and In vivo data support the action of retinoids against dermatophytes, and actual formulations such as topical 0.1% tazarotene gel in the treatment of OM deserve further attention.

3.3. Efficacy of Retinoids against Pneumocystis
Pneumocystis

Pneumocystis jiroveci is an unusual fungus of the genus *Pneumocystis* [44]. This fungus is the causative agent of Pneumocystis pneumonia (PcP), a common opportunistic disease in immunocompromised hosts, such as patients with AIDS [45] and those with other predisposing immune deficiencies [46]. Historically, PcP has been reported mainly in immunocompromised patients. Pereira-Díaz et al. [47] conducted an observational, descriptive transversal study that included all patients admitted in Spain with discharge diagnosis of PcP, registered in the National Health System's Hospital Discharge Records Database of Spain between 2008 and 2012. The results of this first nationwide study in Spain allow a change in the misconception that, after the AIDS pandemic, PcP became an infrequent disease, showing that today it is an emerging problem in immunocompromised patients, including without HIV infection [47]. Lei et al. [48] evaluated the effect of ATRA combined with primaquine in a murine model of pneumocystis infection. Their previous studies of various inflammatory components during PcP found that myeloid-derived suppressor cells (MDSCs) accumulate in the lungs of mice and rats with PcP [48]. ATRA (5 mg/kg/day in 8% DMSO) combined with primaquine (2 mg/kg/day in water) displayed a protective effect against PcP, as the combination of trimethoprim and sulfamethoxazole (TMP, 50 mg/kg/day and SMX, 250 mg/kg/day) induced MDSCs' differentiation into macrophages, which play a key role in the clearance of PcP by recognition, phagocytosis, and degradation of *Pneumocystis* [48–50]. Further studies are needed to evaluate the potential combination of retinoids with standard therapy for *P. jiroveci* infection, also considering it as an emerging problem in immunocompromised patients without HIV infection due to immunosuppressive therapies.

Table 1. Retinoid-based treatments against ungual pathogens.

Fungi	Spp.	Clinical/Experimental Model	Pathological Model	Retinoid	Combination	Results	Reference
Candida	*albicans*	In vitro culture		Retinoid derivatives containing a benzimidazole moiety		Antimicrobial activity MIC 1.56µg/mL	[16]
	albicans	In vitro culture		1 g of Tazarotene 0.1% gel dissolved in 3mL of physiological solution		Fungistatic activity	[12]
	glabrata	In vitro culture		1 g of Tazarotene 0.1% gel dissolved in 3mL of physiological solution		Fungistatic activity	[12]
	krusei	In vitro culture		Retinoid derivatives containing a benzimidazole moiety		Antimicrobial activity MIC 1.56µg/mL	[16]
	Not specified	Human	Chronic hyperplastic candidiasis nystatin-resistant	0.18% isotretinoin applied twice a day for one month		Clinical resolution after one month	[17]
Malassezia	*furfur*	Human	Pityriasis versicolor	Retinoic acid 0.05% cream vs. retinoic acid 0.05% lotion twice daily for 3 weeks		50 patients totally. 33 patients (73.33%) and 3 weeks in 45 patients (97.83%) with a mean of 2.27 weeks	[25]
	Not specified	Human	Pityriasis versicolor	Adapalene gel vs. ketoconazole 2%		Clinical resolution after 4 weeks and 30/40 (75%) presented mycological negative test vs. Ketoconazole cream 2% 28/40 (70%) and 28/40 (70%)	[26]
	Not specified	Human	Pityriasis versicolor	Oral isotretinoin 20 mg/day (0,4 mg/kg/day) for 6 weeks		Clinical resolution after 6 weeks	[28]
Aspergillus	*fumigatus*	structural bioinformatic analysis		ATRA		Competitive inhbitor of the Hsp90 ATP-binding site	[13]
	fumigatus	In vitro colture		ATRA		Fungistatic activity (0.5 and 1 mM) down-regulation of HSP90 mRNa and protein expression; enhances the phagocytosis of macrophages (5 or 10mM)	[13]
	fumigatus	Rat	Invasive pulmonary aspergillosis (IPA)	ATRA, 2 mg/kg i.p. for 6 days	Alone; versus posaconazole; versus vehicol	Reduction in mortality of IPA	[13]
	niger	Human	Onychomycosis	Tazarotene 0.1% gel twice daily for three months	Alone or plus tioconazole (28% nail paint)	Clinical resolution of OM after three months	[32]

Table 1. *Cont.*

Fungi	Spp.	Clinical/Experimental Model	Pathological Model	Retinoid	Combination	Results	Reference
	flavus	Human	Onychomycosis	Tazarotene 0.1% gel twice daily for three months	Alone or plus tioconazole (28% nail paint)	Clinical resolution of OM after three months	[32]
	fumigatus	Murine model	Keratitis	Fenretinide 100 µM subconjunctival injection		Inhibition of neutrophil recruitment and IL-1β production	[34]
Dermatophyte (*Trichophyton*)	*rubrum*	Human	Onychomycosis	Tazarotene 0.1% gel once daily for 12 weeks		Clinical resolution	[12]
	mentagrophytes	Human	Onychomycosis	Tazarotene 0.1% gel once daily for 12 weeks		Clinical resolution	[12]
	verrucosum	In vitro culture		1 g of Tazarotene 0.1% gel dissolved in 3 mL of physiological solution		Fungistatic activity	[12]
	tonsurans	In vitro culture		1 g of Tazarotene 0.1% gel dissolved in 3 mL of physiological solution		Fungistatic activity	[12]
Dermatophyte (*Epidermophyton*)	*floccosum*	Human	Onychomycosis	Tazarotene 0.1% gel once daily for 12 weeks		Fungistatic activity	[12]
Pneumocystis	*jiroveci*	Mice and rats	pneumonia	ATRA 5 mg/kg/day in 8% DMSO	Primaquine 2 mg/kg/day in water	Engaged myeloid-derived suppressor cells	[48]
Fonsecaea	Not specified	Human	Chromoblastomycosis	Acitretin 50 mg/day	Itraconazole and 200 mg/day topical imiquimod for 5 weeks	Clinical resolution	[37]
	Not specified	Human	Chromoblastomycosis	Acitretin 20 mg/day for 5 weeks	Itraconazole and 200 mg/day topical imiquimod for 5 weeks	Clinical resolution	[37]
	monophora	Human	Chromoblastomycosis	Acitretin 20 mg/Kg for 1 month	Itraconazole 200 mg/day for 1 month	Clinical resolution	[38]
Rhodotorula	*mucilaginosa*	Human	Onychomycosis and psoriasis	Oral acitretin 10 mg/day for 8 weeks	Topical cal-cipotriol/betamethasone for 8 weeks	Clinical improvement within 8 weeks	[21]

147

4. Discussion

To date, In vitro and In vivo studies demonstrated a remarkable antifungal efficacy of retinoids against a broad spectrum of opportunistic fungi, as summarized in Table 2.

Table 2. Evaluation of retinoid efficacy against mycosis.

	Tretinoin	Tazarotene	Isotretinoin	Acitretin	Fenretinide	Adapalene
In vitro studiies	*A. fumigatus* *C. albicans* *Microsporum* spp. *Trichophyton* spp. *Epidermophyton* spp.	*C. glabrata* *C. albicans* *T. verrucosum*	*A. fumigatus* *A. niger* *C. albicans*			
In vivo studies	*M. furfur* *A. fumigatus*	*A. niger* *T. rubrum* *T. tonsurans* *T. mentagrophytes* *E. floccosum* *C. albicans* *A. flavus*	*M. furfur* *P. jiroveci*	*F. monophora* *R. mucilaginosa*	*A. fumigatus*	*M. furfur*

In vitro studies have demonstrated the efficacy of ATRA against *Candida albicans*, *Aspergillus fumigatus* and dermatophytes; isotretinoin against *Candida albicans*, *Aspergillus fumigatus* and *Aspergillus niger*; and tazarotene against *Candida albicans*, *Candida glabrata* and *Trychophyton verrucosum*. In vivo studies have reported the clinical efficacy of ATRA against *Malassezia furfur* and *Aspergillus fumigatus*; isotretinoin against *Pneumocystis jiroveci* and *Malassezia furfur*; adapalene against *Malassezia furfur*; fenretinide against *Aspergillus fumigatus*; acitretin against *Fonsecaea monophora* and *Rhodotorula mucilaginosa*; and tazarotene against the dermatophytes *Trychophyton rubrum*, *Trychophyton mentagrophytes*, *Trychophyton tonsurans* and *Epidermophyton floccosum*. The chemical structures of the tested compounds are reported in Figure 2. Notwithstanding the In vivo and In vitro efficacy of retinoids, just one trial has evaluated retinoids in fungal infections (Table S1).

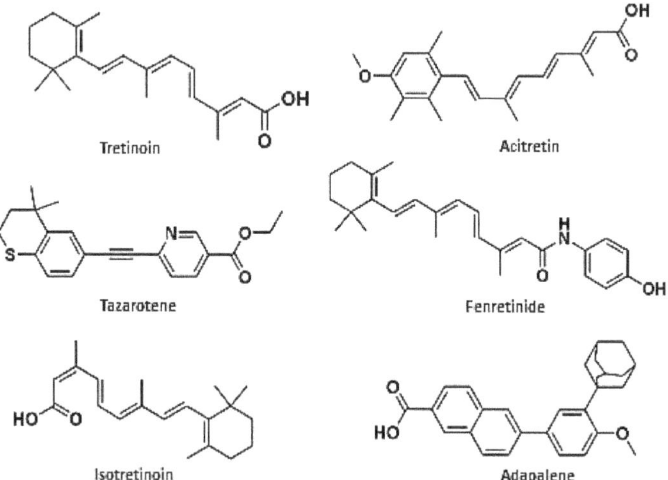

Figure 2. Chemical structures of the active retinoids against mycosis. The sketches were drawn in ChemDraw ®Ultra 7.0. Perkin Elmer Italia S.p.A.

Unmet Needs in Fungal Infections

Existing anti-fungal agents have a variety of limitations, from toxicity to rising levels of resistance in common fungal pathogens, significant drug interactions and sub-optimal efficacy [51]. The rising proportion of the community in Western countries with induced immune deficiencies caused by immunomodulatory therapies and their expanding indications is gradually increasing the burden of fungal diseases. Equally, unlike anti-bacterial therapies, which have exceptionally good response rates in sensitive pathogens, antifungal therapies are not as efficacious, due to poor host immunity, innate fungal properties and drug dosing limitations [52]. Moreover, drugs administered by the enteral or parenteral routes need to reach the blood circulation to be distributed to all apparatus and act on the infected site. To contemplate retinoids in the prevention and treatment of fungal infections, their common adverse events—such as local erythema, liver toxicity, dry skin, irritation, photosensitization and teratogenic risk—must be considered, as they can reduce the therapeutic adherence to a great extent [53]. Furthermore, retinoids' efficacy is strongly limited by several disadvantages, including low water solubility, short lifetime due to the degradation by the cytochrome P450-dependent monooxygenase system, and high sensibility to oxygen, light, and heat. Taken together, all these characteristics drastically reduce bioavailability and, consequently, the therapeutic potential [54]. Consequently, a strategy to reduce side-effects related to free retinoid administration, while increasing their bioavailability and maximizing their therapeutic effects, can be represented by the design and development of advanced formulations. With this aim, several drug delivery systems (DDSs), both local and systemic, loaded with retinoids have been developed. DDSs are carriers able to encapsulate active molecules and deliver them to cell/tissue targets in order to reduce the side effects and achieve an increased therapeutic efficacy compared to formulations containing the drug only in the free form. Among retinoids, all-*trans*-retinoic acid (ATRA) represents the first-choice treatment for several skin diseases, and could be used as a future choice in clinical practice for fungal infections. To avoid the aforementioned adverse effects, recently, Zuccari et al. [55] prepared ATRA-loaded micelles (ATRA-TPGSs), by its encapsulation in D-α-tocopheryl-polyethylene-glycol-succinate (TPGS). Loaded micelles of 12 nm mean diameter were further embedded in a Carbopol®-based gel and applied on porcine skin for topical release evaluation. The nanogel was capable of improving drug absorption through the stratum corneum by an increment of drug concentration in the formulation without the use of organic solvents such as ethanol and propylene glycol. Their new formulation, characterized by low polydispersity, slightly negative zeta potential, and good encapsulation efficiency, highlights the improvement in patient compliance necessary to achieve better therapeutic outcomes [55]. Fungal keratitis is an infection of the cornea caused by fungi as *Fusarium*, *Aspergillus* and *Candida* spp. [56]. Although ATRA has proven to be effective against *Candida* and *Aspegillus* spp., the role of retinoids in ophthalmology is controversial. In the 1980s, ocular surface disease emerged as a potential therapeutic target, since vitamin A deficiency was known to cause epithelial squamous metaplasia and glandular atrophy [56]. To date, no new retinoid ophthalmic formulations for fungal keratitis have been developed. This limitation paves the way for future ophthalmic formulation associated with drugs in the guidelines for ocular infections potentially sensitive to ATRA. As previously reported, pulmonary aspergillosis is a collective term used to refer to a number of conditions caused by Aspergillus spp. infections. Moreover, pulmonary drug delivery offers the advantage of positioning the bioactive molecule in direct contact with the pathological lung epithelia, thus ensuring a rapid onset of the therapeutic response. High local drug concentrations may be easily obtained by pulmonary administration with a concomitant increase in the pharmacological effect but without the side effects elicited by other administration routes [57]. As previously reported, fenretinide has shown In vivo activity against *A. fumigatus* infection. This semisynthetic retinoid shows a more favorable toxicological profile, characterized by minimal systemic toxicity and good tolerability. Unfortunately, the drug's hydrophobic character strongly hinders its aqueous solubilization. Aqueous fenretinide formulations have been obtained by complexation with cyclodextrins [58], encapsulation into nanomicelles [59], or liposomes [60]. Complexation with 2-hydroxypropyl beta-cyclodextrin has increased fenretinide aqueous solubility from 0.017 mg/mL (pure drug) to 2.41 mg/mL (complex). The aqueous formulation of the complexed drug, administered by the parenteral route, was well-tolerated and increased the drug's bioavailability and antitumor activity in mouse models of different tumor types [58]. Nanoencapsulation in phosphatidylcholine-glyceryltributyrate nanomicelles has increased the fenretinide aqueous solubilization up to 3.88 mg/mL (nanoencapsulated drug). The intravenous administration of the nanomicelles in mice bearing tumor xenografts showed enhanced drug bioavailability and antitumor activity [59]. Moreover, high tolerability was demonstrated by the absence of adverse effects after repeated administrations and for long periods. The ability of liposomes to encapsulate both hydrophilic and lipophilic molecules and to be linked to targeting moieties makes them very promising candidates for drug delivery [60]. Due to their biocompatibility, biodegradability and no immunogenicity, liposomes are the first DDSs that have been translated to clinical applications [61]. Therefore, the In vivo tolerability and the ability to provide high fenretinide solubilization levels suggest that complexation with cyclodextrins [58], encapsulation in nanomicelles [59] or liposomes [60] can be a valuable means for the preparation of safe and efficient pulmonary fenretinide formulations. It has been demonstrated that the encapsulation of various lipophilic nutraceuticals/pharmaceuticals/cosmeceuticals inside lipid-based nanocarrier systems can protect from photo/chemical degradation, improve the aqueous solubility, and allow deeper skin penetration of similar active ingredients. Solid lipid nanoparticles (SLNs) [62], nanostructured lipid carriers (NLCs) [63], liposomes [64], niosomes [65], and nanoemulsions (NEs) [66] are examples of lipid-based systems that have been proven to decrease drug degradation, improve drug targeting, and enhance the efficacy of retinoids in the treatment of skin disorders. Among lipid-based DDSs, ATRA-loaded SLNs are currently one of the most studied DDSs, thanks to their scalability, small particle size, good drug protection, and high drug-loading capacity. Despite these improvements, the drug could undergo expulsion and crystallization during storage. To overcome this drawback, NCLs were performed. Nevertheless, up to now, among the various lipid-based formulations designed to deliver retinoid compounds, NE-based drug delivery systems have been identified as the most feasible and economical method of topical therapy for various skin disorders. Through research advancements using homolipids and heterolipids as excipients, NE formulations have gained much attention due to their ability to enhance the topical efficacy of otherwise poorly permeable retinoid compounds. NEs have demonstrated wide compatibility with different retinoid compounds, surfactants, and oil systems. They are also easy to process and manufacture. This has generated further interest in NEs as drug carriers in the development of various topical formulations [66].

5. Conclusions and Future Perspectives

Retinoids have found widespread use in clinical practice, and in recent years, the focus has shifted to their possible antifungal action, both in humans and in experimental In vitro and In vivo models. However, to date, there is a lack of clinical trials that have evaluated the efficacy, tolerability and safety of retinoids against localized or systemic fungal infections. Data from clinical studies show the therapeutic efficacy of retinoids in the treatment of superficial mycoses. In detail, tretinoin and isotretinoin have proven to be effective against *M. furfur*, while tazarotene seems to be the best therapeutic choice for onychomycosis due to dermatophytes. Moreover, tretinoin displays a remarkable antifungal activity both in In vitro and In vivo experimental models against *A. fumigatus* and *C. albicans*, which represent the major opportunistic fungi associated with invasive and life-threatening mycoses in severely immunocompromised patients. Currently, *Candida* biofilm-related infections represent a serious global health threat and are often particularly difficult to treat due to the increasing emergence of multidrug-resistant (MDR) fungal strains. As with bacteria, fungal biofilms are well structured communities in which microorganisms are embedded within self-produced matrix of extracellular polymeric substances (EPSs), composed mainly of polysaccharides, proteins DNA and lipids [67]. Candida is able to produce biofilms on both biotic and abiotic surfaces, such as human tissues during infections and implanted medical devices [68]. Fungal biofilms play a crucial role in the pathogenesis of superficial mucosal and life-threatening systemic mycosis [69]. Embedded within biofilms, the fungal cells are more resistant to antifungal drugs and host immune defense mechanisms than their planktonic counterparts. Therefore, microorganisms which grow in biofilms are able to persist inside the host, causing severe infections, especially in nosocomial settings [70]. Thus, there is an urgent need to develop novel therapeutic strategies to prevent biofilm formation and/or disrupt pre-formed biofilms. In this context, retinoids could represent a novel approach to prevent and reduce biofilm formation. Nowadays, although it has been shown that these agents can exert bactericidal and anti-biofilm activity in bacterial cultures, there are no data about their potential use to counteract biofilm-based fungal infections [71]. In this regard, due to its own fungistatic effect, we hypothesize that ATRA may have great potential against *Candida albicans* biofilm formation by blocking both hyphal extension and budding replication of the yeast cells. Indeed, the hyphal germination is a crucial event in the biofilm formation, contributing to its architectural stability. Furthermore, the fungal cells must reach a critical density to release extracellular molecules that work as auto-inducers to produce an extracellular biofilm matrix. This mechanism, known as quorum sensing (QS), represents another critical step in the development of biofilms. ATRA, by inhibiting the replication of the fungal cells, might also interfere with QS mechanism, thus leading to impaired biofilm formation [72]. Therefore, the potential employment of this molecule could be particularly beneficial in a clinical setting of selected patients, such as individuals undergoing medical devices or organ solid and bone marrow transplantation to prevent systemic mycoses. From this perspective, further studies are needed to assess the potential anti-biofilm activity of this drug. In conclusion, retinoids, either alone or in combination with currently antifungal drugs, could be very promising agents for novel therapeutic or preventive antifungal options, thus overcoming the major clinical hurdle of drug resistance in fungi.

Supplementary Materials: The following are available online at https://www.mdpi.com/article/10.3390/ph14100962/s1, Figure S1: Flow chart reporting the research strategy. Table S1: Clinical trial reporting use of retinoid in fungal infection. Source: https://clinicaltrials.gov/ (accessed on 8 August 2021).

Author Contributions: Conceptualization, T.C., R.G. and E.C.; methodology, T.C., R.G., G.C.; software, T.C., R.G., G.C.; validation, T.C., R.G., G.C., G.Z., P.D.F., S.G., L.B. and E.C.; formal analysis, T.C., R.G., G.Z., and E.C.; investigation, T.C., R.G., G.Z; resources, T.C., R.G. and E.C.; data curation, T.C., R.G., G.Z., L.B. and E.C.; writing—original draft preparation, T.C., R.G., G.C., G.Z., and E.C.; writing—review and editing, T.C., R.G., G.Z., L.B., P.D.F., S.G. and E.C.; visualization, T.C., R.G. and E.C.; supervision, T.C., R.G., G.Z., E.C., P.D.F. and L.B.; project administration, T.C., R.G.,E.C and L.B. All authors have read and agreed to the published version of the manuscript.

Funding: This research received no external funding.

Institutional Review Board Statement: Not applicable.

Informed Consent Statement: Not applicable.

Data Availability Statement: Data sharing not applicable.

Acknowledgments: We thank Denis Mariano for his technical editing assistance.

Conflicts of Interest: The authors declare no conflict of interest.

Abbreviations

ABPA	Allergic Bronchial Pulmonary Aspergillosis
AIDS	Acquired Immunodeficiency Syndrome
AML	Acute Myeloid Leukaemia
APL	Acute Promyelocytic Leukaemia
ATRA	All-Trans Retinoic Acid
CPA	Chronic Pulmonary Aspergillosis
DDSs	Drug Delivery Systems
DMSO	Dimethyl sulfoxide
EPSs	Extracellular Polymeric Substances
IBA	Invasive Bronchial Aspergillosis
IL-1β	Interleukin 1 beta
IFN-γ	Interferon gamma
IPA	Invasive Pulmonary Aspergillosis
HIV	Human Immunodeficiency Virus
HSP90	Heat Shock Protein 90
JNK	c-Jun N-terminal Kinase
LOX-1	Lectin-type Oxidized LDL receptor 1
MDR	Multidrug resistance
MDSCs	Myeloid-Derived Suppressor Cells
MPO	Myeloperoxidase
NEs	Nanoemulsions
NLCs	Nanostructured Lipid Carriers
OM	Onychomycosis
OSI	Onychomycosis Severity Index
PcP	Pneumocystis Pneumonia
qRT-PCR	Real-Time Quantitative Reverse Transcription Polymerase Chain Reaction
QS	Quorum Sensing
RARs	Retinoic Acid Receptors
SAFS	Severe Asthma with Fungal Sensitization
SLNs	Solid Lipid Nanoparticles
SMX	sulfamethoxazole
TMP	trimethoprim
TPGS	D-α-tocopheryl-polyethylene-glycol-succinate

References

1. Enoch, D.A.; Yang, H.; Aliyu, S.H.; Micallef, C. The Changing Epidemiology of Invasive Fungal Infections. *Methods Mol. Biol.* **2017**, *1508*, 17–65. [CrossRef]
2. Mayer, F.L.; Kronstad, J.W. Cryptococcus neoformans. *Trends Microbiol.* **2020**, *28*, 163–164. [CrossRef]
3. Cornely, O.A.; Hoenigl, M.; Lass-Flörl, C.; Chen, S.C.-A.; Kontoyiannis, D.P.; Morrissey, C.O.; Thompson, G.R., III. Mycoses Study Group Education and Research Consortium (MSG-ERC); European Confederation of Medical Mycology (ECMM). Defining breakthrough invasive fungal infection–Position paper of the mycoses study group education and research consortium and the European Confederation of Medical Mycology. *Mycoses* **2019**, *62*, 716–729. [CrossRef]
4. Ibáñez-Martínez, E.; Ruiz-Gaitán, A.; Pemán-García, J. Update on the diagnosis of invasive fungal infection. *Rev. Espanola Quimioter.* **2017**, *30* (Suppl. S1), 16–21.
5. Bongomin, F.; Gago, S.; Oladele, R.O.; Denning, D.W. Global and Multi-National Prevalence of Fungal Diseases—Estimate Precision. *J. Fungi* **2017**, *3*, 57. [CrossRef]
6. Iyer, N.; Vaishnava, S. Vitamin A at the interface of host–commensal–pathogen interactions. *PLoS Pathog.* **2019**, *15*, e1007750. [CrossRef] [PubMed]
7. Gombart, A.F.; Pierre, A.; Maggini, S. A Review of Micronutrients and the Immune System–Working in Harmony to Reduce the Risk of Infection. *Nutrients* **2020**, *12*, 236. [CrossRef] [PubMed]
8. Campione, E.; Cosio, T.; Lanna, C.; Mazzilli, S.; Ventura, A.; Dika, E.; Gaziano, R.; Dattola, A.; Candi, E.; Bianchi, L. Predictive role of vitamin A serum concentration in psoriatic patients treated with IL-17 inhibitors to prevent skin and systemic fungal infections. *J. Pharmacol. Sci.* **2020**, *144*, 52–56. [CrossRef] [PubMed]
9. Campione, E.; Cosio, T.; Lanna, C.; Mazzilli, S.; Dika, E.; Bianchi, L. Clinical efficacy and reflectance confocal microscopy monitoring in moderate-severe skin aging treated with a polyvinyl gel containing retinoic and glycolic acid: An assessor-blinded 1-month study proof-of-concept trial. *J. Cosmet. Dermatol.* **2021**, *20*, 310–315. [CrossRef] [PubMed]

10. Cosio, T.; Di Prete, M.; Gaziano, R.; Lanna, C.; Orlandi, A.; Di Francesco, P.; Bianchi, L.; Campione, E. Trifarotene: A Current Review and Perspectives in Dermatology. *Biomedicines* **2021**, *9*, 237. [CrossRef]
11. Girmenia, C.; Coco, F.L.; Breccia, M.; Latagliata, R.; Spadea, A.; D'Andrea, M.; Gentile, G.; Micozzi, A.; Alimena, G.; Martino, P.; et al. Infectious complications in patients with acute promyelocytic leukaemia treated with the AIDA regimen. *Leukemia* **2003**, *17*, 925–930. [CrossRef] [PubMed]
12. Campione, E.; Paterno, E.J.; Diluvio, L.; Costanza, G.; Bianchi, L.; Carboni, I.; Chimenti, S.; Orlandi, A.; Marino, D.; Favalli, C. Tazarotene as alternative topical treatment for onychomycosis. *Drug Des. Dev. Ther.* **2015**, *9*, 879–886. [CrossRef] [PubMed]
13. Campione, E.; Gaziano, R.; Doldo, E.; Marino, D.; Falconi, M.; Iacovelli, F.; Tagliaferri, D.; Pacello, L.; Bianchi, L.; Lanna, C.; et al. Antifungal Effect of All- trans Retinoic Acid against Aspergillus fumigatus In vitro and in a Pulmonary Aspergillosis In vivo Model. *Antimicrob. Agents Chemother.* **2021**, *65*, e01874-20. [CrossRef] [PubMed]
14. Arendrup, M.C.; Patterson, T.F. Multidrug-Resistant Candida: Epidemiology, Molecular Mechanisms, and Treatment. *J. Infect. Dis.* **2017**, *216* (Suppl. S3), S445–S451. [CrossRef]
15. Colombo, A.L.; De Almeida Júnior, J.N.; Guinea, J. Emerging multidrug-resistant Candida species. *Curr. Opin. Infect. Dis.* **2017**, *30*, 528–538. [CrossRef]
16. Alagoz, Z.A.; Yildiz, S.; Buyukbingol, E. Antimicrobial Activities of Some Tetrahydronaphthalene-Benzimidazole Derivatives. *Chemotherapy* **2007**, *53*, 110–113. [CrossRef] [PubMed]
17. Scardina, G.A.; Ruggieri, A.; Messina, P. Chronic hyperplastic candidosis: A pilot study of the efficacy of 0.18% isotretinoin. *J. Oral Sci.* **2009**, *51*, 407–410. [CrossRef] [PubMed]
18. Ioannou, P.; Vamvoukaki, R.; Samonis, G. Rhodotorula species infections in humans: A systematic review. *Mycoses* **2019**, *62*, 90–100. [CrossRef]
19. Xiao, M.; Chen, S.C.; Kong, F.; Fan, X.; Cheng, J.-W.; Hou, X.; Zhou, M.-L.; Wang, H.; Xu, Y.-C. Five-year China Hospital Invasive Fungal Surveillance Net (CHIF-NET) study of invasive fungal infections caused by noncandidal yeasts: Species distribution and azole susceptibility. *Infect. Drug Resist.* **2018**, *11*, 1659–1667. [CrossRef]
20. Martini, K.; Müller, H.; Huemer, H.P.; Höpfl, R. Nail psoriasis masqueraded by secondary infection withRhodotorula mucilaginosa. *Mycoses* **2013**, *56*, 690–692. [CrossRef]
21. Jarros, I.C.; Veiga, F.F.; Corrêa, J.L.; Barros, I.L.E.; Gadelha, M.C.; Voidaleski, M.F.; Pieralisi, N.; Pedroso, R.B.; Vicente, V.A.; Negri, M.; et al. Microbiological and virulence aspects of Rhodotorula mucilaginosa. *EXCLI J.* **2020**, *19*, 687–704. [PubMed]
22. Ge, G.; Li, D.; Mei, H.; Lu, G.; Zheng, H.; Liu, W.; Shi, D. Different toenail onychomycosis due to Rhodotorula mucilaginosa and Candida parapsilosis in an immunocompetent young adult. *Med. Mycol. Case Rep.* **2019**, *24*, 69–71. [CrossRef] [PubMed]
23. Idris, N.F.B.; Huang, G.; Jia, Q.; Yuan, L.; Li, Y.; Tu, Z. Mixed Infection of Toe Nail Caused by Trichosporon asahii and Rhodotorula mucilaginosa. *Mycopathologia* **2020**, *185*, 373–376. [CrossRef]
24. Theelen, B.; Cafarchia, C.; Gaitanis, G.; Bassukas, I.D.; Boekhout, T.; Dawson, T.L. Malassezia ecology, pathophysiology, and treatment. *Med. Mycol.* **2018**, *56* (Suppl. S1), S10–S25. [CrossRef] [PubMed]
25. Handojo, I.; Subagjo, B.; Hadi, S. The effect of topical retinoic acid (Airol) in the treatment of tinea versicolor. *Southeast Asian J. Trop. Med. Public Health* **1977**, *8*, 93–98.
26. Shi, T.-W.; Ren, X.-K.; Yu, H.-X.; Tang, Y.-B. Roles of Adapalene in the Treatment of Pityriasis Versicolor. *Dermatology* **2012**, *224*, 184–188. [CrossRef] [PubMed]
27. Rusu, A.; Tanase, C.; Pascu, G.-A.; Todoran, N. Recent Advances Regarding the Therapeutic Potential of Adapalene. *Pharmaceuticals* **2020**, *13*, 217. [CrossRef]
28. Yazici, S.; Baskan, E.B.; Saricaoglu, H. Long-term remission of recurren pityriasis versicolor with short-term systemic isotretinoin therapy. *J. Dermatol. Cosmetol.* **2018**, *2*, 1. [CrossRef]
29. Quindós, G. Epidemiología de las micosis invasoras: Un paisaje en continuo cambio [Epidemiology of invasive mycoses: A landscape in continuous change]. *Rev. Iberoam. Micol.* **2018**, *35*, 171–178. [CrossRef]
30. Gago, S.; Overton, N.L.D.; Ben-Ghazzi, N.; Novak-Frazer, L.; Read, N.D.; Denning, D.W.; Bowyer, P. Lung colonization by Aspergillus fumigatus is controlled by ZNF77. *Nat. Commun.* **2018**, *9*, 3835. [CrossRef]
31. Latgé, J.-P.; Chamilos, G. Aspergillus fumigatus and Aspergillosis in 2019. *Clin. Microbiol. Rev.* **2019**, *33*, e00140-18. [CrossRef] [PubMed]
32. El-Salam, S.S.A.; Omar, G.A.; Mahmoud, M.T.; Said, M. Comparative study between the effect of topical tazarotene 0.1 gel alone vs. its combination with tioconazole nail paint in treatment of onychomycosis. *Dermatol. Ther.* **2020**, *33*, e14333. [CrossRef] [PubMed]
33. Carratù, M.R.; Marasco, C.; Mangialardi, G.; Vacca, A. Retinoids: Novel immunomodulators and tumour-suppressive agents? *Br. J. Pharmacol.* **2012**, *167*, 483–492. [CrossRef] [PubMed]
34. Zhao, W.; Che, C.; Liu, K.; Zhang, J.; Jiang, N.; Yuan, K.; Zhao, G. Fenretinide Inhibits Neutrophil Recruitment and IL-1β Production in Aspergillus fumigatus Keratitis. *Cornea* **2018**, *37*, 1579–1585. [CrossRef] [PubMed]
35. Krzyściak, P.M.; Pindycka-Piaszczyńska, M.; Piaszczyński, M. Chromoblastomycosis. *Adv. Dermatol. Allergol.* **2014**, *31*, 310–321. [CrossRef]
36. De Brito, A.C.; Bittencourt, M.D.J.S. Chromoblastomycosis: An etiological, epidemiological, clinical, diagnostic, and treatment update. *An. Bras. Dermatol.* **2018**, *93*, 495–506. [CrossRef]

37. Belda, W.; Criado, P.R.; Passero, L.F.D. Case Report: Treatment of Chromoblastomycosis with Combinations including Acitretin: A Report of Two Cases. *Am. J. Trop. Med. Hyg.* **2020**, *103*, 1852–1854. [CrossRef]
38. Bao, F.; Wang, Q.; Yu, C.; Shang, P.; Sun, L.; Zhou, G.; Wu, M.; Zhang, F. Case Report: Successful Treatment of Chromoblastomycosis Caused by Fonsecaea monophora in a Patient with Psoriasis Using Itraconazole and Acitretin. *Am. J. Trop. Med. Hyg.* **2018**, *99*, 124–126. [CrossRef] [PubMed]
39. Bitew, A. Dermatophytosis: Prevalence of Dermatophytes and Non-Dermatophyte Fungi from Patients Attending Arsho Advanced Medical Laboratory, Addis Ababa, Ethiopia. *Dermatol. Res. Pract.* **2018**, *2018*, 8164757. [CrossRef] [PubMed]
40. Méhul, B.; De Coi, N.; Grundt, P.; Genette, A.; Voegel, J.J.; Monod, M. Detection ofTrichophyton rubrumandTrichophyton interdigitale in onychomycosis using monoclonal antibodies against Sub6 (Tri r 2). *Mycoses* **2019**, *62*, 32–40. [CrossRef]
41. Kassir, M.; Karagaiah, P.; Sonthalia, S.; Katsambas, A.; Galadari, H.; Gupta, M.; Lotti, T.; Wollina, U.; Abdelmaksoud, A.; Grabbe, S.; et al. Selective RAR agonists for acne vulgaris: A narrative review. *J. Cosmet. Dermatol.* **2020**, *19*, 1278–1283. [CrossRef] [PubMed]
42. Wolf, J.E. Potential anti-inflammatory effects of topical retinoids and retinoid analogues. *Adv. Ther.* **2002**, *19*, 109–118. [CrossRef] [PubMed]
43. Gaziano, R.; Campione, E.; Iacovelli, F.; Marino, D.; Pica, F.; Di Francesco, P.; Aquaro, S.; Menichini, F.; Falconi, M.; Bianchi, L. Antifungal activity of *Cardiospermum halicacabum* L. (Sapindaceae) against *Trichophyton rubrum* occurs through molecular interaction with fungal Hsp90. *Drug Des. Dev. Ther.* **2018**, *12*, 2185–2193. [CrossRef] [PubMed]
44. Fishman, J.A. Pneumocystis jiroveci. *Semin. Respir. Crit. Care Med.* **2020**, *41*, 141–157. [CrossRef] [PubMed]
45. Gingerich, A.; Norris, K.; Mousa, J. *Pneumocystis* pneumonia: Immunity, Vaccines and Treatments. *Pathogens* **2021**, *10*, 236. [CrossRef]
46. Kato, H.; Samukawa, S.; Takahashi, H.; Nakajima, H. Diagnosis and treatment of Pneumocystis jirovecii pneumonia in HIV-infected or non-HIV-infected patients—difficulties in diagnosis and adverse effects of trimethoprim-sulfamethoxazole. *J. Infect. Chemother.* **2019**, *25*, 920–924. [CrossRef]
47. Pereira-Díaz, E.; Moreno-Verdejo, F.; De La Horra, C.; Guerrero, J.A.; Calderón, E.J.; Medrano, F.J. Changing Trends in the Epidemiology and Risk Factors of Pneumocystis Pneumonia in Spain. *Front. Public Health* **2019**, *7*, 275. [CrossRef]
48. Lei, G.-S.; Zhang, C.; Shao, S.; Jung, H.-W.; Durant, P.J.; Lee, C.-H. All-Trans Retinoic Acid in Combination with Primaquine Clears Pneumocystis Infection. *PLoS ONE* **2013**, *8*, e53479. [CrossRef]
49. Lee, J.-M.; Seo, J.-H.; Kim, Y.-J.; Kim, Y.-S.; Ko, H.-J.; Kang, C.-Y. The restoration of myeloid-derived suppressor cells as functional antigen-presenting cells by NKT cell help and all-trans-retinoic acid treatment. *Int. J. Cancer* **2012**, *131*, 741–751. [CrossRef]
50. Mirza, N.; Fishman, M.; Fricke, I.; Dunn, M.; Neuger, A.M.; Frost, T.J.; Lush, R.M.; Antonia, S.; Gabrilovich, D.I. All-trans-Retinoic Acid Improves Differentiation of Myeloid Cells and Immune Response in Cancer Patients. *Cancer Res.* **2006**, *66*, 9299–9307. [CrossRef]
51. Wiederhold, N.P. Antifungal resistance: Current trends and future strategies to combat. *Infect. Drug Resist.* **2017**, *10*, 249–259. [CrossRef]
52. Waterer, G. Advances in anti-fungal therapies. *Mycopathologia* **2021**, 1–8. [CrossRef]
53. Khalil, S.; Bardawil, T.; Stephan, C.; Darwiche, N.; Abbas, O.; Kibbi, A.G.; Nemer, G.; Kurban, M. Retinoids: A journey from the molecular structures and mechanisms of action to clinical uses in dermatology and adverse effects. *J. Dermatol. Treat.* **2017**, *28*, 684–696. [CrossRef]
54. Ferreira, R.; Napoli, J.; Enver, T.; Bernardino, L.; Ferreira, L. Advances and challenges in retinoid delivery systems in regenerative and therapeutic medicine. *Nat. Commun.* **2020**, *11*, 4265. [CrossRef]
55. Zuccari, G.; Baldassari, S.; Alfei, S.; Marengo, B.; Valenti, G.; Domenicotti, C.; Ailuno, G.; Villa, C.; Marchitto, L.; Caviglioli, G. D-α-Tocopherol-Based Micelles for Successful Encapsulation of Retinoic Acid. *Pharmaceuticals* **2021**, *14*, 212. [CrossRef] [PubMed]
56. Austin, A.; Lietman, T.; Rose-Nussbaumer, J. Update on the Management of Infectious Keratitis. *Ophthalmology* **2017**, *124*, 1678–1689. [CrossRef]
57. Kaur, R.; Kaur, R.; Singh, C.; Kaur, S.; Goyal, A.K.; Singh, K.K.; Singh, B. Inhalational Drug Delivery in Pulmonary Aspergillosis. *Crit. Rev. Ther. Drug Carr. Syst.* **2019**, *36*, 183–217. [CrossRef] [PubMed]
58. Orienti, I.; Francescangeli, F.; De Angelis, M.L.; Fecchi, K.; Bongiorno-Borbone, L.; Signore, M.; Peschiaroli, A.; Boe, A.; Bruselles, A.; Costantino, A.; et al. A new bioavailable fenretinide formulation with antiproliferative, antimetabolic, and cytotoxic effects on solid tumors. *Cell Death Dis.* **2019**, *10*, 529. [CrossRef] [PubMed]
59. Orienti, I.; Nguyen, F.; Guan, P.; Kolla, V.; Calonghi, N.; Farruggia, G.; Chorny, M.; Brodeur, G.M. A Novel Nanomicellar Combination of Fenretinide and Lenalidomide Shows Marked Antitumor Activity in a Neuroblastoma Xenograft Model. *Drug Des. Dev. Ther.* **2019**, *13*, 4305–4319. [CrossRef] [PubMed]
60. Di Paolo, D.; Pastorino, F.; Zuccari, G.; Caffa, I.; Loi, M.; Marimpietri, D.; Brignole, C.; Perri, P.; Cilli, M.; Nico, B.; et al. Enhanced anti-tumor and anti-angiogenic efficacy of a novel liposomal fenretinide on human neuroblastoma. *J. Control. Release* **2013**, *170*, 445–451. [CrossRef]
61. Bulbake, U.; Doppalapudi, S.; Kommineni, N.; Khan, W. Liposomal Formulations in Clinical Use: An Updated Review. *Pharmaceutics* **2017**, *9*, 12. [CrossRef]
62. Charoenputtakhun, P.; Opanasopit, P.; Rojanarata, T.; Ngawhirunpat, T. All-trans retinoic acid-loaded lipid nanoparticles as a transdermal drug delivery carrier. *Pharm. Dev. Technol.* **2013**, *19*, 164–172. [CrossRef]

63. Raza, K.; Singh, B.; Lohan, S.; Sharma, G.; Negi, P.; Yachha, Y.; Katare, O.P. Nano-lipoidal carriers of tretinoin with enhanced percutaneous absorption, photostability, biocompatibility and anti-psoriatic activity. *Int. J. Pharm.* **2013**, *456*, 65–72. [CrossRef]
64. Sinico, C.; Manconi, M.; Peppi, M.; Lai, F.; Valenti, D.; Fadda, A.M. Liposomes as carriers for dermal delivery of tretinoin: In vitro evaluation of drug permeation and vesicle–skin interaction. *J. Control. Release* **2005**, *103*, 123–136. [CrossRef]
65. Manca, M.L.; Manconi, M.; Nacher, A.; Carbone, C.; Valenti, D.; Maccioni, A.M.; Sinico, C.; Fadda, A.M. Development of novel diolein–niosomes for cutaneous delivery of tretinoin: Influence of formulation and In vitro assessment. *Int. J. Pharm.* **2014**, *477*, 176–186. [CrossRef]
66. AlGahtani, M.S.; Ahmad, M.Z.; Ahmad, J. Nanoemulgel for Improved Topical Delivery of Retinyl Palmitate: Formulation Design and Stability Evaluation. *Nanomaterials* **2020**, *10*, 848. [CrossRef]
67. Íñigo, M.; Del Pozo, J.L. Fungal biofilms: From bench to bedside. *Rev. Espanola Quimioter.* **2019**, *31* (Suppl. S1), 35–38.
68. Costa-Orlandi, C.B.; Sardi, J.C.O.; Pitangui, N.S.; De Oliveira, H.C.; Scorzoni, L.; Galeane, M.C.; Medina-Alarcón, K.P.; Melo, W.C.M.A.; Marcelino, M.Y.; Braz, J.D.; et al. Fungal Biofilms and Polymicrobial Diseases. *J. Fungi* **2017**, *3*, 22. [CrossRef] [PubMed]
69. Ben-Ami, R. Treatment of Invasive Candidiasis: A Narrative Review. *J. Fungi* **2018**, *4*, 97. [CrossRef] [PubMed]
70. Kowalski, C.H.; Morelli, K.A.; Schultz, D.; Nadell, C.D.; Cramer, R.A. Fungal biofilm architecture produces hypoxic microenvironments that drive antifungal resistance. *Proc. Natl. Acad. Sci. USA* **2020**, *117*, 22473–22483. [CrossRef] [PubMed]
71. Tan, F.; She, P.; Zhou, L.; Liu, Y.; Chen, L.; Luo, Z.; Wu, Y. Bactericidal and Anti-biofilm Activity of the Retinoid Compound CD437 Against Enterococcus faecalis. *Front. Microbiol.* **2019**, *10*, 2301. [CrossRef] [PubMed]
72. Padder, S.A.; Prasad, R.; Shah, A.H. Quorum sensing: A less known mode of communication among fungi. *Microbiol. Res.* **2018**, *210*, 51–58. [CrossRef] [PubMed]

pharmaceuticals

Article

Design, Synthesis and Anticandidal Evaluation of Indazole and Pyrazole Derivatives

Karen Rodríguez-Villar [1], Alicia Hernández-Campos [2], Lilián Yépez-Mulia [3],
Teresita del Rosario Sainz-Espuñes [4], Olivia Soria-Arteche [4], Juan Francisco Palacios-Espinosa [4],
Francisco Cortés-Benítez [4], Martha Leyte-Lugo [5], Bárbara Varela-Petrissans [4],
Edgar A. Quintana-Salazar [4] and Jaime Pérez-Villanueva [4,*]

[1] Doctorado en Ciencias Biológicas y de la Salud, Universidad Autónoma Metropolitana (UAM),
Ciudad de México 04960, Mexico; qkarenrodv@hotmail.com
[2] Departamento de Farmacia, Facultad de Química, Universidad Nacional Autónoma de México (UNAM),
Ciudad de México 04510, Mexico; hercam@unam.mx
[3] Unidad de Investigación Médica en Enfermedades Infecciosas y Parasitarias, UMAE Hospital de Pediatría,
Centro Médico Siglo XXI, Instituto Mexicano del Seguro Social, Ciudad de México 06720, Mexico;
lilianyepez@yahoo.com
[4] Departamento de Sistemas Biológicos, División de Ciencias Biológicas y de la Salud,
Universidad Autónoma Metropolitana-Xochimilco (UAM-X), Ciudad de México 04960, Mexico;
trsainz@correo.xoc.uam.mx (T.d.R.S.-E.); soriao@correo.xoc.uam.mx (O.S.-A.);
jpalacios@correo.xoc.uam.mx (J.F.P.-E.); jcortesb@correo.xoc.uam.mx (F.C.-B.); bvra5302@gmail.com (B.V.-P.);
edgarqsl2811@gmail.com (E.A.Q.-S.)
[5] Catedrático CONACyT Comisionado al Departamento de Sistemas Biológicos,
División de Ciencias Biológicas y de la Salud, Universidad Autónoma Metropolitana-Xochimilco (UAM-X),
Ciudad de México 04960, Mexico; mleyte@correo.xoc.uam.mx
* Correspondence: jpvillanueva@correo.xoc.uam.mx; Tel.: +52-5-54-83-72-59; Fax: +52-5-55-94-79-29

Citation: Rodríguez-Villar, K.;
Hernández-Campos, A.; Yépez-Mulia,
L.; Sainz-Espuñes, T.d.R.; Soria-
Arteche, O.; Palacios-Espinosa, J.F.;
Cortés-Benítez, F.; Leyte-Lugo, M.;
Varela-Petrissans, B.; Quintana-
Salazar, E.A.; et al. Design, Synthesis
and Anticandidal Evaluation of
Indazole and Pyrazole Derivatives.
Pharmaceuticals **2021**, *14*, 176.
https://doi.org/10.3390/ph14030176

Academic Editor: Jong Heon Kim

Received: 24 December 2020
Accepted: 20 February 2021
Published: 24 February 2021

Publisher's Note: MDPI stays neutral
with regard to jurisdictional claims in
published maps and institutional affiliations.

Abstract: Candidiasis, caused by yeasts of the genus Candida, is the second cause of superficial and mucosal infections and the fourth cause of bloodstream infections. Although some antifungal drugs to treat candidiasis are available, resistant strains to current therapies are emerging. Therefore, the search for new candicidal compounds is certainly a priority. In this regard, a series of indazole and pyrazole derivatives were designed in this work, employing bioisosteric replacement, homologation, and molecular simplification as new anticandidal agents. Compounds were synthesized and evaluated against *C. albicans*, *C. glabrata*, and *C. tropicalis* strains. The series of 3-phenyl-1*H*-indazole moiety (**10a–i**) demonstrated to have the best broad anticandidal activity. Particularly, compound **10g**, with *N*,*N*-diethylcarboxamide substituent, was the most active against *C. albicans* and both miconazole susceptible and resistant *C. glabrata* species. Therefore, the 3-phenyl-1*H*-indazole scaffold represents an opportunity for the development of new anticandidal agents with a new chemotype.

Keywords: anticandidal activity; indazole; pyrazole; 3-phenyl-1*H*-indazole; drug design

1. Introduction

Candidiasis, caused by yeasts of the genus *Candida*, is the second cause of superficial and mucosal infections and the fourth cause of bloodstream infections [1,2]. *Candida* species are normal inhabitants of the oropharynx, gastrointestinal tract, and vagina in humans. However, these species are classified as opportunistic and can change from harmless to pathogenic upon variation of the host environment by physiological or non-physiological changes [1,3].

One of the most frequent mucosal infections is vulvovaginal candidiasis (VVC), which affects millions of women every year and is considered an important public health problem. An estimated 70–75% of women will be affected at some point in their lifetimes, of which approximately 40–50% of initially infected women will have two episodes, and 5–10% of them will develop recurrent VVC [3]. Indeed, this disease is associated with enhanced susceptibility to human immunodeficiency virus (HIV) infection, pregnancy complications, and increased risks of stillbirth or neonatal death. Moreover, when VVC is left untreated, many complications have been associated, such as pelvic inflammatory disease, infertility, ectopic pregnancy, pelvic abscess, spontaneous abortion, and menstrual disorders [3,4]. On the other hand, invasive candidiasis is of greater concern because it is the most common fungal disease among hospitalized patients associated with a 40% mortality rate [2,5,6]. According to estimates, invasive candidiasis affects more than 250,000 people globally every year, and it is the cause of more than 50,000 deaths [5]. In fact, the Centers for Disease Control and Prevention (CDC) estimated that approximately 25,000 cases of candidemia occur in the USA every year [7].

Candida albicans is the main etiologic species associated with VVC and invasive candidiasis globally [2,4,5]. However, in the last decades, other *Candida* species such as *C. glabrata*, *C. parapsilosis*, *C. tropicalis*, *C. krusei*, and lastly *C. auris*, have also been demonstrated to cause VVC and invasive candidiasis [2,3,6,7]. The conventional treatments for *Candida* infections are limited to amphotericin B (deoxycholate and various lipid formulations), echinocandins (anidulafungin, caspofungin, and micafungin), azoles (fluconazole, itraconazole, and voriconazole) and flucytosine (5-FC) [6–9]. Nowadays, the preferred treatment for mucosal and invasive candidiasis is fluconazole, which is active against most *Candida* species and is as effective as amphotericin B, but with fewer side effects. Nevertheless, fluconazole has limited activity against *C. glabrata* and *C. krusei*, and reports of fluconazole-resistant strains have increased [2,5,9].

Indazole is an important heterocyclic scaffold in medicinal chemistry since it is associated with a broad range of biological activities, e.g., anti-inflammatory, antiprotozoal, antihypertensive, anticancer, antitumor, antifungal, antibacterial, anti-HIV, antiplatelet, and others [10–17]. Previous studies by our group showed that some 2,3-diphenil-2*H*-indazole derivatives substituted with methyl ester or carboxylic acid groups (compounds **3a–d**; Figure 1) have in vitro activity against *C. albicans* and *C. glabrata*. Particularly, compounds **3a** and **3c** showed a minimum inhibitory concentration (MIC) of 3.807 mM and 15.227 mM against *C. albicans* and *C. glabrata*, respectively [17]. However, the information available regarding the structural requirements for anticandidal activity is still limited. Therefore, it is necessary to design and synthesize new indazole derivatives to gain knowledge about the structural modifications required to improve the candicidal activity. As part of our efforts in the search for candicidal compounds, initially, fourteen 2,3-diphenyl-2*H*-indazole derivatives (series 1; Figure 1) were designed by bioisosteric replacement and homologation. It is worth mentioning that esters (e.g., **3a** and **3c**) can be easily hydrolyzed to carboxylic acids and are usually considered prodrugs; therefore, we proposed a replacement by amides that are a similar functional group, but slightly more stable. Different amines were considered to explore the effect of bulky substituents in the activity. Also, fourteen derivatives, which include 3,5-1*H*-pyrazole, 2-phenyl-2*H*-indazole and 3-phenyl-1*H*-indazole, were designed employing molecular simplification strategies by removing rings from the original 2,3-diphenil-2*H*-indazole scaffold (series 2, Figure 1). This strategy has been used to reduce structural complexity, improve physicochemical properties, and to find the minimum molecular features that are necessary for the anticandidal activity. Compounds proposed were synthesized and tested in vitro against four *Candida* strains. Then, taking advantage of the results revealed by the biological evaluations, the third series of six 3-phenyl-1*H*-indazole carboxamides were synthesized and tested.

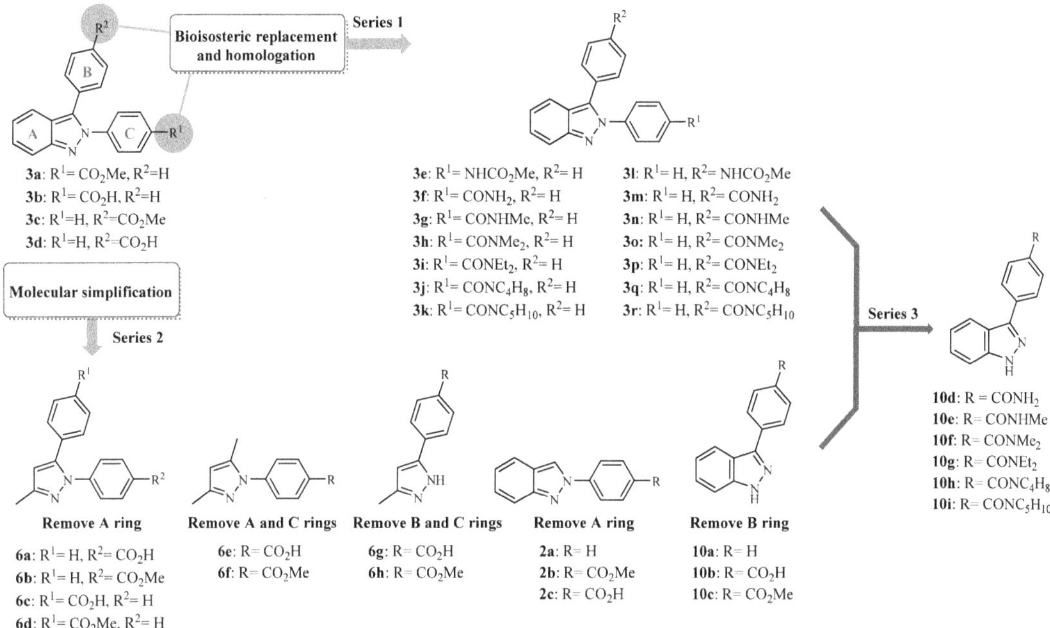

Figure 1. Compounds designed by bioisosteric replacement, homologation, and molecular simplification.

2. Results and Discussion

2.1. Synthesis of 2-Phenyl-2H-Indazole and 2,3-Diphenyl-2H-Indazole Derivatives

The synthesis of 2,3-diphenyl-2*H*-indazole derivatives **3a–r** was carried out as outlined in Scheme 1. 2-Phenyl-2*H*-indazoles (**2a**, **2b**, **2d**, and **2e**) were synthesized by the Cadogan reaction starting from 2-nitrobenzaldehyde (**1**) and the appropriate *p*-substituted aniline under reflux with ethanol to obtain the Schiff's base, which was then reduced as well as cyclized with P(OEt)₃ [17,18]. Compound **2c** was obtained by basic hydrolysis of **2b**. Most of the 2,3-diphenyl-2*H*-indazole derivatives were prepared by a palladium-catalyzed arylation of the corresponding 2-phenyl-2*H*-indazole with substituted halobenzene as previously reported [17,19], whereas compound **3e** was synthesized by a Suzuki-Miyaura coupling of methyl (4-(3-bromo-2*H*-indazol-2-yl)phenyl)carbamate with phenylboronic acid [20]. Esters **3a** and **3c** were hydrolyzed with NaOH to give **3b** and **3d**, respectively, in good yields. Additionally, carboxylic acids **3b** and **3d** were converted to acyl chlorides and then treated with the corresponding amine yielding the carboxamides **3g–k** and **3n–r**. Compounds **3f** and **3m** were prepared by a palladium-catalyzed arylation with iodobenzene or 4-halobenzonitrile followed by acidic conversion to the appropriate carboxamide. Yields for 2-phenyl-2*H*-indazole derivatives (**2a–e**) were moderate to high. The palladium-catalyzed arylation yields were slightly lower than that of previously reported data [17,19]. Compound **3e** could not be obtained by this method; instead, a selective bromination of **2d** at position 3, followed by a Suzuki-Miyaura coupling reaction, was applied to give the product in moderate yield (58%).

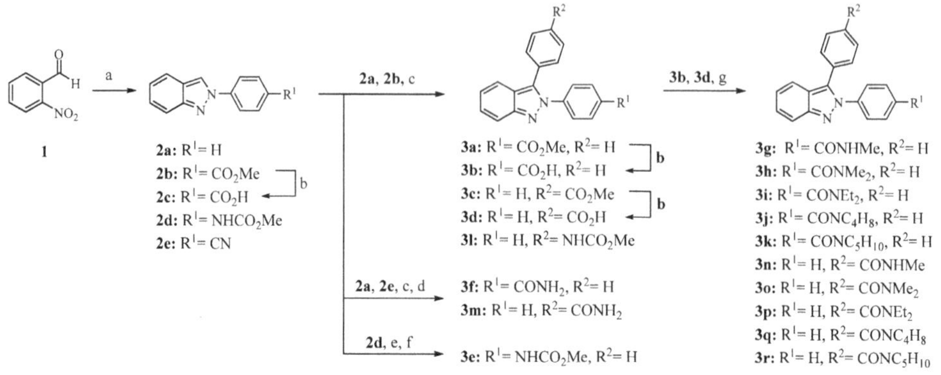

Reagents and conditions: (a) 1. $R^1C_6H_4NH_2$, EtOH, reflux; 2. $P(OEt)_3$, 150°C; (b) NaOH, MeOH/H_2O, reflux; (c) $R^2C_6H_4I$, PPh$_3$, Ag$_2$CO$_3$, Pd(dppf)Cl$_2$•DCM, H_2O, 55 °C; (d) H_2SO_4, rt; (e) Br$_2$ 1M, AcOH, rt; (f) Phenylboronic acid, Na$_2$CO$_3$, Pd(PPh$_3$)$_4$, DME/H_2O (3:1), MW at 155 °C; (g) 1. SOCl$_2$, benzene, reflux; 2. Amine, rt.

Scheme 1. Synthesis of 2*H*-indazole derivatives **2a–e** and **3a–r**.

2.2. Synthesis of 1H-Pyrazole and 1H-Indazole Derivatives

The synthesis of 3,5-disubstituted pyrazole derivatives is displayed in Scheme 2. Pyrazoles **6a**, **6e**, and **6h** were prepared by cyclocondensation of the 1,3-dicarbonyl compounds (**4a–c**) and 4-hydrazinylbenzoic acid (**5a**) or hydrazine (**5b**) with diluted H_2SO_4 in MeOH at room temperature (Scheme 2, Method A) [21]. Compounds **6b** and **6f** were obtained by Fischer–Speier esterification from carboxylic acids **6a** and **6e**, respectively, whereas hydrolysis of **6h** produced the carboxylic acid **6g** in good yields. Compound **6c** was synthesized through a two-step sequence starting from 5-methyl-2-phenyl-2,4-dihydro-3*H*-pyrazol-3-one (**7**), which was first converted to 5-chloro-pyrazole intermediary (**7a**) and then coupled with 4-carbomethoxyphenylboronic acid pinacol ester under microwave irradiation (Scheme 2, Method B) [20]. *O*-Methylation of **6c** with methyl iodide at room temperature yielded the ester **6d**. It is important to mention that all pyrazole derivatives were obtained with moderate to good yields following the proposed chemical synthesis.

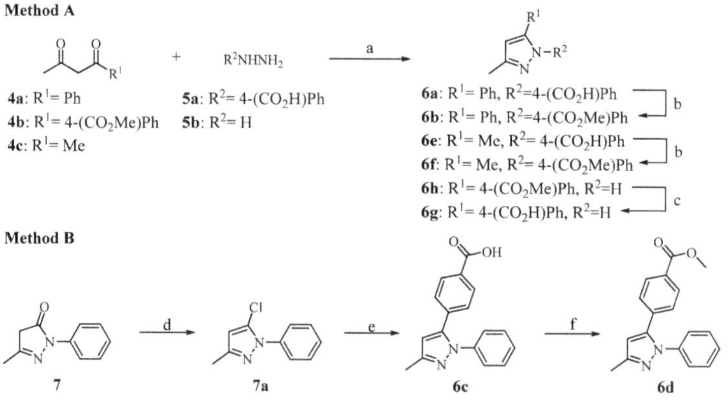

Reagents and conditions: (a) 5% H_2SO_4 in MeOH, rt; (b) 5% H_2SO_4 in MeOH, reflux; (c) 60% H_2SO_4, reflux; (d) POCl$_3$, reflux; (e) 4-Methoxycarbonylphenylboronic acid pinacol ester, Pd(PPh$_3$)$_4$, Na$_2$CO$_3$, MeCN/H_2O (4:1), MW at 175 °C; (f) MeI, Na$_2$CO$_3$, DMF, rt.

Scheme 2. Synthesis of pyrazole derivatives **6a–g**.

Finally, a slight modification of the method reported by Huff et al. was performed, since 3-bromo-1*H*-indazole (**8**) was coupled under microwave irradiation with phenylboronic acid (**9a**) or 4-methoxycarbonylphenylboronic acid pinacol ester (**9b**) to afford the 3-phenyl-1*H*-indazole derivatives **10a** and **10b**, respectively (Scheme 3) [22]. Fischer–Speier esterification of compound **10b** gave **10c** in good yield.

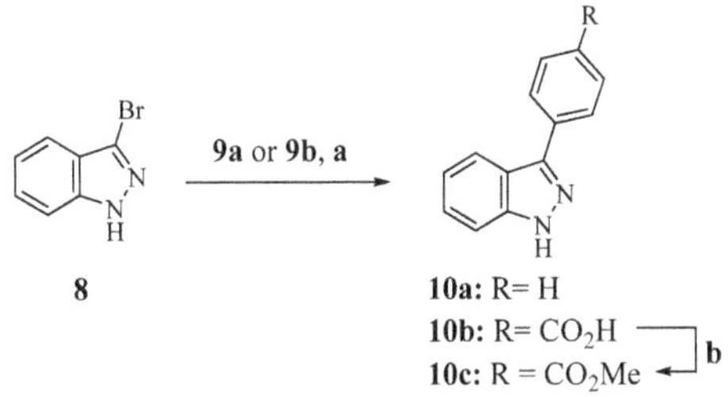

8 → **9a** or **9b, a** →

10a: R= H
10b: R= CO$_2$H
10c: R = CO$_2$Me — **b**

Reagents and conditions:
(a) Phenylboronic acid or 4-methoxycarbonylphenylboronic acid pinacol ester, Pd(AcO)$_2$, Na$_2$CO$_3$, PPh$_3$, H$_2$O/n-propanol (1:3), 150 °C, MW;
(b) 5% H$_2$SO$_4$ in MeOH, reflux.

Scheme 3. Synthesis of 3-phenyl-1*H*-indazole derivatives **10a–c**. All synthesized compounds were characterized by ^{1}H NMR and ^{13}C NMR spectra and the new structures were also characterized by mass spectrometry. The nuclear magnetic resonance and mass spectra of compounds can be found in Figures S2–S71 in the Supplementary Materials.

2.3. Anticandidal Activity

Some structural changes were performed to gain knowledge about the features that would improve the anticandidal activity of 2,3-diphenyl-2*H*-indazole derivatives. The methyl ester group of **3a** and **3c**, and the carboxylic acid group of **3b** and **3d** were replaced by methyl carbamate (**3e**, **3l**), as well as carboxamide and *N*-substituted carboxamides (**3f–k** and **3m–r**). The anticandidal activity of these new derivatives was tested against *C. albicans*, miconazole susceptible and resistant *C. glabrata* (ATCC 90030 and 32554, respectively) and *C. tropicalis* (ATCC 750) using the cylinder-plate method [23], which results are presented in Table 1. Overall, *N*-substituted carboxamides **3h–k** and **3n–r** showed activity against *C. albicans* at 1 mM. Particularly, compound **3j** showed the maximum inhibition of *C. albicans* growth at 10 mM, but it also had activity at 1 mM. It also showed weak activity against miconazole-resistant *G. glabrata* at 10 mM. Compounds **3h** and **3p** were active against *C. albicans*, and miconazole-resistant *C. glabrata* species at 1 mM. Regarding carboxamide derivatives, compound **3f** was active just against *C. tropicalis* at 10 mM, while compound **3m** had weak anticandidal activity against *C. albicans* and miconazole susceptible *C. glabrata* at 10 mM. Concerning methyl carbamate derivatives **3e** and **3l**, they did not show activity against *C. albicans*, but **3e** was active against *C. tropicalis* at 10 mM and **3l** had activity against susceptible and resistant miconazole *C. glabrata* at 10 mM. All these results suggest that compounds **3h** and **3p** (*N,N*-dimethyl and *N,N*-diethyl carboxamide, respectively) had the best candicidal activity against *C. albicans* and miconazole resistant *C. glabrata* followed by **3j** and **3r** (pyrrolidine and piperidine carboxamide, respectively). It is worth noticing that these compounds demonstrated better activity against the miconazole-resistant *C. glabrata* than to the susceptible *C. glabrata*.

Table 1. In vitro growth inhibition of 2,3-diphenyl-2H-indazoles, 2-phenyl-2H-indazole, 3,5-1H-pyrazole, and 3-phenyl-1H-indazole derivatives against *Candida* species.

ID	R[1]	R[2]	*C. albicans*			*C. glabrata*			*C. glabrata* [1]			*C. tropicalis*
			[ ] mM									
			10	1	0.1	10	1	0.1	10	1	0.1	10
3a	CO_2Me	H	+	–	–	–	–	–	–	–	–	–
3b	CO_2H	H	+	–	–	+	–	–	+	–	–	+ +
3c	H	CO_2Me	–	–	–	–	–	–	–	–	–	–
3d	H	CO_2H	+	–	–	–	–	–	+	–	–	–
3e	$NHCO_2Me$	H	–	–	–	–	–	–	–	–	–	+ +
3f	$CONH_2$	H	–	–	–	–	–	–	–	–	–	+ +
3g	CONHMe	H	–	–	–	+	–	–	–	–	–	–
3h	$CONMe_2$	H	+	+	–	–	–	–	+	+	–	–
3i	$CONEt_2$	H	+	+	–	–	–	–	–	–	–	–
3j	$CONC_4H_8$	H	+ +	+	–	–	–	–	+	–	–	–
3k	$CONC_5H_{10}$	H	+	+	–	–	–	–	–	–	–	–
3l	H	$NHCO_2Me$	–	–	–	+	–	–	+	–	–	–
3m	H	$CONH_2$	+	–	–	+	–	–	–	–	–	–
3n	H	CONHMe	+	+	–	–	–	–	–	–	–	–
3o	H	$CONMe_2$	+	+	–	–	–	–	–	–	–	–
3p	H	$CONEt_2$	+	+	–	–	–	–	+	+	–	–
3q	H	$CONC_4H_8$	+	+	–	–	–	–	–	–	–	–
3r	H	$CONC_5H_{10}$	+	+	–	–	–	–	+	–	–	–
6a	Ph	4-(CO_2H)Ph	–	–	–	–	–	–	+	+	–	–
6b	Ph	4-(CO_2Me)Ph	+	+	–	+	–	–	+	+	–	–
6c	4-(CO_2H)Ph	Ph	+	+	–	–	–	–	+	+	–	–
6d	4-(CO_2Me)Ph	Ph	+	+	–	–	–	–	+	+	–	–
6e	Me	4-(CO_2H)Ph	–	–	–	–	–	–	+	+	–	–
6f	Me	4-(CO_2Me)Ph	+	+	–	+	+	–	+	+	–	+ +
6g	4-(CO_2H)Ph	H	+	+	–	–	–	–	–	–	–	–
6h	4-(CO_2Me)Ph	H	+	+	–	–	–	–	–	–	–	–
2a	H	–	–	+	+	–	–	–	–	–	–	–
2b	CO_2H	–	–	+	+	+	+[2]	+ +	–	–	–	–
2c	CO_2Me	–	–	–	+	–	+	–	–	+ +	+ +	–
10a	H	–	–	+ + +	+ +	+	+ +	+	–	+ + +	+ +	+	–
10b	CO_2H	–	–	+ +	+ +	–	+	+	+	–	–	–	–
10c	CO_2Me	–	–	+ + +	+ + +	+ +	+ +	+	+	–[2]	+ +	–	–
V [3]			–			–			–			–
M [4]			+ + +			+ + +			+			+ +
F [5]			+ +			–			+ + +			+

[1] Miconazole resistant strain; [2] compound precipitation was observed at 10 mM. (–) inactive (<12 mm); (+) weakly active (13–16 mm); (+ +) moderately active (17–21 mm); (+ + +) strongly active (22–29 mm). [3] Vehicle; [4] Miconazole; [5] Fluconazole.

On the other hand, to understand the effect of the molecular simplification of the indazole nucleus on the anticandidal activity 3,5-1H-pyrazole (**6a–h**), 2-phenyl-1H-indazole (**2a–c**) and 3-phenyl-1H-indazole derivatives (**10a–c**) were tested against *Candida* strains. Their anticandidal activity is presented in Table 1. Noteworthy, most of the 3,5-1H-pyrazole derivatives (**6b–d** and **6f**) showed anticandidal activity at lower concentrations than those of 2,3-diphenyl indazole analogs (**3a–d**). These were active against *C. albicans* and miconazole-resistant *C. glabrata* at 1 mM. Compounds **6a** and **6e** showed only activity against miconazole-resistant *C. glabrata* (at 1 mM), and **6g** and **6h** were only active against *C. albicans* at 1 mM. It is worth noticing that the pyrazole **6f** exhibited broad growth inhibition against *C. albicans*, miconazole susceptible and resistant *C. glabrata-strains* and *C. tropicalis*.

Regarding the activity of 2-phenyl-2H-indazole (**2a–c**) and 3-phenyl-1H-indazole (**10a–c**) derivatives, these compounds showed anticandidal activity at lower concentrations than 2,3-diphenyl indazole and 3,5-1H-pyrazole derivatives. Compound **2b** had the best activity against *C. albicans* and miconazole susceptible *C. glabrata* strain at 100 µM and 1 mM respectively, while **2c** had the best activity against miconazole resistant *C. glabrata* at 1 mM. The 3-phenyl-1*H*-indazole compounds, **10a** and **10c**, displayed broad activity against *C. albicans*, miconazole susceptible and resistant *C. glabrata*. Notably, the 3-phenylindazole **10a** had activity against *C. albicans* and the miconazole resistant *C. glabrata* strain at 100 µM, and to miconazole susceptible *C. glabrata* at 1 mM, whereas **10c** had the best inhibition against *C. albicans* and miconazole susceptible *C. glabrata* at 100 µM. Concerning **10b**, it showed activity against *C. albicans* at 1 mM and miconazole susceptible *C. glabrata* at 100 µM. Additionally, to discard the DMSO effect in the growth inhibition, a control of solvent was included in all assays (Figure S1, supporting material).

According to these results, principally structure-activity relationship features derived from compounds tested are exhibited in Figure 2. The replacement of the carbonyl group by *N*-substituted carboxamides on 2,3-diphenyl-2*H*-indazole nucleus led to an increase in the growth inhibitory activity against *C. albicans*. The molecular simplification of 2,3-diphenylindazole improved the antifungal activity against *Candida* strains tested. The 3,5-disubstituted pyrazoles displayed better growth inhibition than the 2,3-diphenylindazole analogs against *C. albicans* and miconazole resistant *C. glabrata*. At the same time, 2-phenyl-2*H*-indazoles were active against *C. albicans* and both *C. glabrata* strains. It is important to emphasize that 3-phenyl-1*H*-indazole derivatives showed the highest antifungal activity against all *Candida* strains tested except for *C. tropicalis*.

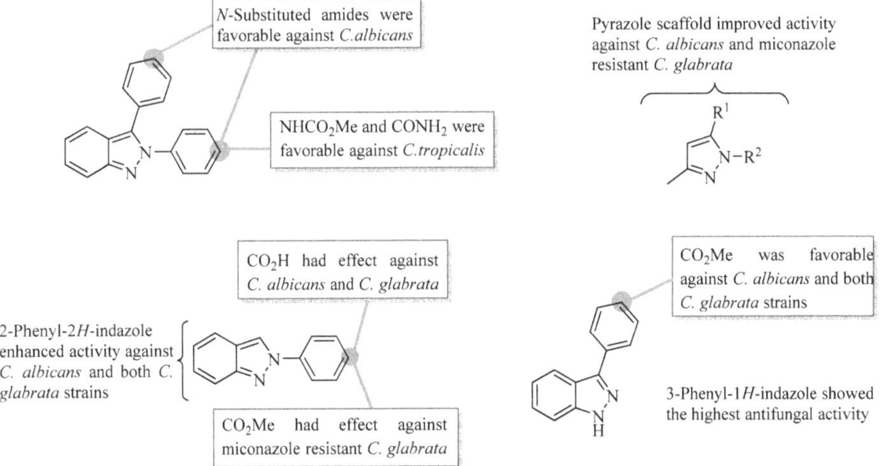

Figure 2. Principally structure-activity relationship features derived from compounds tested.

It is important to mention that prototype compound **10a**, which represents the nucleus for the most active compounds, was also tested on three representative human cell lines employing the MTT method [17]. The results showed 100% of cellular viability on HeLa (human cervix), K562 (human chronic myelogenous leukemia) and SW620 (human colon) cell lines after 48h of compound exposition at 50 µM.

2.4. Synthesis and Determination of Minimum Inhibitory Concentration (MIC) of 3-Phenyl-1H-Indazole Derivatives

Considering all the data obtained about 2,3-diphenyl indazole amides, as well as molecular simplification used to select the best scaffold, six new 3-phenyl-1*H*-indazole derivatives bearing small or bulky amides were synthesized (**10d–i**). These compounds were synthetized through the Suzuki-Miyaura reaction of coupling 3-bromo-1*H*-indazole (**8**) with several 4-phenylboronic acid pinacol esters (**9c–h**) (Scheme 4) [22]. Yields of carboxamide derivatives were lower compared to their unsubstituted or carboxylic acid analogs. The above indicates that phenylboronic acid pinacol esters with a carboxamide group are less reactive.

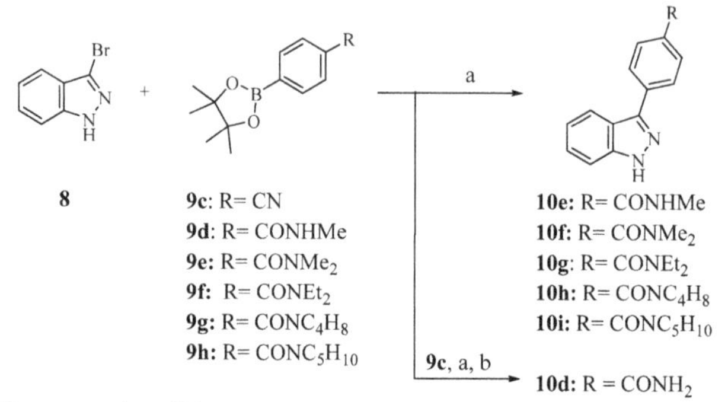

Reagents and conditions:
(a) Pd(AcO)$_2$, Na$_2$CO$_3$, PPh$_3$, H$_2$O/n-propanol (1:3), MW at 150 °C; (b) H$_2$SO$_4$, rt.

Scheme 4. Synthesis of carboxamide 3-phenyl-1*H*-indazole derivatives.

The in vitro anticandidal assays against *C. albicans* and *C. glabrata* strains of 3-phenyl-1*H*-indazole derivates **10a–i** were carried out following the cylinder-plate method previously described [23]. The minimum inhibitory concentration (MIC) values were determined and are shown in Table 2. It is noteworthy that fluconazole and miconazole were used as reference drugs, and in addition, the purity of the compounds was determined by quantitative NMR (purity > 95%) to ensure that the observed MIC values were caused by the reported compounds (Table S1 and Figure S72, supporting material). Most of the amides kept their activity against all *Candida* strains, similar to compounds **10a–c**. Indeed, some of them showed a slight increase in anticandidal activity. From this series, compounds **10c, 10g**, and **10i** had the best activity against *C. albicans*. In contrast, **10f** is the most active compound against both miconazole susceptible and resistant *C. glabrata*. In addition, **10g** was active against miconazole-resistant *C. glabrata*. It is important to emphasize that compounds with bulky carboxamides (**10g–i**) had better anticandidal activity against *C. albicans* than small carboxamides (**10d–f**). These results indicated that the carboxamide group on the 3-phenyl-1*H*-indazole scaffold kept antifungal activity against tested *C. albicans* and *C. glabrata* strains. The MIC determination of the compounds tested showed to be less active than the reference drugs. However, the antifungal activity of the proposed compounds is still relevant and can be considered as leads for the development of new anticandidal agents.

Table 2. Minimum Inhibitory Concentration (μM) of 3-phenyl-1*H*-indazole derivatives.

ID	R	*C. albicans*	*C. glabrata*	*C. glabrata* [1]
10a	H	100	150	100
10b	CO$_2$H	200	100	–
10c	CO$_2$Me	75	100	200
10d	CONH$_2$	150	150	100
10e	CONHMe	150	100	100
10f	CONMe$_2$	150	75	75
10g	CONEt$_2$	75	100	75
10h	CONC$_4$H$_8$	100	100	150
10i	CONC$_5$H$_{10}$	75	100	125
Fluconazole		50	–	50
Miconazole		<50	<50	–

[1] Miconazole resistant; (–) inactive.

3. Materials and Methods

3.1. Chemistry

All chemicals and starting materials were obtained from Sigma-Aldrich (Toluca, MC, Mexico). Reactions were monitored by TLC on 0.2 mm percolated silica gel 60 F254 plates (Merck, Darmstadt, Germany) and visualized by irradiation with a UV lamp (Upland, CA, USA). Silica gel 60 (70–230 mesh, Düren, Germany) was used for column chromatography. Melting points were determined in open capillary tubes with a Büchi M-565 melting point apparatus (Flawil, Switzerland). Microwave-assisted reactions were carried out in a monomodal reactor equipped with a hydraulic pressure sensing device and an infrared temperature-sensor (Anton Parr Monowave 300, Anton Parr, Graz, Austria). ^{1}H-NMR and ^{13}C-NMR spectra were measured with an Agilent DD2 spectrometer (Santa Clara, CA, USA), operating at 600 MHz and 151 MHz for ^{1}H and ^{13}C, respectively. Chemical shifts are given in parts per million relatives to tetramethylsilane (TMS, $\delta = 0$); J values are given in Hz. Splitting patterns are expressed as follow: s, singlet; d, doublet; t, triplet; q, quartet; dd, doublet of doublet; dt, doublet of triplete; ddd, doublet of doublets of doublets; m, multiplet; bs, broad singlet. High-resolution mass spectra were recorded on a Bruker ESI/APCI-TOF, MicroTOF-II-Focus spectrometer (Billerica, MA, USA) by electrospray ionization (ESI) and low-resolution mass spectra were recorded on a Water Xevo TQ-MS by ESI. The purity of compounds **10a–i** was determined by quantitative NMR spectroscopy (qNMR) using sodium 2,2-dimethyl- 2-silapentane-5-sulfonate (DSS) as internal standard (Cambridge Isotope Laboratories, Inc., Tewksbury, MA, USA). All compounds were named using the automatic name generator tool implemented in ChemDraw 19.1.1.21 software (PerkinElmer, Waltham, MA, USA), according to IUPAC rules.

3.2. General Procedure for the Synthesis

3.2.1. 2-Phenyl-2*H*-Indazole Derivatives (**2a**, **2b**, **2d**, and **2e**)

2-Phenyl-2*H*-indazole derivatives were synthesized employing a slight modification of the Cadogan method [17,18]. A mixture of 2-nitrobenzaldehyde (200 mg, 1.32 mmol) and aniline or the corresponding substituted aniline (1.32 mmol) was dissolved in ethanol (10 mL) and heated at reflux for 2 h. Then, the solvent was removed under vacuum to give the appropriate 1-(2-nitrophenyl)-*N*-phenylmethanimine. This same reaction was carried out at room temperature to obtain the methyl 4-((2-nitrobenzylidene)amino)benzoate. Later, triethyl phosphite (3.96 mmol) was added and heated at 150 °C employing an oil bath for 2 h under N$_2$ atmosphere, until the starting material was consumed. After cooling, the mixture was treated with 20 mL of 5% hydrogen peroxide solution, and the product was extracted with ethyl acetate (3 × 15 mL). The combined organic layers were dried over Na$_2$SO$_4$ and concentrated in vacuo. The crude product was purified by column chromatography using hexane/ethyl acetate (80:20). Only compound **2b** was isolated by vacuum filtration and washed with cold ethanol to give a pure compound.

2-Phenyl-2*H*-indazole (**2a**). White solid (60% yield); m.p.: 81.2–81.6 °C (lit. [18]: 81–82 °C); the spectroscopic data matched previously reported data [17,24]: ^{1}H NMR (600 MHz, CDCl$_3$) δ 8.40 (d, J = 0.9 Hz, 1H), 7.91–7.88 (m, 2H), 7.79 (dd, J = 8.8, 0.9 Hz, 1H), 7.70 (dt, J = 8.5, 1.0 Hz, 1H), 7.54–7.50 (m, 2H), 7.41–7.37 (m, 1H), 7.32 (ddd, J = 8.8, 6.6, 1.0 Hz, 1H), 7.11 (ddd, J = 8.4, 6.6, 0.7 Hz, 1H); ^{13}C NMR (151 MHz, CDCl$_3$): δ 149.78, 140.52, 129.54, 127.88, 126.81, 122.76, 122.44, 120.99, 120.39, 120.37, 117.94.

Methyl 4-(2*H*-indazol-2-yl) benzoate (**2b**). After completion of the reaction, the product was filtrated and washed with cold MeOH. White solid (38% yield); m.p.: 185.8–186.2 °C (lit. [25]: 186–187 °C); the spectroscopic data matched previously reported data [17]: ^{1}H NMR (600 MHz, CDCl$_3$) δ 8.47 (d, J = 0.7 Hz, 1H), 8.22–8.18 (m, 2H), 8.02–7.99 (m, 2H), 7.77 (dd, J = 8.8, 0.8 Hz, 1H), 7.69 (d, J = 8.5 Hz, 1H), 7.33 (ddd, J = 8.8, 6.6, 1.0 Hz, 1H), 7.14–7.10 (m, 1H), 3.95 (s, 3H); ^{13}C NMR (151 MHz, CDCl$_3$) δ 166.19, 150.19, 143.64, 131.16, 129.27, 127.45, 123.01, 122.98, 120.47, 120.26, 118.06, 52.33.

Methyl (4-(2*H*-indazol-2-yl)phenyl)carbamate (**2d**). Pale orange solid (17% yield); m.p.: 163.7–165.0 °C; ^{1}H NMR (600 MHz, CDCl$_3$) δ 8.46 (bs, 1H), 8.38 (d, J = 0.9 Hz, 1H),

7.83–7.79 (m, 2H), 7.76 (dd, *J* = 8.8, 0.9 Hz, 1H), 7.70 (d, *J* = 8.4 Hz, 1H), 7.64 (d, *J* = 7.6 Hz, 2H), 7.31 (ddd, *J* = 8.7, 6.6, 1.1 Hz, 1H), 7.10 (ddd, *J* = 8.4, 6.6, 0.8 Hz, 1H), 3.80 (s, 3H); ^{13}C NMR (151 MHz, CDCl$_3$) δ 154.27, 149.48, 138.52, 135.46, 126.59, 122.62, 122.20, 121.44, 120.30, 120.25, 119.20, 117.62, 52.17; MS [M+H]$^+$ *m/z* 268.12.

4-(2*H*-indazol-2-yl)benzonitrile (**2e**). Yellow solid (19% yield); m.p.; 162.0–163.5 °C; (lit. [24]: 163.4–164.6 °C); the spectroscopic data matched previously reported data [24]: ^{1}H NMR (600 MHz, CDCl$_3$) δ 8.47 (d, *J* = 0.9 Hz, 1H), 8.10–8.05 (m, 2H), 7.85–7.80 (m, 2H), 7.76 (dd, *J* = 8.8, 0.9 Hz, 1H), 7.70 (dt, *J* = 8.5, 1.0 Hz, 1H), 7.35 (ddd, *J* = 8.8, 6.6, 1.1 Hz, 1H), 7.14 (ddd, *J* = 8.5, 6.6, 0.8 Hz, 1H); ^{13}C NMR (151 MHz, CDCl$_3$) δ 150.37, 143.33, 133.69, 127.87, 123.36, 123.14, 120.85, 120.49, 120.37, 118.13, 118.08, 111.18; MS [M + H]$^+$ *m/z* 220.11.

3.2.2. 4-(2*H*-Indazol-2-yl) Benzoic Acid (**2c**)

Methyl 4-(2*H*-indazol-2-yl) benzoate (300 mg, 1.26 mmol) was dissolved in methanol (10 mL) and an aqueous solution of NaOH (3.77 mmol in 5 mL of water) was added. The reaction mixture was heated under reflux for 5 h. After completion, the reaction mixture was cooled on an ice bath and acidified to pH 1 with HCl to induce precipitation. The solid was filtered under vacuum and dried.

4-(2*H*-Indazol-2-yl) benzoic acid (**2c**). White solid (96% yield); m.p.: 288.3–288.5 °C (lit. [25]: 286–288 °C); the spectroscopic data matched previously reported data [17]: ^{1}H NMR (600 MHz, DMSO-*d6*) δ 13.14 (bs, 1H), 9.22 (s, 1H), 8.26 (d, *J* = 8.6 Hz, 2H), 8.15 (d, *J* = 8.6 Hz, 2H), 7.79 (d, *J* = 8.4 Hz, 1H), 7.74 (d, *J* = 8.7 Hz, 1H), 7.41–7.30 (m, 1H), 7.19–7.08 (m, 1H); ^{13}C NMR (151 MHz, DMSO-*d6*) δ 166.47, 149.23, 142.84, 130.83, 129.65, 127.28, 122.55, 122.44, 122.03, 120.99, 119.86, 117.49.

3.2.3. 2,3-Diphenyl-2*H*-Indazole Derivatives (**3a**, **3c**, and **3l**)

2,3-Diphenyl-2*H*-indazole derivatives **3a**, **3c**, and **3l** were synthesized by a palladium-catalyzed arylation previously described by Ohnmacht et al. [19]. Compounds **3a** and **3c** were synthesized using the appropriate 2-phenyl-2*H*-indazole and the substituted 4-iodobenzene, whereas compound **3l** was synthesized from 2-phenyl-2*H*-indazole and methyl(4-bromophenyl) carbamate.

Methyl 4-(3-phenyl-2*H*-indazol-2-yl) benzoate (**3a**). Pale yellow solid (40% yield); m.p.: 152.4–154.9 °C (lit. [17]: 152.4–154.9 °C); the spectroscopic data matched previously reported data [17]: ^{1}H NMR (600 MHz, CDCl$_3$) δ 8.09–8.03 (m, 2H), 7.80 (d, *J* = 8.8 Hz, 1H), 7.69 (d, *J* = 8.5 Hz, 1H), 7.57–7.49 (m, 2H), 7.41 (dt, *J* = 3.7, 1.2 Hz, 3H), 7.35 (dd, *J* = 7.5, 2.0 Hz, 3H), 7.15 (ddd, *J* = 8.5, 6.6, 0.8 Hz, 1H), 3.93 (s, 3H); ^{13}C NMR (151 MHz, CDCl$_3$) δ 166.19, 149.33, 143.75, 135.68, 130.40, 129.68, 129.61, 128.96, 128.64, 127.44, 125.68, 122.88, 122.06, 120.52, 117.79, 52.32.

Methyl 4-(2-phenyl-2*H*-indazol-3-yl) benzoate (**3c**). Pale yellow solid (76% yield): m.p.: 164.5–166.3 °C (lit. [17]: 164.5–166.3 °C) the spectroscopic data matched previously reported data [17,26]: ^{1}H NMR (600 MHz, CDCl$_3$) δ 8.09–8.03 (m, 2H), 7.82 (d, *J* = 8.8 Hz, 1H), 7.72 (dt, *J* = 8.5, 0.9 Hz, 1H), 7.46–7.36 (m, 8H), 7.19 (ddd, *J* = 8.5, 6.5, 0.6 Hz, 1H), 3.93 (s, 3H); ^{13}C NMR (151 MHz, CDCl$_3$) δ 166.55, 149.08, 139.99, 134.37, 134.13, 129.97, 129.66, 129.49, 129.18, 128.59, 127.14, 126.04, 123.18, 121.90, 120.09, 118.02, 52.29.

Methyl (4-(2-phenyl-2*H*-indazol-3-yl)phenyl)carbamate (**3l**). Pale brown solid (26% yield): m.p.: 197.9–199.7 °C; ^{1}H NMR (600 MHz, DMSO-*d$_6$*) δ 9.85 (s, 1H), 7.70 (ddt, *J* = 24.0, 8.5, 0.9 Hz, 2H), 7.53 (d, *J* = 8.6 Hz, 2H), 7.50–7.41 (m, 5H), 7.37 (ddd, *J* = 8.7, 6.6, 1.1 Hz, 1H), 7.30–7.26 (m, 2H), 7.14 (ddd, *J* = 8.5, 6.5, 0.8 Hz, 1H), 3.68 (s, 3H); ^{13}C NMR (151 MHz, DMSO-*d$_6$*) δ 154.36, 148.61, 140.36, 139.82, 135.37, 130.46, 129.52, 128.89, 127.25, 126.45, 123.41, 122.67, 121.46, 120.95, 118.66, 117.78, 52.20; MS [M+H]$^+$ *m/z* 344.1381.

3.2.4. 2,3-Diphenyl-2*H*-Indazole Derivatives (**3b** and **3d**)

Employing the hydrolysis method 3.2.2 described above, compounds **3b** and **3d** were prepared from their esters **3a** or **3c**, respectively.

4-(3-Phenyl-2*H*-indazol-2-yl) benzoic acid (**3b**). White solid (70% yield); m.p.: 129.2–130.1 °C (lit. [17]: 129.2–130.1 °C); the spectroscopic data matched previously reported data [17]: ^{1}H NMR (600 MHz, DMSO-d_6) δ 8.01 (d, *J* = 8.6 Hz, 2H), 7.77 (d, *J* = 8.8 Hz, 1H), 7.69 (d, *J* = 8.5 Hz, 1H), 7.57 (d, *J* = 8.6 Hz, 2H), 7.44 (dddd, *J* = 11.9, 7.6, 5.4, 3.9 Hz, 6H), 7.21–7.15 (m, 1H); ^{13}C NMR (151 MHz, DMSO-d_6) δ 166.49, 148.52, 143.08, 135.26, 130.53, 130.15, 129.52, 129.03, 128.95, 128.71, 127.26, 125.99, 122.81, 121.38, 120.40, 117.49.

4-(2-Phenyl-2*H*-indazol-3-yl) benzoic acid (**3d**). White solid (88% yield); mp: 296.2–298.2 °C (lit. [17]: 296.2–298.2 °C); the spectroscopic data matched previously reported data [17]: ^{1}H NMR (600 MHz, DMSO-d_6) δ δ 7.92 (d, *J* = 8.2 Hz, 2H), 7.75 (d, *J* = 8.7 Hz, 1H), 7.71 (d, *J* = 8.5 Hz, 1H), 7.50–7.41 (m, 5H), 7.38 (ddd, *J* = 8.7, 6.6, 0.9 Hz, 1H), 7.28 (d, *J* = 8.3 Hz, 2H), 7.19–7.13 (m, 1H); ^{13}C NMR (151 MHz, DMSO-d_6) δ 169.13, 148.11, 140.15, 139.79, 135.10, 129.32, 129.17, 128.98, 128.37, 128.26, 126.77, 125.88, 122.34, 121.02, 120.38, 117.29.

3.2.5. 2,3-Diphenyl-2*H*-Indazole Derivatives (**3f** and **3m**)

The proper 2-phenyl-2*H*-indazole (**2a** or **2e**) and 4-bromobenzonitrile or 4-iodobenzene were reacted employing the previously described method by Ohnmacht et al. [19] to give the benzonitrile derivative. Then, the intermediate was dissolved and stirred in H_2SO_4 (1 mL) at room temperature overnight. After completion, the mixture was poured into ice-water (15 mL) to induce precipitation. The solid was filtered under vacuum and dried.

4-(3-Phenyl-2*H*-indazol-2-yl)benzamide (**3f**). White solid (57% yield): m.p.: 221.6–222.8 °C; ^{1}H NMR (600 MHz, DMSO-d_6) δ 8.09 (s, 1H), 7.97–7.92 (m, 2H), 7.76 (dt, *J* = 8.8, 0.9 Hz, 1H), 7.68 (dt, *J* = 8.5, 1.0 Hz, 1H), 7.55–7.37 (m, 9H), 7.18 (ddd, *J* = 8.5, 6.6, 0.8 Hz, 1H). ^{13}C NMR (151 MHz, DMSO-d_6) δ 167.30, 148.85, 142.33, 135.65, 134.40, 129.97, 129.50, 129.45, 129.09, 128.77, 127.59, 126.14, 123.16, 121.78, 120.81, 117.91; MS [M+H]$^+$ *m/z* 314.1294.

4-(2-Phenyl-2*H*-indazol-3-yl)benzamide (**3m**). Pale yellow solid (34% yield): m.p.: 230.6–232.7 °C; ^{1}H NMR (600 MHz, DMSO-d_6) δ 8.05 (s, 1H), 7.93 (d, *J* = 8.3 Hz, 2H), 7.77 (d, *J* = 8.7 Hz, 1H), 7.72 (d, *J* = 8.5 Hz, 1H), 7.47 (ddd, *J* = 15.3, 11.3, 7.9 Hz, 8H), 7.40 (dd, *J* = 11.7, 3.7 Hz, 1H), 7.20 (dd, *J* = 8.1, 6.9 Hz, 1H); ^{13}C NMR (151 MHz, DMSO-d_6) δ 167.63, 148.69, 140.16, 134.64, 134.28, 132.31, 129.70, 129.64, 129.16, 128.40, 127.41, 126.57, 123.30, 121.71, 120.68, 117.96; MS [M+H]$^+$ *m/z* 314.1281.

3.2.6. Methyl (4-(3-Phenyl-2*H*-indazol-2-yl)phenyl) Carbamate (**3e**)

Methyl (4-(phenyl-2*H*-indazole-2-yl)phenyl) carbamate **2d** (500 mg, 1.88 mmol) was dissolved in acetic acid (5 mL). Bromine (1.8 mL of 1 M solution in acetic acid) was slowly added at 0–5 °C and then led to warm at room temperature and stirred overnight. After completion of the reaction, ice-water was added and the solid formed was filtered and dried under vacuum. The crude intermediate (0.5 mmol) was treated with phenylboronic acid (0.55 mmol), Na_2CO_3 (1.5 mmol), Pd(PPh$_3$)$_4$ (0.01 mmol) and 3 mL of DME/water (3:1). The mixture was heated under microwave irradiation at 155 °C for 30 min [20]. After cooling, the solvent was removed under vacuum and the obtained product was purified by column chromatography using hexane/ethyl acetate (60:40).

Methyl (4-(3-phenyl-2*H*-indazol-2-yl)phenyl)carbamate (**3e**). White solid (58% yield): m.p.: 201.5–202.5 °C; ^{1}H NMR (600 MHz, DMSO-d_6) δ 9.91 (s, 1H), 7.73 (dt, *J* = 8.8, 0.9 Hz, 1H), 7.67 (dt, *J* = 8.5, 1.0 Hz, 1H), 7.53 (d, *J* = 8.9 Hz, 2H), 7.48–7.44 (m, 2H), 7.43–7.39 (m, 1H), 7.39–7.31 (m, 5H), 7.15 (ddd, *J* = 8.5, 6.6, 0.8 Hz, 1H), 3.69 (s, 3H); ^{13}C NRM (151 MHz, DMSO-d_6) δ 154.39, 148.48, 139.76, 135.26, 134.59, 129.87, 129.71, 129.32, 128.83, 127.16, 127.02, 122.81, 121.45, 120.71, 118.52, 117.77, 52.25; MS [M+H]$^+$ *m/z* 344.1391.

3.2.7. 2,3-Diphenyl-2*H*-Indazole Carboxamides (**3g–k** and **3n–r**)

Method A: To a solution of carboxylic acids **3b** or **3d** (250 mg, 0.8 mmol) in benzene (5 mL), SOCl$_2$ (0.35 mL, 4.8 mmol) was added. The mixture was heated at 70 °C for 4 h. After completion, the excess of SOCl$_2$ was distilled-off at reduced pressure (3 × 5 mL of benzene) to give the acyl chloride intermediate and then stirred with an excess of the adequate amine (16 mmol) at room temperature for 30 min. The mixture was poured in methanol and heated until solids were dissolved. Water was added to induces the precipitation of the compound. The formed solid was separated by vacuum filtration. Finally, the crude product was purified by column chromatography using hexane/ethyl acetate (80:20).

N,N-Diethyl-4-(3-phenyl-2*H*-indazol-2-yl)benzamide (**3i**). Pale yellow solid (80% yield); m.p.: 136.5–138.0 °C. ^{1}H NMR (600 MHz, CDCl$_3$) δ 7.79 (d, *J* = 8.8 Hz, 1H), 7.71 (d, *J* = 8.5 Hz, 1H), 7.52–7.48 (m, 2H), 7.48–7.35 (m, 8H), 7.15 (ddd, *J* = 8.3, 6.5, 0.5 Hz, 1H), 3.55 (s, 2H), 3.24 (s, 2H), 1.25 (s, 3H), 1.09 (s, 3H); ^{13}C NMR (151 MHz, CDCl$_3$) δ 170.25, 149.18, 140.78, 137.00, 135.61, 129.71, 129.57, 128.99, 128.57, 127.29, 127.21, 125.93, 122.73, 121.91, 120.59, 117.72, 43.35, 39.47, 14.21, 12.90; MS [M+H]$^+$ *m/z* 370.23.

(4-(3-Phenyl-2*H*-indazol-2-yl)phenyl)(pyrrolidin-1-yl)methanone (**3j**). Pale brown solid (40% yield); m.p.: 186.1–186.2 °C; ^{1}H NMR (600 MHz, CDCl$_3$) δ 7.79 (d, *J* = 8.8 Hz, 1H), 7.71 (d, *J* = 8.5 Hz, 1H), 7.61–7.52 (m, 2H), 7.52–7.45 (m, 2H), 7.45–7.34 (m, 6H), 7.18–7.12 (m, 1H), 3.65 (t, *J* = 7.0 Hz, 2H), 3.40 (t, *J* = 6.6 Hz, 2H), 2.00–1.93 (m, 2H), 1.92–1.85 (m, 2H); ^{13}C NMR (151 MHz, CDCl$_3$) δ 168.60, 149.15, 141.22, 136.84, 135.56, 129.66, 129.56, 128.95, 128.54, 127.97, 127.26, 125.70, 122.71, 121.90, 120.54, 117.70, 49.56, 46.29, 26.40, 24.41; MS [M+H]$^+$ *m/z* 368.24.

(4-(3-Phenyl-2*H*-indazol-2-yl)phenyl)(piperidin-1-yl)methanone (**3k**). White solid (73% yield); m.p.: 136.0–138.0 °C; ^{1}H NMR (600 MHz, CDCl$_3$) δ 7.81 (d, *J* = 8.8 Hz, 1H), 7.73 (d, *J* = 8.5 Hz, 1H), 7.48–7.34 (m, 10H), 7.20–7.14 (m, 1H), 3.71 (bs, 2H), 3.38 (bs, 2H), 1.85–1.57 (bs, 6H); ^{13}C NMR (151 MHz, CDCl$_3$) 169.56, 149.00, 140.02, 136.12, 134.41, 131.01, 129.56, 129.13, 128.45, 127.41, 127.04, 126.02, 122.85, 121.81, 120.22, 117.89, 48.77, 43.20, 26.59, 25.54, 24.54; MS [M+H]$^+$ *m/z* 382.26.

N,N-Diethyl-4-(2-phenyl-2*H*-indazol-3-yl)benzamide (**3p**). White solid (88 % yield); m.p.: 148.4–149.6 °C; ^{1}H NMR (600 MHz, CDCl$_3$) δ 7.81 (d, *J* = 8.8 Hz, 1H), 7.73 (d, *J* = 8.5 Hz, 1H), 7.49–7.36 (m, 10H), 7.19–7.14 (m, 1H), 3.56 (bs, 2H), 3.29 (bs, 2H), 1.26 (bs, 3H), 1.14 (bs, 3H); ^{13}C NMR (151 MHz, CDCl$_3$) δ 170.57, 149.03, 140.06, 136.93, 134.47, 130.79, 129.59, 129.13, 128.47, 127.06, 126.92, 126.05, 122.86, 121.80, 120.24, 117.92, 43.39, 39.41, 14.25, 12.89; MS [M+H]$^+$ *m/z* 370.26.

(4-(2-Phenyl-2*H*-indazol-3-yl)phenyl)(pyrrolidin-1-yl)methanone (**3q**). White solid (37% yield); m.p.: 167.0–168.8 °C; ^{1}H NMR (600 MHz, CDCl$_3$) δ 7.81 (d, *J* = 8.8 Hz, 1H), 7.72 (dt, *J* = 8.5, 0.8 Hz, 1H), 7.56 (d, *J* = 8.3 Hz, 2H), 7.46–7.36 (m, 8H), 7.17 (ddd, *J* = 8.4, 6.5, 0.7 Hz, 1H), 3.66 (t, *J* = 7.0 Hz, 2H), 3.46 (t, *J* = 6.6 Hz, 2H), 2.01–1.95 (m, 2H), 1.94–1.88 (m, 2H); ^{13}C NMR (151 MHz, CDCl$_3$) δ 168.95, 149.03, 140.04, 136.82, 134.46, 131.39, 129.44, 129.15, 128.49, 127.68, 127.09, 126.04, 122.90, 121.84, 120.23, 117.91, 49.60, 46.33, 26.46, 24.43; MS [M+H]$^+$ *m/z* 368.24.

(4-(2-Phenyl-2*H*-indazol-3-yl)phenyl)(piperidin-1-yl)methanone (**3r**). White solid (90% yiled); m.p.: 198.2–199.2 °C; ^{1}H NMR (600 MHz, CDCl$_3$) δ 7.81 (d, *J* = 8.8 Hz, 1H), 7.73 (d, *J* = 8.5 Hz, 1H), 7.55–7.28 (m, 10H), 7.20–7.14 (m, 1H), 3.72 (bs, 2H), 3.38 (bs, 2H), 1.69 (bs, 6H).^{13}C NMR (151 MHz, CDCl$_3$) δ 169.56, 149.00, 140.02, 136.12, 134.41, 131.01, 129.56, 129.13, 128.45, 127.41, 127.04, 126.02, 122.85, 121.81, 120.22, 117.89, 48.77, 43.22, 26.57, 25.59, 24.54; MS [M+H]$^+$ *m/z* 382.26.

Method B: The acyl chlorides were synthesized employing the same procedure as described in method A. Next, the crude acyl chloride was mixed with methylamine hydrochloride or *N,N*-dimethylamine hydrochloride (1 mmol) and dissolved in anhydrous CH$_2$Cl$_2$ (3 mL). Then, triethylamine (1.0 mmol) was slowly added to the mixture and stirred at room temperature for 2 h. After completion, the mixture was concentrated in vacuo. The product was purified by column chromatography using hexane/ethyl acetate (70:30).

N-Methyl-4-(3-phenyl-2*H*-indazol-2-yl)benzamide (**3g**). Pale brown solid (43% yield); m.p.:194.3–195.4 °C. ^{1}H NMR (600 MHz, CDCl$_3$) δ 7.78 (dt, *J* = 8.9, 0.9 Hz, 1H), 7.77–7.74 (m, 2H), 7.70 (dt, *J* = 8.5, 0.9 Hz, 1H), 7.53–7.43 (m, 2H), 7.43–7.30 (m, 6H), 7.15 (ddd, *J* = 8.5, 6.5, 0.8 Hz, 1H), 6.38 (d, *J* = 4.3 Hz, 1H), 3.01 (d, *J* = 4.8 Hz, 3H); ^{13}C NMR (151 MHz, CDCl$_3$)

δ 167.20, 149.22, 142.42, 135.66, 134.18, 129.65, 129.53, 128.94, 128.60, 127.68, 127.40, 125.88, 122.81, 121.95, 120.53, 117.66, 26.89; MS [M+H]$^+$ *m/z* 328.1439.

N,N-Dimethyl-4-(3-phenyl-2*H*-indazol-2-yl)benzamide (**3h**). Pale brown solid (70% yield); m.p.:153.0–154.5 °C. ^{1}H NMR (600 MHz, DMSO-d_6) δ 7.75 (d, *J* = 8.7 Hz, 1H), 7.68 (d, *J* = 8.5 Hz, 1H), 7.53–7.35 (m, 10H), 7.20–7.14 (m, 1H), 2.99 (s, 3H), 2.91 (s, 3H); ^{13}C NMR (151 MHz, DMSO-d_6) δ 169.52, 148.78, 140.70, 136.78, 135.56, 129.95, 129.48, 129.43, 129.08, 128.23, 127.53, 126.32, 123.12, 121.67, 120.82, 117.89, 35.18; MS [M+H]$^+$ *m/z* 342.24.

N-Methyl-4-(2-phenyl-2*H*-indazol-3-yl)benzamide (**3n**). Pale brown solid (62% yield); m.p.: 235.7–236.6 °C; ^{1}H NMR (600 MHz, DMSO-d_6) δ 8.52 (d, *J* = 4.5 Hz, 1H), 7.89–7.86 (m, 2H), 7.79–7.75 (m, 1H), 7.72 (dd, *J* = 8.5, 1.0 Hz, 1H), 7.51–7.44 (m, 7H), 7.40 (ddd, *J* = 8.6, 6.6, 0.9 Hz, 1H), 7.23–7.17 (m, 1H), 2.79 (d, *J* = 4.6 Hz, 3H); ^{13}C NMR (151 MHz, DMSO-d_6) 166.40, 148.69, 140.16, 134.62. 134.54, 132.10, 129.76, 129.63, 129.14, 127.98, 127.41, 126.56, 123.29, 121.69, 120.67, 117.97, 26.72; MS [M+H]$^+$ *m/z* 328.1442.

N,N-Dimethyl-4-(2-phenyl-2*H*-indazol-3-yl)benzamide (**3o**). White solid (82% yield); m.p.: 166.2–168.0 °C; ^{1}H NMR (600 MHz, DMSO-d_6) 7.76 (dt, *J* = 8.7, 0.7 Hz, 1H), 7.72 (dt, *J* = 8.5, 0.8 Hz, 1H), 7.51–7.44 (m, 7H), 7.43–7.38 (m, 3H), 7.19 (ddd, *J* = 8.5, 6.6, 0.7 Hz, 1H), 2.99 (s, 3H), 2.92 (s, 3H); ^{13}C NMR (151 MHz, DMSO-d_6) 174.63, 153.43, 144.91, 141.38, 139.46, 135.24, 134.48, 134.37, 133.89, 132.67, 132.15, 131.28, 127.94, 126.42, 125.46, 122.70, 39.94; MS [M+H]$^+$ *m/z* 342.21.

3.2.8. Synthesis of Pyrazole Derivatives (**6a**, **6e**, and **6h**)

To a solution of 1,3-dicarbonyl compound (2 mmol) in 5 mL of a 5% methanolic solution of H_2SO_4, hydrazine hydrate or 4-hydrazinylbenzoic acid (2 mmol) was added. The reaction mixture was stirred at room temperature for 2–5 h and then cooled on an ice bath. To induce a complete precipitation, water (2–5 mL) was added. The resulting solid was separated in vacuo and dried [21].

4-(3-Methyl-5-phenyl-1*H*-pyrazol-1-yl)benzoic acid (**6a**). White solid (91% yield); m.p.: 126.3–128.3 °C; the spectroscopic data matched previously reported data [27] ^{1}H NMR (600 MHz, CDCl$_3$) δ 8.06–8.01 (m, 2H), 7.40–7.36 (m, 2H), 7.36–7.31 (m, 3H), 7.25–7.20 (m, 2H), 6.34 (s, 1H), 3.49 (s, 1H), 2.41 (s, 3H); ^{13}C NMR (151 MHz CDCl$_3$) δ 170.56, 150.50, 144.12, 144.10, 130.96, 130.43, 128.73, 128.66, 128.56, 127.58, 124.29, 109.03, 13.53; MS [M+H]$^+$ *m/z* 279.17.

4-(3,5-Dimethyl-1*H*-pyrazol-1-yl)benzoic acid (**6e**). White solid (90% yield); m.p.: 160.7–161.8 °C; the spectroscopic data matched previously reported data [28]: ^{1}H NMR (600 MHz, CDCl$_3$) 8.19 (d, *J* = 8.7 Hz, 2H), 7.59 (d, *J* = 8.7 Hz, 2H), 6.05 (s, 1H), 2.39 (d, *J* = 0.6 Hz, 3H), 2.33 (s, 3H); RMN ^{13}C (151 MHz, CDCl$_3$) δ 170.33, 150.06, 143.94, 139.73, 131.15, 127.74; MS [M+H]$^+$ *m/z* 217.10.

Methyl 4-(3-methyl-1*H*-pyrazol-5-yl)benzoate (**6h**). Yellow solid (82% yield); m.p.: 188.0–191 °C; ^{1}H NMR (600 MHz, DMSO-d_6) δ 12.80 (bs, 1H), 7.97 (d, *J* = 8.5 Hz, 2H), 7.90 (d, *J* = 8.4 Hz, 2H), 6.56 (s, 1H), 3.86 (s, 3H), 2.27 (s, 3H); ^{13}C NMR (151 MHz, DMSO-d_6) δ 166.48, 130.06, 128.42, 125.38, 125.09, 102.41, 52.47, 11.22; MS [M+H]$^+$ *m/z* 217.10.

3.2.9. Pyrazole Derivatives (**6b** and **6f**)

A solution of carboxylic acid (**6a** or **6e**, 1.1 mmol) in 5 mL of a 5% methanolic solution of H_2SO_4 was heated under reflux for 2 h. When the reaction was completed, the mixture was neutralized with a 10% aqueous solution of NaHCO$_3$. The resulting solid was separated under vacuum filtration and dried.

Methyl 4-(3-methyl-5-phenyl-1*H*-pyrazol-1-yl)benzoate (**6b**). White solid (84% yield); m.p.: 100.2–101.6 °C; ^{1}H NMR (600 MHz, CDCl$_3$) δ 7.97 (d, *J* = 8.8 Hz, 2H), 7.37–7.33 (m, 2H), 7.33–7.29 (m, 3H), 7.24–7.19 (m, 2H), 6.33 (s, 1H), 3.90 (s, 3H), 2.39 (s, 3H). ^{13}C NMR (151 MHz, CDCl$_3$) δ 166.38, 150.28, 143.95, 143.63, 130.51, 130.29, 128.67, 128.56, 128.41, 128.18, 124.19, 108.84, 52.14, 13.56; MS [M+H]$^+$ *m/z* 293.19.

Methyl 4-(3,5-dimethyl-1*H*-pyrazol-1-yl)benzoate (**6f**). White solid (73% yield); m.p.: 60.2–61.2 °C; ^{1}H NMR (600 MHz, CDCl$_3$) 8.12 (d, *J* = 8.8 Hz, 2H), 7.55 (d, *J* = 8.8 Hz, 2H),

6.03 (s, 1H), 3.94 (s, 3H), 2.37 (d, J = 0.6 Hz, 3H), 2.30 (s, 3H); ^{13}C NMR (151 MHz, CDCl$_3$) 166.38, 149.87, 143.61, 139.53, 130.52, 128.25, 123.54, 108.16, 52.19, 13.48, 12.78; MS [M+H]$^+$ m/z 231.13.

3.2.10. 4-(3-Methyl-1H-pyrazol-5-yl) Benzoic Acid (**6g**)

Methyl 4-(3-methyl-1H-pyrazol-5-yl) benzoate **6h** (250 mg, 1.1 mmol) was dissolved in a 5 mL of a 60% aqueous solution of H$_2$SO$_4$ and heated under reflux for 2 h. After completion, the reaction was poured into 10 mL of iced water. The resulting solid was filtered in vacuo, washed with cold water and dried.

4-(3-Methyl-1H-pyrazol-5-yl)benzoic acid (**6g**). Yellow solid (95% yield); m.p.: 285–286 °C; ^{1}H NMR (600 MHz, DMSO-d_6) δ 7.96 (d, J = 8.5 Hz, 2H), 7.88 (d, J = 8.5 Hz, 2H), 6.55 (d, J = 0.7 Hz, 1H), 2.29–2.26 (m, 3H); ^{13}C NMR (151 MHz, DMSO-d_6) δ 167.55, 148.45, 141.87, 137.76, 130.20, 129.65, 125.27, 102.40, 11.40; MS [M+H]$^+$ m/z 203.13.

3.2.11. 4-(3-Methyl-1-phenyl-1H-pyrazol-5-yl) Benzoic Acid (**6c**)

5-Methyl-2-phenyl-2,4-dihydro-3H-pyrazol-3-one (1.0 g, 5.74 mmol) was dissolved in POCl$_3$ (1.5 mL) and heated under reflux for 5 h. The mixture was cooled on an ice bath and quenched with 10 mL of water. The product was extracted with ethyl acetate (3 × 10 mL), the combined organic layers were dried over Na$_2$SO$_4$ and concentrated in vacuo. The resulting oily product (200 mg, 1.0 mmol) was mixed with 4-carbomethoxyphenylboronic acid pinacol ester (300 mg, 1.1 mmol), Pd(PPh$_3$)$_4$ (116 mg, 0.1 mmol), Na$_2$CO$_3$ (424 mg, 4 mmol) and 5 mL of MeCN/H$_2$O (4:1). The reaction was heated under microwave irradiation at 175 °C for 20 min twice. Then, the solvent was removed in vacuo. The resulting crude product was purified by column chromatography using hexane/ethyl acetate (80:20).

4-(3-Methyl-1-phenyl-1H-pyrazol-5-yl)benzoic acid (**6c**). Pale brown solid (55% yield); m.p.: 221.0–223.3 °C; ^{1}H NMR (600 MHz, CDCl$_3$) δ 8.05–7.97 (m, 2H), 7.37–7.32 (m, 2H), 7.32–7.28 (m, 3H), 7.27 (dd, J = 8.4, 1.4 Hz, 2H), 6.41 (s, 1H), 2.41 (s, 3H).; ^{13}C NMR (151 MHz, CDCl$_3$) δ 170.85, 149.72, 142.48, 139.73, 135.69, 130.23, 129.04, 128.77, 128.47, 127.56, 125.24, 108.41, 13.46; MS [M+H]$^+$ m/z 279.17.

3.2.12. Methyl 4-(3-Methyl-1-phenyl-1H-pyrazol-5-yl) Benzoate (**6d**)

A mixture of 4-(3-methyl-1-phenyl-1H-pyrazol-5-yl) benzoic acid **6c** (200 mg, 0.7 mmol) and Na$_2$CO$_3$ (75 mg, 0.7 mmol), DMF (2 mL) and water (0.5 mL) was stirred for 15 min at room temperature. Afterward, methyl iodide (0.7 mmol) was added to the mixture and stirred at room temperature for 1 h. Then, the reaction was poured into water (10 mL) and extracted with ethyl acetate (3 × 5 mL). The combined organic layers were dried over Na$_2$SO$_4$ and concentrated under vacuum. The crude product was purified by recrystallization from ethanol.

Methyl 4-(3-methyl-1-phenyl-1H-pyrazol-5-yl)benzoate (**6d**). White solid (82% yield); 92.0–94.3 °C; the spectroscopic data matched previously reported data [29] ^{1}H NMR (600 MHz, CDCl$_3$) 7.95 (d, J = 8.6 Hz, 2H), 7.36–7.31 (m, 2H), 7.31–7.23 (m, 5H), 6.39 (s, 1H), 3.91 (s, 3H), 2.39 (s, 3H); ^{13}C NMR (151 MHz, CDCl$_3$) 166.57, 149.64, 142.54, 139.87, 135.02, 129.63, 129.47, 128.99, 128.40, 127.42, 125.17, 108.28, 52.18, 13.52; MS [M+H]$^+$ m/z 293.15.

3.2.13. Synthesis of 3-Phenyl-1H-Indazoles (**10a**, **10b**, and **10d–i**)

A mixture of 3-bromo-1H-indazole (200 mg, 1.01 mmol), phenylboronic acid or appropriate aryl boronic acid pinacol ester (1.11 mmol), Na$_2$CO$_3$ (128 mg, 1.21 mmol), PPh$_3$ (8 mg, 0.03 mmol), Pd(OAc)$_2$ (2.2 mg, 0.01 mmol) and 5 mL of n-propanol/H$_2$O (3:1) was heated under microwave irradiation at 150 °C for 20 min. The solvent was removed in vacuo and the resulting residue was purified by column chromatography using hexane/ethyl acetate (50:50).

3-Phenyl-1H-indazole (**10a**). The product was purified using column chromatography and hexane/ethyl acetate (80:20) as mobile phase. White solid (80% yield). m.p.:

105.2–106.8 °C [lit. [30]: 106–107 °C]; the spectroscopic data matched previously reported data [31,32]: ^{1}H NMR (600 MHz, CDCl$_3$) δ 11.44 (bs, 1H), 8.05–7.99 (m, 3H), 7.56–7.51 (m, 2H), 7.47–7.42 (m, 1H), 7.37–7.33 (m, 1H), 7.29–7.25 (m, 1H), 7.23–7.19 (m, 1H); ^{13}C NMR (151 MHz, CDCl$_3$) δ 145.74, 141.68, 133.56, 128.92, 128.16, 127.71, 126.77, 121.33, 121.08, 120.95, 110.21; Purity (qNMR, % w/w): 98.89 ± 2.29.

4-(1*H*-Indazol-3-yl)benzoic acid (**10b**). White solid (82% yield); m.p.: 283.5–286 °C; the spectroscopic data matched previously reported data [32]: ^{1}H NMR (600 MHz, DMSO-d_6) δ 13.52 (bs, 1H), 13.05 (bs, 1H), 8.19–8.14 (m, 3H), 8.12–8.08 (m, 2H), 7.64 (d, J = 8.4 Hz, 1H), 7.48–7.42 (m, 1H), 7.26 (ddd, J = 7.8, 6.8, 0.8 Hz, 1H); ^{13}C NMR (151 MHz, DMSO-d_6) δ 172.33, 148.26, 147.20, 146.82, 143.12, 135.17, 134.68, 131.70, 131.48, 126.67, 125.81, 125.76, 125.28, 116.01; Purity (qNMR, % w/w): 96.48 ± 0.42.

4-(1*H*-Indazol-3-yl)benzamide (**10d**). Pale brown solid (45% yield); m.p.: 253.0–255.5 °C; ^{1}H NMR (600 MHz, DMSO-d_6) δ 13.43 (s, 1H), 8.14 (d, J = 8.3 Hz, 1H), 8.12–8.09 (m, 3H), 8.04 (d, J = 8.5 Hz, 2H), 7.63 (d, J = 8.4 Hz, 1H), 7.49–7.41 (m, 2H), 7.25 (ddd, J = 7.9, 6.9, 0.7 Hz, 1H); ^{13}C NMR (151 MHz, DMSO-d_6) δ 168.04, 142.74, 142.03, 136.86, 133.54, 128.59, 126.71, 126.68, 121.80, 121.08, 120.52, 111.16; MS [M+H]$^+$ m/z 238.14; Purity (qNMR, % w/w): 95.23±0.71.

4-(1*H*-Indazol-3-yl)-*N*-methylbenzamide (**10e**). White solid (34% yield); m.p.: 239.5–241.0 °C; ^{1}H NMR (600 MHz, DMSO-d_6) δ 13.41 (s, 1H), 8.55 (d, J = 4.5 Hz, 1H), 8.14 (d, J = 8.2 Hz, 1H), 8.10 (d, J = 8.4 Hz, 2H), 7.99 (d, J = 8.4 Hz, 2H), 7.62 (d, J = 8.4 Hz, 1H), 7.47–7.41 (m, 1H), 7.28–7.22 (m, 1H), 3.38 (s, 3H); ^{13}C NMR (151 MHz DMSO-d_6) δ 166.17, 142.17, 141.47, 136.12, 133.22, 127.59, 126.22, 126.13, 121.23, 120.54, 119.96, 110.60, 26.19; MS [M+H]$^+$ m/z 252.17; Purity (qNMR, % w/w): 96.59 ± 0.80.

4-(1*H*-Indazol-3-yl)-*N,N*-dimethylbenzamide (**10f**). White solid (48% yield); m.p.: 160.5–163.0 °C; the spectroscopic data matched previously reported data [12]: ^{1}H NMR (600 MHz, DMSO-d_6) δ 13.38 (s, 1H), 8.12 (d, J = 8.3 Hz, 1H), 8.07 (d, J = 8.3 Hz, 2H), 7.62 (d, J = 8.4 Hz, 1H), 7.56 (d, J = 8.3 Hz, 2H), 7.46–7.40 (m, 1H), 7.24 (ddd, J = 7.9, 6.8, 0.8 Hz, 1H), 3.01 (d, J = 16.2 Hz, 6H); ^{13}C NMR (151 MHz, DMSO-d_6) δ 169.76, 142.30, 141.47, 135.30, 134.58, 127.60, 126.27, 126.12, 121.16, 120.47, 119.92, 110.60, 34.71; MS [M+H]$^+$ m/z: 266.20; Purity (qNMR, % w/w): 97.56 ± 1.07.

N,N-Diethyl-4-(1*H*-indazol-3-yl)benzamide (**10g**). White solid (53% yield); m.p.: 259.3–261.5 °C; ^{1}H NMR (600 MHz, DMSO-d_6) δ 13.37 (s, 1H), 8.12 (d, J = 8.2 Hz, 1H), 8.07 (d, J = 8.3 Hz, 2H), 7.62 (d, J = 8.4 Hz, 1H), 7.50 (d, J = 8.3 Hz, 2H), 7.46–7.40 (m, 1H), 7.27–7.21 (m, 1H), 3.46 (s, 2H), 3.27 (s, 2H), 1.16 (bs,3H), 1.10 (bs, 3H); ^{13}C NMR (151 MHz, DMSO-d_6) δ 169.62, 142.31, 141.47, 136.17, 134.25, 126.71, 126.42, 126.11, 121.14, 120.47, 119.90, 110.59, 42.78, 13.98, 12.74; MS [M+H]$^+$ m/z 294.23; Purity (qNMR, % w/w): 96.68 ± 0.33.

(4-(1*H*-Indazol-3-yl)phenyl)(pyrrolidin-1-yl)methanone (**10h**). White solid (21% yield); m.p.: 201.0–202.3 °C; ^{1}H NMR (600 MHz, DMSO-d_6) δ 13.39 (s, 1H), 8.12 (d, J = 8.3 Hz, 1H), 8.07 (dd, J = 6.6, 1.8 Hz, 2H), 7.70–7.66 (m, 2H), 7.62 (d, J = 8.4 Hz, 1H), 7.46–7.40 (m, 1H), 7.24 (ddd, J = 7.9, 6.9, 0.8 Hz, 1H), 3.52–3.46 (m, 4H), 1.91–1.81 (m, 4H); ^{13}C NMR (151 MHz DMSO-d_6) δ 167.79, 142.28, 141.47, 135.97, 135.04, 134.91, 127.70, 126.17, 126.11, 121.18, 120.46, 119.93, 110.60, 48.88, 45.91, 25.93, 23.82; MS [M+H]$^+$ m/z 292.20; Purity (qNMR, % w/w): 99.12 ± 1.25.

(4-(1*H*-indazol-3-yl)phenyl)(piperidin-1-yl)methanone (**10i**). White solid (42% yield); m.p.: 193.5–195.5 °C; ^{1}H NMR (600 MHz, DMSO-d_6) δ 13.38 (s, 1H), 8.12 (d, J = 8.2 Hz, 1H), 8.07 (d, J = 8.2 Hz, 2H), 7.62 (d, J = 8.6 Hz, 1H), 7.52 (d, J = 8.2 Hz, 2H), 7.46–7.39 (m, 1H), 7.27–7.20 (m, 1H), 3.61 (bs, 2H), 3.36 (bs, 2H), 1.65–14.5 (m, 6H); ^{13}C NMR (151 MHz, DMSO-d_6) δ 169.08, 142.84, 142.00, 135.86, 135.06, 127.83, 126.94, 126.66, 121.69, 121.01, 120.45, 111.13, 48.56, 42.84, 26.47, 25.74, 24.51; MS [M+H]$^+$ m/z 306.23; Purity (qNMR, % w/w): 97.32 ± 0.62.

3.2.14. Methyl 4-(1*H*-Indazol-3-yl) Benzoate (**10c**)

Employing the esterification method 3.2.9. described above, compound **10c** was synthesized from acid derivative **10b**.

Methyl 4-(1*H*-indazol-3-yl)benzoate (**10c**). Yellow solid (94% yield); m.p.: 205.5–208.0 °C; the spectroscopic data matched previously reported data [16]: ^{1}H NMR (600 MHz, DMSO-d_6) δ 13.52 (s, 1H), 8.19 (d, *J* = 8.4 Hz, 2H), 8.16 (d, *J* = 8.1 Hz, 1H), 8.11 (d, *J* = 8.4 Hz, 2H), 7.64 (d, *J* = 8.4 Hz, 1H), 7.45 (dd, *J* = 11.3, 3.8 Hz, 1H), 7.27 (dd, *J* = 11.2, 3.9 Hz, 1H), 3.90 (s, 3H); ^{13}C NMR (151 MHz DMSO-d_6) δ 165.95, 141.78, 141.53, 138.24, 129.73, 128.20, 126.54, 126.24, 121.46, 120.47, 120.00, 110.74, 52.07; Purity (qNMR, % *w/w*): 96.76 ± 1.06.

3.3. Anticandidal Activity Assays

3.3.1. Agar Diffusion Method

Susceptibility assays were carried out using the cylinder-plate method (agar diffusion method) described in the general methods of analysis MGA 0100 of the Mexican Pharmacopeia [23]. *Candida* strains from the American Type Culture Collection (ATCC) were used: *Candida albicans* (18,804), *Candida glabrata* (90,030, susceptible to miconazole), *Candida glabrata* (32,554, resistant to miconazole), and *Candida tropicalis* (750). Compounds tested, fluconazole and miconazole were properly weighed and solubilized in dimethyl sulfoxide (DMSO) at 100%. The microorganisms were grown and maintained on Sabouraud dextrose broth (SB, Bioxon, Mexico) for 24 h at 35 ± 2 °C. A base layer was prepared into Petri dishes (100 × 20 mm) employing 10 mL of a mixture of agar SB with *Candida* suspension adjusted to 0.5 McFarland standard. After solidification, a second layer was formed by the addition of 10 mL of agar SB. In each plate, five stainless steel cylinders of uniform size (8 × 6 × 10 mm) were placed on the surface and filled with 100 μL of a solution of the compounds at 10, 1 and 0.1 mM. Fluconazole (1 mM) and miconazole (1 mM) were used as positive controls, and vehicle as a negative control. The plates were incubated at 35 ± 2 °C for 24 h. The degree of effectiveness was measured by determining the zone of inhibition in millimeters produced by the tested compounds and considering the effect produced by DMSO.

3.3.2. Determination of Minimum Inhibitory Concentration (MIC) of the Most Effective Compounds

Minimum inhibitory concentration is defined as the lowest concentration of compounds against *Candida* strains that inhibit their growth after 24 h of incubation. The 3-phenyl-1*H*-indazole derivatives, which showed antimicrobial activity at the least concentration tested (100 μM), were evaluated against *Candida albicans* (18,804), *Candida glabrata* (90,030, susceptible to miconazole), *Candida glabrata* (32,554, resistant to miconazole). The cylinder-plate method and different concentrations of tested compounds (50, 75, 100, 125, 150, and 200 μM) were used for the assay. Fluconazole and miconazole were used as positive controls and vehicle (DMSO) as negative control. Experiments were carried out in duplicate and repeated three times.

3.4. Cytotoxicity Assays

HeLa (human cervix), K562 (human chronic myelogenous leukemia), and SW620 (human colon) cells were grown in DMEM (Invitrogen Corporation, Carlsbad, CA, USA) supplemented with 10% FBS (BioWest, Riverside, MO, USA) and maintained in standard culture conditions (37 °C, 95% humidity, and 5% CO_2). Cells were grown to a density of 80% and then were harvested using sterile PBS/EDTA (pH 7.4) before starting every experiment. Cells were seeded in 96-well plates (7000 cells/well in 200 μL of DMEM). After 24 h the cells were exposed to test compounds dissolved in 0.5 % DMSO and diluted with DMEM at 50 μM, to reach 250 μL in the well. The exposure time was 48 h, and then viability was determined by MTT assay. The absorbance of formazan was determined for each well and its viability was related to the vehicle (100%) [17].

3.5. Quantitative Nuclear Magnetic Resonance Spectroscopy

The purity of compounds **10a–i** was determined by qNMR using sodium 2,2-dimethyl-2-silapentane-5-sulfonate (DSS, 97%, CIL) as an internal standard. The 4-bromobenzonitrile (99%, Sigma-Aldrich) was used for the validation of the method. As solvent deuterated dimethyl sulfoxide-d6 (DMSO-d_6, >99.96%, CIL) was used.

Samples were prepared as a mixture of compounds tested and DSS was dissolved in DMSO-d_6. The measurements were carried out with an Agilent DD2 spectrometer operating at 600 MHz. In general, the experiments were acquired employing the parameters: 90° pulse, an acquisition time of 7 s, relaxation delay of 60 s, digital resolution of 0.8 Hz, a spectral window of 15 ppm, and a total of 64 scans. Phase and baseline corrections were done automatically using the software MestReNova v.6.0.2. The signal integration was done in automatic mode. For determination of purity were used separated signals (H-5, H-6, or NH) in the aromatic region and three signals of DSS (CH_2) in the aliphatic region. The percentage of purity was calculated by Equation (1):

$$P_{Sample} = \frac{I_{Sample}}{I_{Std}} \times \frac{N_{Std}}{N_{Sample}} \times \frac{M_{Sample}}{M_{Std}} \times \frac{m_{Std}}{m_{Sample}} \times P_{Std} \tag{1}$$

where I is the integrated area, N is the number of spins, M is the molar mass, m is the gravimetric weight, and P is the purity in % w/w [33,34].

4. Conclusions

In the present study, several indazole and pyrazole derivatives were designed and synthesized by bioisosteric replacement, homologation, and molecular simplification. Besides, a new series of derivatives was designed by combining the 3-phenyl-1*H*-indazole moiety with carboxamide substituents. Taking the activity data from the series of the 3-phenyl-1*H*-indazole moiety, compounds **10c**, **10g**, and **10i** had the best effect against *C. albicans*. In comparison, **10f** and **10g** were active against both miconazole susceptible and resistant *C. glabrata* strains. In particular, compound **10g** was demonstrated to have the best anticandidal activity against *Candida* strains tested. According to the data, the molecular simplification along with *p*-carbonyl substituents was crucial for the anticandidal activity. Therefore, the 3-phenyl-1*H*-indazole nucleus is a new scaffold for anticandidal agents that has not been reported to date. In light of these results, new research could be conducted to explore other substituents attached to the 3-phenyl group to generate more information about the structural requirements needed to improve the anticandidal activity.

Supplementary Materials: The following are available online at https://www.mdpi.com/1424-8247/14/3/176/s1, Figure S1: Determination of the inhibition zone for compound 2b tested at 10, 1, and 0.1 mM against *C. albicans*, Figures S2–S41: ^{1}H NMR and ^{13}C NMR spectra of compounds 2a–e, 3a–r, 6a–h and 10a–i, Figures S42–S71: MS spectra of compounds 2d, 2e, 3e–r, 6a–h and 10d–i, Table S1: Percentage of purity of compounds 10a–i by qNMR used an internal standard and Figure S72: Spectra of qNMR for 4-bromobenzonitrile (purity > 99%) as a reference compound.

Author Contributions: Conceptualization, K.R.-V., A.H.-C., L.Y.-M., T.d.R.S.-E., and J.P.-V.; funding acquisition, O.S.-A. and J.P.-V.; investigation, K.R.-V., B.V.-P., E.A.Q.-S., and J.P.-V.; methodology, K.R.-V., A.H.-C., L.Y.-M., T.d.R.S.-E., and J.P.-V.; project administration, O.S.-A. and J.P.-V.; resources, O.S.-A., J.F.P.-E., and F.C.-B.; supervision, A.H.-C., L.Y.-M., T.d.R.S.-E., M.L.-L., and J.P.-V.; writing—original draft, K.R.-V.; writing—review and editing, A.H.-C., L.Y.-M., O.S.-A., J.F.P.-E., F.C.-B., M.L.-L., and J.P.-V. All authors have read and agreed to the published version of the manuscript.

Funding: This research was funded by CONACyT (grants CB-2015-01-258554 and 1238).

Institutional Review Board Statement: Not applicable.

Informed Consent Statement: Not applicable.

Data Availability Statement: The data presented in this study are available in supplementary material.

Acknowledgments: K.R.-V. thanks CONACyT for the Ph.D. scholarship awarded 538233/291415 to study in the program of Doctorado en Ciencias Biológicas y de la Salud at UAM. The authors would like to express thanks to Ernesto Sánchez Mendoza, Mónica A. Rincón Guevara, and Miguel Cortés Gines for the analytical support, and Pharmometrica research center for the support with mass spectra determinations.

Conflicts of Interest: The authors declare no conflict of interest.

References

1. Brown, G.D.; Denning, D.W.; Gow, N.A.R.; Levitz, S.M.; Netea, M.G.; White, T.C. Hidden Killers: Human Fungal Infections. *Sci. Transl. Med.* **2012**, *4*, 165rv13. [CrossRef] [PubMed]
2. Bongomin, F.; Gago, S.; Oladele, R.O.; Denning, D.W. Global and Multi-National Prevalence of Fungal Diseases—Estimate Precision. *J. Fungi* **2017**, *3*, 57. [CrossRef] [PubMed]
3. Gonçalves, B.; Ferreira, C.; Alves, C.T.; Henriques, M.; Azeredo, J.; Silva, S. Vulvovaginal candidiasis: Epidemiology, microbiology and risk factors. *Crit. Rev. Microbiol.* **2015**, *42*, 905–927. [CrossRef]
4. Centers for Disease Control and Prevention. Sexually transmitted diseases treatment guidelines, 2015. MMWR. Recommendations and reports: Morbidity and mortality weekly report. *Recomm. Rep.* **2015**, *64*, 1–137.
5. Kullberg, B.J.; Arendrup, M.C. Invasive Candidiasis. *N. Engl. J. Med.* **2015**, *373*, 1445–1456. [CrossRef] [PubMed]
6. Pfaller, M.A.; Diekema, D.J. Epidemiology of Invasive Candidiasis: A Persistent Public Health Problem. *Clin. Microbiol. Rev.* **2007**, *20*, 133–163. [CrossRef]
7. Centers for Disease Control and Prevention. Invasive Candidiasis. Available online: https://www.cdc.gov/fungal/diseases/candidiasis/invasive/statistics.html (accessed on 1 January 2020).
8. Fisher, M.C.; Hawkins, N.J.; Sanglard, D.; Gurr, S.J. Worldwide emergence of resistance to antifungal drugs challenges human health and food security. *Science* **2018**, *360*, 739–742. [CrossRef]
9. Pappas, P.G.; Kauffman, C.A.; Andes, D.R.; Clancy, C.J.; Marr, K.A.; Ostrosky-Zeichner, L.; Reboli, A.C.; Schuster, M.G.; Vazquez, J.A.; Walsh, T.J.; et al. Clinical Practice Guideline for the Management of Candidiasis: 2016 Update by the Infectious Diseases Society of America. *Clin. Infect. Dis.* **2016**, *62*, e1–e50. [CrossRef]
10. Thangadurai, A.; Minu, M.; Wakode, S.; Agrawal, S.; Narasimhan, B. Indazole: A medicinally important heterocyclic moiety. *Med. Chem. Res.* **2012**, *21*, 1509–1523. [CrossRef]
11. Gaikwad, D.D.; Chapolikar, A.D.; Devkate, C.G.; Warad, K.D.; Tayade, A.P.; Pawar, R.P.; Domb, A.J. Synthesis of indazole motifs and their medicinal importance: An overview. *Eur. J. Med. Chem.* **2015**, *90*, 707–731. [CrossRef] [PubMed]
12. Reddy, G.; Ivaturi, K.; Allaka, T. Design, Synthesis and Docking Studies of New Indazole Derivatives as Potent Cytotoxic and Antibacterial Agents. *Indian J. Heterocycl. Chem.* **2019**, *28*, 467–476.
13. Lipunova, G.N.; Nosova, E.V.; Charushin, V.N.; Chupakhin, O.N. Fluorine-containing indazoles: Synthesis and biological activity. *J. Fluor. Chem.* **2016**, *192*, 1–21. [CrossRef]
14. Vega, M.C.; Rolón, M.; Montero-Torres, A.; Fonseca-Berzal, C.; Escario, J.A.; Gómez-Barrio, A.; Gálvez, J.; Marrero-Ponce, Y.; Arán, V.J. Synthesis, biological evaluation and chemometric analysis of indazole derivatives. 1,2-Disubstituted 5-nitroindazolinones, new prototypes of antichagasic drug. *Eur. J. Med. Chem.* **2012**, *58*, 214–227. [CrossRef]
15. Park, J.S.; A Yu, K.; Kang, T.H.; Kim, S.; Suh, Y.-G. Discovery of novel indazole-linked triazoles as antifungal agents. *Bioorganic Med. Chem. Lett.* **2007**, *17*, 3486–3490. [CrossRef]
16. Chen, H.-S.; Kuo, S.-C.; Teng, C.-M.; Lee, F.-Y.; Wang, J.-P.; Lee, Y.-C.; Kuo, C.-W.; Huang, C.-C.; Wu, C.-C.; Huang, L.-J. Synthesis and antiplatelet activity of ethyl 4-(1-benzyl-1H-indazol-3-yl)benzoate (YD-3) derivatives. *Bioorganic Med. Chem.* **2008**, *16*, 1262–1278. [CrossRef]
17. Pérez-Villanueva, J.; Yépez-Mulia, L.; González-Sánchez, I.; Palacios-Espinosa, J.F.; Soria-Arteche, O.; Sainz-Espuñes, T.R.; Cerbón, M.A.; Rodríguez-Villar, K.; Rodríguez-Vicente, A.K.; Cortés-Gines, M.; et al. Synthesis and Biological Evaluation of 2H-Indazole Derivatives: Towards Antimicrobial and Anti-Inflammatory Dual Agents. *Molecules* **2017**, *22*, 1864. [CrossRef] [PubMed]
18. 2-PHENYLINDAZOLE. *Org. Synth.* **1968**, *48*, 113. [CrossRef]
19. Ohnmacht, S.A.; Culshaw, A.J.; Greaney, M.F. Direct Arylations of 2H-Indazoles On Water. *Org. Lett.* **2010**, *12*, 224–226. [CrossRef]
20. Kümmerle, A.E.; Schmitt, M.; Cardozo, S.V.S.; Lugnier, C.; Villa, P.; Lopes, A.B.; Romeiro, N.C.; Justiniano, H.; Martins, M.A.; Fraga, C.A.M.; et al. Design, Synthesis, and Pharmacological Evaluation ofN-Acylhydrazones and Novel Conformationally Constrained Compounds as Selective and Potent Orally Active Phosphodiesterase-4 Inhibitors. *J. Med. Chem.* **2012**, *55*, 7525–7545. [CrossRef] [PubMed]
21. Wang, Z.-X.; Qin, H.-L. Solventless syntheses of pyrazole derivativesElectronic supplementary information (ESI) available: Analytical and spectroscopic data. *Green Chem.* **2004**, *6*, 90–92. [CrossRef]
22. Huff, B.E.; Koenig, T.M.; Mitchell, D.; Staszak, M.A. Synthesis of Unsymmetrical Biaryls Using a Modified Suzuki Cross-Coupling: 4-Biphenylcarboxaldehyde. *Org. Synth.* **2003**, *75*, 53. [CrossRef]
23. Secretaria de Salud, Comisión permanente de la Farmacopea de los Estados Unidos Mexicanos. Valoración microbiológica de antibióticos, MGA 0100. In *Farmacopea de los Estados Unidos Mexicanos FEUM*, 12th ed.; FEUM: Ciudad de México, México, 2018.
24. Haag, B.; Peng, Z.; Knochel, P. Preparation of Polyfunctional Indazoles and Heteroarylazo Compounds Using Highly Functionalized Zinc Reagents. *Org. Lett.* **2009**, *11*, 4270–4273. [CrossRef]

25. Zhou, D.; Sze, J.Y.-C.; Gross, J.L.; Robichaud, A.J. Preparation of Azacyclylbenzamide Derivatives as Histamine-3 Antago-nists for Treating CNS Disorders. U.S. Patent 20080293771A1, 27 November 2008.
26. Geng, X.; Wang, C. Rhenium-Catalyzed [4 + 1] Annulation of Azobenzenes and Aldehydes via Isolable Cyclic Rhenium(I) Complexes. *Org. Lett.* **2015**, *17*, 2434–2437. [CrossRef]
27. Vickerstaffe, E. The Development and Application of Automated Multi-Step Polymer Assisted Solution Phase Synthesis for the Preparation of Biologically Active Compound Arrays. Ph.D. Thesis, University of Cambridge, Cambridge, UK, 2004.
28. Porcelloni, M.; D'Andrea, P.; Rossi, C.; Sisto, A.; Ettorre, A.; Madami, A.; Altamura, M.; Giuliani, S.; Meini, S.; Fattori, D. α,α-Cyclopentaneglycine Dipeptides Capped with Biaryls as Tachykinin NK2Receptor Antagonists. *ChemMedChem* **2008**, *3*, 1048–1060. [CrossRef] [PubMed]
29. Sar, D.; Bag, R.; Yashmeen, A.; Bag, S.S.; Punniyamurthy, T. Synthesis of Functionalized Pyrazoles via Vanadium-Catalyzed C–N Dehydrogenative Cross-Coupling and Fluorescence Switch-On Sensing of BSA Protein. *Org. Lett.* **2015**, *17*, 5308–5311. [CrossRef]
30. Croce, P.D.; La Rosa, C. A Convenient Synthesis of Indazoles. *Synthesis* **1984**, *1984*, 982–983. [CrossRef]
31. Counceller, C.M.; Eichman, C.C.; Wray, B.C.; Stambuli, J.P. A Practical, Metal-Free Synthesis of 1H-Indazoles. *Org. Lett.* **2008**, *10*, 1021–1023. [CrossRef] [PubMed]
32. Shamsabadi, A.; Chudasama, V. A facile route to 1H- and 2H-indazoles from readily accessible acyl hydrazides by exploiting a novel aryne-based molecular rearrangement. *Chem. Commun.* **2018**, *54*, 11180–11183. [CrossRef]
33. Malz, F.; Jancke, H. Validation of quantitative NMR. *J. Pharm. Biomed. Anal.* **2005**, *38*, 813–823. [CrossRef]
34. Schoenberger, T.; Bernstein, M.; Mestrelab, U.K.; Braouet, C.; Ron Crouch, J.; Suematsu, T.; Guillou, C.; Klare, H.; Cologne, B.W.Z.; Bernie, O.; et al. Guideline for QNMR Analysis. DWG-NMR-001, ENFSI. 2019. Available online: http://enfsi.eu/wp-content/uploads/2017/06/qNMR-Guideline_version001.pdf (accessed on 17 February 2021).

pharmaceuticals

Review

Augmenting Azoles with Drug Synergy to Expand the Antifungal Toolbox

Aidan Kane and Dee A. Carter *

School of Life and Environmental Sciences and Sydney ID, University of Sydney,
Camperdown, NSW 2006, Australia; aidan.kane@sydney.edu.au
* Correspondence: dee.carter@sydney.edu.au; Tel.: +61-2-9351-5383

Abstract: Fungal infections impact the lives of at least 12 million people every year, killing over 1.5 million. Wide-spread use of fungicides and prophylactic antifungal therapy have driven resistance in many serious fungal pathogens, and there is an urgent need to expand the current antifungal arsenal. Recent research has focused on improving azoles, our most successful class of antifungals, by looking for synergistic interactions with secondary compounds. Synergists can co-operate with azoles by targeting steps in related pathways, or they may act on mechanisms related to resistance such as active efflux or on totally disparate pathways or processes. A variety of sources of potential synergists have been explored, including pre-existing antimicrobials, pharmaceuticals approved for other uses, bioactive natural compounds and phytochemicals, and novel synthetic compounds. Synergy can successfully widen the antifungal spectrum, decrease inhibitory dosages, reduce toxicity, and prevent the development of resistance. This review highlights the diversity of mechanisms that have been exploited for the purposes of azole synergy and demonstrates that synergy remains a promising approach for meeting the urgent need for novel antifungal strategies.

Keywords: antifungal; azole; synergy; mycosis; resistance; *Candida*; dermatophytes; natural products

Citation: Kane, A.; Carter, D.A. Augmenting Azoles with Drug Synergy to Expand the Antifungal Toolbox. *Pharmaceuticals* **2022**, *15*, 482. https://doi.org/10.3390/ph15040482

Academic Editor: Jong Heon Kim, Luisa W. Cheng and Kirkwood Land

Received: 15 March 2022
Accepted: 26 March 2022
Published: 14 April 2022

1. Introduction

1.1. The Burden of Fungal Disease

Fungal pathogens present an ever-increasing threat to global health. An estimated 1.5 million people are killed by fungal infections every year, and the incidence of several serious mycoses is growing [1,2]. It is likely that the global fungal burden is underestimated, as several invasive fungal infections are under-reported in the developed world due to their association with other predisposing illnesses [3–5]. Australian clinics have recently seen a near-doubling of systemic candidaemia caused by drug-resistant *Candida glabrata*, with increasing rates of invasive candidaemia seen in Europe and the USA [6–8]. *Candida auris*, an emerging yeast pathogen infamous for its high tolerance to most important antifungals, has been the cause of several recent outbreaks, both before and during the COVID-19 pandemic [9–12]. Drug-resistant biofilms of various species of *Candida* have become increasingly responsible for fatal nosocomial infections [13]. Lethal infections with *Aspergillus fumigatus* and *Cryptococcus* sp. also remain a pressing concern, together causing an estimated 400,000 deaths per year, with chronic pulmonary aspergillosis severely affecting close to 3 million people [2]. Morbidity of non-life-threatening topical fungal infections is also increasing at an alarming rate. Cutaneous dermatophytosis affects 12 to 13 million people a year, and nail infections are extraordinarily difficult to treat, with less than 13% of cases fully resolved [14,15]. Decreasing susceptibility to topical antifungals has been observed in *Exophiala dermatitidis* and *Malassezia* sp. [15–18].

The increased incidence of emerging systemic, superficial, and cutaneous fungal pathogens has increased the demand for novel antifungal medications. However, despite the advances over the past four decades in bringing azoles and echinocandins to market, the currently available antifungals still operate via a limited number of mechanisms

(Figure 1) and unfortunately all have significant problems, including toxicity, difficulty of administration, limited bioavailability and efficacy, and an often high cost [19–21].

Current therapies are also being increasingly compromised by antifungal resistance. Widespread over-use of agricultural fungicides appears to be driving cross-resistance in a range of pathogenic moulds and yeasts, and the use of prophylactic and over-the-counter azoles for superficial infections can promote the acquisition of resistance [22–25]. While recent innovations in anti-retroviral therapy have dramatically reduced the rates of fungal infection associated with HIV-AIDS, the case fatality rates of invasive fungal infections have remained constant [2,26,27]. Due to its propensity to up-regulate active efflux pumps, *C. glabrata* is an emerging cause of recurrent candidiasis, and multiple-drug resistance has been observed in an alarming proportion of clinical isolates of *C. auris* [28–30]. Clearly, there is an urgent and unmet need for new approaches to treat fungal infections [31].

1.2. Azole Antifungals

Arguably the most successful class of drug in the antifungal toolbox is the azoles. Azoles operate by binding to and inhibiting lanosterol 14α-demethylase, a fungal cytochrome P450 enzyme essential to the biosynthesis of ergosterol (Figure 1) [32]. First-generation triazoles, such as itraconazole and fluconazole, have low host toxicity, and the more important azoles are administered orally and have excellent pharmacodynamic properties [33]. Itraconazole and the second-generation triazole voriconazole have high respiratory bioavailability and fluconazole has superb neural pharmacokinetics, making it well suited as a maintenance therapy for cryptococcal meningitis. A promising new azole prodrug, isavuconazonium, was approved by the FDA in 2015, while a new generation of tetrazoles is currently in clinical and agrochemical development [34–39]. Older azoles like fluconazole are cheap, off-patent and widely available, including in developing countries where they are needed the most. The currently available prescription and over-the-counter azoles, and the mycoses for which they are commonly indicated, are detailed in Table 1.

Azoles are a vital tool in the fight against fungal infection. It has been projected that the antifungal market will be USD 13 billion by 2026, with roughly 2.4 billion of that allocated to azoles: USD 1.4 billion in oral and intravenous drugs, and USD 1 billion in topical and over-the-counter solutions [40,41]; however, azoles have limitations. Many filamentous fungal pathogens are not sensitive to azoles, particularly fluconazole [33]. Second-generation triazoles like voriconazole can be toxic to the host [42]. Azoles are often used prophylactically in neutropenic and transplant patients, but this can encourage the development of resistance [23,43]. Azole-based fungicides are used in agriculture on a massive scale, and cross-resistance between these and azole antifungals is seen with alarming frequency [22]. *Candida auris* particularly represents an escalating threat to the usefulness of azoles, as it is frequently highly azole-tolerant and often multi-drug resistant [44].

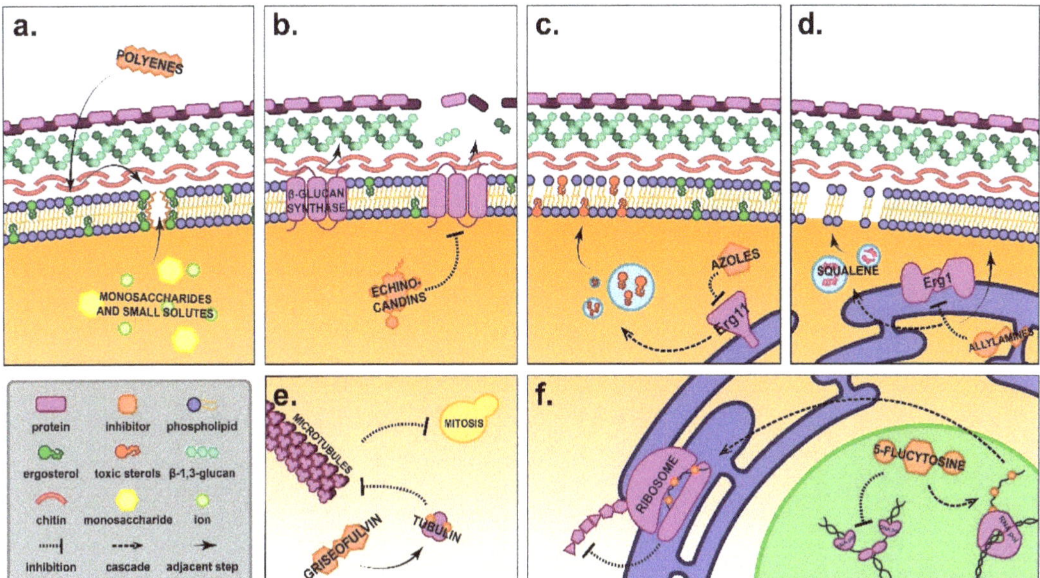

Figure 1. Mechanisms of action of commercially available antifungals. (**a**) Polyenes bind to ergosterol in the plasma membrane, forming pores that permit the efflux of vital small solutes like potassium ions and simple sugars [45]. (**b**) Echinocandins operate by inhibiting 1,3-β-D-glucan synthase in the fungal membrane, depriving the cell wall of glucans and therefore its structural integrity [46]. (**c**) Azoles inhibit Erg11 (lanosterol 14α-demethylase) preventing the biosynthesis of ergosterol and resulting in a build-up of toxic methyl-sterols that incorporate into the plasma membrane. The result is a loss of membrane structure and inhibition of growth [47]. (**d**) Allylamines inhibit ergosterol biosynthesis by antagonising squalene epoxidase Erg1, which converts squalene to squalene epoxide. As well as preventing the biosynthesis of ergosterol, this results in the build-up of squalene, which is deposited into lipid vesicles that disrupt the plasma membrane [48]. (**e**) Griseofulvin binds to tubulin in the fungal cell, preventing the formation of microtubules and arresting mitosis [49]. (**f**) 5-flucytosine is a pyrimidine analogue that is converted to 5-fluorouracil inside the cell. This fluoridated nucleotide is incorporated into mRNA, halting ribosomal processing and inhibiting protein translation. 5-fluorouracil also antagonises Cdc21, or thymidylate synthase, preventing the biosynthesis of thymidine nucleotides and inhibiting DNA synthesis [50].

There are various mechanisms of resistance to azole antifungals. Mutations in *ERG11* (also known as *CYP51*), which encodes the target enzyme lanosterol 14α-demethylase, can prevent enzyme binding, or *ERG11* expression can be up-regulated via changes to its promoter or regulating transcription factors or via gene duplication and aneuploidy [51]. Azoles can be actively excluded from the cell via membrane bound ABC and MFS efflux pumps [52–54], which can be increased in expression via aneuploidy or by alterations to transcription factors leading to constitutive expression. Mutations in ERG3 have also been found to increase resistance, thought to be via prevention of the build-up of toxic intermediates [55–61]. Frequently, these resistance mechanisms lead to cross-resistance across azole drug types, and although the research community is working to derive new azole antifungals by modelling lanosterol 14α-demethylase crystal structures [62–64], the increased resistance seen in clinical strains can undermine their use as a monotherapy.

Table 1. Currently available azole antifungals and associated mycoses.

Class	Application	Azole	Brand	Mycosis	Notes	Ref.
Imidazole	Topical	butoconazole	Gynazole-1, Mycelex-3	uncomplicated and recurrent vaginal candidiasis		[65]
		climbazole	Squaphane, Pitiren	dandruff and seborrhoeic dermatitis caused by *Malassezia* sp.		[66]
		clotrimazole [†]	Lotrimin	oral and vaginal candidiasis, and tinea versicolor, cruris and pedis	WHO Essential Medicine	[66,67]
		eberconazole	Ebernet	cutaneous candidiasis and dermatophytosis	Approved in EU in 2015	[68]
		econazole	Spectrazole, Ecostatin	tinea pedis and cruris, vaginal candidiasis	Also repels clothes moths	[69]
		flutrimazole	Flusporan, Topiderm	cutaneous dermatophytosis including tinea pedis		[70]
		isoconazole	Icaden, Travogen	tinea pedis and vaginal candidiasis	Effective against Gram-positive bacteria	[71]
		ketoconazole [†]	Nizoral	seborrhoeic dermatitis, dandruff, tinea and cutaneous candidiasis	Also systemic	[72]
		luliconazole	Luzu	tinea pedis and cruris and other dermatophytoses	FDA-approved in 2013	[73]
		miconazole [†]	Monistat, Desenex	dermatophytosis and cutaneous, oral and vaginal candidiasis	WHO Essential Medicine	[74]
		oxiconazole	Oxistat, Oxizole	dermatophytoses and cutaneous candidiasis		[75]
		sertaconazole	Ertaczo, Dermofix	tinea pedis and vaginal candidiasis	Also anti-inflammatory and anti-pruritic	[76,77]
		sulconazole	Exelderm	dermatophytoses	Also anti-carpet beetle	[78,79]
		tioconazole	Vagistat-1	onychomycosis, dermatophytoses and vaginal candidiasis	Also called thioconazole	[80]
	Systemic	ketoconazole	Nizoral (oral)	mycoses caused by *Candida, Histoplasma* and *Coccidioides*	Systemic use for extreme cases only	[81]
Triazole	Topical	efinaconazole	Jublia, Clenafin	onychomycosis	Low cure rate, but higher than other drugs	[82]
		fluconazole [†]	Diflucan	dermatophytoses and cutaneous candidiasis	WHO Essential Medicine, more commonly systemic	[21]
		terconazole	Terazol	acute and chronic vaginal candidiasis		[83]
	Systemic	fluconazole [†]	Diflucan	candidiasis, cryptococcosis, histoplasmosis, blastomycosis	WHO Essential Medicine, oral or intravenous	[21]
		fosfluconazole	Prodif	prophylaxis in the immunocompromised	Fluconazole prodrug	[84,85]
		fosravuconazole	Nailin	onychomycosis	Ravuconazole prodrug	[86]
		isavuconazonium	Cresemba	mucormycosis and invasive aspergillosis	Isavuconazole prodrug	[35,87]
		itraconazole [†]	Sporanox, Orungal	aspergillosis, histoplasmosis, coccidioidomycosis and blastomycosis	WHO Essential Medicine	[48,88]
		posaconazole	Noxafil, Posanol	invasive candidiasis, aspergilosis, mucormycosis and scedosporiosis	FDA-approved in 2006	[89]
		voriconazole [†]	Vfend	aspergillosis, candidiasis, penicilliosis, histoplasmosis and fusariosis	WHO Essential Medicine	[90]

[†] Most commonly prescribed azole antifungals. Other azole antifungals no longer on the market include ravuconazole, a triazole similar to voriconazole which was discontinued after Phase-III clinical trials and the thiazole abafungin, which is no longer available [91].

1.3. Antimicrobial Synergy

Recent expiries in patent protection of early-generation azoles have led to generic alternatives for voriconazole, fluconazole, posaconazole and efinaconazole entering the market and have made space for innovations. There is a particular increasing interest in enhancing the antifungal activity of current azoles using drug synergy.

Synergy occurs when two compounds produce an increased inhibitory effect beyond what would be expected by adding the effects of the compounds individually [92]. Significant synergy is determined by the Fractional Inhibitory Concentration Index (*FICI*), which is calculated as the sum of the ratios of the Minimum Inhibitory Concentration (*MIC*) of the drugs when used alone and together, according to the following equation:

$$FICI = \frac{MIC\ of\ drug\ A\ in\ combination}{MIC\ of\ drug\ A\ alone} + \frac{MIC\ of\ drug\ B\ in\ combination}{MIC\ of\ drug\ B\ alone}$$

When the *FICI* of two drugs is ≤ 0.5 their interaction is considered synergistic, and when this is >4 it is considered antagonistic. An *FICI* between 0.5 and 4 is considered indifferent [92].

Antimicrobial synergy can overcome resistance and lower inhibitory dosages to within clinically achievable levels [93,94]. Synergy can expand the spectra of activity of the individual drugs, making azoles a viable option against pathogens that may be otherwise resistant. Antimicrobial synergy can also make otherwise fungistatic drugs fungicidal, including many azoles [95,96].

Antimicrobial synergy has been exploited in the clinic for years for the treatment of HIV and malaria [97,98], and the most successful induction therapy for cryptococcal meningitis is a synergistic combination of amphotericin B and 5-flucytosine. Synergy between two drugs can be potentiated by their co-operation on multiple enzymes belonging to the same pathway; for example, sulfamethoxazole and trimethoprim are antibacterial agents that target the folic acid biosynthesis pathway at two different sites to synergistically improve both the spectrum and potency of bacterial inhibition [99]. Synergy can also be produced between drugs that inhibit different processes: for example, amphotericin B damages the fungal plasma membrane and 5-flucytosine interrupts translation and replication [100,101].

1.4. Aims and Scope of This Review

The mechanism of action of azoles is well understood (Figure 1) and various synergists have been described that operate on related (Figure 2) or quite disparate (Figure 3) pathways or mechanisms. In this review, we describe recent studies that have reported azole synergists. There are several excellent earlier reviews that have considered aspects of azole synergy [52,102,103]; here we provide an update with a particular focus on developments made over the past six years. Table 2 provides an overview of these findings presented as a heatmap, with the extent of synergy demonstrated between commonly used azoles and secondary agents in blue, and the spectrum of fungi found to be affected in yellow. Below we explore each of the different classes of synergists and their potential as combination therapies.

Table 2. Interactions between azoles and synergists and their spectrum of activity described in published studies.

Category	Synergist	Azole Synergy							Synergy in % Strains Tested						Notes [1]	Ref.
		Fluconazole	Itraconazole	Voriconazole	Isavuconazole	Posaconazole	Ketoconazole	Miconazole	*C. albicans*	AR *Candida*	*A. fumigatus*	AR *Aspergillus*	Dermatophytes	Other		
Antifungals	terbinafine														Incl. *Scedosporium* sp. and *Pythium* sp.	[104–112]
	caspofungin														Dep. on mechanism of resistance	[113–116]
	anidulafungin															[114,117,118]
	micafungin														Incl. *C. auris*	[115,116,118]
	natamycin														Incl. *Fusarium* sp.	[119]
	ciclopirox															[120]
	flucytosine														Incl. *C. auris*	[121]
	voriconazole														Limited synergy in the Mucorales	[122]
	amorolfine														Onychomycosis clinical trials	[123–126]
	K20														Also clotrimazole, incl. *C. neoformans*	[127]
	oxadiazole pept.															[128]
Antibacterials	sulfamethoxazole														Incl. *C. auris*	[129,130]
	sulfa- antibiotics														Some effective vs. biofilms and in vivo	[130]
	doxycycline														Effective against biofilms and in vivo	[131–133]
	tigecycolline														Incl. *Fusarium* sp., limited anti-biofilm	[132,134,135]
	minocycline														Incl. *C. neoformans*, *Scedosporium* sp.	[133,136–139]
	gentamicin														Effective against biofilms	[140]
	linezolid														Little synergy, but reduced dosage	[141]
	polymyxin B														Incl. *Rhodotorula* and *Lichtheimia* sp.	[142]
	colistin														Incl. *C. auris*	[96,143,144]
Antiparasitics	pyrvinium pam.															[145,146]
	chloroquine															[147]
	artemisinins														Effective against biofilms	[146]
	INK128														Incl. *Fusarium* and *Exophiala* sp.	[148]
	mefloquine														Incl. *C. neoformans*	[149]
Antivirals	saquinavir														Incl. *Histoplasma capsulatum*	[150]
	ritonavir														Incl. *Histoplasma capsulatum*	[150]
	adamantanamine														Switch from fungistatic to fungicidal	[151]
	ribavirin														Effective against biofilms and in vivo	[152]

Table 2. *Cont.*

Category	Synergist	Azole Synergy							Synergy in % Strains Tested						Notes [1]	Ref.
		Fluconazole	Itraconazole	Voriconazole	Isavuconazole	Posaconazole	Ketoconazole	Miconazole	*C. albicans*	AR *Candida*	*A. fumigatus*	AR *Aspergillus*	Dermatophytes	Other		
	lopinavir														Incl. *C. auris*	[153]
								Efflux Inhibitors								
Calcium Inhibitors	tetrandrine														Effective in vivo	[154–157]
	verapamil														Dep. on mechanism of resistance	[158,159]
Other	eucalyptal D														Natural product	[160]
	dodenoic acid															[161]
	azoffluxin														Incl. *C. auris*	[162]
	ospemifeme														Incl. *C. neoformans* and *C. auris*	[163]
	phialocephalarin															[164]
	palmarumycin P3															[164]
	geraniol														Effective in vivo, natural product	[165]
								Repurposed Drugs								
Statins	lovastatin														Incl. *Rhizopus* sp.	[166–168]
	atorvastatin														Incl. *Rhizopus* sp., *Cryptococcus* sp.	[166,167,169]
	fluvastatin														Incl. *Rhizopus* sp.	[166,167]
	simvastatin														Incl. *Rhizopus* sp., *Cryptococcus* sp.	[166,167,170]
	pitavastatin															[171]
Bisphosphonates	risedronate														Incl. *Cryptococcus* sp.	[172]
	alendronate														Incl. *Cryptococcus* sp.	[172]
	zoledronate														Incl. *Cryptococcus* sp.	[172]
Immunomodulators	promethazine															[173,174]
	terfenadine														Effective against biofilms	[175]
	ebastine														Effective against biofilms	[175]
	dexamethasone														Effective against biofilms	[176]
	budesonide														Effective in vivo	[177]
	methotrexate															[178]
Psychoactives	bromperidol															[179]
	fluoxetine														Effective in vivo, *C. albicans* only	[180]
	haloperidol															[173]
	sertraline														Incl. *Trichosporon asahii*	[181]
Calcineurin Inhibitors	cyclosporine														Incl. *S. cerevisiae*	[182–187]
	tacrolimus														Incl. *S. cerevisiae*	[183,188–193]

Table 2. *Cont.*

Category	Synergist	Azole Synergy							Synergy in % Strains Tested						Notes [1]	Ref.
		Fluconazole	Itraconazole	Voriconazole	Isavuconazole	Posaconazole	Ketoconazole	Miconazole	*C. albicans*	AR *Candida*	*A. fumigatus*	AR *Aspergillus*	Dermatophytes	Other		
	B2														Piperidone derivative, incl. *C. neoformans*	[257]
	H1-J10														Novel HSP90/HDAC inhibitors	[258]
	L1-C2														Novel lipopeptides	[259]
	AZD8055														Novel TOR inhibitor	[260]
	KEY:		extremely strong synergy												synergistic in all strains tested	
			strong synergy												synergistic in >20% strains tested	
			weak synergy												synergistic in <20% strains tested	
			borderline synergy												not synergistic in any strains tested	

[1] "Incl." refers to species included under "Other".

2. Synergy between Azoles and Currently Available Antimicrobials

Existing antimicrobial pharmaceuticals that have been proven safe or tolerable for humans and have approval from regulatory bodies such as the FDA make an attractive starting point for antifungal synergy. Theoretically, combining antimicrobials might broaden their activity spectrum to include pathogens that are not susceptible to either drug as a monotherapy [261]. Although combining azoles with other classes of antifungals could be expected to often give synergy, certain antibacterial, antiparasitic and antiviral drugs have also been shown to interact synergistically.

2.1. Azole-Antifungal Synergy

Numerous antifungals, including azoles, allylamines and morpholines, target different points of the ergosterol biosynthesis pathway, as summarised in Figure 2. Given that the mechanisms of action of these drugs are often closely related, they are prime targets for investigation as potential azole synergists. Terbinafine is an allylamine that has been investigated across a broad array of azoles and various fungal and oomycete pathogens and was found to synergise with some systemic azoles (Table 2), particularly against azole-resistant *C. albicans* isolates. A combined treatment strategy using terbinafine and fluconazole was also shown to resolve persistent oropharyngeal thrush in a clinical setting [104–106]. For other species of *Candida*, however, terbinafine-azole synergy is weaker and only seen in azole-sensitive isolates, although the combination shifts inhibition from fungistatic to fungicidal [107]. Terbinafine and azoles have proven effective against clinical isolates of *Scedosporium prolificans*, a hard-to-treat pathogen of the lungs, sinuses and brain, at clinically achievable concentrations [108,109]. For other pathogens, terbinafine–azole synergy is narrow in spectrum, with only a very few isolates of *Aspergillus*, azole-resistant non-*albicans Candida* species, azole-resistant dermatophytes and *Pythium insidiosum*, (a fungus-like oomycete that is the cause of often-fatal pythiosis) affected [106,107,110,111,262,263].

Antifungals from the echinocandin class, which act by weakening the cell wall, were considered potentially attractive as azole synergists when they first became available; however, for most combinations, the pairs either synergise strongly but in a select subset of isolates, or work across a global spectrum of isolates but with only weakly synergistic interactions (Table 2) [113–117]. One particularly promising pair is micafungin and voriconazole, which strongly inhibits azole- and echinocandin-resistant *Candida auris*, bringing the required dosages of both to well within clinically achievable levels [115]. In a recent survey, anidulafungin, caspofungin and micafungin all displayed strong synergy with isavuconazole in *C. auris* [118]. Given the extreme level of resistance to azoles demonstrated by many *C. auris* isolates and the threat of emerging resistance to echinocandins, which are currently the front-line antifungal for *C. auris* infection, these combinations may warrant immediate clinical application [264–266].

Broadly speaking, most other market antifungals have demonstrated weak or no synergy with azoles. The topical polyene natamycin weakly synergises with voriconazole against a narrow range of azole-susceptible pathogens, while ciclopirox synergises strongly with itraconazole but only in a small number of dermatophytes [119,120]. Flucytosine and even voriconazole have been tested with other azoles but were indifferent or only weakly synergistic for various fungal species, including *C. auris* and pathogens in the order Mucorales [121,122]. Amorolfine, however, which acts on the ergosterol biosynthesis pathway subsequent to azoles (Figure 2) displayed more promise, synergising with systemic and topical azoles against dermatophytes in vitro, and demonstrating efficacy in an open randomised clinical trial of notoriously refractive onychomycosis [123–125]. While promising, this combined treatment strategy for topical fungal infections is yet to make it to market.

A few recently developed novel antifungals that are not yet available commercially appear promising as azole synergists. K20, a novel fungus-specific amphiphilic aminoglycoside that inhibits *Fusarium* spp. and a variety of yeast pathogens, displayed strong synergy with a wide variety of systemic azoles for almost all *Candida* isolates tested [127,267].

Oxadiazole-tagged macrocyclic peptides, which are capable of inhibiting azole-resistant *C. glabrata* and *C. tropicalis* strains also interacted synergistically with fluconazole, but with less synergy and a narrower spectrum of activity than K20 (Table 2) [128].

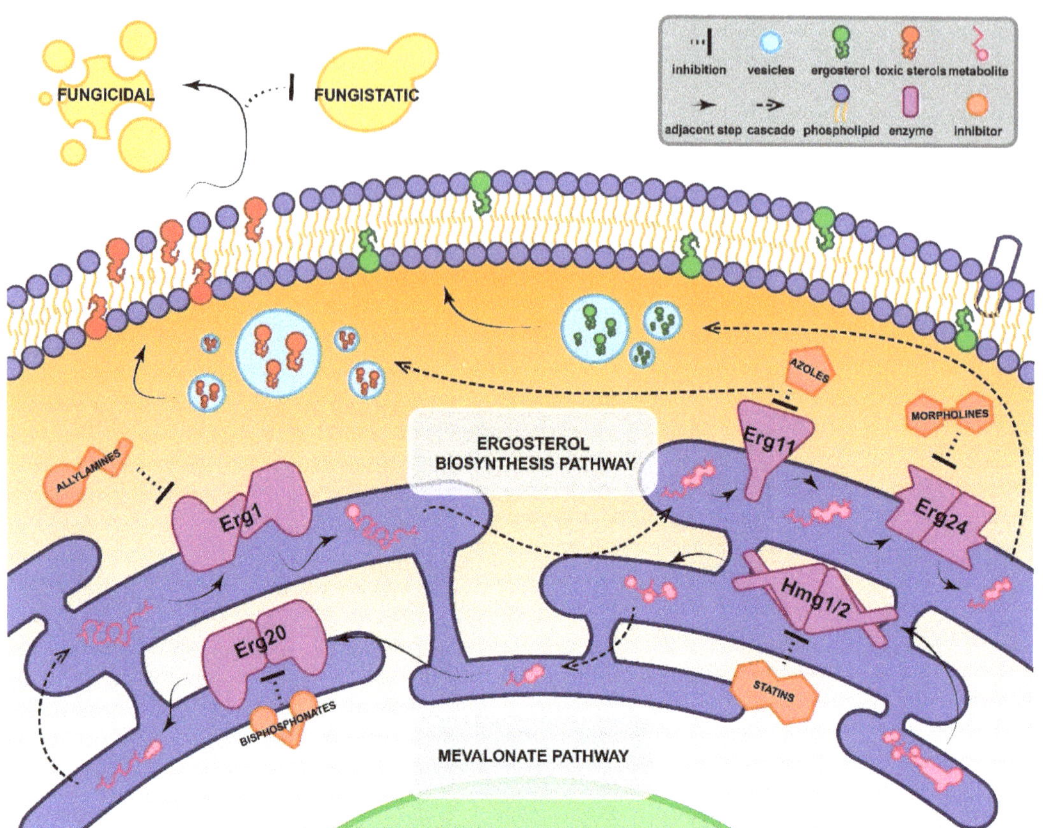

Figure 2. Proposed mechanism of synergy between azoles and inhibitors that operate on the ergosterol biosynthesis and mevalonate pathways. In fungi, the synthesis of ergosterol occurs primarily in the endoplasmic reticulum, with the final product packaged into vesicles to be incorporated into the membrane [268]. Azole drugs inhibit Erg11, preventing lanosterol from being converted into dimethyltrienol and leading to the build-up of toxic methyl sterols. These are incorporated into the membrane instead of mature ergosterol, causing a loss of membrane structure and an inability to divide and resulting in the fungistatic arrest of growth [47]. Synergistic inhibitors co-operate at points up- and down-stream of Erg11, increasing the generation of toxic ergosterol precursors and other terpene-derived metabolites. The mevalonate pathway, upstream from the ergosterol biosynthesis pathway, is responsible for the biosynthesis of squalene, a precursor to all fungal membrane sterols. Statins like atorvastatin inhibit the HMG-CoA reductases Hmg1 and Hmg2, which are responsible for the production of mevalonate from HMG-CoA [269]. Further downstream, bisphosphonates like zoledronate inhibit farnesyl pyrophosphate synthase, or Erg20, which catalyses the production of farnesyl pyrophosphate from dimethylallyl pyrophosphate [270]. In the ergosterol biosynthesis pathway, squalene is converted into squalene epoxide by Erg1, a squalene epoxidase, which can be inhibited by allylamines like terbinafine [271]. Downstream from Erg11, dimethyltrienol is converted again into dimethylzymosterol by Erg24, a sterol reductase that is inhibited by morpholine antifungals such as amorolfine [47,272]. The resulting destabilisation of the cell membrane means synergy can often produce a fungicidal effect in the pathogen.

2.2. Azole–Antibacterial Synergy

Among the antibacterials, the sulfa-based drugs have shown the greatest potential to date for azole synergy (Table 2). These inhibit folic acid biosynthesis and several, including the widely available antibacterial sulfamethoxazole, have displayed strong synergy with azoles in most azole-resistant *Candida* isolates, including isolates of *C. auris* [129,130]. Tetracycline-based antibacterial agents have demonstrated synergy with azoles specifically against azole-resistant pathogens (Table 2). Doxycycline, tigecycline and minocycline could all potentially be used to improve therapy with azoles for persistent cases of candidiasis, aspergillosis and fusariosis [131–137]. Minocycline in particular has recently demonstrated potential as an itraconazole synergist in *C. neoformans* and *Scedosporium* sp. [138,139]. This synergy has been consistently reproduced in in vivo infection models and against sessile forms of *Candida*, with some demonstrating efficacy against biofilms. Colistin is a last-resort antibiotic that has recently displayed promising synergy in vitro with isavuconazole in *Aspergillus* spp. and *Candida auris,* and in vivo in *Candida albicans* [93,96,143,144,252,253].

Other antibacterial compounds have failed to display any real potential as combined antifungal therapies. Gentamicin was synergistic with fluconazole in biofilms produced by some resistant *Candida* species. Linezolid failed to produce true synergy with any azoles tested but it did reduce the required dosage of both drugs to a clinically achievable level for a limited spectrum of fungi (Table 2) [140,141]. While reducing the dose is one of the main goals when developing novel combination treatments, in the wake of more viable therapeutic leads it seems unlikely that drugs like linezolid will see further development. Furthermore, the in vivo use of voriconazole and clarithromycin, another antibacterial found to be synergistic in vitro, was found to cause acute kidney damage, illustrating the potential for undesirable consequences when exploiting antimicrobial synergy [273].

2.3. Azole–Antiparasitic Synergy

Only a limited number of antiparasitic compounds have been explored for azole synergy in recent years, but some of them show promise. Chloroquine and artemisinin interacted synergistically with azoles against azole-resistant *Candida* strains, while pyrvinium pamoate–azole synergy was observed in a broad suite of dermatophytes [145–148]. Mefloquine and related compounds displayed limited synergy in *C. neoformans,* but did potentiate a strong fungicidal effect when combined with fungistatic fluconazole [149].

2.4. Azole–Antiviral Synergy

Antivirals have recently been investigated as potential antifungal synergists, with the anti-retrovirals showing the most promise. Saquinavir and ritonavir were effective at synergistically cooperating with azoles against *Histoplasma*, a systemic fungal pathogen [150]. Other antivirals like ribavirin and 2-adamantanine have shown promise at treating biofilms of azole-resistant *C. albicans* and potentiating the antifungal activity of azoles from fungistatic to fungicidal, respectively [151,152]. Lopinavir is an antiviral that shows extremely strong synergy with voriconazole in a majority of azole-resistant *C. auris* strains, and certainly warrants further investigation [153].

3. Active Efflux Modulators

The active, ATP-dependent transport of antimicrobial compounds out of the fungal cell is one of the most concerning mechanisms of antifungal resistance emerging today. It is the principal mechanism of azole tolerance in *C. glabrata* and in a significant portion of *C. auris* and azole-resistant *C. albicans* isolates [44,274,275]. There is also evidence that active efflux is responsible for azole resistance in some *Aspergillus* species and dermatophytes [276,277]. Due to the importance of pumps for resistance in bacterial pathogens and malarial parasites, inhibition of active efflux is a popular target in antimicrobial discovery and the development of combined antifungal treatments [278,279].

Most efflux pump inhibitors that have been investigated as azole synergists interact directly with the membrane-bound pump. Several drugs that affect the intracellular homeostasis of calcium by blocking calcium channels have been found to have some affinity for membrane-bound transporters, particularly tetrandrine and verapamil. Traditionally used as immunomodulators and vasodilators, these have produced antifungal synergy with a variety of azoles. Tetrandrine combined with posaconazole has been proposed as an effective solution for persistent and temporary candidiasis [154–159,280]. Eucalyptal D and dodecenoic acid are essential oil extracts that have been shown to directly antagonise ABC transporters [160,161]. These exhibited strong synergy with fluconazole and itraconazole, respectively, in azole-resistant *Candida.* Azoffluxin, an oxindole Cdr1 inhibitor, has been shown to synergise strongly with fluconazole in all non-clade III *C. auris* and azole-resistant *C. albicans* strains, both in vitro and in vivo [162]. Other receptor antagonists may be cross-reactive to membrane-bound transporters; for example, ospemifeme is a promising therapeutic lead that has a broad spectrum of activity and synergises very potently with fluconazole (Figure 3g) [159,163].

Several efflux inhibitors that show promising synergy with azoles do not directly bind pump proteins but interfere with efflux through other mechanisms. Palmarumycin P3 and phialocephalarin B are naturally occurring quinone derivatives that synergise with fluconazole in pump-dependent, azole-resistant *C. albicans*. It appears that the mechanism of synergy is due to the ability of these derivatives to directly inhibit nuclear transcription factors, modulating the expression of the principal pump-coding gene *MDR1* [164]. Geraniol is a unique synergist that causes the localisation of pumps to become dysregulated, preventing them from being incorporated into the membrane and resulting in weak synergy [165]. The controversial anti-cancer drug ponatinib is able to inhibit active efflux in multiple yeast pathogens by interfering with the proton motive force, producing strong synergy with fluconazole in all strains tested [196]. Other potential co-drugs like cationic triphenylphosphonium have been shown to improve the activity of efflux pump inhibitors, suggesting the potential development of a triple-drug antifungal treatment strategy [281].

Many other compounds currently being investigated for their ability to synergise with azoles promote the downstream inhibition of efflux, despite no direct action on the pumps themselves or their expression. Haloperidol and promethazine are repurposed drugs that have displayed the ability to modulate active efflux. Both show great promise as azole adjuvants for the treatment of cutaneous mycoses and dandruff [18,173,174]. Thymol and carvacrol are terpenoids extracted from thyme leaves that interact synergistically with azoles and inhibit efflux in azole-resistant *Candida,* including *C. auris* [211,212,282]. Numerous other natural products with synergistic potential have displayed similar indirect anti-efflux properties [203,211,223,227,233].

It should be noted that fungal drug efflux pumps have an extremely broad spectrum of substrates that they transport [283]. Given this, many of the compounds discussed in this review may be substrates of efflux pumps that compete with azoles, preventing the transport of the azole and prolonging its effect. This competition may contribute to the synergy observed between some drug pairs where the underlying mechanism is currently unknown.

4. Repurposing Other Pharmaceuticals

Repurposing existing drugs can short-cut the process of drug development and regulatory approval. In recent years, a popular approach in drug discovery has been to screen libraries of existing compounds for novel repurposed uses, including antifungal activity. The resulting "hits" may have completely different mechanisms of action to any other currently available antifungal, opening new avenues for drug design. Many of these repurposed antimicrobials are also tested for their capacity as antifungal adjuvants [284,285].

4.1. Statins

Statins are common anti-cholesterol medications and until recently were considered one of the most promising routes for antifungal discovery. Statins operate on HMG-CoA reductase in the mevalonate pathway, upstream from azole-targeted demethylases

(Figure 2) [269]. Unfortunately, however, as demonstrated in Table 2, synergy between azoles and statins is often minor or bordering on indifferent. In addition, while it might appear that some statin–azole pairings have a decent spectrum of activity, most studies have tested only 1–3 strains per species. Their applicability to a wide variety of mycoses is therefore difficult to gauge [166–170]. Where synergy has been observed, the mechanism was shown to be primarily driven by the co-operation of the drug pairs on the same pathway (Figure 2) [286]. Newer statins such as pitavastatin have been found to exhibit extremely strong antifungal synergy with fluconazole in azole-resistant *Candida*, which may reignite future interest [171]. Statins can unfortunately have adverse off-target effects and drug interactions that range from unpleasant to lethal, especially for already vulnerable mycosis patients. These include diabetes, liver cirrhosis, irreparable damage to skeletal muscle and sexual dysfunction [287,288].

4.2. Bisphosphonates

Bisphosphonates are anti-osteoporosis drugs and show promising synergy with fluconazole. Like statins, bisphosphonates operate on the mevalonate pathway where they target farnesyl pyrophosphate synthase (or Erg11; Figure 2) [289]. Of the bisphosphonates tested, zoledronate resulted in strong synergy across numerous strains of *Cryptococcus* tested in vitro and in an in vivo model and significantly limited the development of antifungal resistance [172]. Due to their propensity to bind to bone mineral and their implication in osteonecrosis, market bisphosphonates have limited applicability for the treatment of invasive mycoses [289,290]; however, their antifungal synergy and potent immunostimulatory properties make them attractive lead compounds for further development [291].

4.3. Repurposing Miscellaneous Pharmaceuticals

Although most fail to produce strong, broad-spectrum synergy and may have undesirable anti-inflammatory or immunosuppressive effects, certain immunomodulators could be useful as lead compounds for development as synergists to treat *Candida* biofilms [175–178,292]. The antihistamine promethazine appears potentially attractive as a novel topical anti-dandruff and anti-tinea treatment, as it synergised strongly with azoles in all strains of dermatophytes tested [18,173,174].

Some psychoactive drugs have displayed limited synergy with azoles along a narrow spectrum of activity. Bromperidol and fluoxetine show limited potential as azole adjuvants for the treatment of candidiasis, while the commonly prescribed antidepressant sertraline synergised weakly with azoles in the opportunistic yeast *Trichosporon* [179–181]. Haloperidol is an antipsychotic that may be more promising as a topical combined treatment due to its strong synergy with fluconazole and itraconazole in many strains of dermatophytes [18,173].

Two of the most attractive groups of compounds for developing new synergists are calcineurin inhibitors and calcium channel blockers. These drugs modulate calcium ion homeostasis, which is vital for cellular signalling, and have been considered promising antifungal leads for a variety of diverse pathogens for more than a decade [184–186,293]. Calcium channel blockers have also been shown to further enhance the synergy between fluconazole and doxycycline in a series of three-way checkerboards [131]. The calcium channel blockers tetrandrine and verapamil are discussed above in the context of their role in efflux, but their ability to disturb calcium homeostasis is likely also part of their antifungal effect. Other inhibitors like cyclosporine and tacrolimus target calcineurin and calmodulin, a calcium-activated complex responsible for the up-regulation of genes related to growth and the fungal stress response (Figure 3c) [18,182–185,187,294]. Tacrolimus specifically produces significant synergy for the majority of dermatophyte species, while cyclosporine reliably and potently interacts with fluconazole in *C. albicans* [183,188,189]. It should also be noted that calcineurin inhibitors like tacrolimus directly affect the ATPase activity of efflux pumps in addition to acting on calcineurin, thereby both directly and indirectly inhibiting active efflux [295].

The growing drug repurposing initiative has yielded synergists with novel mechanisms of action with significant therapeutic potential, as illustrated in Figure 3. Fungal membrane proton pumps are vital enzymes, generating the membrane potential required for membrane-bound transporter function and the uptake of nutrients required for ATP synthesis, thereby enabling ATP-dependent drug efflux [52]. A wide variety of proton pump inhibitors have been shown to produce strong synergy with fluconazole in azole-resistant *Candida* isolates [194]. HSP90 (HSP82 in yeast) inhibitors like geldanamycin and ganetespib repress the fungal response to stress, reducing the survival of yeasts during azole-induced inhibition (Figure 3a) [195,197,209]. Histone deacetylase inhibitors have been popular in the development of antifungal therapies but surprisingly few have been tested for synergy with azoles. One exception is givinostat, a potential anti-aspergillosis treatment when paired with posaconazole [198]. Lonafarnib inhibits farnesyltransferase, preventing vital post-translational modifications of fungal proteins and producing moderate azole synergy in *Aspergillus* (Figure 3b) [199]. Other recently discovered azole synergists with more limited utility inhibit superoxide dismutase, chemosensitising *C. albicans* to oxidative stressors induced by azole treatment (Figure 3f) [200]. Others such as DIBI, lactoferrin, D-penicillamine and EDTA chelate ions vital for enzymatic function, resulting in dysregulated apoptosis in the cell, but these result in only a weak or narrow-spectrum antifungal synergy (Figure 3d) [201,202,245,253].

Some repurposed pharmaceuticals interact synergistically with azoles through a mechanism that has yet to be fully elucidated. Licofelone failed to make it through clinical trials as an anti-arthritic but was shown to abrogate azole resistance in biofilms of *C. albicans* [203]. Phenylbutyrate is an aromatic fatty acid used to treat hyperammonaemia and has been shown to weakly synergise with various systemic azoles against resistant *Candida* species [204]. Other repurposed inhibitors including sedatives, antiseptics, diuretics and analgesics have displayed some promising synergy with azoles against a narrow spectrum of strains that may warrant further exploration [205–208,210].

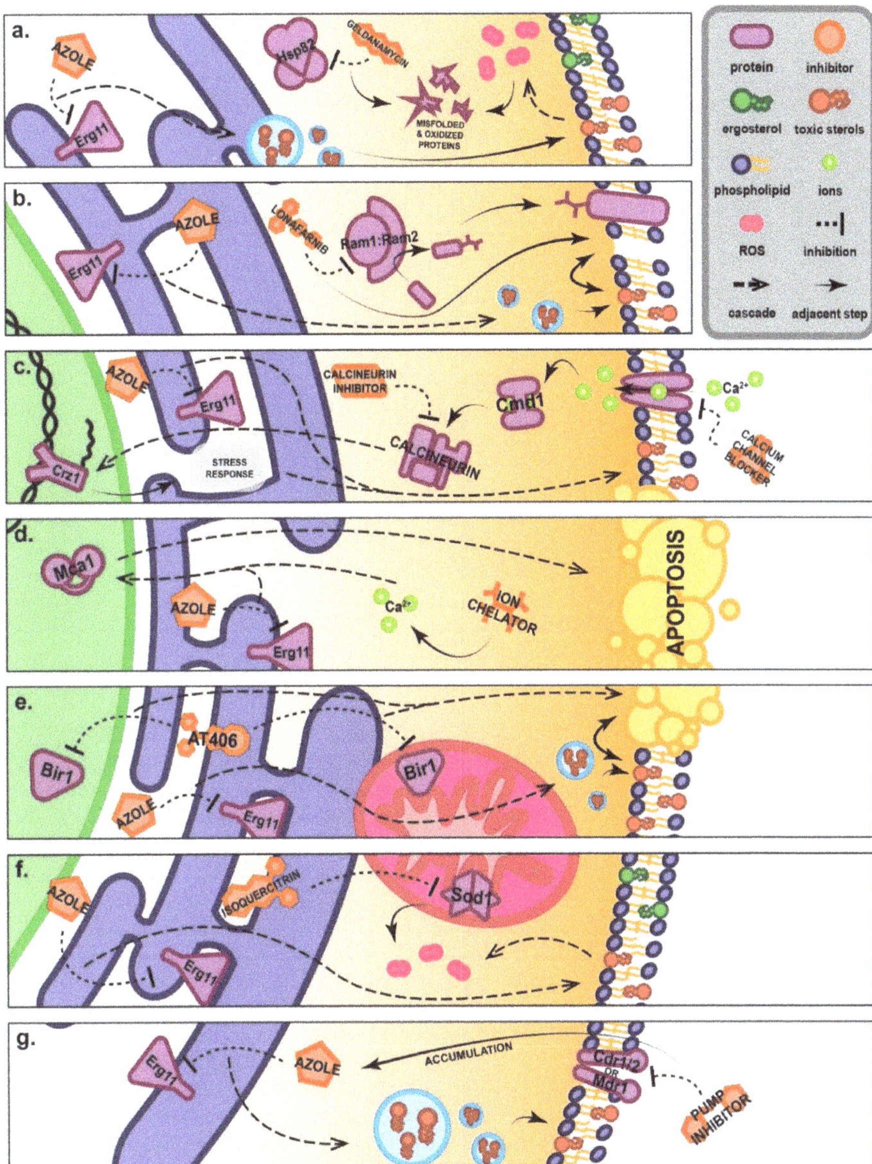

Figure 3. Proposed novel mechanisms of synergy between azoles and inhibitors that operate on entirely separate pathways. (**a**) HSP82 inhibitors like geldanamycin prevent the association of Hsp82 with proteins, inhibiting proper folding of nascent proteins and degradation of senescent proteins. Accumulation of toxic oxygen radicals results in oxidation of proteins, which would ordinarily be degraded. HSP82 inhibitor–azole synergy therefore appears to rise from the accumulation of oxidatively damaged, toxic proteins [195]. (**b**) Inhibition of protein farnesylation by farnesyltransferase (Ram1:Ram2) inhibitors such as lonafarnib results in reduced translocation of membrane-bound proteins. This decline in the population of membrane proteins combines synergistically with the azole-induced build-up of toxic sterols, resulting in increased membrane instability [199]. (**c**) Calcium

channel blockers and calcineurin inhibitors prevent the activation of the calcineurin complex by calmodulin (Cmd1). This results in an inability of calcineurin to dephosphorylate Crz1, which would ordinarily mobilise it to the nucleus. Crz1 is a transcription factor responsible for the regulation of several stress-related genes. Calcium channel blockers and calcineurin inhibitors therefore impair the cellular stress response, sensitising the cell to the antifungal effect of azoles [182,280,296]. (**d**) Ion chelators like DIBI and D-penicillamine bind to ions and disrupt cellular ion homeostasis. Evidence suggests that it is the disturbance of calcium homeostasis that results in the promotion of metacaspase (Mca1)-dependent apoptosis when paired with azoles, synergistically enhancing the fungicidal effect [202,253]. (**e**) AT406 is an antagonist of the inhibition of apoptosis proteins (IAPs) such as Bir1, which is present in both the mitochondrion and the nucleus. There is evidence that membrane weakness due to toxic sterol build-up improves the pro-apoptotic effects of AT406 [256]. (**f**) Some novel synergists, such as isoquercitrin, have demonstrated the ability to inhibit mitochondrial superoxide dismutase, Sod1. Sod1 becomes unable to neutralise harmful reactive oxygen species that accumulate during azole treatment, resulting in rapid accumulation of radicals and potentiating a toxic oxidative effect [200]. (**g**) Direct and indirect inhibitors of both ABC transporters such as Cdr1 and MFS transporters such as Mdr1 prevent the active efflux of toxic compounds such as fluconazole out of the cell, resulting in an accumulation of the drug and extending its antifungal effect. In turn, the destabilised membrane may reduce or prevent incorporation of transmembrane proteins including pumps, further reducing the efflux capabilities of the cell [173,183,240].

5. Azole Synergy with Natural Products

Nature has long been the source of novel compounds, including important anti-cancer and antimicrobial chemotherapies [297]. Table 3 lists naturally produced synergists that have been included in this review alongside their original biological source.

Some essential oils and essential oil extracts have shown strong potential as lead compounds, but most are only effective at concentrations that are not clinically achievable or are active only in particular isolates. Thymol and carvacrol are well-characterised terpenoids from thyme and oregano essential oils (Table 3) and have been found to synergise with fluconazole in wild-type *C. albicans* and have displayed some activity in *C. auris* biofilms [211,212,282]. Acetophenone is a small ketone present in many foods that showed promise as a topical antifungal when paired with ketoconazole [213]. Osthole from tonka bean oil and houttuyfonate from fish mint oil both displayed excellent synergy with azoles in azole-resistant species of *Candida* [214,215]. Oridonin is a staple of traditional Chinese medicine that has displayed strong synergy with common azoles in resistant isolates of *Candida* [222]. Menthol extracted from mint synergised well with itraconazole, but only for a fraction of the *Candida* strains tested [216]. Glabridin from liquorice root synergised with voriconazole in all *A. fumigatus* strains tested [220]. Several oil extracts displayed excellent anti-biofilm activity when combined with fluconazole, including chito-oligosaccharides, tyrosol from olive oil, allyl isothiocyanate from mustard oil and butylphthalide from celery oil [217–219,221]. Some crude essential oils have also been investigated for potential synergy, including oils from Indian frankincense, tea tree, sea buckthorn and guava leaf. While some promising anti-*Candida* and anti-dermatophyte synergy was observed, these crude oils are too complex to be called therapeutic leads, and more refined bioactive fractions need to be identified [223–226].

Alkaloid and terpenoid metabolites make up many of the most promising naturally derived synergists. The alkaloid phytochemicals berberine and palmatine are particularly attractive, displaying synergy with fluconazole in both planktonic and biofilm forms of nearly all strains tested in the *Candida* genus, *Saccharomyces cerevisiae* and *Talaromyces marneffei* [227–234]. Harmine is an alkaloid that interacts extremely synergistically with multiple systemic azoles, but only in a fraction of the strains tested, while guttiferone, a terpenoid, synergises less acutely but with more total strain coverage, particularly in non-*albicans* species of *Candida* [235,236]. Berberine and guttiferone are two of the very few phytochemicals actually taken past the point of a therapeutic lead, undergoing chemical modifications and development into more synergistic novel derivatives in further studies [236,250,298]. Farnesol is an isoprenoid that displays limited or no synergy in planktonic forms of *C. auris* but is highly synergistic against its biofilms. This contrasts with asiatic acid, a terpenoid that synergises with fluconazole only in planktonic *Candida* or in vivo, but not in biofilms [238]. Farnesol may, thus, be a promising tool for fighting highly problematic biofilms of *C. auris* [237].

Phenol derivatives are an important class of organic phytochemicals, often vital for mediating the plant response to stress [299]. Many phenolics have been shown to have antimicrobial activity, and several may be potential synergistic azole co-drugs [300]. Both the lignan magnolol and the diphenol diorcinol have proven effective against every *Candida* strain tested, with the former synergising well with fluconazole and the latter not technically synergising but sharply decreasing the required inhibitory dosage, bringing it to well within a clinically achievable concentration [239,240]. In contrast, proanthocyanidin, a plant polyphenol, synergised with fluconazole very weakly and in only a small number of the azole-resistant non-*albicans Candida* isolates tested [241]. Other phenolics show greater promise, particularly for the treatment of cutaneous or oropharyngeal candidiasis. Epigallocatechin gallate and asarones both synergise with topical azoles, with the former particularly effective against *Candida* biofilms [242,243]. Pyrogallol is a phenol that synergises with various market antifungals by inhibiting active efflux in both azole-susceptible and azole-resistant *Candida* [244]. Catechol, while unable to inhibit *Candida* itself, potentiates the antifungal activity of azoles and polyenes, and prevents up-regulation of virulence-associated genes. Curiously, catechol did not reduce the viability of *Candida* biofilms, but did reduce their hydrophobicity [301].

Antimicrobial peptides are found in cells from all taxonomic Kingdoms and are vital for the defence against infection, potentially synergising with other antimicrobials [302]. Synergy has been seen with the milk protein lactoferrin with azoles in resistant isolates of *Candida*, but not *Cryptococcus* [201,245]. Beauvericin is an antibiotic and insecticidal peptide derivative called a depsipeptide that synergises well with fluconazole. With only a limited group of azole-susceptible strains of *C. albicans* tested, however, it remains unclear whether it can be called a truly promising therapeutic lead [246].

Table 3. Bioactive natural products named in this review and their original biological source(s).

Type	Synergist	Common Name	Latin Name	Ref.
		Source		
Essential Oil Extracts	thymol	thyme, ajwain, wild bergamot	*Thymus vulgaris, Trachyspermum ammi, Monarda fistulosa*	[211]
	carvacrol	oregano, thyme, marjoram	*Origanum vulgare, Thymus vulgaris, Origanum majorana*	[211]
	acetophenone	apple, apricot, beef, cheese, croton	*Malus domestica, Prunus armeniaca, Bos taurus, Croton* sp.	[213]
	osthole	snowparsley, wild celery, shishiudo	*Cnidium monnieri, Angelica archangelica, Angelica pubescens*	[214]
	houttuyfonate	fish mint	*Houttuynia cordata*	[215]
	menthol	wild mint, peppermint	*Mentha arvensis, Mentha piperita*	[216]
	tyrosol	olive, argan	*Olea europaea, Argania spinosa*	[217]
	allyl isothiocyanate	mustard, radish, horseradish, wasabi	*Sinapis alba, Raphanus raphanistrum, Armoracia rusticana*	[218]
	butylphthalide	celery	*Apium graveolens*	[219]
	glabridin	liquorice	*Glycyrrhiza glabra*	[220]
	oridonin	blushred	*Rabdosia rubescens*	[222]
Crude Essential Oils	sea-buckthorn	sea-buckthorn	*Hippophae rhamnoides*	[223]
	guava leaf	guava, pineapple guava	*Psidium guajava, Acca sellowiana*	[224]
	frankincense	Indian frankincense	*Boswellia serrata*	[225]
	TTO	tea tree	*Melaleuca alternifolia*	[226]

Table 3. *Cont.*

		Source		
Type	**Synergist**	**Common Name**	**Latin Name**	**Ref.**
Alkaloids	berberine	barberry, tree turmeric, prickly poppy	*Berberis vulgaris, Berberis aristate, Argemone mexicana*	[228–232]
	palmatine	Amur cork tree, yanhusuo	*Phellodendron amurense, Corydalis yanhusuo*	[233,234]
	harmine	wild rue, ayahuasca	*Peganum harmala, Banisteriopsis caapi*	[235]
Other terpenoids	guttiferone	boarwood root	*Symphonia globulifera*	[236]
	farnesol	plants, animals and fungi		[237]
Other phenols	magnolol	Chinese magnolia, southern magnolia	*Magnolia officinalis, Magnolia grandiflora*	[239]
	diorcinol	fungal symbiont	*Epichloe bromicola*	[240]
	proanthocyanidin	pine bark, cranberries, grape seeds	*Pinus* sp., *Vaccinium oxycoccus, Vitis labrusca*	[241]
	epigallocatechin	black tea, white tea, green tea	*Camellia sinensis*	[242]
	asarone	sweet flag, wild ginger	*Acorus calamus, Asarum* sp.	[243]
Peptides	lactoferrin	bovine and human (milk, mucous)	*Bos taurus, Homo sapiens*	[201,245]
	beauvericin	white muscardine, *Fusarium*	*Beauveria bassiana, Fusarium* sp.	[246]

6. Azole Synergy with Novel Compounds

Numerous new compounds that have been shown to have good antifungal activity as a monotherapy have also been found to synergize with azoles. Synergy is often inconsistent for azole-susceptible and azole-resistant strains, however. Consistent with the cross-resistance observed between azoles, novel azole derivatives are generally effective only in sensitive yeasts and not in resistant ones [247,248]. Novel azoles conjugated with triphenylphosphonium cations have displayed improved mitochondrial targeting and have shown an improved fungicidal effect when combined with Hsp90 inhibitors [303].

Several of the more promising novel compounds currently under investigation are chemically modified derivatives of promising therapeutic leads. Derivatives of the aforementioned isoquinolone and phthalazine, natural metabolites berberine, piperidol, caffeic acid and guttiferone, the anti-inflammatory celecoxib, and phenylpentanol all demonstrated strong synergy with fluconazole in a significant portion of *Candida* strains tested, and in particular in azole-resistant strains [171,236,249–251,254,257]. Other promising novel compounds like beta-glucan synthase inhibitor SCY-078, ion chelator DIBI, TOR inhibitor AZD8055, efflux modulators, and a group of novel antifungal chalcones have all proven highly synergistic against azole-resistant *C. albicans*, *C. glabrata* and other *Candida* species with reduced sensitivity to azoles [173,252,253,255,260]. A group of novel ultra-short cationic lipopeptides were not able to fully synergise with fluconazole, but did interact additively [259].

Two promising classes of compounds that have been developed in the past year are "dual-inhibitors", which are single compounds that are designed to attack more than one druggable target at once. One group of these that was strongly antifungal contains a piperazine moiety, allowing it to inhibit 14α-demethylase, and a zinc binding group to inhibit HDAC [304]. Another used fluconazole conjugated with COX inhibitors and was able to consistently inhibit pathogenic *Candida* [305]. As these dual inhibitors are single compounds, they cannot be considered truly synergistic; however, a class of novel Hsp90/HDAC dual inhibitors has displayed strong synergy with fluconazole in azole-resistant *Candida* [258].

An interesting novel antifungal mechanism of action is the promotion of dysregulated apoptosis in the fungal cell. Control of apoptosis in yeasts is partially governed by the regulation of Inhibitors of Apoptosis Proteins (IAPs), which prevent progression to cell death [306]. A new IAP antagonist, AT406, promotes apoptosis in *C. albicans* and the outbreak pathogen *Exophiala dermatitidis*, sensitising the cell to the oxidative stress produced by the azoles (Figure 3e) [256]. This is a truly novel mode of fungicidal action that may be a source of an entirely new class of azole synergists.

7. Conclusions

With the rise of opportunistic and emerging fungal pathogens and increasing rates of antifungal resistance, there is an urgent need for new therapies in our antifungal toolbox. As this review has shown, a reliable path to success is to improve commonly used azole antifungals with synergists, and there is substantial diversity in the compounds and approaches that have yielded synergy. Although high-throughput screening has become a popular method for discovering new therapeutic agents, a rational, target-based approach to drug discovery may yield more reliable and effective therapeutic leads. Compounds co-operating with azoles on the ergosterol and mevalonate biosynthesis pathways have displayed consistent synergy in various fungal pathogens. Rational drug design, building on a known mechanism of action or starting with already approved drugs with known pharmacokinetic data, may take newly developed drugs into market more rapidly. On the other hand, hypothesis-free drug screening initiatives may yield novel synergies that would otherwise be undiscovered, opening new avenues for drug design.

There are several gaps in current studies that await further research. From Table 2, there is a clear focus on developing combined treatment strategies to combat *Candida*, particularly azole-resistant clinical isolates, and with a few notable exceptions there is a paucity of data exploring azole synergy in *Cryptococcus* and filamentous fungi. There is also a focus on improving the systemic azoles, particularly fluconazole, while for topical pathogens where oral bioavailability is not required, more could be gained by exploring other azoles. Finally, any translation of synergy into clinical use must deal with the issue of co-administration of two (or more) compounds. New systems that package drugs into nanoparticle delivery systems or co-crystallise compounds into a single formulation, may enable the development of single-dose synergistic treatments [307,308]. There are currently no azole-based antifungal combinations used to treat mycoses, but the need for new treatments and the threats to azoles from intrinsic and acquired resistance make drug synergy an increasingly attractive avenue for antifungal development.

Author Contributions: Conceptualization, A.K. and D.A.C.; investigation, A.K.; resources, D.A.C.; writing—original draft preparation, A.K.; writing—review and editing, D.A.C.; visualization, A.K.; supervision, D.A.C.; project administration, D.A.C.; funding acquisition, D.A.C. All authors have read and agreed to the published version of the manuscript.

Funding: This research received no external funding.

Institutional Review Board Statement: Not applicable.

Informed Consent Statement: Not applicable.

Data Availability Statement: Data sharing not applicable.

Acknowledgments: The authors would like to extend their gratitude to Kenya Fernandes for her assistance proof-reading the final draft of this review.

Conflicts of Interest: The authors declare no conflict of interest.

References

1. Enoch, D.A.; Yang, H.; Aliyu, S.H.; Micallef, C. The Changing Epidemiology of Invasive Fungal Infections. *Methods Mol. Biol.* **2017**, *1508*, 17–65. [CrossRef] [PubMed]
2. Bongomin, F.; Gago, S.; Oladele, R.; Denning, D. Global and Multi-National Prevalence of Fungal Diseases—Estimate Precision. *J. Fungi* **2017**, *3*, 57. [CrossRef] [PubMed]
3. Schelenz, S.; Barnes, R.A.; Kibbler, C.C.; Jones, B.L.; Denning, D.W. Standards of Care for Patients with Invasive Fungal Infections within the United Kingdom: A National Audit. *J. Infect.* **2009**, *58*, 145–153. [CrossRef]
4. Bitar, D.; Lortholary, O.; le Strat, Y.; Nicolau, J.; Coignard, B.; Tattevin, P.; Che, D.; Dromer, F. Population-Based Analysis of Invasive Fungal Infections, France, 2001-2010. *Emerg. Infect. Dis.* **2014**, *20*, 1149–1155. [CrossRef] [PubMed]

5. Rees, J.R.; Pinner, R.W.; Hajjeh, R.A.; Brandt, M.E.; Reingold, A.L. The Epidemiological Features of Invasive Mycotic Infections in the San Francisco Bay Area, 1992-1993: Results of Population-Based Laboratory Active Surveillance. *Clin. Infect. Dis.* **1998**, *27*, 1138–1150. [CrossRef] [PubMed]
6. Chapman, B.; Slavin, M.; Marriott, D.; Halliday, C.; Kidd, S.E.; Arthur, I.; Bak, N.; Heath, C.; Kennedy, K.; Morrissey, C.O.; et al. Changing Epidemiology of Candidaemia in Australia. *J. Antimicrob. Chemother.* **2017**, *72*, 1103–1108. [CrossRef]
7. Tsay, S.V.; Mu, Y.; Williams, S.; Epson, E.; Nadle, J.; Bamberg, W.M.; Barter, D.M.; Johnston, H.L.; Farley, M.M.; Harb, S.; et al. Burden of Candidemia in the United States, 2017. *Clin. Infect. Dis.* **2020**, *71*, E449–E453. [CrossRef]
8. Astvad, K.M.T.; Johansen, H.K.; Røder, B.L.; Rosenvinge, F.S.; Knudsen, J.D.; Lemming, L.; Schønheyder, H.C.; Hare, R.K.; Kristensen, L.; Nielsen, L.; et al. Update from a 12-Year Nationwide Fungemia Surveillance: Increasing Intrinsic and Acquired Resistance Causes Concern. *J. Clin. Microbiol.* **2018**, *56*, e01564-17. [CrossRef]
9. Eyre, D.W.; Sheppard, A.E.; Madder, H.; Moir, I.; Moroney, R.; Quan, T.P.; Griffiths, D.; George, S.; Butcher, L.; Morgan, M.; et al. A *Candida auris* Outbreak and Its Control in an Intensive Care Setting. *N. Engl. J. Med.* **2018**, *379*, 1322–1331. [CrossRef]
10. Ruiz-Gaitán, A.; Moret, A.M.; Tasias-Pitarch, M.; Aleixandre-López, A.I.; Martínez-Morel, H.; Calabuig, E.; Salavert-Lletí, M.; Ramírez, P.; López-Hontangas, J.L.; Hagen, F.; et al. An Outbreak Due to *Candida auris* with Prolonged Colonisation and Candidaemia in a Tertiary Care European Hospital. *Mycoses* **2018**, *61*, 498–505. [CrossRef]
11. Schelenz, S.; Hagen, F.; Rhodes, J.L.; Abdolrasouli, A.; Chowdhary, A.; Hall, A.; Ryan, L.; Shackleton, J.; Trimlett, R.; Meis, J.F.; et al. First Hospital Outbreak of the Globally Emerging *Candida auris* in a European Hospital. *Antimicrob. Resist. Infect. Control.* **2016**, *5*, 35. [CrossRef] [PubMed]
12. Prestel, C.; Anderson, E.; Forsberg, K.; Lyman, M.; de Perio, M.A.; Kuhar, D.; Edwards, K.; Rivera, M.; Shugart, A.; Walters, M.; et al. *Candida auris* Outbreak in a COVID-19 Specialty Care Unit—Florida, July–August 2020. *MMWR Morb. Mortal. Wkly. Rep.* **2021**, *70*, 56–57. [CrossRef] [PubMed]
13. Douglas, L.J. *Candida* Biofilms and Their Role in Infection. *Trends Microbiol.* **2003**, *11*, 30–36. [CrossRef]
14. Seebacher, C.; Bouchara, J.-P.; Mignon, B. Updates on the Epidemiology of Dermatophyte Infections. *Mycopathologia* **2008**, *166*, 335–352. [CrossRef]
15. Gupta, A.K.; Paquet, M.; Simpson, F.C. Therapies for the Treatment of Onychomycosis. *Clin. Dermatol.* **2013**, *31*, 544–554. [CrossRef]
16. Kirchhoff, L.; Olsowski, M.; Rath, P.-M.; Steinmann, J. *Exophiala dermatitidis*: Key Issues of an Opportunistic Fungal Pathogen. *Virulence* **2019**, *10*, 984–998. [CrossRef]
17. Cafarchia, C.; Iatta, R.; Immediato, D.; Puttilli, M.R.; Otranto, D. Azole Susceptibility of *Malassezia pachydermatis* and *Malassezia furfur* and Tentative Epidemiological Cut-off Values. *Med. Mycol.* **2015**, *53*, 743–748. [CrossRef]
18. Iatta, R.; Puttilli, M.R.; Immediato, D.; Otranto, D.; Cafarchia, C. The Role of Drug Efflux Pumps in *Malassezia pachydermatis* and *Malassezia furfur* Defence against Azoles. *Mycoses* **2017**, *60*, 178–182. [CrossRef]
19. Moen, M.D.; Lyseng-Williamson, K.A.; Scott, L.J. Liposomal Amphotericin B. *Drugs* **2009**, *69*, 361–392. [CrossRef]
20. Kneale, M.; Bartholomew, J.S.; Davies, E.; Denning, D.W. Global Access to Antifungal Therapy and Its Variable Cost. *J. Antimicrob. Chemother.* **2016**, *71*, 3599–3606. [CrossRef]
21. Brammer, K.W.; Farrow, P.R.; Faulkner, J.K. Pharmacokinetics and Tissue Penetration of Fluconazole in Humans. *Clin. Infect. Dis.* **1990**, *12*, S318–S326. [CrossRef] [PubMed]
22. Verweij, P.E.; Snelders, E.; Kema, G.H.; Mellado, E.; Melchers, W.J. Azole Resistance in *Aspergillus fumigatus*: A Side-Effect of Environmental Fungicide Use? *Lancet Infect. Dis.* **2009**, *9*, 789–795. [CrossRef]
23. Goldman, M.; Cloud, G.A.; Smedema, M.; Lemonte, A.; Connolly, P.; Mckinsey, D.S.; Kauffman, C.A.; Moskovitz, B.; Wheat, L.J.; Flanigan, C.; et al. Does Long-Term Itraconazole Prophylaxis Result in In Vitro Azole Resistance in Mucosal *Candida albicans* Isolates from Persons with Advanced Human Immunodeficiency Virus Infection? *Antimicrob. Agents Chemother.* **2000**, *44*, 1585–1587. [CrossRef]
24. Ruggero, M.A.; Topal, J.E. Development of Echinocandin-Resistant *Candida albicans* Candidemia Following Brief Prophylactic Exposure to Micafungin Therapy. *Transpl. Infect. Dis.* **2014**, *16*, 469–472. [CrossRef] [PubMed]
25. Bastos, R.W.; Carneiro, H.C.S.; Oliveira, L.V.N.; Rocha, K.M.; Freitas, G.J.C.; Costa, M.C.; Magalhães, T.F.F.; Carvalho, V.S.D.; Rocha, C.E.; Ferreira, G.F.; et al. Environmental Triazole Induces Cross-Resistance to Clinical Drugs and Affects Morphophysiology and Virulence of *Cryptococcus gattii* and *C. neoformans*. *Antimicrob. Agents Chemother.* **2018**, *62*, e01179-17. [CrossRef]
26. Bamba, S.; Lortholary, O.; Sawadogo, A.; Millogo, A.; Guiguemdé, R.T.; Bretagne, S. Decreasing Incidence of Cryptococcal Meningitis in West Africa in the Era of Highly Active Antiretroviral Therapy. *AIDS* **2012**, *26*, 1039–1041. [CrossRef]
27. D'Arminio Monforte, A.; Sabin, C.A.; Phillips, A.; Sterne, J.; May, M.; Justice, A.; Dabis, F.; Grabar, S.; Ledergerber, B.; Gill, J.; et al. The Changing Incidence of AIDS Events in Patients Receiving Highly Active Antiretroviral Therapy. *Arch. Intern. Med.* **2005**, *165*, 416–423. [CrossRef]
28. Bennett, J.E.; Izumikawa, K.; Marr, K.A. Mechanism of Increased Fluconazole Resistance in *Candida glabrata* during Prophylaxis. *Antimicrob. Agents Chemother.* **2004**, *48*, 1773–1777. [CrossRef]
29. Rodrigues, C.F.; Silva, S.; Henriques, M. *Candida glabrata*: A Review of Its Features and Resistance. *Eur. J. Clin. Microbiol. Infect. Dis.* **2014**, *33*, 673–688. [CrossRef]
30. Chowdhary, A.; Sharma, C.; Meis, J.F. *Candida auris*: A Rapidly Emerging Cause of Hospital-Acquired Multidrug-Resistant Fungal Infections Globally. *PLOS Pathog.* **2017**, *13*, e1006290. [CrossRef]

31. Denning, D.W.; Bromley, M. How to Bolster the Antifungal Pipeline. *Science (1979)* **2015**, *347*, 1414–1416. [CrossRef] [PubMed]
32. Odds, F.C.; Brown, A.J.P.; Gow, N.A.R. Antifungal Agents: Mechanisms of Action. *Trends Microbiol.* **2003**, *11*, 272–279. [CrossRef]
33. Allen, D.; Wilson, D.; Drew, R.; Perfect, J. Azole Antifungals: 35 Years of Invasive Fungal Infection Management. *Expert Rev. Anti-Infect. Ther.* **2015**, *13*, 787–798. [CrossRef] [PubMed]
34. Andes, D.; Kovanda, L.; Desai, A.; Kitt, T.; Zhao, M.; Walsh, T.J. Isavuconazole Concentration in Real-World Practice: Consistency with Results from Clinical Trials. *Antimicrob. Agents Chemother.* **2018**, *62*, e00585-18. [CrossRef] [PubMed]
35. Miceli, M.H.; Kauffman, C.A. Isavuconazole: A New Broad-Spectrum Triazole Antifungal Agent. *Clin. Infect. Dis.* **2015**, *61*, 1558–1565. [CrossRef] [PubMed]
36. Brand, S.R.; Degenhardt, T.P.; Person, K.; Sobel, J.D.; Nyirjesy, P.; Schotzinger, R.J.; Tavakkol, A. A Phase 2, Randomized, Double-Blind, Placebo-Controlled, Dose-Ranging Study to Evaluate the Efficacy and Safety of Orally Administered VT-1161 in the Treatment of Recurrent Vulvovaginal Candidiasis. *Am. J. Obstet. Gynecol.* **2018**, *218*, 624.e1–624.e9. [CrossRef]
37. Wiederhold, N.P.; Xu, X.; Wang, A.; Najvar, L.K.; Garvey, E.P.; Ottinger, E.A.; Alimardanov, A.; Cradock, J.; Behnke, M.; Hoekstra, W.J.; et al. In Vivo Efficacy of VT-1129 against Experimental Cryptococcal Meningitis with the Use of a Loading Dose-Maintenance Dose Administration Strategy. *Antimicrob. Agents Chemother.* **2018**, *62*, e01315-18. [CrossRef]
38. Monk, B.C.; Keniya, M.V. Roles for Structural Biology in the Discovery of Drugs and Agrochemicals Targeting Sterol 14α-Demethylases. *J. Fungi* **2021**, *7*, 67. [CrossRef]
39. Hargrove, T.Y.; Garvey, E.P.; Hoekstra, W.J.; Yates, C.M.; Wawrzak, Z.; Rachakonda, G.; Villalta, F.; Lepesheva, G.I. Crystal Structure of the New Investigational Drug Candidate VT-1598 in Complex with *Aspergillus fumigatus* Sterol 14α-Demethylase Provides Insights into Its Broad-Spectrum Antifungal Activity. *Antimicrob. Agents Chemother.* **2017**, *61*, e00570-17. [CrossRef]
40. Antifungal Drugs Market Size 2019 | Demand | Industry Forecast. Available online: https://www.reportsanddata.com/report-detail/antifungal-drugs-market (accessed on 8 March 2021).
41. Global Antifungal Drugs Market Report 2020 | Orbis Research. Available online: https://www.orbisresearch.com/reports/index/global-antifungal-drugs-market-report-2020 (accessed on 8 March 2021).
42. Zonios, D.; Yamazaki, H.; Murayama, N.; Natarajan, V.; Palmore, T.; Childs, R.; Skinner, J.; Bennett, J.E. Voriconazole Metabolism, Toxicity, and the Effect of Cytochrome P450 2C19 Genotype. *J. Infect. Dis.* **2014**, *209*, 1941–1948. [CrossRef]
43. Mann, P.A.; McNicholas, P.M.; Chau, A.S.; Patel, R.; Mendrick, C.; Ullmann, A.J.; Cornely, O.A.; Patino, H.; Black, T.A. Impact of Antifungal Prophylaxis on Colonization and Azole Susceptibility of *Candida* Species. *Antimicrob. Agents Chemother.* **2009**, *53*, 5026–5034. [CrossRef] [PubMed]
44. Chaabane, F.; Graf, A.; Jequier, L.; Coste, A.T. Review on Antifungal Resistance Mechanisms in the Emerging Pathogen *Candida auris*. *Front. Microbiol.* **2019**, *10*, 2788. [CrossRef]
45. Brajtburg, J.; Powderly, W.G.; Kobayashi, G.S.; Medoff, G. Amphotericin B: Current Understanding of Mechanisms of Action. *Antimicrob. Agents Chemother.* **1990**, *34*, 183–188. [CrossRef] [PubMed]
46. Perlin, D.S. Current Perspectives on Echinocandin Class Drugs. *Future Microbiol.* **2011**, *6*, 441–457. [CrossRef] [PubMed]
47. Georgopapadakou, N.H. Antifungals: Mechanism of Action and Resistance, Established and Novel Drugs. *Curr. Opin. Microbiol.* **1998**, *1*, 547–557. [CrossRef]
48. Leyden, J. Pharmacokinetics and Pharmacology of Terbinafine and Itraconazole. *J. Am. Acad. Dermatol.* **1998**, *38*, S42–S47. [CrossRef]
49. de Carli, L.; Larizza, L. Griseofulvin. *Mutat. Res./Rev. Genet. Toxicol.* **1988**, *195*, 91–126. [CrossRef]
50. Vermes, A. Flucytosine: A Review of Its Pharmacology, Clinical Indications, Pharmacokinetics, Toxicity and Drug Interactions. *J. Antimicrob. Chemother.* **2000**, *46*, 171–179. [CrossRef]
51. Heilmann, C.J.; Schneider, S.; Barker, K.S.; Rogers, P.D.; Morschhäuser, J. An A643T Mutation in the Transcription Factor Upc2p Causes Constitutive ERG11 Upregulation and Increased Fluconazole Resistance in *Candida albicans*. *Antimicrob. Agents Chemother.* **2010**, *54*, 353–359. [CrossRef]
52. Holmes, A.R.; Cardno, T.S.; Strouse, J.J.; Ivnitski-Steele, I.; Keniya, M.V.; Lackovic, K.; Monk, B.C.; Sklar, L.A.; Cannon, R.D. Targeting Efflux Pumps to Overcome Antifungal Drug Resistance. *Future Med. Chem.* **2016**, *8*, 1485–1501. [CrossRef]
53. Cannon, R.D.; Lamping, E.; Holmes, A.R.; Niimi, K.; Baret, P.V.; Keniya, M.V.; Tanabe, K.; Niimi, M.; Goffeau, A.; Monk, B.C. Efflux-Mediated Antifungal Drug Resistance. *Clin. Microbiol. Rev.* **2009**, *22*, 291–321. [CrossRef] [PubMed]
54. Lamping, E.; Monk, B.C.; Niimi, K.; Holmes, A.R.; Tsao, S.; Tanabe, K.; Niimi, M.; Uehara, Y.; Cannon, R.D. Characterization of Three Classes of Membrane Proteins Involved in Fungal Azole Resistance by Functional Hyperexpression in *Saccharomyces cerevisiae*. *Eukaryot. Cell* **2007**, *6*, 1150–1165. [CrossRef] [PubMed]
55. Selmecki, A.; Forche, A.; Berman, J. Aneuploidy and Isochromosome Formation in Drug-Resistant *Candida albicans*. *Science (1979)* **2006**, *313*, 367–370. [CrossRef] [PubMed]
56. Selmecki, A.M.; Dulmage, K.; Cowen, L.E.; Anderson, J.B.; Berman, J. Acquisition of Aneuploidy Provides Increased Fitness during the Evolution of Antifungal Drug Resistance. *PLoS Genet.* **2009**, *5*, e1000705. [CrossRef] [PubMed]
57. Lockhart, S.R.; Frade, J.P.; Etienne, K.A.; Pfaller, M.A.; Diekema, D.J.; Balajee, S.A. Azole Resistance in *Aspergillus fumigatus* Isolates from the ARTEMIS Global Surveillance Study Is Primarily Due to the TR/L98H Mutation in the Cyp51A Gene. *Antimicrob. Agents Chemother.* **2011**, *55*, 4465–4468. [CrossRef]
58. Gonzalez-Jimenez, I.; Lucio, J.; Amich, J.; Cuesta, I.; Arroyo, R.S.; Alcazar-Fuoli, L.; Mellado, E. A CYP51b Mutation Contributes to Azole Resistance in *Aspergillus fumigatus*. *J. Fungi* **2020**, *6*, 315. [CrossRef]

59. Leonardelli, F.; Macedo, D.; Dudiuk, C.; Cabeza, M.S.; Gamarra, S.; Garcia-Effron, G. *Aspergillus fumigatus* Intrinsic Fluconazole Resistance Is Due to the Naturally Occurring T301I Substitution in Cyp51Ap. *Antimicrob. Agents Chemother.* **2016**, *60*, 5420–5426. [CrossRef]

60. Vermitsky, J.P.; Edlind, T.D. Azole Resistance in *Candida glabrata*: Coordinate Upregulation of Multidrug Transporters and Evidence for a Pdr1-like Transcription Factor. *Antimicrob. Agents Chemother.* **2004**, *48*, 3773–3781. [CrossRef]

61. Hagiwara, D.; Miura, D.; Shimizu, K.; Paul, S.; Ohba, A.; Gonoi, T.; Watanabe, A.; Kamei, K.; Shintani, T.; Moye-Rowley, W.S.; et al. A Novel Zn2-Cys6 Transcription Factor AtrR Plays a Key Role in an Azole Resistance Mechanism of *Aspergillus fumigatus* by Co-Regulating Cyp51A and Cdr1B Expressions. *PLoS Pathog.* **2017**, *13*, e1006096. [CrossRef]

62. Keniya, M.V.; Sabherwal, M.; Wilson, R.K.; Woods, M.A.; Sagatova, A.A.; Tyndall, J.D.A.; Monk, B.C. Crystal Structures of Full-Length Lanosterol 14α-Demethylases of Prominent Fungal Pathogens *Candida albicans* and *Candida glabrata* Provide Tools for Antifungal Discovery. *Antimicrob. Agents Chemother.* **2018**, *62*, e01134-18. [CrossRef]

63. Sheng, C.; Miao, Z.; Ji, H.; Yao, J.; Wang, W.; Che, X.; Dong, G.; Lü, J.; Guo, W.; Zhang, W. Three-Dimensional Model of Lanosterol 14α-Demethylase from *Cryptococcus neoformans*: Active-Site Characterization and Insights into Azole Binding. *Antimicrob. Agents Chemother.* **2009**, *53*, 3487–3495. [CrossRef] [PubMed]

64. Lewis, R.E.; Kontoyiannis, D.P. Rationale for Combination Antifungal Therapy. *Pharmacotherapy* **2001**, *21*, 149S–164S. [CrossRef] [PubMed]

65. Droegemueller, W.; Adamson, D.G.; Brown, D.; Cibley, L.; Fleury, F.; Lepage, M.E.; Henzl, M.R. Three-Day Treatment with Butoconazole Nitrate for Vulvovaginal Candidiasis. *Obstet. Gynecol.* **1984**, *64*, 530–534. [PubMed]

66. Schmidt, A. In Vitro Activity of Climbazole, Clotrimazole and Silver-Sulphadiazine against Isolates of *Malassezia pachydermatis*. *J. Vet. Med. Ser. B* **1997**, *44*, 193–197. [CrossRef]

67. Mendling, W.; Plempel, M. Vaginal Secretion Levels after 6 Days, 3 Days and 1 Day of Treatment with 100, 200 and 500 Mg Vaginal Tablets of Clotrimazole and Their Therapeutic Efficacy. *Chemotherapy* **1982**, *28*, 43–47. [CrossRef]

68. Fernández-Torres, B.; Inza, I.; Guarro, J. In Vitro Activities of the New Antifungal Drug Eberconazole and Three Other Topical Agents against 200 Strains of Dermatophytes. *J. Clin. Microbiol.* **2003**, *41*, 5209–5211. [CrossRef]

69. Thienpont, D.; van Cutsem, J.; van Nueten, J.M. Biological and Toxicological Properties of Econazole, a Broad Spectrum Antimycotic. *Arzneim.-Forsch./Drug Res.* **1975**, *25*, 224–230.

70. Gerven, F.; van Odds, F.C. The Anti-*Malassezia furfur* Activity In Vitro and in Experimental Dermatitis of Six Imidazole Antifungal Agents: Bifonazole, Clotrimazole, Flutrimazole, Ketoconazole, Miconazole and Sertaconazole. *Mycoses* **1995**, *38*, 389–393. [CrossRef]

71. Veraldi, S. Isoconazole Nitrate: A Unique Broad-Spectrum Antimicrobial Azole Effective in the Treatment of Dermatomycoses, Both as Monotherapy and in Combination with Corticosteroids. *Mycoses* **2013**, *56*, 3–15. [CrossRef]

72. Green, C.A.; Farr, P.M.; Shuster, S. Treatment of Seborrhoeic Dermatitis with Ketoconazole: II. Response of Seborrhoeic Dermatitis of the Face, Scalp and Trunk to Topical Ketoconazole. *Br. J. Dermatol.* **1987**, *116*, 217–221. [CrossRef]

73. Khanna, D.; Bharti, S. Luliconazole for the Treatment of Fungal Infections: An Evidence-Based Review. *Core Evid.* **2014**, *9*, 113. [CrossRef] [PubMed]

74. van Cutsem, J.M.; Thienpont, D. Miconazole, a Broad-Spectrum Antimycotic Agent with Antibacterial Activity. *Chemotherapy* **1972**, *17*, 392–404. [CrossRef] [PubMed]

75. Polak, A. Oxiconazole, a New Imidazole Derivative. Evaluation of Antifungal Activity In Vitro and In Vivo. *Arzneim.-Forsch./Drug Res.* **1982**, *32*, 17–24.

76. Liebel, F.; Lyte, P.; Garay, M.; Babad, J.; Southall, M.D. Anti-Inflammatory and Anti-Itch Activity of Sertaconazole Nitrate. *Arch. Dermatol. Res.* **2006**, *298*, 191–199. [CrossRef] [PubMed]

77. Carrillo-Muñoz, A.J.; Giusiano, G.; Ezkurra, P.A.; Quindós, G. Sertaconazole: Updated Review of a Topical Antifungal Agent. *Expert Rev. Anti-Infect. Ther.* **2005**, *3*, 333–342. [CrossRef]

78. Benfield, P.; Stephen, P. Sulconazole: A Review of Its Antimicrobial Activity and Therapeutic Use in Superficial Dermatomycoses. *Drugs* **1988**, *35*, 143–153. [CrossRef]

79. Sunderland, M.R.; Cruickshank, R.H.; Leighs, S.J. The Efficacy of Antifungal Azole and Antiprotozoal Compounds in Protection of Wool from Keratin-Digesting Insect Larvae. *Text. Res. J.* **2014**, *84*, 924–931. [CrossRef]

80. Jevons, S.; Gymer, G.E.; Brammer, K.W.; Cox, D.A.; Leeming, M.R. Antifungal Activity of Tioconazole (UK-20,349), a New Imidazole Derivative. *Antimicrob. Agents Chemother.* **1979**, *15*, 597–602. [CrossRef]

81. Dismukes, W.E.; Stamm, A.M.; Graybill, J.R.; Craven, P.C.; Stevens, D.A.; Stiller, R.L.; Sarosi, G.A.; Medoff, G.; Gregg, C.R.; Gallis, H.A.; et al. Treatment of Systemic Mycoses with Ketoconazole: Emphasis on Toxicity and Clinical Response in 52 Patients. National Institute of Allergy and Infectious Diseases Collaborative Antifungal Study. *Ann. Intern. Med.* **1983**, *98*, 13–20. [CrossRef]

82. Elewski, B.E.; Rich, P.; Pollak, R.; Pariser, D.M.; Watanabe, S.; Senda, H.; Ieda, C.; Smith, K.; Pillai, R.; Ramakrishna, T.; et al. Efinaconazole 10% Solution in the Treatment of Toenail Onychomycosis: Two Phase III Multicenter, Randomized, Double-Blind Studies. *J. Am. Acad. Dermatol.* **2013**, *68*, 600–608. [CrossRef]

83. van Cutsem, J.; van Gerven, F.; Zaman, R.; Janssen, P.A.J. Terconazole—A New Broad-Spectrum Antifungal. *Chemotherapy* **1983**, *29*, 322–331. [CrossRef]

84. Sobue, S.; Tan, K.; Layton, G.; Eve, M.; Sanderson, J.B. Pharmacokinetics of Fosfluconazole and Fluconazole Following Multiple Intravenous Administration of Fosfluconazole in Healthy Male Volunteers. *Br. J. Clin. Pharmacol.* **2004**, *58*, 20–25. [CrossRef] [PubMed]

85. Takahashi, D.; Nakamura, T.; Shigematsu, R.; Matsui, M.; Araki, S.; Kubo, K.; Sato, H.; Shirahata, A. Fosfluconazole for Antifungal Prophylaxis in Very Low Birth Weight Infants. *Int. J. Pediatr.* **2009**, *2009*, 274768. [CrossRef] [PubMed]

86. Watanabe, S.; Tsubouchi, I.; Okubo, A. Efficacy and Safety of Fosravuconazole L-Lysine Ethanolate, a Novel Oral Triazole Antifungal Agent, for the Treatment of Onychomycosis: A Multicenter, Double-Blind, Randomized Phase III Study. *J. Dermatol.* **2018**, *45*, 1151–1159. [CrossRef] [PubMed]

87. Marty, F.M.; Ostrosky-Zeichner, L.; Cornely, O.A.; Mullane, K.M.; Perfect, J.R.; Thompson, G.R.; Alangaden, G.J.; Brown, J.M.; Fredricks, D.N.; Heinz, W.J.; et al. Isavuconazole Treatment for Mucormycosis: A Single-Arm Open-Label Trial and Case-Control Analysis. *Lancet Infect. Dis.* **2016**, *16*, 828–837. [CrossRef]

88. Odds, F.C. Itraconazole—A New Oral Antifungal Agent with a Very Broad Spectrum of Activity in Superficial and Systemic Mycoses. *J. Dermatol. Sci.* **1993**, *5*, 65–72. [CrossRef]

89. Torres, H.A.; Hachem, R.Y.; Chemaly, R.F.; Kontoyiannis, D.P.; Raad, I.I. Posaconazole: A Broad-Spectrum Triazole Antifungal. *Lancet Infect. Dis.* **2005**, *5*, 775–785. [CrossRef]

90. Hoffman, H.L.; Rathbun, R.C. Review of the Safety and Efficacy of Voriconazole. *Expert Opin. Investig. Drugs* **2002**, *11*, 409–429. [CrossRef]

91. BioCentury—Abasol Abafungin Regulatory Update. Available online: https://www.biocentury.com/article/126706/abasol-abafungin-regulatory-update (accessed on 16 March 2021).

92. Odds, F.C. Synergy, Antagonism, and What the Chequerboard Puts between Them. *J. Antimicrob. Chemother.* **2003**, *52*, 1. [CrossRef]

93. Xu, X.; Xu, L.; Yuan, G.; Wang, Y.; Qu, Y.; Zhou, M. Synergistic Combination of Two Antimicrobial Agents Closing Each Other's Mutant Selection Windows to Prevent Antimicrobial Resistance. *Sci. Rep.* **2018**, *8*, 7237. [CrossRef]

94. Coates, A.R.M.; Hu, Y.; Holt, J.; Yeh, P. Antibiotic Combination Therapy against Resistant Bacterial Infections: Synergy, Rejuvenation and Resistance Reduction. *Expert Rev. Anti-Infect. Ther.* **2020**, *18*, 5–15. [CrossRef] [PubMed]

95. Zubko, E.I.; Zubko, M.K. Co-Operative Inhibitory Effects of Hydrogen Peroxide and Iodine against Bacterial and Yeast Species. *BMC Res. Notes* **2013**, *6*, 272. [CrossRef] [PubMed]

96. Bibi, M.; Murphy, S.; Benhamou, R.I.; Rosenberg, A.; Ulman, A.; Bicanic, T.; Fridman, M.; Berman, J. Combining Colistin and Fluconazole Synergistically Increases Fungal Membrane Permeability and Antifungal Cidality. *ACS Infect. Dis.* **2021**, *7*, 377–389. [CrossRef] [PubMed]

97. Trickey, A.; May, M.T.; Vehreschild, J.J.; Obel, N.; Gill, M.J.; Crane, H.M.; Boesecke, C.; Patterson, S.; Grabar, S.; Cazanave, C.; et al. Survival of HIV-Positive Patients Starting Antiretroviral Therapy between 1996 and 2013: A Collaborative Analysis of Cohort Studies. *Lancet HIV* **2017**, *4*, e349–e356. [CrossRef]

98. Bell, A. Antimalarial Drug Synergism and Antagonism: Mechanistic and Clinical Significance. *FEMS Microbiol. Lett.* **2005**, *253*, 171–184. [CrossRef] [PubMed]

99. Hitchings, G.H. Mechanism of Action of Trimethoprim-Sulfamethoxazole. *J. Infect. Dis.* **1973**, *128*, S433–S436. [CrossRef]

100. Bennett, J.E.; Dismukes, W.E.; Duma, R.J.; Medoff, G.; Sande, M.A.; Gallis, H.; Leonard, J.; Fields, B.T.; Bradshaw, M.; Haywood, H.; et al. A Comparison of Amphotericin B Alone and Combined with Flucytosine in the Treatment of Cryptoccal Meningitis. *N. Engl. J. Med.* **1979**, *301*, 126–131. [CrossRef]

101. Stamm, A.M.; Diasio, R.B.; Dismukes, W.E.; Shadomy, S.; Cloud, G.A.; Bowles, C.A.; Karam, G.H.; Espinel-Ingroff, A. Toxicity of Amphotericin B plus Flucytosine in 194 Patients with Cryptococcal Meningitis. *Am. J. Med.* **1987**, *83*, 236–242. [CrossRef]

102. Campitelli, M.; Zeineddine, N.; Samaha, G.; Maslak, S. Combination Antifungal Therapy: A Review of Current Data. *J. Clin. Med. Res.* **2017**, *9*, 451–456. [CrossRef]

103. Johnson, M.D.; MacDougall, C.; Ostrosky-Zeichner, L.; Perfect, J.R.; Rex, J.H. Combination Antifungal Therapy. *Antimicrob. Agents Chemother.* **2004**, *48*, 693–715. [CrossRef]

104. Weig, M.; Müller, F.M.C. Synergism of Voriconazole and Terbinafine against *Candida albicans* Isolates from Human Immunodeficiency Virus-Infected Patients with Oropharyngeal Candidiasis. *Antimicrob. Agents Chemother.* **2001**, *45*, 966–968. [CrossRef] [PubMed]

105. Ghannoum, M.A.; Elewski, B. Successful Treatment of Fluconazole-Resistant Oropharyngeal Candidiasis by a Combination of Fluconazole and Terbinafine. *Clin. Diagn. Lab. Immunol.* **1999**, *6*, 921–923. [CrossRef] [PubMed]

106. Perea, S.; Gonzalez, G.; Fothergill, A.W.; Sutton, D.A.; Rinaldi, M.G. In Vitro Activities of Terbinafine in Combination with Fluconazole, Itraconazole, Voriconazole, and Posaconazole against Clinical Isolates of *Candida glabrata* with Decreased Susceptibility to Azoles. *J. Clin. Microbiol.* **2002**, *40*, 1831–1833. [CrossRef]

107. Cantón, E.; Pemán, J.; Gobernado, M.; Viudes, A.; Espinel-Ingroff, A. Synergistic Activities of Fluconazole and Voriconazole with Terbinafine against Four *Candida* Species Determined by Checkerboard, Time-Kill, and Etest Methods. *Antimicrob. Agents Chemother.* **2005**, *49*, 1593–1596. [CrossRef] [PubMed]

108. Meletiadis, J.; Mouton, J.W.; Rodriguez-Tudela, J.L.; Meis, J.F.G.M.; Verweij, P.E. In Vitro Interaction of Terbinafine with Itraconazole against Clinical Isolates of *Scedosporium prolificans*. *Antimicrob. Agents Chemother.* **2000**, *44*, 470–472. [CrossRef] [PubMed]

109. Meletiadis, J.; Mouton, J.W.; Meis, J.F.G.M.; Verweij, P.E. In Vitro Drug Interaction Modeling of Combinations of Azoles with Terbinafine against Clinical *Scedosporium prolificans* Isolates. *Antimicrob. Agents Chemother.* **2003**, *47*, 106–117. [CrossRef]
110. Ryder, N.S.; Leitner, I. Synergistic Interaction of Terbinafine with Triazoles or Amphotericin B against *Aspergillus* Species. *Med. Mycol.* **2001**, *39*, 91–95. [CrossRef]
111. Bidaud, A.L.; Schwarz, P.; Chowdhary, A.; Dannaoui, E. In Vitro Antifungal Combination of Terbinafine with Itraconazole against Isolates of Trichophyton Species. *Antimicrob. Agents Chemother.* **2021**, *66*, e0144921. [CrossRef]
112. Barchiesi, F.; Di Francesco, L.F.; Scalise, G. In Vitro Activities of Terbinafine in Combination with Fluconazole and Itraconazole against Isolates of *Candida albicans* with Reduced Susceptibility to Azoles. *Antimicrob. Agents Chemother.* **1997**, *41*, 1812–1814. [CrossRef]
113. Mavridou, E.; Meletiadis, J.; Rijs, A.; Mouton, J.W.; Verweij, P.E. The Strength of Synergistic Interaction between Posaconazole and Caspofungin Depends on the Underlying Azole Resistance Mechanism of *Aspergillus fumigatus*. *Antimicrob. Agents Chemother.* **2015**, *59*, 1738–1744. [CrossRef]
114. Buil, J.B.; Brüggemann, R.J.M.; Denardi, L.B.; Melchers, W.J.G.; Verweij, P.E. In Vitro Interaction of Isavuconazole and Anidulafungin against Azole-Susceptible and Azole-Resistant *Aspergillus fumigatus* Isolates. *J. Antimicrob. Chemother.* **2020**, *75*, 2582–2586. [CrossRef] [PubMed]
115. Fakhim, H.; Chowdhary, A.; Prakash, A.; Vaezi, A.; Dannaoui, E.; Meis, J.F.; Badali, H. In Vitro Interactions of Echinocandins with Triazoles against Multidrug-Resistant *Candida auris*. *Antimicrob. Agents Chemother.* **2017**, *61*, e01056-17. [CrossRef] [PubMed]
116. Fakhim, H.; Vaezi, A.; Dannaoui, E.; Sharma, C.; Mousavi, B.; Chowdhary, A.; Meis, J.F.; Badali, H. In Vitro Combination of Voriconazole with Micafungin against Azole-Resistant Clinical Isolates of *Aspergillus fumigatus* from Different Geographical Regions. *Diagn. Microbiol. Infect. Dis.* **2018**, *91*, 266–268. [CrossRef] [PubMed]
117. Pfaller, M.A.; Messer, S.A.; Deshpande, L.M.; Rhomberg, P.R.; Utt, E.A.; Castanheira, M. Evaluation of Synergistic Activity of Isavuconazole or Voriconazole plus Anidulafungin and the Occurrence and Genetic Characterization of *Candida auris* Detected in a Surveillance Program. *Antimicrob. Agents Chemother.* **2021**, *65*, e02031-20. [CrossRef]
118. Caballero, U.; Kim, S.; Eraso, E.; Quindós, G.; Vozmediano, V.; Schmidt, S.; Jauregizar, N. In Vitro Synergistic Interactions of Isavuconazole and Echinocandins against *Candida auris*. *Antibiotics* **2021**, *10*, 355. [CrossRef]
119. Sradhanjali, S.; Yein, B.; Sharma, S.; Das, S. In Vitro Synergy of Natamycin and Voriconazole against Clinical Isolates of *Fusarium*, *Candida*, *Aspergillus* and *Curvularia* spp. *Br. J. Ophthalmol.* **2018**, *102*, 142–145. [CrossRef]
120. Gupta, A.K.; Kohli, Y. In Vitro Susceptibility Testing of Ciclopirox, Terbinafine, Ketoconazole and Itraconazole against Dermatophytes and Nondermatophytes, and In Vitro Evaluation of Combination Antifungal Activity. *Br. J. Dermatol.* **2003**, *149*, 296–305. [CrossRef]
121. Bidaud, A.L.; Botterel, F.; Chowdhary, A.; Dannaouia, E. In Vitro Antifungal Combination of Flucytosine with Amphotericin B, Voriconazole, or Micafungin against *Candida auris* Shows No Antagonism. *Antimicrob. Agents Chemother.* **2019**, *63*, e01393-19. [CrossRef]
122. Macedo, D.; Leonardelli, F.; Dudiuk, C.; Vitale, R.G.; del Valle, E.; Giusiano, G.; Gamarra, S.; Garcia-Effron, G. In Vitro and In Vivo Evaluation of Voriconazole-Containing Antifungal Combinations against *Mucorales* Using a *Galleria mellonella* Model of Mucormycosis. *J. Fungi* **2019**, *5*, 5. [CrossRef]
123. Polak, A. Combination of Amorolfine with Various Antifungal Drugs in Dermatophytosis. *Mycoses* **2009**, *36*, 43–49. [CrossRef]
124. Tamura, T.; Asahara, M.; Yamamoto, M.; Yamaura, M.; Matsumura, M.; Goto, K.; Rezaei-Matehkolaei, A.; Mirhendi, H.; Makimura, M.; Makimura, K. In Vitro Susceptibility of Dermatomycoses Agents to Six Antifungal Drugs and Evaluation by Fractional Inhibitory Concentration Index of Combined Effects of Amorolfine and Itraconazole in Dermatophytes. *Microbiol. Immunol.* **2014**, *58*, 1–8. [CrossRef] [PubMed]
125. Lecha, M. Amorolfine and Itraconazole Combination for Severe Toenail Onychomycosis; Results of an Open Randomized Trial in Spain. *Br. J. Dermatol.* **2008**, *145*, 21–26. [CrossRef]
126. Laurent, A.; Monod, M. Production of *Trichophyton rubrum* Microspores in Large Quantities and Its Application to Evaluate Amorolfine/Azole Compound Interactions In Vitro. *Mycoses* **2017**, *60*, 581–586. [CrossRef] [PubMed]
127. Shrestha, S.K.; Grilley, M.; Anderson, T.; Dhiman, C.; Oblad, J.; Chang, C.-W.T.; Sorensen, K.N.; Takemoto, J.Y. In Vitro Antifungal Synergy between Amphiphilic Aminoglycoside K20 and Azoles against *Candida* Species and *Cryptococcus neoformans*. *Med. Mycol.* **2015**, *53*, 837–844. [CrossRef] [PubMed]
128. Revie, N.M.; Robbins, N.; Whitesell, L.; Frost, J.R.; Appavoo, S.D.; Yudin, A.K.; Cowen, L.E. Oxadiazole-Containing Macrocyclic Peptides Potentiate Azole Activity against Pathogenic *Candida* Species. *mSphere* **2020**, *5*, e00256-20. [CrossRef]
129. Eldesouky, H.E.; Li, X.; Abutaleb, N.S.; Mohammad, H.; Seleem, M.N. Synergistic Interactions of Sulfamethoxazole and Azole Antifungal Drugs against Emerging Multidrug-Resistant *Candida auris*. *Int. J. Antimicrob. Agents* **2018**, *52*, 754–761. [CrossRef]
130. Eldesouky, H.E.; Mayhoub, A.; Hazbun, T.R.; Seleema, M.N. Reversal of Azole Resistance in *Candida albicans* by Sulfa Antibacterial Drugs. *Antimicrob. Agents Chemother.* **2018**, *62*, e00701-17. [CrossRef]
131. Gao, Y.; Zhang, C.; Lu, C.; Liu, P.; Li, Y.; Li, H.; Sun, S. Synergistic Effect of Doxycycline and Fluconazole against *Candida albicans* Biofilms and the Impact of Calcium Channel Blockers. *FEMS Yeast Res.* **2013**, *13*, 453–462. [CrossRef]
132. Hooper, R.W.; Ashcraft, D.S.; Pankey, G.A. In Vitro Synergy with Fluconazole plus Doxycycline or Tigecycline against Clinical *Candida glabrata* Isolates. *Med. Mycol.* **2019**, *57*, 122–126. [CrossRef]

133. Gu, W.; Yu, Q.; Yu, C.; Sun, S. In Vivo Activity of Fluconazole/Tetracycline Combinations in *Galleria mellonella* with Resistant *Candida albicans* Infection. *J. Glob. Antimicrob. Resist.* **2018**, *13*, 74–80. [CrossRef]

134. Venturini, T.P.; Rossato, L.; Chassot, F.; Keller, J.T.; Piasentin, F.B.; Santurio, J.M.; Alves, S.H. In Vitro Synergistic Combinations of Pentamidine, Polymyxin B, Tigecycline and Tobramycin with Antifungal Agents against *Fusarium* spp. *J. Med. Microbiol.* **2016**, *65*, 770–774. [CrossRef] [PubMed]

135. Hacioglu, M.; Tan, A.S.B.; Dosler, S.; Inan, N.; Otuk, G. In Vitro Activities of Antifungals Alone and in Combination with Tigecycline against *Candida albicans* Biofilms. *PeerJ* **2018**, *2018*, e5263. [CrossRef] [PubMed]

136. Gao, L.; Sun, Y.; Yuan, M.; Li, M.; Zeng, T. In Vitro and In Vivo Study on the Synergistic Effect of Minocycline and Azoles against Pathogenic Fungi. *Antimicrob. Agents Chemother.* **2020**, *64*, e00290-20. [CrossRef] [PubMed]

137. Tan, J.; Jiang, S.; Tan, L.; Shi, H.; Yang, L.; Sun, Y.; Wang, X. Antifungal Activity of Minocycline and Azoles Against Fluconazole-Resistant *Candida* Species. *Front. Microbiol.* **2021**, *12*, 1185. [CrossRef]

138. Tan, L.; Shi, H.; Chen, M.; Wang, Z.; Yao, Z.; Sun, Y. In Vitro Synergistic Effect of Minocycline Combined with Antifungals against *Cryptococcus neoformans*. *J. Med. Mycol.* **2022**, *32*, 101227. [CrossRef]

139. Yang, F.; Sun, Y.; Lu, Q. The Synergistic Effect of Minocycline and Azole Antifungal Drugs against *Scedosporium* and *Lomentospora* Species. *BMC Microbiol.* **2022**, *22*, 21. [CrossRef]

140. Lu, M.; Yu, C.; Cui, X.; Shi, J.; Yuan, L.; Sun, S. Gentamicin Synergises with Azoles against Drug-Resistant *Candida albicans*. *Int. J. Antimicrob. Agents* **2018**, *51*, 107–114. [CrossRef]

141. Lu, M.; Yang, X.; Yu, C.; Gong, Y.; Yuan, L.; Hao, L.; Sun, S. Linezolid in Combination With Azoles Induced Synergistic Effects Against *Candida albicans* and Protected *Galleria mellonella* Against Experimental Candidiasis. *Front. Microbiol.* **2019**, *9*, 3142. [CrossRef]

142. Yousfi, H.; Ranque, S.; Rolain, J.M.; Bittar, F. In Vitro Polymyxin Activity against Clinical Multidrug-Resistant Fungi. *Antimicrob. Resist. Infect. Control* **2019**, *8*, 66. [CrossRef]

143. Schwarz, P.; Djenontin, E.; Dannaoui, E. Colistin and Isavuconazole Interact Synergistically In Vitro against *Aspergillus nidulans* and *Aspergillus niger*. *Microorganisms* **2020**, *8*, 1447. [CrossRef]

144. Schwarz, P.; Bidaud, A.-L.; Dannaoui, E. In Vitro Synergy of Isavuconazole in Combination with Colistin against *Candida auris*. *Sci. Rep.* **2020**, *10*, 21448. [CrossRef] [PubMed]

145. Gao, L.; Sun, Y.; He, C.; Zeng, T.; Li, M. Synergy between Pyrvinium Pamoate and Azoles against *Exophiala dermatitidis*. *Antimicrob. Agents Chemother.* **2018**, *62*, e02361-17. [CrossRef] [PubMed]

146. de Cremer, K.; Lanckacker, E.; Cools, T.L.; Bax, M.; de Brucker, K.; Cos, P.; Cammue, B.P.A.; Thevissen, K. Artemisinins, New Miconazole Potentiators Resulting in Increased Activity against *Candida albicans* Biofilms. *Antimicrob. Agents Chemother.* **2015**, *59*, 421–426. [CrossRef] [PubMed]

147. Li, Y.; Wan, Z.; Liu, W.; Li, R. Synergistic Activity of Chloroquine with Fluconazole against Fluconazole-Resistant Isolates of *Candida* Species. *Antimicrob. Agents Chemother.* **2015**, *59*, 1365–1369. [CrossRef] [PubMed]

148. Gao, L.; Sun, Y.; He, C.; Li, M.; Zeng, T.; Lu, Q. INK128 Exhibits Synergy with Azoles against *Exophiala* spp. and *Fusarium* spp. *Front. Microbiol.* **2016**, *7*, 1658. [CrossRef]

149. Montoya, M.C.; Beattie, S.; Alden, K.M.; Krysan, D.J. Derivatives of the Antimalarial Drug Mefloquine Are Broad-Spectrum Antifungal Molecules with Activity against Drug-Resistant Clinical Isolates. *Antimicrob. Agents Chemother.* **2020**, *64*, e02331-19. [CrossRef]

150. Brilhante, R.S.N.; Caetano, É.P.; Riello, G.B.; de M. Guedes, G.M.; de Souza Collares Maia Castelo-Branco, D.; Fechine, M.A.B.; de Oliveira, J.S.; de Camargo, Z.P.; de Mesquita, J.R.L.; Monteiro, A.J.; et al. Antiretroviral Drugs Saquinavir and Ritonavir Reduce Inhibitory Concentration Values of Itraconazole against *Histoplasma capsulatum* Strains In Vitro. *Braz. J. Infect. Dis.* **2016**, *20*, 155–159. [CrossRef]

151. LaFleur, M.D.; Sun, L.; Lister, I.; Keating, J.; Nantel, A.; Long, L.; Ghannoum, M.; North, J.; Lee, R.E.; Coleman, K.; et al. Potentiation of Azole Antifungals by 2-Adamantanamine. *Antimicrob. Agents Chemother.* **2013**, *57*, 3585–3592. [CrossRef]

152. Zhang, M.; Yan, H.; Lu, M.; Wang, D.; Sun, S. Antifungal Activity of Ribavirin Used Alone or in Combination with Fluconazole against *Candida albicans* Is Mediated by Reduced Virulence. *Int. J. Antimicrob. Agents* **2020**, *55*, 105804. [CrossRef]

153. Eldesouky, H.E.; Salama, E.A.; Lanman, N.A.; Hazbun, T.R.; Seleem, M.N. Potent Synergistic Interactions between Lopinavir and Azole Antifungal Drugs against Emerging Multidrug-Resistant *Candida auris*. *Antimicrob. Agents Chemother.* **2020**, *65*, e00684-20. [CrossRef]

154. Zhao, Y.J.; Liu, W.D.; Shen, Y.N.; Li, D.M.; Zhu, K.J.; Zhang, H. The Efflux Pump Inhibitor Tetrandrine Exhibits Synergism with Fluconazole or Voriconazole against *Candida parapsilosis*. *Mol. Biol. Rep.* **2019**, *46*, 5867–5874. [CrossRef] [PubMed]

155. Li, S.-X.; Song, Y.-J.; Jiang, L.; Zhao, Y.-J.; Guo, H.; Li, D.-M.; Zhu, K.-J.; Zhang, H. Synergistic Effects of Tetrandrine with Posaconazole Against *Aspergillus fumigatus*. *Microb. Drug Resist.* **2017**, *23*, 674–681. [CrossRef] [PubMed]

156. Shi, J.; Li, S.; Gao, A.; Zhu, K.; Zhang, H. Tetrandrine Enhances the Antifungal Activity of Fluconazole in a Murine Model of Disseminated Candidiasis. *Phytomedicine* **2018**, *46*, 21–31. [CrossRef] [PubMed]

157. Li, S.X.; Song, Y.J.; Zhang, L.L.; Shi, J.P.; Ma, Z.L.; Guo, H.; Dong, H.Y.; Li, Y.M.; Zhang, H. An In Vitro and In Vivo Study on the Synergistic Effect and Mechanism of Itraconazole or Voriconazole Alone and in Combination with Tetrandrine against *Aspergillus fumigatus*. *J. Med. Microbiol.* **2015**, *64*, 1008–1020. [CrossRef] [PubMed]

158. Zeng, Q.; Zhang, Z.; Chen, P.; Long, N.; Lu, L.; Sang, H. In Vitro and In Vivo Efficacy of a Synergistic Combination of Itraconazole and Verapamil Against *Aspergillus fumigatus*. *Front. Microbiol.* **2019**, *10*, 1266. [CrossRef]
159. Afeltra, J.; Vitale, R.G.; Mouton, J.W.; Verweij, P.E. Potent Synergistic In Vitro Interaction between Nonantimicrobial Membrane-Active Compounds and Itraconazole against Clinical Isolates of *Aspergillus fumigatus* Resistant to Itraconazole. *Antimicrob. Agents Chemother.* **2004**, *48*, 1335–1343. [CrossRef]
160. Xu, J.; Liu, R.; Sun, F.; An, L.; Shang, Z.; Kong, L.; Yang, M. Eucalyptal D Enhances the Antifungal Effect of Fluconazole on Fluconazole-Resistant *Candida albicans* by Competitively Inhibiting Efflux Pump. *Front. Cell. Infect. Microbiol.* **2019**, *9*, 211. [CrossRef]
161. Yang, D.L.; Hu, Y.L.; Yin, Z.X.; Zeng, G.S.; Li, D.; Zhang, Y.Q.; Xu, Z.H.; Guan, X.M.; Weng, L.X.; Wang, L.H. Cis-2-Dodecenoic Acid Mediates Its Synergistic Effect with Triazoles by Interfering with Efflux Pumps in Fluconazole-Resistant *Candida albicans*. *Biomed. Environ. Sci.* **2019**, *32*, 199–209. [CrossRef]
162. Iyer, K.R.; Camara, K.; Daniel-Ivad, M.; Trilles, R.; Pimentel-Elardo, S.M.; Fossen, J.L.; Marchillo, K.; Liu, Z.; Singh, S.; Muñoz, J.F.; et al. An Oxindole Efflux Inhibitor Potentiates Azoles and Impairs Virulence in the Fungal Pathogen *Candida auris*. *Nat. Commun.* **2020**, *11*, 6429. [CrossRef]
163. Eldesouky, H.E.; Salama, E.A.; Hazbun, T.R.; Mayhoub, A.S.; Seleem, M.N. Ospemifene Displays Broad-Spectrum Synergistic Interactions with Itraconazole through Potent Interference with Fungal Efflux Activities. *Sci. Rep.* **2020**, *10*, 6089. [CrossRef]
164. Xie, F.; Chang, W.; Zhang, M.; Li, Y.; Li, W.; Shi, H.; Zheng, S.; Lou, H. Quinone Derivatives Isolated from the Endolichenic Fungus *Phialocephala fortinii* Are Mdr1 Modulators That Combat Azole Resistance in *Candida albicans*. *Sci. Rep.* **2016**, *6*, 33687. [CrossRef] [PubMed]
165. Singh, S.; Fatima, Z.; Ahmad, K.; Hameed, S. Fungicidal Action of Geraniol against *Candida albicans* Is Potentiated by Abrogated CaCdr1p Drug Efflux and Fluconazole Synergism. *PLoS ONE* **2018**, *13*, e0203079. [CrossRef] [PubMed]
166. Nyilasi, I.; Kocsubé, S.; Krizsán, K.; Galgóczy, L.; Pesti, M.; Papp, T.; Vágvölgyi, C. In Vitro Synergistic Interactions of the Effects of Various Statins and Azoles against Some Clinically Important Fungi. *FEMS Microbiol. Lett.* **2010**, *307*, 175–184. [CrossRef] [PubMed]
167. Nyilasi, I.; Kocsubé, S.; Krizsán, K.; Galgóczy, L.; Papp, T.; Pesti, M.; Nagy, K.; Vágvölgyi, C. Susceptibility of Clinically Important Dermatophytes against Statins and Different Statin-Antifungal Combinations. *Med. Mycol.* **2013**, *52*, 140–148. [CrossRef]
168. Zhou, Y.; Yang, H.; Zhou, X.; Luo, H.; Tang, F.; Yang, J.; Alterovitz, G.; Cheng, L.; Ren, B. Lovastatin Synergizes with Itraconazole against Planktonic Cells and Biofilms of *Candida albicans* through the Regulation on Ergosterol Biosynthesis Pathway. *Appl. Microbiol. Biotechnol.* **2018**, *102*, 5255–5264. [CrossRef] [PubMed]
169. de Q. Ribeiro, N.; Costa, M.C.; Magalhães, T.F.F.; Carneiro, H.C.S.; Oliveira, L.V.; Fontes, A.C.L.; Santos, J.R.A.; Ferreira, G.F.; de S. Araujo, G.R.; Alves, V.; et al. Atorvastatin as a Promising Anticryptococcal Agent. *Int. J. Antimicrob. Agents* **2017**, *49*, 695–702. [CrossRef] [PubMed]
170. Brilhante, R.S.; Caetano, E.P.; Oliveira, J.S.; Castelo-Branco, D.; Souza, E.R.; Alencar, L.P.; Cordeiro, R.; Bandeira, T.; Sidrim, J.J.; Rocha, M.F. Simvastatin Inhibits Planktonic Cells and Biofilms of *Candida* and *Cryptococcus* Species. *Braz. J. Infect. Dis.* **2015**, *19*, 459–465. [CrossRef] [PubMed]
171. Eldesouky, H.E.; Salama, E.A.; Li, X.; Hazbun, T.R.; Mayhoub, A.S.; Seleem, M.N. Repurposing Approach Identifies Pitavastatin as a Potent Azole Chemosensitizing Agent Effective against Azole-Resistant *Candida* Species. *Sci. Rep.* **2020**, *10*, 7525. [CrossRef] [PubMed]
172. Kane, A.; Campbell, L.; Ky, D.; Hibbs, D.; Carter, D. The Antifungal and Synergistic Effect of Bisphosphonates in *Cryptococcus*. *Antimicrob. Agents Chemother.* **2021**, *65*, e01753-20. [CrossRef]
173. Aneke, C.I.; Rhimi, W.; Otranto, D.; Cafarchia, C. Synergistic Effects of Efflux Pump Modulators on the Azole Antifungal Susceptibility of *Microsporum canis*. *Mycopathologia* **2020**, *185*, 279–288. [CrossRef]
174. Brilhante, R.S.N.; de Oliveira, J.S.; de Jesus Evangelista, A.J.; Pereira, V.S.; Alencar, L.P.; de Souza Collares Maia Castelo-Branco, D.; Câmara, L.M.C.; de Lima-Neto, R.G.; de A. Cordeiro, R.; Sidrim, J.J.C.; et al. In Vitro Effects of Promethazine on Cell Morphology and Structure and Mitochondrial Activity of Azole-Resistant *Candida tropicalis*. *Med. Mycol.* **2017**, *56*, 1012–1022. [CrossRef] [PubMed]
175. Dennis, E.K.; Garneau-Tsodikova, S. Synergistic Combinations of Azoles and Antihistamines against *Candida* Species In Vitro. *Med. Mycol.* **2019**, *57*, 874–884. [CrossRef] [PubMed]
176. Sun, W.; Wang, D.; Yu, C.; Huang, X.; Li, X.; Sun, S. Strong Synergism of Dexamethasone in Combination with Fluconazole against Resistant *Candida albicans* Mediated by Inhibiting Drug Efflux and Reducing Virulence. *Int. J. Antimicrob. Agents* **2017**, *50*, 399–405. [CrossRef] [PubMed]
177. Li, X.; Yu, C.; Huang, X.; Sun, S. Synergistic Effects and Mechanisms of Budesonide in Combination with Fluconazole against Resistant *Candida albicans*. *PLoS ONE* **2016**, *11*, e0168936. [CrossRef] [PubMed]
178. Yang, J.; Gao, L.; Yu, P.; Kosgey, J.C.; Jia, L.; Fang, Y.; Xiong, J.; Zhang, F. In Vitro Synergy of Azole Antifungals and Methotrexate against *Candida albicans*. *Life Sci.* **2019**, *235*, 116827. [CrossRef] [PubMed]
179. Holbrook, S.Y.L.; Garzan, A.; Dennis, E.K.; Shrestha, S.K.; Garneau-Tsodikova, S. Repurposing Antipsychotic Drugs into Antifungal Agents: Synergistic Combinations of Azoles and Bromperidol Derivatives in the Treatment of Various Fungal Infections. *Eur. J. Med. Chem.* **2017**, *139*, 12–21. [CrossRef]

180. Gu, W.; Guo, D.; Zhang, L.; Xu, D.; Sun, S. The Synergistic Effect of Azoles and Fluoxetine against Resistant *Candida albicans* Strains Is Attributed to Attenuating Fungal Virulence. *Antimicrob. Agents Chemother.* **2016**, *60*, 6179–6188. [CrossRef]
181. Cong, L.; Liao, Y.; Yang, S.; Yang, R. In Vitro Antifungal Activity of Sertraline and Synergistic Effects in Combination with Antifungal Drugs against Planktonic Forms and Biofilms of Clinical *Trichosporon asahii* Isolates. *PLoS ONE* **2016**, *11*, e0167903. [CrossRef]
182. Jia, W.; Zhang, H.; Li, C.; Li, G.; Liu, X.; Wei, J. The Calcineruin Inhibitor Cyclosporine a Synergistically Enhances the Susceptibility of *Candida albicans* Biofilms to Fluconazole by Multiple Mechanisms. *BMC Microbiol.* **2016**, *16*, 113. [CrossRef]
183. Tome, M.; Zupan, J.; Tomičić, Z.; Matos, T.; Raspor, P. Synergistic and Antagonistic Effects of Immunomodulatory Drugs on the Action of Antifungals against *Candida glabrata* and *Saccharomyces cerevisiae*. *PeerJ* **2018**, *2018*, e4999. [CrossRef]
184. Onyewu, C.; Blankenship, J.R.; Del Poeta, M.; Heitman, J. Ergosterol Biosynthesis Inhibitors Become Fungicidal When Combined with Calcineurin Inhibitors against *Candida albicans, Candida glabrata,* and *Candida krusei*. *Antimicrob. Agents Chemother.* **2003**, *47*, 956–964. [CrossRef] [PubMed]
185. Uppuluri, P.; Nett, J.; Heitman, J.; Andes, D. Synergistic Effect of Calcineurin Inhibitors and Fluconazole against *Candida albicans* Biofilms. *Antimicrob. Agents Chemother.* **2008**, *52*, 1127–1132. [CrossRef] [PubMed]
186. Steinbach, W.J.; Schell, W.A.; Blankenship, J.R.; Onyewu, C.; Heitman, J.; Perfect, J.R. In Vitro Interactions between Antifungals and Immunosuppressants against *Aspergillus fumigatus*. *Antimicrob. Agents Chemother.* **2004**, *48*, 1664–1669. [CrossRef] [PubMed]
187. Marchetti, O.; Moreillon, P.; Glauser, M.P.; Bille, J.; Sanglard, D. Potent Synergism of the Combination of Fluconazole and Cyclosporine in *Candida albicans*. *Antimicrob. Agents Chemother.* **2000**, *44*, 2373–2381. [CrossRef]
188. Gao, L.; Sun, Y.; He, C.; Zeng, T.; Li, M. Synergistic Effects of Tacrolimus and Azoles against *Exophiala dermatitidis*. *Antimicrob. Agents Chemother.* **2017**, *61*, e00948-17. [CrossRef]
189. Gao, L.; Sun, Y. In Vitro Interactions of Antifungal Agents and Tacrolimus against *Aspergillus* Biofilms. *Antimicrob. Agents Chemother.* **2015**, *59*, 7097–7099. [CrossRef]
190. Denardi, L.B.; Mario, D.A.N.; Loreto, É.S.; Santurio, J.M.; Alves, S.H. Synergistic Effects of Tacrolimus and Azole Antifungal Compounds in Fluconazole-Susceptible and Fluconazole-Resistant *Candida glabrata* Isolates. *Braz. J. Microbiol.* **2015**, *46*, 125–129. [CrossRef]
191. Kubiça, T.F.; Denardi, L.B.; Azevedo, M.I.; Oliveira, V.; Severo, L.C.; Santurio, J.M.; Alves, S.H. Antifungal Activities of Tacrolimus in Combination with Antifungal Agents against Fluconazole-Susceptible and Fluconazole-Resistant *Trichosporon asahii* Isolates. *Braz. J. Infect. Dis.* **2016**, *20*, 539–545. [CrossRef]
192. Borba-Santos, L.P.; Reis de Sá, L.F.; Ramos, J.A.; Rodrigues, A.M.; de Camargo, Z.P.; Rozental, S.; Ferreira-Pereira, A. Tacrolimus Increases the Effectiveness of Itraconazole and Fluconazole against *Sporothrix* spp. *Front. Microbiol.* **2017**, *8*, 1759. [CrossRef]
193. Zhang, J.; Tan, J.; Yang, L.; He, Y. Tacrolimus, Not Triamcinolone Acetonide, Interacts Synergistically with Itraconazole, Terbinafine, Bifonazole, and Amorolfine against Clinical Dermatophyte Isolates. *J. De Mycol. Med.* **2018**, *28*, 612–616. [CrossRef]
194. Lu, M.; Yan, H.; Yu, C.; Yuan, L.; Sun, S. Proton Pump Inhibitors Act Synergistically with Fluconazole against Resistant *Candida albicans*. *Sci. Rep.* **2020**, *10*, 498. [CrossRef] [PubMed]
195. Jia, C.; Zhang, J.; Zhuge, Y.; Xu, K.; Liu, J.; Wang, J.; Li, L.; Chu, M. Synergistic Effects of Geldanamycin with Fluconazole Are Associated with Reactive Oxygen Species in *Candida tropicalis* Resistant to Azoles and Amphotericin B. *Free Radic. Res.* **2019**, *53*, 618–628. [CrossRef] [PubMed]
196. Liu, L.; Jiang, T.; Zhou, J.; Mei, Y.; Li, J.; Tan, J.; Wei, L.; Li, J.; Peng, Y.; Chen, C.; et al. Repurposing the FDA-Approved Anticancer Agent Ponatinib as a Fluconazole Potentiator by Suppression of Multidrug Efflux and Pma1 Expression in a Broad Spectrum of Yeast Species. *Microb. Biotechnol.* **2022**, *15*, 482–498. [CrossRef] [PubMed]
197. Li, L.P.; An, M.M.; Shen, H.; Huang, X.; Yao, X.; Liu, J.; Zhu, F.; Zhang, S.Q.; Chen, S.M.; He, L.J.; et al. The Non-Geldanamycin Hsp90 Inhibitors Enhanced the Antifungal Activity of Fluconazole. *Am. J. Transl. Res.* **2015**, *7*, 2589–2602.
198. Sun, Y.; Gao, L.; He, C.; Wu, Q.; Li, M.; Zeng, T. Givinostat Exhibits In Vitro Synergy with Posaconazole against *Aspergillus* spp. *Med. Mycol.* **2016**, *55*, myw131. [CrossRef]
199. Qiao, J.; Sun, Y.; Gao, L.; He, C.; Zheng, W. Lonafarnib Synergizes with Azoles against *Aspergillus* spp. and *Exophiala* spp. *Med. Mycol.* **2018**, *56*, 452–457. [CrossRef]
200. Kim, S.; Woo, E.-R.; Lee, D.G. Synergistic Antifungal Activity of Isoquercitrin: Apoptosis and Membrane Permeabilization Related to Reactive Oxygen Species in *Candida albicans*. *IUBMB Life* **2019**, *71*, 283–292. [CrossRef]
201. Lai, Y.-W.; Campbell, L.T.; Wilkins, M.R.; Pang, C.N.I.; Chen, S.; Carter, D.A. Synergy and Antagonism between Iron Chelators and Antifungal Drugs in *Cryptococcus*. *Int. J. Antimicrob. Agents* **2016**, *48*, 388–394. [CrossRef]
202. Li, Y.; Jiao, P.; Li, Y.; Gong, Y.; Chen, X.; Sun, S. The Synergistic Antifungal Effect and Potential Mechanism of D-Penicillamine Combined With Fluconazole Against *Candida albicans*. *Front. Microbiol.* **2019**, *10*, 2853. [CrossRef]
203. Liu, X.; Li, T.; Wang, D.; Yang, Y.; Sun, W.; Liu, J.; Sun, S. Synergistic Antifungal Effect of Fluconazole Combined with Licofelone against Resistant *Candida albicans*. *Front. Microbiol.* **2017**, *8*, 2101. [CrossRef]
204. Sun, W.; Zhang, L.; Lu, X.; Feng, L.; Sun, S. The Synergistic Antifungal Effects of Sodium Phenylbutyrate Combined with Azoles against *Candida albicans* via the Regulation of the Ras–CAMP–PKA Signalling Pathway and Virulence. *Can. J. Microbiol.* **2019**, *65*, 105–115. [CrossRef] [PubMed]
205. Gao, L.; Sun, Y.; He, C.; Li, M.; Zeng, T. In Vitro Interactions between 17-AAG and Azoles against *Exophiala dermatitidis*. *Mycoses* **2018**, *61*, 853–856. [CrossRef] [PubMed]

206. de Andrade Neto, J.B.; da Silva, C.R.; Barroso, F.D.; do Amaral Valente Sá, L.G.; de Sousa Campos, R.; S Aires do Nascimento, F.B.; Sampaio, L.S.; da Silva, A.R.; da Silva, L.J.; de Sá Carneiro, I.; et al. Synergistic Effects of Ketamine and Azole Derivatives on *Candida* Spp. Resistance to Fluconazole. *Future Microbiol.* **2020**, *15*, 177–188. [CrossRef] [PubMed]

207. Ahangarkani, F.; Khodavaisy, S.; Mahmoudi, S.; Shokohi, T.; Rezai, M.S.; Fakhim, H.; Dannaoui, E.; Faraji, S.; Chowdhary, A.; Meis, J.F.; et al. Indifferent Effect of Nonsteroidal Anti-Inflammatory Drugs (NSAIDs) Combined with Fluconazole against Multidrug-Resistant *Candida auris. Curr. Med. Mycol.* **2019**, *5*, 26–30. [CrossRef] [PubMed]

208. Hao, W.; Wang, Y.; Xi, Y.; Yang, Z.; Zhang, H.; Ge, X. Activity of Chlorhexidine Acetate in Combination with Fluconazole against Suspensions and Biofilms of *Candida auris. J. Infect. Chemother.* **2022**, *28*, 29–34. [CrossRef] [PubMed]

209. Yuan, R.; Tu, J.; Sheng, C.; Chen, X.; Liu, N. Effects of Hsp90 Inhibitor Ganetespib on Inhibition of Azole-Resistant *Candida albicans. Front. Microbiol.* **2021**, *12*, 1280. [CrossRef]

210. Vu, K.; Blumwald, E.; Gelli, A. The Antifungal Activity of HMA, an Amiloride Analog and Inhibitor of Na+/H+ Exchangers. *Front. Microbiol.* **2021**, *12*, 1055. [CrossRef]

211. Ahmad, A.; Khan, A.; Manzoor, N. Reversal of Efflux Mediated Antifungal Resistance Underlies Synergistic Activity of Two Monoterpenes with Fluconazole. *Eur. J. Pharm. Sci.* **2013**, *48*, 80–86. [CrossRef]

212. Shaban, S.; Patel, M.; Ahmad, A. Improved Efficacy of Antifungal Drugs in Combination with Monoterpene Phenols against *Candida auris. Sci. Rep.* **2020**, *10*, 1162. [CrossRef]

213. de Aguiar, F.L.L.; de Morais, S.M.; dos Santos, H.S.; Albuquerque, M.R.J.R.; Bandeira, P.N.; de Brito, E.H.S.; Rocha, M.F.G.; dos Santos Fontenelle, R.O. Antifungal Activity and Synergistic Effect of Acetophenones Isolated from Species *Croton* against Dermatophytes and Yeasts. *J. Med. Plants Res.* **2016**, *10*, 216–222. [CrossRef]

214. Li, D.D.; Chai, D.; Huang, X.W.; Guan, S.X.; Du, J.; Zhang, H.Y.; Sun, Y.; Jiang, Y.Y. Potent In Vitro Synergism of Fluconazole and Osthole against Fluconazole-Resistant *Candida albicans. Antimicrob. Agents Chemother.* **2017**, *61*, e00436-17. [CrossRef] [PubMed]

215. Shao, J.; Cui, Y.; Zhang, M.; Wang, T.; Wu, D.; Wang, C. Synergistic In Vitro Activity of Sodium Houttuyfonate with Fluconazole against Clinical *Candida albicans* Strains under Planktonic Growing Conditions. *Pharm. Biol.* **2017**, *55*, 355–359. [CrossRef] [PubMed]

216. Sharifzadeh, A.; Khosravi, A.R.; Shokri, H.; Tari, P.S. Synergistic Anticandidal Activity of Menthol in Combination with Itraconazole and Nystatin against Clinical *Candida glabrata* and *Candida krusei* Isolates. *Microb. Pathog.* **2017**, *107*, 390–396. [CrossRef] [PubMed]

217. de A. Cordeiro, R.; Teixeira, C.E.C.; Brilhante, R.S.N.; Castelo-Branco, D.S.C.M.; Alencar, L.P.; de Oliveira, J.S.; Monteiro, A.J.; Bandeira, T.J.P.G.; Sidrim, J.J.C.; Moreira, J.L.B.; et al. Exogenous Tyrosol Inhibits Planktonic Cells and Biofilms of *Candida* Species and Enhances Their Susceptibility to Antifungals. *FEMS Yeast Res.* **2015**, *15*, 12. [CrossRef]

218. Raut, J.S.; Bansode, B.S.; Jadhav, A.K.; Karuppayil, S.M. Activity of Allyl Isothiocyanate and Its Synergy with Fluconazole against *Candida albicans* Biofilms. *J. Microbiol. Biotechnol.* **2017**, *27*, 685–693. [CrossRef]

219. Gong, Y.; Liu, W.; Huang, X.; Hao, L.; Li, Y.; Sun, S. Antifungal Activity and Potential Mechanism of N-Butylphthalide Alone and in Combination with Fluconazole against *Candida albicans. Front. Microbiol.* **2019**, *10*, 1461. [CrossRef]

220. Nabili, M.; Aslani, N.; Shokohi, T.; Hedayati, M.T.; Hassanmoghadam, F.; Moazeni, M. In Vitro Interaction between Glabridin and Voriconazole against *Aspergillus fumigatus* Isolates. *Rev. Iberoam. De Micol.* **2021**, *38*, 145–147. [CrossRef]

221. Ganan, M.; Lorentzen, S.B.; Gaustad, P.; Sørlie, M. Synergistic Antifungal Activity of Chito-Oligosaccharides and Commercial Antifungals on Biofilms of Clinical *Candida* Isolates. *J. Fungi* **2021**, *7*, 718. [CrossRef]

222. Chen, H.; Li, H.; Duan, C.; Song, C.; Peng, Z.; Shi, W. Reversal of Azole Resistance in *Candida albicans* by Oridonin. *J. Glob. Antimicrob. Resist.* **2021**, *24*, 296–302. [CrossRef]

223. Sadowska, B.; Budzyńska, A.; Stochmal, A.; Żuchowski, J.; Różalska, B. Novel Properties of *Hippophae Rhamnoides* L. Twig and Leaf Extracts—Anti-Virulence Action and Synergy with Antifungals Studied In Vitro on *Candida* Spp. Model. *Microb. Pathog.* **2017**, *107*, 372–379. [CrossRef]

224. da Gabriella, G.; Pippi, B.; Dalla Lana, D.F.; Amaral, A.P.S.; Teixeira, M.L.; de Souza, K.C.B.; Fuentefria, A.M. Reversal of Fluconazole Resistance Induced by a Synergistic Effect with *Acca sellowiana* in *Candida glabrata* Strains. *Pharm. Biol.* **2016**, *54*, 2410–2419. [CrossRef]

225. Sadhasivam, S.; Palanivel, S.; Ghosh, S. Synergistic Antimicrobial Activity of *Boswellia serrata* Roxb. Ex Colebr. (Burseraceae) Essential Oil with Various Azoles against Pathogens Associated with Skin, Scalp and Nail Infections. *Lett. Appl. Microbiol.* **2016**, *63*, 495–501. [CrossRef] [PubMed]

226. Roana, J.; Mandras, N.; Scalas, D.; Campagna, P.; Tullio, V. Antifungal Activity of *Melaleuca alternifolia* Essential Oil (TTO) and Its Synergy with Itraconazole or Ketoconazole against *Trichophyton rubrum. Molecules* **2021**, *26*, 461. [CrossRef] [PubMed]

227. Yang, Z.; Wang, Q.; Ma, K.; Shi, P.; Liu, W.; Huang, Z. Fluconazole Inhibits Cellular Ergosterol Synthesis to Confer Synergism with Berberine against Yeast Cells. *J. Glob. Antimicrob. Resist.* **2018**, *13*, 125–130. [CrossRef]

228. Li, D.D.; Xu, Y.; Zhang, D.Z.; Quan, H.; Mylonakis, E.; Hu, D.D.; Li, M.B.; Zhao, L.X.; Zhu, L.H.; Wang, Y.; et al. Fluconazole Assists Berberine to Kill Fluconazole-Resistant *Candida albicans. Antimicrob. Agents Chemother.* **2013**, *57*, 6016–6027. [CrossRef]

229. Wei, G.X.; Xu, X.; Wu, C.D. In Vitro Synergism between Berberine and Miconazole against Planktonic and Biofilm *Candida* Cultures. *Arch. Oral Biol.* **2011**, *56*, 565–572. [CrossRef]

230. Iwazaki, R.S.; Endo, E.H.; Ueda-Nakamura, T.; Nakamura, C.V.; Garcia, L.B.; Filho, B.P.D. In Vitro Antifungal Activity of the Berberine and Its Synergism with Fluconazole. *Int. J. Gen. Mol. Microbiol.* **2010**, *97*, 201–205. [CrossRef]

231. Quan, H.; Cao, Y.Y.; Xu, Z.; Zhao, J.X.; Gao, P.H.; Qin, X.F.; Jiang, Y.Y. Potent In Vitro Synergism of Fluconazole and Berberine Chloride against Clinical Isolates of *Candida albicans* Resistant to Fluconazole. *Antimicrob. Agents Chemother.* **2006**, *50*, 1096–1099. [CrossRef]

232. Luo, H.; Pan, K.-S.; Luo, X.-L.; Zheng, D.-Y.; Andrianopoulos, A.; Wen, L.-M.; Zheng, Y.-Q.; Guo, J.; Huang, C.-Y.; Li, X.-Y.; et al. In Vitro Susceptibility of Berberine Combined with Antifungal Agents Against the Yeast Form of *Talaromyces marneffei*. *Mycopathologia* **2019**, *184*, 295–301. [CrossRef]

233. Wang, T.; Shao, J.; Da, W.; Li, Q.; Shi, G.; Wu, D.; Wang, C. Strong Synergism of Palmatine and Fluconazole/Itraconazole Against Planktonic and Biofilm Cells of *Candida* Species and Efflux-Associated Antifungal Mechanism. *Front. Microbiol.* **2018**, *9*, 2892. [CrossRef]

234. Campos, R.D.; da Silva, C.R.; de Andrade Neto, J.B.; Sampaio, L.S.; Aires do Nascimento, F.B.S.; de Moraes, M.O.; Cavalcanti, B.C.; Magalhães, H.I.F.; Gomes, A.O.; Lobo, M.D.P.; et al. Antifungal Activity of Palmatine against Strains of *Candida* Spp. Resistant to Azoles in Planktonic Cells and Biofilm. *Int. J. Curr. Microbiol. Appl. Sci.* **2018**, *7*, 3657–3669. [CrossRef]

235. Li, X.; Wu, X.; Gao, Y.; Hao, L. Synergistic Effects and Mechanisms of Combined Treatment With Harmine Hydrochloride and Azoles for Resistant *Candida albicans*. *Front. Microbiol.* **2019**, *10*, 2295. [CrossRef] [PubMed]

236. Ribeiro de Carvalho, R.; Chaves Silva, N.; Cusinato, M.; Tranches Dias, K.S.; dos Santos, M.H.; Viegas Junior, C.; Silva, G.; Tranches Dias, A.L. Promising Synergistic Activity of Fluconazole with Bioactive Guttiferone-A and Derivatives against Non-*albicans Candida* Species. *J. De Mycol. Med.* **2018**, *28*, 645–650. [CrossRef] [PubMed]

237. Nagy, F.; Vitális, E.; Jakab, Á.; Borman, A.M.; Forgács, L.; Tóth, Z.; Majoros, L.; Kovács, R. In Vitro and In Vivo Effect of Exogenous Farnesol Exposure Against *Candida auris*. *Front. Microbiol.* **2020**, *11*, 957. [CrossRef] [PubMed]

238. Wang, Y.; Lu, C.; Zhao, X.; Wang, D.; Liu, Y.; Sun, S. Antifungal Activity and Potential Mechanism of Asiatic Acid Alone and in Combination with Fluconazole against *Candida albicans*. *Biomed. Pharmacother.* **2021**, *139*, 111568. [CrossRef] [PubMed]

239. Sun, L.-M.; Liao, K.; Liang, S.; Yu, P.-H.; Wang, D.-Y. Synergistic Activity of Magnolol with Azoles and Its Possible Antifungal Mechanism against *Candida albicans*. *J. Appl. Microbiol.* **2015**, *118*, 826–838. [CrossRef]

240. Li, Y.; Chang, W.; Zhang, M.; Li, X.; Jiao, Y.; Lou, H. Synergistic and Drug-Resistant Reversing Effects of Diorcinol D Combined with Fluconazole against *Candida albicans*. *FEMS Yeast Res.* **2015**, *15*, 1. [CrossRef]

241. Moraes, R.C.; Carvalho, A.R.; Lana, A.J.D.; Kaiser, S.; Pippi, B.; Fuentefria, A.M.; Ortega, G.G. In Vitro Synergism of a Water Insoluble Fraction of *Uncaria tomentosa* Combined with Fluconazole and Terbinafine against Resistant Non-*Candida albicans* Isolates. *Pharm. Biol.* **2017**, *55*, 406–415. [CrossRef]

242. Behbehani, J.M.; Irshad, M.; Shreaz, S.; Karched, M. Synergistic Effects of Tea Polyphenol Epigallocatechin 3-O-Gallate and Azole Drugs against Oral *Candida* Isolates. *J. De Mycol. Med.* **2019**, *29*, 158–167. [CrossRef]

243. Kumar, S.N.; Aravind, S.R.; Sreelekha, T.T.; Jacob, J.; Kumar, B.S.D. Asarones from *Acorus calamus* in Combination with Azoles and Amphotericin B: A Novel Synergistic Combination to Compete Against Human Pathogenic *Candida* Species In Vitro. *Appl. Biochem. Biotechnol.* **2015**, *175*, 3683–3695. [CrossRef]

244. Yao, D.; Zhang, G.; Chen, W.; Chen, J.; Li, Z.; Zheng, X.; Yin, H.; Hu, X. Pyrogallol and Fluconazole Interact Synergistically In Vitro against *Candida glabrata* through an Efflux-Associated Mechanism. *Antimicrob. Agents Chemother.* **2021**, *65*, e0010021. [CrossRef] [PubMed]

245. Kobayashi, T.; Kakeya, H.; Miyazaki, T.; Izumikawa, K.; Yanagihara, K.; Ohno, H.; Yamamoto, Y.; Tashiro, T.; Kohno, S. Synergistic Antifungal Effect of Lactoferrin with Azole Antifungals against *Candida albicans* and a Proposal for a New Treatment Method for Invasive Candidiasis. *Jpn. J. Infect. Dis.* **2011**, *64*, 292–296. [CrossRef] [PubMed]

246. Shekhar-Guturja, T.; Tebung, W.A.; Mount, H.; Liu, N.; Köhler, J.R.; Whiteway, M.; Cowen, L.E. Beauvericin Potentiates Azole Activity via Inhibition of Multidrug Efflux, Blocks *Candida albicans* Morphogenesis, and Is Effluxed via Yor1 and Circuitry Controlled by Zcf29. *Antimicrob. Agents Chemother.* **2016**, *60*, 7468–7480. [CrossRef] [PubMed]

247. Fakhim, H.; Emami, S.; Vaezi, A.; Hashemi, S.M.; Faeli, L.; Diba, K.; Dannaoui, E.; Badali, H. In Vitro Activities of Novel Azole Compounds ATTAF-1 and ATTAF-2 against Fluconazole-Susceptible and-Resistant Isolates of *Candida* Species. *Antimicrob. Agents Chemother.* **2017**, *61*, e01106-16. [CrossRef] [PubMed]

248. Shi, C.; Liu, C.; Liu, J.; Wang, Y.; Li, J.; Xiang, M. Anti-*Candida* Activity of New Azole Derivatives Alone and in Combination with Fluconazole. *Mycopathologia* **2015**, *180*, 203–207. [CrossRef]

249. Mood, A.D.; Premachandra, I.D.U.A.; Hiew, S.; Wang, F.; Scott, K.A.; Oldenhuis, N.J.; Liu, H.; Van Vranken, D.L. Potent Antifungal Synergy of Phthalazinone and Isoquinolones with Azoles Against *Candida albicans*. *ACS Med. Chem. Lett.* **2017**, *8*, 168–173. [CrossRef]

250. Li, L.P.; Liu, W.; Liu, H.; Zhu, F.; Zhang, D.Z.; Shen, H.; Xu, Z.; Qi, Y.P.; Zhang, S.Q.; Chen, S.M.; et al. Synergistic Antifungal Activity of Berberine Derivative B-7b and Fluconazole. *PLoS ONE* **2015**, *10*, e0126393. [CrossRef]

251. Koselny, K.; Green, J.; DiDone, L.; Halterman, J.P.; Fothergill, A.W.; Wiederhold, N.P.; Patterson, T.F.; Cushion, M.T.; Rappelye, C.; Wellington, M.; et al. The Celecoxib Derivative AR-12 Has Broad-Spectrum Antifungal Activity In Vitro and Improves the Activity of Fluconazole in a Murine Model of Cryptococcosis. *Antimicrob. Agents Chemother.* **2016**, *60*, 7115–7127. [CrossRef]

252. Ghannoum, M.; Long, L.; Larkin, E.L.; Isham, N.; Sherif, R.; Borroto-Esoda, K.; Barat, S.; Angulo, D. Evaluation of the Antifungal Activity of the Novel Oral Glucan Synthase Inhibitor SCY-078, Singly and in Combination, for the Treatment of Invasive Aspergillosis. *Antimicrob. Agents Chemother.* **2018**, *62*, e00244-18. [CrossRef]

253. Savage, K.A.; del Carmen Parquet, M.; Allan, D.S.; Davidson, R.J.; Holbein, B.E.; Lilly, E.A.; Fidel, P.L. Iron Restriction to Clinical Isolates of *Candida albicans* by the Novel Chelator Dibi Inhibits Growth and Increases Sensitivity to Azoles In Vitro and In Vivo in a Murine Model of Experimental Vaginitis. *Antimicrob. Agents Chemother.* **2018**, *62*, e02576-17. [CrossRef]

254. Dai, L.; Zang, C.; Tian, S.; Liu, W.; Tan, S.; Cai, Z.; Ni, T.; An, M.; Li, R.; Gao, Y.; et al. Design, Synthesis, and Evaluation of Caffeic Acid Amides as Synergists to Sensitize Fluconazole-Resistant *Candida albicans* to Fluconazole. *Bioorganic Med. Chem. Lett.* **2015**, *25*, 34–37. [CrossRef] [PubMed]

255. Wang, Y.H.; Dong, H.H.; Zhao, F.; Wang, J.; Yan, F.; Jiang, Y.Y.; Jin, Y.S. The Synthesis and Synergistic Antifungal Effects of Chalcones against Drug Resistant *Candida albicans*. *Bioorganic Med. Chem. Lett.* **2016**, *26*, 3098–3102. [CrossRef] [PubMed]

256. Sun, Y.; Gao, L.; He, C.; Li, M.; Zeng, T. In Vitro Interactions between IAP Antagonist AT406 and Azoles against Planktonic Cells and Biofilms of Pathogenic Fungi *Candida albicans* and *Exophiala dermatitidis*. *Med. Mycol.* **2018**, *56*, 1045–1049. [CrossRef] [PubMed]

257. Yang, W.; Tu, J.; Ji, C.; Li, Z.; Han, G.; Liu, N.; Li, J.; Sheng, C. Discovery of Piperidol Derivatives for Combinational Treatment of Azole-Resistant Candidiasis. *ACS Infect. Dis.* **2021**, *7*, 650–660. [CrossRef]

258. Li, C.; Tu, J.; Han, G.; Liu, N.; Sheng, C. Heat Shock Protein 90 (Hsp90)/Histone Deacetylase (HDAC) Dual Inhibitors for the Treatment of Azoles-Resistant *Candida albicans*. *Eur. J. Med. Chem.* **2022**, *227*, 113961. [CrossRef]

259. Czechowicz, P.; Neubauer, D.; Nowicka, J.; Kamysz, W.; Gościniak, G. Antifungal Activity of Linear and Disulfide-Cyclized Ultrashort Cationic Lipopeptides Alone and in Combination with Fluconazole against Vulvovaginal *Candida* spp. *Pharmaceutics* **2021**, *13*, 1589. [CrossRef]

260. Sun, Y.; Tan, L.; Yao, Z.; Gao, L.; Yang, J.; Zeng, T. In Vitro and In Vivo Interactions of TOR Inhibitor AZD8055 and Azoles against Pathogenic Fungi. *Microbiol. Spectr.* **2022**, *10*, e0200721. [CrossRef]

261. Chang, Y.-L.; Yu, S.-J.; Heitman, J.; Wellington, M.; Chen, Y.-L. New Facets of Antifungal Therapy. *Virulence* **2017**, *8*, 222–236. [CrossRef]

262. Cavalheiro, A.S.; Maboni, G.; de Azevedo, M.I.; Argenta, J.S.; Pereira, D.I.B.; Spader, T.B.; Alves, S.H.; Santurio, J.M. In Vitro Activity of Terbinafine Combined with Caspofungin and Azoles against *Pythium insidiosum*. *Antimicrob. Agents Chemother.* **2009**, *53*, 2136–2138. [CrossRef]

263. Argenta, J.S.; Santurio, J.M.; Alves, S.H.; Pereira, D.I.B.; Cavalheiro, A.S.; Spanamberg, A.; Ferreiro, L. In Vitro Activities of Voriconazole, Itraconazole, and Terbinafine Alone or in Combination against *Pythium insidiosum* Isolates from Brazil. *Antimicrob. Agents Chemother.* **2008**, *52*, 767–769. [CrossRef]

264. Pappas, P.G.; Kauffman, C.A.; Andes, D.R.; Clancy, C.J.; Marr, K.A.; Ostrosky-Zeichner, L.; Reboli, A.C.; Schuster, M.G.; Vazquez, J.A.; Walsh, T.J.; et al. Clinical Practice Guideline for the Management of Candidiasis: 2016 Update by the Infectious Diseases Society of America. *Clin. Infect. Dis.* **2016**, *62*, e1–e50. [CrossRef] [PubMed]

265. Kordalewska, M.; Lee, A.; Park, S.; Berrio, I.; Chowdhary, A.; Zhao, Y.; Perlin, D.S. Understanding Echinocandin Resistance in the Emerging Pathogen *Candida auris*. *Antimicrob. Agents Chemother.* **2018**, *62*, e00238-18. [CrossRef] [PubMed]

266. Pristov, K.E.; Ghannoum, M.A. Resistance of *Candida* to Azoles and Echinocandins Worldwide. *Clin. Microbiol. Infect.* **2019**, *25*, 792–798. [CrossRef] [PubMed]

267. Shrestha, S.K.; Chang, C.-W.T.; Meissner, N.; Oblad, J.; Shrestha, J.P.; Sorensen, K.N.; Grilley, M.M.; Takemoto, J.Y. Antifungal Amphiphilic Aminoglycoside K20: Bioactivities and Mechanism of Action. *Front. Microbiol.* **2014**, *5*, 671. [CrossRef] [PubMed]

268. Pichler, H.; Gaigg, B.; Hrastnik, C.; Achleitner, G.; Kohlwein, S.D.; Zellnig, G.; Perktold, A.; Daum, G. A Subfraction of the Yeast Endoplasmic Reticulum Associates with the Plasma Membrane and Has a High Capacity to Synthesize Lipids. *Eur. J. Biochem.* **2001**, *268*, 2351–2361. [CrossRef]

269. Istvan, E.S.; Deisenhofer, J. Structural Mechanism for Statin Inhibition of HMG-CoA Reductase. *Science (1979)* **2001**, *292*, 1160–1164. [CrossRef]

270. van Beek, E.; Pieterman, E.; Cohen, L.; Löwik, C.; Papapoulos, S. Farnesyl Pyrophosphate Synthase Is the Molecular Target of Nitrogen-Containing Bisphosphonates. *Biochem. Biophys. Res. Commun.* **1999**, *264*, 108–111. [CrossRef]

271. Balfour, J.A.; Faulds, D. Terbinafine: A Review of Its Pharmacodynamic and Pharmacokinetic Properties, and Therapeutic Potential in Superficial Mycoses. *Drugs* **1992**, *43*, 259–284. [CrossRef]

272. Polak, A. Preclinical Data and Mode of Action of Amorolfine. *Dermatology* **1992**, *184*, 3–7. [CrossRef]

273. Mishima, E.; Maruyama, K.; Nakazawa, T.; Abe, T.; Ito, S. Acute Kidney Injury from Excessive Potentiation of Calcium-Channel Blocker via Synergistic CYP3A4 Inhibition by Clarithromycin Plus Voriconazole. *Intern. Med.* **2017**, *56*, 1687–1690. [CrossRef]

274. Miyazaki, H.; Miyazaki, Y.; Geber, A.; Parkinson, T.; Hitchcock, C.; Falconer, D.J.; Ward, D.J.; Marsden, K.; Bennett, J.E. Fluconazole Resistance Associated with Drug Efflux and Increased Transcription of a Drug Transporter Gene, PDH1, in *Candida glabrata*. *Antimicrob. Agents Chemother.* **1998**, *42*, 1695–1701. [CrossRef] [PubMed]

275. Albertson, G.D.; Niimi, M.; Cannon, R.D.; Jenkinson, H.F. Multiple Efflux Mechanisms Are Involved in *Candida albicans* Fluconazole Resistance. *Antimicrob. Agents Chemother.* **1996**, *40*, 2835–2841. [CrossRef] [PubMed]

276. Fraczek, M.G.; Bromley, M.; Buied, A.; Moore, C.B.; Rajendran, R.; Rautemaa, R.; Ramage, G.; Denning, D.W.; Bowyer, P. The Cdr1B Efflux Transporter Is Associated with Non-Cyp51a-Mediated Itraconazole Resistance in *Aspergillus Fumigatus*. *J. Antimicrob. Chemother.* **2013**, *68*, 1486–1496. [CrossRef] [PubMed]

277. Martinez-Rossi, N.M.; Peres, N.T.A.; Rossi, A. Antifungal Resistance Mechanisms in Dermatophytes. *Mycopathologia* **2008**, *166*, 369–383. [CrossRef]

278. Webber, M.A. The Importance of Efflux Pumps in Bacterial Antibiotic Resistance. *J. Antimicrob. Chemother.* **2003**, *51*, 9–11. [CrossRef]

279. Sanchez, C.P.; McLean, J.E.; Rohrbach, P.; Fidock, D.A.; Stein, W.D.; Lanzer, M. Evidence for a Pfcrt-Associated Chloroquine Efflux System in the Human Malarial Parasite *Plasmodium falciparum*. *Biochemistry* **2005**, *44*, 9862–9870. [CrossRef]

280. Liu, S.; Yue, L.; Gu, W.; Li, X.; Zhang, L.; Sun, S. Synergistic Effect of Fluconazole and Calcium Channel Blockers against Resistant *Candida albicans*. *PLoS ONE* **2016**, *11*, e0150859. [CrossRef]

281. Chang, W.; Liu, J.; Zhang, M.; Shi, H.; Zheng, S.; Jin, X.; Gao, Y.; Wang, S.; Ji, A.; Lou, H. Efflux Pump-Mediated Resistance to Antifungal Compounds Can Be Prevented by Conjugation with Triphenylphosphonium Cation. *Nat. Commun.* **2018**, *9*, 5102. [CrossRef]

282. Miranda-Cadena, K.; Marcos-Arias, C.; Mateo, E.; Aguirre-Urizar, J.M.; Quindós, G.; Eraso, E. In Vitro Activities of Carvacrol, Cinnamaldehyde and Thymol against *Candida* Biofilms. *Biomed. Pharmacother.* **2021**, *143*, 112218. [CrossRef]

283. Prasad, R.; Rawal, M.K. Efflux Pump Proteins in Antifungal Resistance. *Front. Pharmacol.* **2014**, *5*, 202. [CrossRef]

284. Perfect, J.R. The Antifungal Pipeline: A Reality Check. *Nat. Rev. Drug Discov.* **2017**, *16*, 603–616. [CrossRef] [PubMed]

285. Pushpakom, S.; Iorio, F.; Eyers, P.A.; Escott, K.J.; Hopper, S.; Wells, A.; Doig, A.; Guilliams, T.; Latimer, J.; McNamee, C.; et al. Drug Repurposing: Progress, Challenges and Recommendations. *Nat. Rev. Drug Discov.* **2019**, *18*, 41–58. [CrossRef] [PubMed]

286. Cabral, M.E.; Figueroa, L.I.C.; Fariña, J.I. Synergistic Antifungal Activity of Statin-Azole Associations as Witnessed by *Saccharomyces cerevisiae*- and *Candida utilis*-Bioassays and Ergosterol Quantification. *Rev. Iberoam. De Micol.* **2013**, *30*, 31–38. [CrossRef] [PubMed]

287. Thompson, P.D.; Panza, G.; Zaleski, A.; Taylor, B. Statin-Associated Side Effects. *J. Am. Coll. Cardiol.* **2016**, *67*, 2395–2410. [CrossRef]

288. Shaukat, A.; Benekli, M.; Vladutiu, G.D.; Slack, J.L.; Wetzler, M.; Baer, M.R. Simvastatin–Fluconazole Causing Rhabdomyolysis. *Ann. Pharmacother.* **2003**, *37*, 1032–1035. [CrossRef]

289. Drake, M.T.; Clarke, B.L.; Khosla, S. Bisphosphonates: Mechanism of Action and Role in Clinical Practice. *Mayo Clin. Proc.* **2008**, *83*, 1032–1045. [CrossRef]

290. Ruggiero, S.L.; Mehrotra, B.; Rosenberg, T.J.; Engroff, S.L. Osteonecrosis of the Jaws Associated with the Use of Bisphosphonates: A Review of 63 Cases. *J. Oral Maxillofac. Surg.* **2004**, *62*, 527–534. [CrossRef]

291. Cabillic, F.; Toutirais, O.; Lavoué, V.; de La Pintière, C.T.; Daniel, P.; Rioux-Leclerc, N.; Turlin, B.; Mönkkönen, H.; Mönkkönen, J.; Boudjema, K.; et al. Aminobisphosphonate-Pretreated Dendritic Cells Trigger Successful Vγ9Vδ2 T Cell Amplification for Immunotherapy in Advanced Cancer Patients. *Cancer Immunol. Immunother.* **2010**, *59*, 1611–1619. [CrossRef]

292. Singh, N. Antifungal Prophylaxis for Solid Organ Transplant Recipients: Seeking Clarity amidst Controversy. *Clin. Infect. Dis.* **2000**, *31*, 545–553. [CrossRef]

293. Shirazi, F.; Kontoyiannis, D.P. The Calcineurin Pathway Inhibitor Tacrolimus Enhances the In Vitro Activity of Azoles against Mucorales via Apoptosis. *Eukaryot. Cell* **2013**, *12*, 1225–1234. [CrossRef]

294. Cyert, M.S. Calcineurin Signaling in *Saccharomyces cerevisiae*: How Yeast Go Crazy in Response to Stress. *Biochem. Biophys. Res. Commun.* **2003**, *311*, 1143–1150. [CrossRef]

295. Tegos, G.P.; Haynes, M.; Jacob Strouse, J.; Md. T. Khan, M.; Bologa, C.G.; Oprea, T.I.; Sklar, L.A. Microbial Efflux Pump Inhibition: Tactics and Strategies. *Curr. Pharm. Des.* **2012**, *17*, 1291–1302. [CrossRef] [PubMed]

296. Stathopoulos-Gerontides, A.; Guo, J.J.; Cyert, M.S. Yeast Calcineurin Regulates Nuclear Localization of the Crz1p Transcription Factor through Dephosphorylation. *Genes Dev.* **1999**, *13*, 798–803. [CrossRef] [PubMed]

297. Harvey, A.L.; Edrada-Ebel, R.; Quinn, R.J. The Re-Emergence of Natural Products for Drug Discovery in the Genomics Era. *Nat. Rev. Drug Discov.* **2015**, *14*, 111–129. [CrossRef] [PubMed]

298. Xu, Y.; Wang, Y.; Yan, L.; Liang, R.M.; di Dai, B.; Tang, R.J.; Gao, P.H.; Jiang, Y.Y. Proteomic Analysis Reveals a Synergistic Mechanism of Fluconazole and Berberine against Fluconazole-Resistant *Candida albicans*: Endogenous ROS Augmentation. *J. Proteome Res.* **2009**, *8*, 5296–5304. [CrossRef] [PubMed]

299. Dangles, O. Antioxidant Activity of Plant Phenols: Chemical Mechanisms and Biological Significance. *Curr. Org. Chem.* **2012**, *16*, 692–714. [CrossRef]

300. Bouarab-Chibane, L.; Forquet, V.; Lantéri, P.; Clément, Y.; Léonard-Akkari, L.; Oulahal, N.; Degraeve, P.; Bordes, C. Antibacterial Properties of Polyphenols: Characterization and QSAR (Quantitative Structure-Activity Relationship) Models. *Front. Microbiol.* **2019**, *10*, 829. [CrossRef]

301. Jothi, R.; Sangavi, R.; Kumar, P.; Pandian, S.K.; Gowrishankar, S. Catechol Thwarts Virulent Dimorphism in *Candida albicans* and Potentiates the Antifungal Efficacy of Azoles and Polyenes. *Sci. Rep.* **2021**, *11*, 21049. [CrossRef]

302. Bahar, A.; Ren, D. Antimicrobial Peptides. *Pharmaceuticals* **2013**, *6*, 1543–1575. [CrossRef]

303. Wang, X.; Liu, J.; Chen, J.; Zhang, M.; Tian, C.; Peng, X.; Li, G.; Chang, W.; Lou, H. Azole-Triphenylphosphonium Conjugates Combat Antifungal Resistance and Alleviate the Development of Drug-Resistance. *Bioorganic Chem.* **2021**, *110*, 104771. [CrossRef]

304. Zhu, T.; Chen, X.; Li, C.; Tu, J.; Liu, N.; Xu, D.; Sheng, C. Lanosterol 14α-Demethylase (CYP51)/Histone Deacetylase (HDAC) Dual Inhibitors for Treatment of *Candida tropicalis* and *Cryptococcus neoformans* Infections. *Eur. J. Med. Chem.* **2021**, *221*, 113524. [CrossRef] [PubMed]

305. Elias, R.; Basu, P.; Fridman, M. Fluconazole-COX Inhibitor Hybrids: A Dual-Acting Class of Antifungal Azoles. *J. Med. Chem.* **2022**, *65*, 2361–2373. [CrossRef] [PubMed]

306. Uren, A.G.; Beilharz, T.; O'Connell, M.J.; Bugg, S.J.; van Driel, R.; Vaux, D.L.; Lithgow, T. Role for Yeast Inhibitor of Apoptosis (IAP)-like Proteins in Cell Division. *Proc. Natl. Acad. Sci. USA* **1999**, *96*, 10170–10175. [CrossRef] [PubMed]
307. Scorzoni, L.; de Paula e Silva, A.C.A.; Marcos, C.M.; Assato, P.A.; de Melo, W.C.M.A.; de Oliveira, H.C.; Costa-Orlandi, C.B.; Mendes-Giannini, M.J.S.; Fusco-Almeida, A.M. Antifungal Therapy: New Advances in the Understanding and Treatment of Mycosis. *Front. Microbiol.* **2017**, *8*, 36. [CrossRef]
308. Weng, J.; Wong, S.N.; Xu, X.; Xuan, B.; Wang, C.; Chen, R.; Sun, C.C.; Lakerveld, R.; Kwok, P.C.L.; Chow, S.F. Cocrystal Engineering of Itraconazole with Suberic Acid via Rotary Evaporation and Spray Drying. *Cryst. Growth Des.* **2019**, *19*, 2736–2745. [CrossRef]

pharmaceuticals

MDPI

Article

Synthetic Derivatives against Wild-Type and Non-Wild-Type *Sporothrix brasiliensis*: In Vitro and In Silico Analyses

Lais Cavalcanti dos Santos Velasco de Souza [1], Lucas Martins Alcântara [1], Pãmella Antunes de Macêdo-Sales [1], Nathália Faria Reis [1], Débora Sena de Oliveira [1], Ricardo Luiz Dantas Machado [1], Reinaldo Barros Geraldo [2], André Luis Souza dos Santos [3], Vítor Francisco Ferreira [4,5], Daniel Tadeu Gomes Gonzaga [6], Fernando de Carvalho da Silva [7], Helena Carla Castro [2] and Andréa Regina de Souza Baptista [1,*]

1   Center for Microorganisms' Investigation, Fluminense Federal University, Niterói 24020-150, Brazil; laiscavalcanti@id.uff.br (L.C.d.S.V.d.S.); martins_lucas@id.uff.br (L.M.A.); pantunes@id.uff.br (P.A.d.M.-S.); nathaliafariareis@id.uff.br (N.F.R.); dsena@id.uff.br (D.S.d.O.); ricardomachado@id.uff.br (R.L.D.M.)
2   Laboratory of Antibiotics, Biochemistry and Molecular Modeling, Institute of Biology, Fluminense Federal University, Niterói 24210-201, Brazil; reinaldobgeraldo@gmail.com (R.B.G.); hcastrorangel@yahoo.com.br (H.C.C.)
3   Laboratory of Advanced Studies of Emerging and Resistant Microorganisms, Federal University of Rio de Janeiro, Rio de Janeiro 21941-902, Brazil; andre@micro.ufrj.br
4   Department of Pharmaceutical Technology, Faculty of Pharmacy, Graduate Program in Applied Health Sciences, Niterói 24241-000, Brazil; vitorferreira@id.uff.br
5   Department of Pharmaceutical Technology, Faculty of Pharmacy, Fluminense Federal University, Niterói 24241-000, Brazil
6   Pharmacy Unit, State University of the West Zone, Rio de Janeiro 23070-200, Brazil; danieltadeugonzaga@yahoo.com.br
7   Department of Organic Chemistry, Institute of Chemistry, Fluminense Federal University, Niterói 24020-150, Brazil; fcsilva@id.uff.br
*   Correspondence: andrearegina@id.uff.br; Tel.: +55-21-2629-2559

Citation: de Souza, L.C.d.S.V.; Alcântara, L.M.; de Macêdo-Sales, P.A.; Reis, N.F.; de Oliveira, D.S.; Machado, R.L.D.; Geraldo, R.B.; dos Santos, A.L.S.; Ferreira, V.F.; Gonzaga, D.T.G.; et al. Synthetic Derivatives against Wild-Type and Non-Wild-Type *Sporothrix brasiliensis*: In Vitro and In Silico Analyses. *Pharmaceuticals* **2022**, *15*, 55. https://doi.org/10.3390/ph15010055

Academic Editors: Jong Heon Kim, Luisa W. Cheng and Kirkwood Land

Received: 19 November 2021
Accepted: 21 December 2021
Published: 1 January 2022

Publisher's Note: MDPI stays neutral with regard to jurisdictional claims in published maps and institutional affiliations.

Abstract: Recently, the well-known geographically wide distribution of sporotrichosis in Brazil, combined with the difficulties of effective domestic feline treatment, has emphasized the pressing need for new therapeutic alternatives. This work considers a range of synthetic derivatives as potential antifungals against *Sporothrix brasiliensis* isolated from cats from the hyperendemic Brazilian region. Six *S. brasiliensis* isolates from the sporotrichotic lesions of itraconazole responsive or non-responsive domestic cats were studied. The minimum inhibitory concentrations (MICs) of three novel hydrazone derivatives and eleven novel quinone derivatives were determined using the broth microdilution method (M38-A2). In silico tests were also used to predict the pharmacological profile and toxicity parameters of these synthetic derivatives. MICs and MFCs ranged from 1 to >128 μg/mL. The ADMET computational analysis failed to detect toxicity while a good pharmacological predictive profile, with parameters similar to itraconazole, was obtained. Three hydrazone derivatives were particularly promising candidates as antifungal agents against itraconazole-resistant *S. brasiliensis* from the Brazilian hyperendemic region. Since sporotrichosis is a neglected zoonosis currently spreading in Latin America, particularly in Brazil, the present data can contribute to its future control by alternative antifungal drug design against *S. brasiliensis*, the most virulent and prevalent species of the hyperendemic context.

Keywords: sporotrichosis; *Felis catus*; quinones; hydrazones; zoonoses

1. Introduction

Sporotrichosis is a post-traumatic implantation subcutaneous mycosis of worldwide occurrence and increasing incidence, especially in tropical and subtropical regions such as Asia and Latin America. In Brazil, it is a hyperendemic neglected zoonosis and a major public health concern [1,2]. The pathogenic fungi causing this dermatozoonosis, from the genus *Sporothrix*, are thermo-dimorphic and saprophytic microorganisms [2]. *Sporothrix brasiliensis* is the most prevalent species in Brazil, directly related to the cat-transmitted zoonotic route [2,3]. The recent detection of feline and human patients with *S. brasiliensis*-caused sporotrichosis in Argentina [4–6] in addition to reports of zoonotic cases in other countries such as Paraguay and Panamá [7,8]. Suspected or possible cases were detected in Bolivia and Colombia, revealing its potential dissemination across South America [9]. One of the main challenges to be addressed is the occurrence of severe non-responsive forms of this mycosis in cats, in addition to recurrences and reinfections, characterizing zoonotic sporotrichosis as a matter of large relevance to public health [3,10].

Currently, the drug of choice for human and animal sporotrichosis treatment is itraconazole [11,12]. This azole is hepatotoxic, exclusively administered orally, with a high cost for the affected population, most of which live in socially vulnerable areas [13]. Thus, despite being improved, sporotrichosis therapy remains an unsolved problem since the reduction in *S. brasiliensis* sensitivity to this azole has been reported [12,14,15]. These authors described *S. brasiliensis* isolates obtained from patients residing in distinct Brazilian states such as Rio de Janeiro [16–18], São Paulo [16], Minas Gerais [18] and Rio Grande do Sul [12,14,19]. Our research group recently described the unprecedented existence of domestic felines coinfected by wild-type and non-wild-type *S. brasiliensis* in distinct areas of the hyperendemic region [17]. The ability to adapt and manifest the phenomenon of antifungal resistance to conventional antifungals is a reality for *Sporothrix* species originating from human and animal cases. Furthermore, *S. brasiliensis* presents a high ability to display resistance mechanisms, although these have not yet been fully elucidated [15].

The pharmaceutical industry is currently seeking new approaches for developing antimicrobial drugs to mitigate the worldwide burden of fungal diseases. In addition, the discovery and commercialization of new antifungal agents occurs in the long term and is costly. With regard to *Sporothrix* spp. infections, a few synthetic derivatives [20–22] have been investigated and described in the literature. Additionally, the potential correlation between feline therapeutic follow up with the in vitro itraconazole performance against *S. brasiliensis* has rarely been investigated. The present study aimed to determine the potential antifungal properties of a range of synthetic derivatives against wild-type and non-wild-type *S. brasiliensis* from diseased cats of the Brazilian hyperendemic area.

2. Results

2.1. Clinical Epidemiological Data

Table 1 summarizes the clinical epidemiological data of the six diseased felines from which the *S. brasiliensis* isolates were obtained. More than half were male (66.7%) and half of the total were unneutered. The average age of these animals was 3.58 years (SD ± 3.47 years), ranging from one to ten years. The majority were free roaming (83.3%), residing in urban areas with contact with soil and/or plants. Lesions were distributed in different anatomical sites, regardless of disease severity and the extent of itraconazole treatment.

Two domestic felines, the carriers of the WT3 and of the NWT1 isolates, presented more than one episode of sporotrichosis relapse. The NWT1 carrier underwent tail amputation 4 years ago and suffered from a new episode of sporotrichosis in the tail stump at the time of this study. Itraconazole treatment ranged from 2 to more than 120 months, with a dosage of 100 mg/cat/day.

2.2. Antifungal Susceptibility Assay

After the positive initial screening by an adaptation of the diffusion disc method (CLSI M2-A8) [24], eleven quinone derivatives (Q1–Q11) and three hydrazones (H1–H3) were selected for further investigation by the microdilution broth test against *S. brasiliensis* yeast-like forms (Figure 1). The results of the MIC and MFC values of the six clinical isolates and the standard strain against both hydrazone and quinone derivatives as well as itraconazole are shown in Table 2, Table 3 (geometric means to hydrazones) and Table 4.

Table 1. Clinical epidemiological and laboratorial data of domestic cats infected by the six *S. brasiliensis* isolates investigated in the present study.

Clinical/ Epidemiological	Wild Type			Non-Wild Type		
	WT1	WT2	WT3	NWT1	NWT2	NWT3
Age (years)	10	5	2.5	2	1	1
Sex	M	F	M	M	M	F
Castration	Yes	No	No	Yes	No	Yes
Free roaming	No	Yes	Yes	Yes	Yes	Yes
Anatomical site	Abdomen, head, chest and pelvic limb	Head and neck	Head, thoracic and pelvic limb	Base of tail and paw	Head	Pelvic limb
Relapse	No	No	Yes	Yes	No	No
Duration of ITC treatment (months)	60	NI	8	>120	NI	2

WT: wild type; NWT: non-wild type to ITC [23]; M: male; F: female; ITC: itraconazole; NI: not informed.

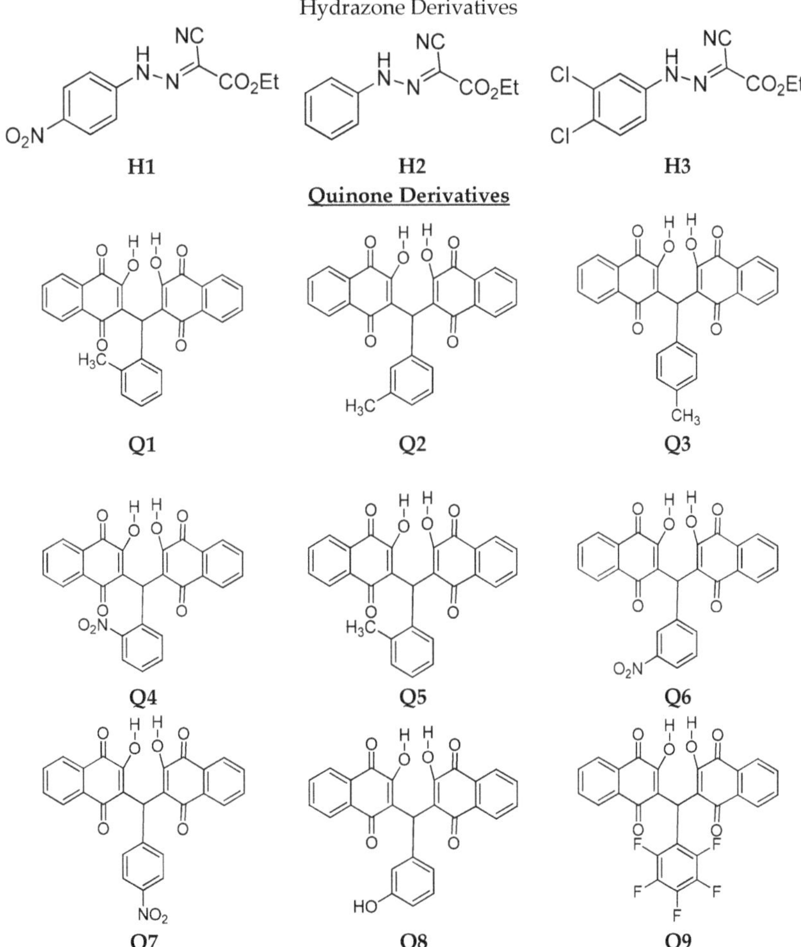

Figure 1. *Cont.*

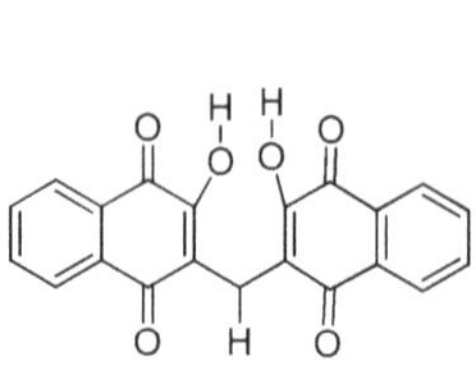

Q10 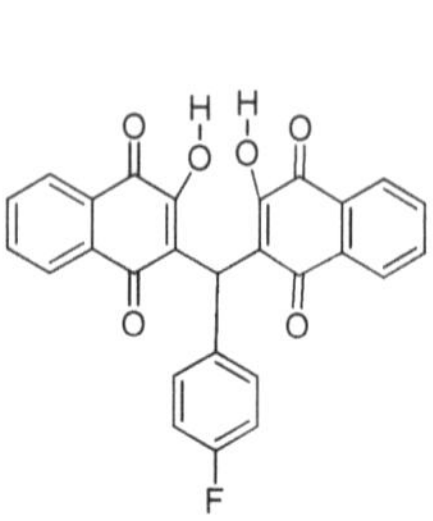   **Q11**

Figure 1. Molecular structures of hydrazone and quinone derivatives.

Table 2. In vitro susceptibility (µg/mL) of the six isolated *Sporothrix brasiliensis* (yeast forms) to the novel hydrazone derivatives.

Azole/Hydrazones		ATCC/Clinical Isolates						
		S. bra	WT 1	WT 2	WT 3	NWT 1	NWT 2	NWT 3
ITC	MIC	2	1	4	1	32	16	8
	MFC	16	8	32	8	>128	128	128
H1	MIC	8	2	4	2	1	8	8
	MFC	4	2	8	2	2	16	8
H2	MIC	8	16	16	8	8	16	8
	MFC	8	16	16	8	16	16	16
H3	MIC	4	2	4	16	1	16	1
	MFC	4	4	8	16	1	16	1

Sbra: *Sporothrix brasiliensis* (ATCC MYA-4823); MIC: minimum inhibitory concentration; MFC: minimum fungicidal concentration.

Table 3. Geometric means generated from the in vitro susceptibility (µg/mL) of the six isolated *Sporothrix brasiliensis* (yeast forms) to novel hydrazone derivatives.

Azole/Hydrazones		ATCC/Clinical Isolates		
		S. bra	WT GM	NWT GM
ITC	MIC	2	2	18.6
	MFC	16	16	128
H1	MIC	8	2.7	5.7
	MFC	4	4	8.7
H2	MIC	8	13.3	10.7
	MFC	8	13.3	16
H3	MIC	4	7.3	6
	MFC	4	9.3	6

Sbra: *Sporothrix brasiliensis* (ATCC MYA-4823); MIC: minimum inhibitory concentration; MFC: minimum fungicidal concentration; GM: geometric mean.

An itraconazole MIC variation between 1 and 32 µg/mL and an MFC variation between 8 and >128 µg/mL against *S. brasiliensis* were observed (Table 2). Isolates obtained from the lesions of long-term treated cats with recurrent episodes presented the highest MICs for itraconazole (NWT1 MIC = 32 µg/mL; NWT2 MIC = 16 µg/mL and NWT3 MIC = 8 µg/mL) as well as the highest MFCs (Table 2).

S. brasiliensis exposure to hydrazones provided MICs varying from 1 to 16 µg/mL (Table 2), while for quinones, the MICs ranged from 32 to >128 µg/mL (Table 4). MICs' and MFCs' geometric means (GMs) for the hydrazone derivatives ranged from 2.7 to 13.3 µg/mL (Table 3). The GMs of the MICs obtained for NWT isolates exposed to the three hydrazones were between a third and a half lower (5.7–10.7 µg/mL) than those obtained for itraconazole (Table 3). In contrast, for the quinone derivatives, higher MICs and MFCs were detected (32–>128 µg/mL; Table 4). The clinical isolate NWT1, obtained from a two-year-old domestic feline irresponsive to itraconazole, showing several episodes of relapse, presented in vitro results suggestive of sensitivity to the Q8 (MIC/MFC = 32 µg/mL; Table 4) as well as to H1, H2 and H3 (MIC = 1–8µg/mL; MFC = 1–16 µg/mL; Table 2).

Table 4. In vitro susceptibility (µg/mL) of the six isolated *Sporothrix brasiliensis* (yeast form) to novel quinone derivatives.

Azole/Quinones		ATCC/Clinical Isolates						
		S. bra	WT 1	WT 2	WT 3	NWT 1	NWT 2	NWT 3
Itraconazole	MIC	2	1	4	1	32	16	8
	MFC	16	8	32	8	>128	128	128
Q1	MIC	>128	32	>128	NA	128	NA	64
	MFC	>128	32	>128	NA	128	NA	64
Q2	MIC	>128	64	NA	NA	>128	NA	64
	MFC	>128	64	NA	NA	>128	NA	128
Q3	MIC	>128	64	NA	NA	>128	NA	128
	MFC	>128	128	NA	NA	>128	NA	128
Q4	MIC	>128	128	NA	NA	>128	>128	128
	MFC	>128	128	NA	NA	>128	>128	128
Q5	MIC	>128	NA	NA	NA	NA	NA	128
	MFC	>128	NA	NA	NA	NA	NA	128
Q6	MIC	NA	128	NA	NA	NA	NA	64
	MFC	NA	128	NA	NA	NA	NA	64
Q7	MIC	NA	128	NA	NA	NA	NA	64
	MFC	NA	128	NA	NA	NA	NA	64
Q8	MIC	32	128	>128	NA	32	>128	64
	MFC	64	128	>128	NA	32	>128	64
Q9	MIC	>128	128	NA	NA	NA	NA	128
	MFC	>128	128	NA	NA	NA	NA	128
Q10	MIC	128	128	NA	NA	NA	NA	128
	MFC	128	128	NA	NA	NA	NA	128
Q11	MIC	128	64	NA	NA	NA	NA	128
	MFC	128	64	NA	NA	NA	NA	128

Sbra: *Sporothrix brasiliensis* (ATCC MYA-4823); MIC: minimum inhibitory concentration; MFC: minimum fungicidal concentration; NA: not analyzed due to the lack of evidence after TSA screening.

The WT1 clinical isolate from the feline treated with itraconazole for 60 months showed MICs compatible with sensitivity after exposure to itraconazole (1 µg/mL), Q1 (32 µg/mL), H1 (2 µg/mL), H2 (16 µg/mL) and H3 (2 µg/mL). Meanwhile, the MFCs were, respectively, Q1 (32 µg/mL), H1 (2 µg/mL), H2 (16 µg/mL) and H3 (4 µg/mL). WT2 and WT3 exposed to hydrazones exhibited the same suggestive phenotype (Table 2). MICs and MFCs of the quinones were also high for the wild-type isolates (Table 4). According to the MFCs, all hydrazones and quinones presented a fungicidal profile.

2.3. The In Silico Toxicological Profile and Pharmacokinetics

Toxicity was predicted as described in Table 5. The oral rat acute toxicity (LD_{50}) of all derivatives and itraconazole ranged from 1.952 to 2.984 mol/kg. In contrast, oral rat chronic toxicity (LOAEL) varied from 0.055 log mg/kg bw/day to 2.974 mg/kg bw/day. Minnow toxicity varied between −6.407 log LC50 and 0.919 log LC50. Q6, Q7, H1 and H3 did not show evidence of a hepatotoxic effect. No hydrazone was a potential inhibitor of HERG I and II channels, while quinones did not seem to inhibit HERG I. The in silico predictions of four toxicological endpoints revealed that Q4, Q11 and itraconazole were immunotoxic, and that Q4, Q5 and H1 were mutagenic. No compound presented any presumed carcinogenic or cytotoxic properties.

Table 5. Toxicological in silico profile of synthetic derivatives.

Compounds	Oral Rat Acute Toxicity (LD50)	Oral Rat Chronic Toxicity (LOAEL)	Minnow Toxicity	HERG I	HERG II	Hepatotoxicity	Immuno-toxicity	Carcino-genicity	Cyto-toxicity	Muta-genicity
	Numeric (mol/kg)	Numeric (log mg/kg_bw/day)	Numeric (log LC 50)	Categorical (Yes/No)			Categorical (Active/Inactive)			
Itraconazole	2.966	0.055	−4.446	No	Yes	Yes	Yes	No	No	No
Q1	1.952	2.322	−2.223	No	Yes	Yes	No	No	No	No
Q2	2.098	2.398	−2.889	No	Yes	Yes	No	No	No	No
Q3	2.245	2.381	−2.098	No	Yes	Yes	No	No	No	No
Q4	2.580	1.549	−1.846	No	Yes	Yes	Yes	No	No	Yes
Q5	2.577	1.581	−1.211	No	Yes	Yes	No	No	No	Yes
Q6	2.972	1.221	0.919	No	Yes	No	No	No	No	No
Q7	2.965	2.426	0.274	No	Yes	No	No	No	No	No
Q8	2.711	2.974	−6.407	No	Yes	Yes	No	No	No	No
Q9	2.844	1.724	−0.550	No	No	Yes	No	No	No	No
Q10	2.905	1.578	−1.852	No	Yes	Yes	No	No	No	No
Q11	2.247	2.597	−2.126	No	Yes	Yes	Yes	No	No	No
H1	2.596	1.380	0.167	No	No	No	No	No	No	Yes
H2	2.549	1.340	0.644	No	No	Yes	No	No	No	No
H3	2.984	1.238	−0.396	No	No	No	No	No	No	No

Itra: itraconazole; HERG I: Type 1 human Ether-a-go-go-related gene; HERG II: Type 2 human Ether-a-go-go-related gene.

The pharmacokinetics profile of THE synthetic derivatives in silico shown in Table S1 summarizes the in silico intestinal absorption analysis of all compounds showing values above 80% (84.81–100%), and most of the derivatives (8/14) exhibited good Caco-2 permeability, similarly to itraconazole. In contrast, Q1, Q7, H2 and H3 were not determined to be P-glycoprotein substrates. No hydrazone derivative was suggested as an inhibitor of P-glycoprotein I/II while Q9 was not assigned as a P-glycoprotein II inhibitor.

The volume of distribution (VDss) showed lower values (−0.807 and −0.169 log L/kg) for almost all derivatives. Likewise, low values were shown for the majority of derivatives with regard to the unbound drug fraction (UF). However, H1, H2, H3 and Q8 compounds presented estimated values ranging from 0.274 to 0.386, above those of itraconazole (Table S1). In addition, H3 was the only derivative with a suggested capability of blood–brain barrier permeability (BBB).

These compounds were also analyzed for interaction with CYP 450 enzymes, as substrates (2D6 and 3A4) or as inhibitors (CYP1A2, CYP2C19, CYP2C9, CYP2D6, CYP3A4). The hydrazone derivatives were not substrates for CYP3A4, and none of them were substrates for CYP2D6. As inhibitors, only the H3 compound showed an inhibitory profile and only against CYP1A2. Q1, Q2, Q3, Q4 and Q10 compounds inhibited CYP2C19,

CYP2C9, CYP2D6 and CYP3A4. The candidates showed total clearance activity ranging from −0.926 log mL/min/kg to 0.786 log mL/min/kg. Similarly, no renal OCT2 substrate activity was found for either classes.

3. Discussion

The present work investigated *S. brasiliensis* isolates obtained from the lesions of domestic cats with sporotrichosis, clinically classified as responsive or non-responsive to treatment with itraconazole. This treatment varied from 2 to 60 months, and as expected, *S. brasiliensis* showed MICs compatible with reduced sensitivity to itraconazole or the "non-wild-type" phenotype, whenever obtained from non-responsive cats, as previously proposed (ESPINEL-INGROFF et al. [23] Almeida-Paes et al. [25]. The cats' mean age of approximately 3 year was in accordance with previous reports on sporotrichosis, as was the predominance of males. Indeed, with this species, sexual maturity increases fighting for females and concomitant exposure to other diseased animals [1–3].

Although itraconazole is the first therapeutic option for human and domestic feline sporotrichosis treatment [12,14,23], broth microdilution cut-off points have not been established. As a consequence, the definition of qualitative parameters of *Sporothrix* spp. response (sensitivity or resistance) to the different antifungal drugs remains a riddle to be solved. In the attempt to clarify this definition, epidemiological cutoff values (ECVs) were suggested [23,25] though not yet fully correlated to the patients' clinical responses [12,14,23]. Recently, Nakasu et al. [12] published a single study in an attempt to draw a parallel between in vitro data and the feline therapeutic follow up.

Recent publications have suggested that ECVs with MICs in the range of 0.5–2 mg/L for itraconazole were considered indicative of *S. brasiliensis* in vitro sensitivity [12,14,23,25]. The present study data on *S. brasiliensis* clinical isolates support this indication since those originating from domestic felines with an effective therapeutic response to the azole presented an in vitro MIC of 2 µg/mL. In contrast, the *S. brasiliensis* obtained from animals showing poor therapeutic response to itraconazole provided MIC values between 8 and 32 µg/mL, and could therefore be considered as "non-wild type" [23,25] or even resistant [12,14,15]. Likewise, in a veterinary hospital in southern Brazil (Pelotas, RS), Nakasu et al. [12] showed that approximately half of the investigated cats were non-responsive to itraconazole, with the corresponding isolates "resistant" to this drug. However, the reported MICs for itraconazole detected in the present investigation were higher than those reported by these authors. This might reflect the fact that the South Region of Brazil is located 1800 km away from the hyperendemic area (Rio de Janeiro, Southeast Brazil) with a lower level of sporotrichosis endemicity [1,3].

Different authors have suggested that *S. brasiliensis* shows reduced sensitivity to azoles [12,14,23,25], underlining the need to distinguish therapeutic alternatives. Recently, quinones became the subject of several studies due to their diverse known biological activities, such as their antibacterial [26,27] and antifungal potential [21,28,29]. These drugs also have antifungal properties against *S. schenckii* [29]. The present study analyzed eleven new quinone derivatives against WT and NWT clinical isolates and the standard strain of *S. brasiliensis*. Quinone derivatives Q1 and Q8 exhibited MIC and MFC values compatible with potential antifungal molecules [28,29]. Indeed, Tandon et al. [29,30], assessing quinone derivatives' antibacterial and antifungal potential, showed MICs between 6.25 and ≥50 µg/mL for distinct *S. schenckii* isolates. Another example of an antifungal naphthoquinone is the substituted α-and-β-2,3-dihydrofuranaphthoquinones which were synthesized and evaluated against the main zoonotic *Sporothrix* species [21]. Two compounds were strongly active against these dimorphic fungi, namely naphthoquinone 1a, (Figure 2) which presented an MIC of 4 mg/mL, and naphthoquinone 1b (Figure 2), which presented an MIC of 2 mg/mL and 4 mg/mL against *S. brasiliensis* and *S. schenckii*, respectively. In the present study, the MICs obtained after the in vitro exposure of *S. brasiliensis* versus quinones were higher than those reported by Ferreira et al. [21], but still compatible with therapeutic potential.

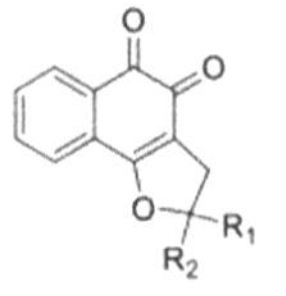

Figure 2. Naphthoquinones which previously presented antifungal activity against *Sporothrix* spp. as published by Janeczko et al. [31].

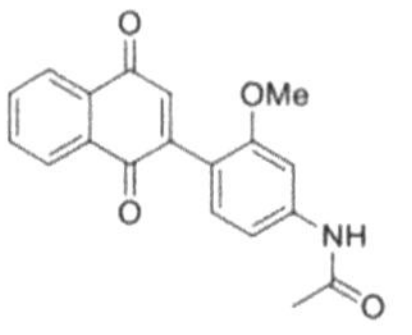

The antifungal activity of naphthoquinones can also be found against other fungi, such as *Candida albicans*. In the work of Janeczko et al. [31], a series of 1,4-naphthoquinones was prepared and tested against *C. albicans*. The compound (2) was the most active with an MIC of 8 mg/mL (Figure 2). The precedents for antifungal activities indicate that this class of compounds has good prospects, especially for hybrids with heterocyclic nuclei. Altogether, these data corroborate the potential of this class of molecules as a source of antifungal drug candidates.

Three hydrazones investigated in the present study also showed potential antifungal activity against different clinical isolates of *S. brasiliensis*. The in vitro exposure to H1 resulted in an MIC of 1 µg/mL against the *S. brasiliensis* NWT1, while the corresponding itraconazole value was 32-fold higher. Remarkably, this isolate infected a cat with a history of prolonged itraconazole-non-responsive sporotrichosis manifested by lesion recurrence at the base of its amputated tail. The standard strain of *S. brasiliensis* (*Sbra*) revealed the same in vitro response to these molecules.

The therapeutic potential for the hydrazones was previously investigated with several described biological activities [32]. Its antimycotic potential against dimorphic fungi, such as *Coccidioides posadasii*, as well as highly relevant yeasts including *Candida* spp. and *Trichosporon asahii*, in both community and hospital-acquired infections [32–35], were reported. Previously, Cordeiro et al. [34] described hydrazone MICs ranging from 2 to 250 µg/mL against *Coccidioides posadasii*, higher than those reported for *S. brasiliensis* in the present study, as compatible with the further investment of hydrazones as candidate drugs against this dimorphic fungus. Furthermore, distinct *Candida* species challenged this in vitro growth in the presence of different hydrazone derivatives resulting in MICs ranging from 0.25 to 128 µg/mL [35]. Previously, Casanova et al. [33] proposed that an MIC value between 8 and 32 µg/mL could be considered an encouraging result after the in vitro testing of new hydrazones against *Candida parapsilosis* and *Trichosporon asahii*. Our results indicate the strong potential of future investment in hydrazones for controlling fungal infections.

Hydrazones are present in different aryl groups which confer antifungal activity [36,37]. Abu-Melha et al. [37] performed tests with a one bis aryl hydrazine containing a thiazole ring (Hydrazone 03; Figure 3) against *C. albicans*. This compound resulted in MICs of 0.18 µg/mL and 3.18 µg/mL against azole-sensitive and azole-resistant *C. albicans*, respectively [37]. Carvalho et al. [36] evaluated a series of hydrazones containing the nucleus 7-chloroquinoline. The hydrazone shown in Figure 3 (hydrazone 04) stood out with a high percentage of inhibition against *C. albicans* (71% growth inhibition at a concentration of 12.5 mg/mL).

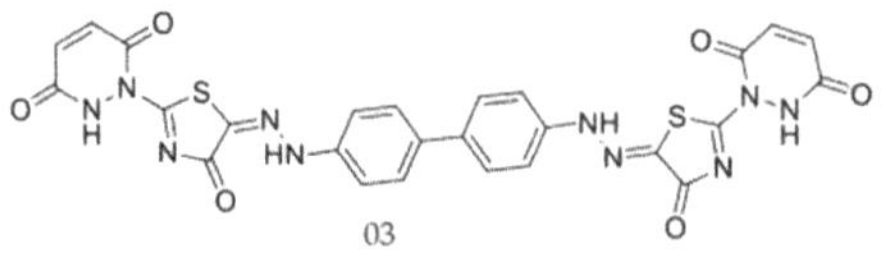

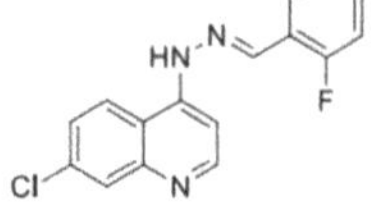

Figure 3. Aryl hydrazones which previously presented antifungal activity against *Candida albicans*, as published by Carvalho et al. and Abu-Melha et al. [36,37].

Hydrazone and quinone derivatives were submitted to in silico ADME/Tox property analyses. Absorption is the process whereby the drug candidate moves from the point of administration (extravascular site) to the blood (systemic circulation). The values above 90%, present an optimal intestinal profile [38]. In the present work, all tested derivatives revealed good theoretical absorption in the intestine, similarly to itraconazole. Moreover, some quinone derivatives and itraconazole presented permeability to Caco-2 cells [39]. The Caco-2 in silico assay is a useful feature to analyze the passive absorption in the intestine, so the hydrophobic parameters shows the most relevance in this relationship [40]. Thus, parameters such as hydrophobic (LogP), molecular polar surface area (TPSA), molecular weight and donor/acceptor of hydrogen may influence Caco-2 results. Therefore, due to the hydrophilic profile of the quinolone derivatives such as Q4, Q5 and Q11 and of the hydrazone derivatives (H1–H3), these did not show permeability profiles to the in silico Caco-2 assay.

One last absorption parameter was analyzed by employing the P-glycoprotein transporters' substrate or inhibitor [41]. The non-linear absorption kinetics of P-glycoprotein substrates has been reported due to the saturation transporter-mediated efflux activity, promoting the inhibition of intestinal P-glycoprotein resulting in significant effects of drug interaction (DDI) [42,43]. The results suggested that Q1, Q7, H2 and H3 derivative compounds were not P-glycoprotein substrates suggesting a good in silico oral bioavailability for all analyzed derivatives to be further explored.

The volume of distribution (VDss), fraction unbound (human) and blood–brain barrier membrane (BBB) permeability were also analyzed, as a distribution parameter for all the studied compounds. The VDss indicates the theoretical volume that a total dose would need to be uniformly distributed in the plasma to obtain the same concentration observed in blood plasma. All compounds showed lower values of VDss, suggesting that all these prototypes need low concentrations to remain in blood plasma. A fraction of the unbound drug of all hydrazones is higher than that of itraconazole. These results indicate that these groups of molecules have a good predictive distribution because the unbound form of the drug is responsible for exercising pharmacological activity [44]. The blood–brain barrier (BBB) was analyzed and according to the SwissADME calculation, only the H3 derivative could pass through the blood–brain barrier.

Enzymatic metabolism indicates the chemical biotransformation of a drug in the body, which plays a vital role in converting drug compounds [43]. CYP enzymes represent the most studied phase I drug-metabolizing enzymes and are also implicated in drug–drug interactions (DDIs) mediated by drug inhibition [45]. Although none of the derivatives were substrates for the tested Cytochrome P450 isozyme 2D6, all quinones are substrates of 3A4, such as itraconazole. Additionally, among these derivatives, only H3 acts as an inhibitor for CYP1A2. Interestingly, CYP1A2 is one main xenobiotic-metabolizing enzyme in humans, and a recent study associated this enzyme with the bioactivation of procarcinogens, including 4-(methylnitrosamino)-1-(3-pyridyl)-1-butanone (NNK), a tobacco-specific and potent pulmonary carcinogen [43,45].

Clearance is a constant that describes the relationship between drug concentration in the body and is important in determining the elimination of the drug [46]. Apart from hydrazone derivatives, all other new compounds exhibited lower values, including itraconazole. These metabolic and excretion results indicated that, theoretically, the quinone derivatives present a hepatic metabolization, while the hydrazone derivatives present a renal metabolization.

Toxicity has been a significant concern to the safety of drug candidates. Hepatotoxicity is still one of the major problems of drug toxicity [47]. Consequently, only Q6, Q7, H1 and H3 derivatives showed no predicted hepatotoxicity. Concerning the in silico toxicity test with minnows, an equivalent lethal concentration value (LC$_{50}$), representing the concentration of a molecule necessary to cause the death of 50% of experimentally tested Fathead minnows, LC$_{50}$ < 0.5 mM (i.e., log LC$_{50}$ < −0.3) is regarded as causing acute toxicity. Q6, Q7, H1 and H2 derivatives do not seem to present acute toxicity in the minnow test. All quinones—apart from Q8—and hydrazones proved to be less toxic than itraconazole. No hydrazone was potentially an inhibitor of HERG I and II channels [48], while quinones did not seem to inhibit HERG I, avoiding fatal heart-beat problems and short QT syndrome.

The oral rat acute toxicity expresses the compound toxic potential in terms of lethal dosage values (LD_{50} in mol/kg). These chronic toxicity studies aim to detect the lowest doses of a compound that will cause the lowest observed adverse effect levels (LOAELs), and the treatment period and exposure time of the compound must also be considered [49]. Log LOAEL predicted values for the tested compounds suggest that a larger dose of each compound must be used to induce adverse effects, indicating safety for these compounds compared to itraconazole.

ProTox-II proposes to classify drugs into several different steps, including a toxicological (immunotoxicity model), genotoxicological (cytotoxicity, mutagenicity and carcinogenicity model) endpoints, and the toxicity of a specific protein target [50,51]. The in silico results revealed that all compounds do not present cytotoxicity and carcinogenicity. In general, theoretical toxicity studies show that the derivatives were less toxic than the antifungal of choice.

4. Materials and Methods

4.1. Fungal Isolates

Six clinical *S. brasiliensis* isolates were obtained from the lesions of cats with laboratory-confirmed sporotrichosis. The clinical evaluations were performed by veterinarians, and the six cats' lesions, suspected to be from sporotrichosis, were later confirmed by mycological culture isolation and the genotyping of *S. brasiliensis*; all cats resided in the hyperendemic region of Rio de Janeiro, Brazil, as previously described [3]. These were selected from the Center for Microorganisms' Investigation fungal collection, based on the itraconazole in vitro results [52–54] following the criteria proposed by Espinel-Ingroff et al. [23], as wild type (MIC $\leq$ 2µg/mL) or non-wild type (MIC $\geq$ 4µg/mL). These were further designated as "WT1", "WT2", "WT3" and "NWT1", "NWT2", "NWT3", respectively. The reference strain *S. brasiliensis* ATCC MYA-4823 (*Sbra*) was included in all experiments.

4.2. Growth Conditions

The clinical isolates were maintained by cryopreservation in its yeast phase at -20 °C in the fungus collection of the Center for Microorganism's Investigation (CIM) until its reactivation for the conduction of the experiments. For reactivation, cryotubes were defrosted with subsequent replication on Sabouraud agar 2% dextrose (Becton, Dickinson, and Company—BD, Franklin Lakes, NJ, USA) and incubated at room temperature for five days for microbial growth in the form of conidia.

4.3. Synthetic Derivatives

Seventy-eight novel chemical derivatives of seven different classes (pyrazoles, pyrazolones, quinolones, naphthoquinones, hydrazones, n-phthalimides and quinones) were screened by disk diffusion antimicrobial sensitivity test (adapted Kirby–Bauer methodology) [55].

4.4. Antifungal Susceptibility Assays

Susceptibility testing was performed according to the standardized broth microdilution technique described by the CLSI in documents M38-A2 and M27-A3 [52,53] for yeast-like cells and conidia. The antifungal used as the experimental control was itraconazole.

Minimum fungicidal concentrations (MFCs) were obtained from subcultures on Petri dishes, including Sabouraud agar 2% dextrose (SDA; Becton, Dickinson and Company—BD, Franklin Lakes, NJ, USA), filamentous phase and brain heart infusion (BHI; Becton, Dickinson and Company—BD, Franklin Lakes, NJ, USA), yeast phase, with 30 µL of the MIC cell suspension. In plates, they were incubated at 25 °C for five days (conidia) and at 37 °C for seven days (yeast). After reading the number of colonies, the MFC was established as the lowest derivative concentration capable of eliminating 99.9% of the fungal growth [16].

Prototypes providing MICs under $\leq$32 µg/mL against *S. brasiliensis* were considered to be promising candidates for future antifungal drug development. Three independent experiments were performed for all assays.

4.5. In Silico Toxicity and Pharmacological Profiles

In silico pharmacokinetic properties and toxicity estimations (ADMET) were evaluated using pkCSM—pharmacokinetic web server (http://biosig.unimelb.edu.au/pkcsm/ (accessed on 21 June 2021) [49], OSIRIS Property Explorer [56] and SwissADME webserver [57]. These results were compared with the itraconazole profile. The evaluated theoretical pharmacokinetic properties were absorption, distribution, metabolism and excretion. Absorption suggests a theoretical human intestinal absorption due to the Caco-2 permeability, P-glycoprotein substrate and P-glycoprotein I/II inhibition. For distribution, the theoretical steady-state volume of distribution was analyzed, the blood–brain barrier penetration (BBBP) and the fraction unbound to serum proteins in humans. The metabolism analyses were based on the relationships with cytochrome P450 (CYP) enzymes: the inhibition of CYP1A2, CYP3A4, CYP2C19, CYP2C9 and CYP2D6, and as the substrate of CYP3A4 and CYP2D6. Lastly, excretion parameters were assessed by the total theoretical clearance and renal OCT2 substrate. Toxicological analyses comprise toxicity target, hepatotoxicity, hERGI/II inhibitors and the toxicity end points such as carcinogenic, mutagenic, immunotoxic and cytotoxic parameters.

5. Conclusions

This original work was based on the premise of finding new potent and promising compounds with antifungal activity, based on the fact that antifungal armamentarium is very limited and many fungal species are resistant to clinically available drugs. Thus, our original aim was to evaluate the possible antifungal action of novel compounds, including those from the hydrazone and quinone classes, which are well-recognized molecules with several biological properties such as antimicrobial potential. Three hydrazone derivatives (H1, H2 and H3) showed good in vitro and in silico performances against the yeast pathogenic phase of the dimorphic fungi *S. brasiliensis*, compared with those of other promising antifungal drug candidates. These findings are particularly relevant if the results against itraconazole-resistant isolates, originally from non-responsive cats, are to be considered. Furthermore, although understanding the mechanisms of action and the antifungal drug targets are important, future studies of the mechanisms involved in *S. brasiliensis* drug resistance are necessary. Our results set the grounds for the discovery of novel promising compounds which can be used as a platform for the synthesis of more potent derivatives. Finally, since sporotrichosis is a neglected zoonosis currently spreading in Brazil and Latin America, the present data can contribute to its future control by alternative antifungal drug design against *S. brasiliensis*—the most virulent and prevalent species of the hyperendemic scenario.

Supplementary Materials: The following are available online at https://www.mdpi.com/article/10.3390/ph15010055/s1, Table S1: In silico pharmacokinetics profile of the hydrazone and quinone synthetic derivatives.

Author Contributions: Conceptualization, A.R.d.S.B., P.A.d.M.-S., R.L.D.M., A.L.S.d.S. and H.C.C.; data curation, A.R.d.S.B. and L.C.d.S.V.d.S.; formal analysis, A.R.d.S.B., L.C.d.S.V.d.S., D.S.d.O., R.B.G. and A.L.S.d.S.; funding acquisition, A.R.d.S.B. and R.L.D.M.; investigation, L.C.d.S.V.d.S., L.M.A. and N.F.R.; methodology, A.R.d.S.B., P.A.d.M.-S. and H.C.C.; project administration, A.R.d.S.B.; resources, A.R.d.S.B., D.S.d.O., R.L.D.M., R.B.G., A.L.S.d.S., V.F.F., D.T.G.G. and F.d.C.d.S.; supervision, A.R.d.S.B. and H.C.C.; validation, L.C.d.S.V.d.S., L.M.A. and N.F.R.; visualization, L.C.d.S.V.d.S. and P.A.d.M.-S.; writing—original draft, L.C.d.S.V.d.S. and N.F.R.; writing—review and editing, A.R.d.S.B., L.C.d.S.V.d.S., L.M.A., P.A.d.M.-S., R.L.D.M., R.B.G., V.F.F., D.T.G.G., F.d.C.d.S. and H.C.C. All authors have read and agreed to the published version of the manuscript.

Funding: This study was supported by grants from the Brazilian Agencies: Conselho Nacional de Desenvolvimento Científico e Tecnológico (PIBIC-CNPq-UFF, Brazil); Fundação de Amparo à Pesquisa no Estado do Rio de Janeiro, Brazil (FAPERJ-E-26/103.198/2011; E-26/010.001882/2014), Coordenação de Aperfeiçoamento de Pessoal de Nível Superior, Brazil (CAPES)—Financial Code 001 and PROEX-MEC. R.L.D.M., A.L.S.S. and V.F.F. are research fellows of CNPq (PQ-CNPq).

Institutional Review Board Statement: This study was approved by and conducted according to the norms of the Ethics Committee on Animal Use by the Fluminense Federal University, Rio de Janeiro/BR (CEUA-UFF, protocol number 208/2012; 13 December 2012 and protocol number 7561040518; 14 June 2018).

Informed Consent Statement: Informed consent was obtained from all the tutors of the domestic feline involved in the study.

Data Availability Statement: All data are contained in the main text of the article.

Acknowledgments: The authors would like to thank the veterinarian collaborators and cat owners for allowing their animals to be part of this study. We are grateful to Norman Ratcliffe for editing and suggestions for improvements of the manuscript.

Conflicts of Interest: The authors declare no conflict of interest.

References

1. Gremião, I.D.F.; Miranda, L.H.M.; Reis, E.G.; Rodrigues, A.M.; Pereira, S.A. Zoonotic Epidemic of Sporotrichosis: Cat to Human Transmission. *PLoS Pathog.* **2017**, *13*, e1006077. [CrossRef] [PubMed]
2. Gremião, I.D.F.; Rocha, E.M.D.S.D.; Montenegro, H.; Carneiro, A.J.B.; Xavier, M.O.; de Farias, M.R.; Monti, F.; Mansho, W.; Pereira, R.H.D.M.A.; Pereira, S.A.; et al. Guideline for the management of feline sporotrichosis caused by Sporothrix brasiliensis and literature revision. *Braz. J. Microbiol.* **2021**, *52*, 107–124. [CrossRef] [PubMed]
3. Macêdo-Sales, P.A.; Souto, S.R.L.S.; Destefani, C.A.; Lucena, R.P.; Machado, R.L.D.; Pinto, M.R.; Rodrigues, A.M.; Lopes-Bezerra, L.M.; Rocha, E.M.S.; Baptista, A.R.S. Domestic feline contribution in the transmission of Sporothrix in Rio de Janeiro State, Brazil: A comparison between infected and non-infected populations. *BMC Vet. Res.* **2018**, *14*, 19. [CrossRef] [PubMed]
4. Etchecopaz, A.; Scarpa, M.; Mas, J.; Cuestas, M.L. Sporothrix brasiliensis: A growing hazard in the Northern area of Buenos Aires Province? *Rev. Argent. Microbiol.* **2020**, *52*, 350–351. [CrossRef]
5. Etchecopaz, A.N.; Lanza, N.; Toscanini, M.A.; Devoto, T.B.; Pola, S.J.; Daneri, G.L.; Iovannitti, C.A.; Cuestas, M.L. Sporotrichosis caused by Sporothrix brasiliensis in Argentina: Case report, molecular identification and in vitro susceptibility pattern to antifungal drugs. *J. Mycol. Med.* **2020**, *30*, 100908. [CrossRef] [PubMed]
6. Etchecopaz, A.; Toscanini, M.; Gisbert, A.; Mas, J.; Scarpa, M.; Iovannitti, C.; Bendezú, K.; Nusblat, A.; Iachini, R.; Cuestas, M. Sporothrix brasiliensis: A Review of an Emerging South American Fungal Pathogen, Its Related Disease, Presentation and Spread in Argentina. *J. Fungi* **2021**, *7*, 170. [CrossRef] [PubMed]
7. Rios, M.E.; Suarez, J.; Moreno, J.; Vallee, J.; Moreno, J.P. Zoonotic Sporotrichosis Related to Cat Contact: First Case Report from Panama in Central America. *Cureus* **2018**, *10*, e2906. [CrossRef] [PubMed]
8. García Duarte, J.M.; Wattiez Acosta, V.R.; Fornerón Viera, P.M.L.; Aldama Caballero, A.; Gorostiaga Matiauda, G.A.; Rivelli de Oddone, V.B.; Pereira Brunell, J.G. Esporotricosis trasmitida por gato doméstico. Reporte de un caso familiar. *Rev. Nac.* **2017**, *9*, 67–76. [CrossRef]
9. Rossow, J.A.; Queiroz-Telles, F.; Caceres, D.H.; Beer, K.D.; Jackson, B.R.; Pereira, J.G.; Ferreira Gremião, I.D.; Pereira, S.A. A One Health Approach to Combatting Sporothrix brasiliensis: Narrative Review of an Emerging Zoonotic Fungal Pathogen in South America. *J. Fungi* **2020**, *6*, 247. [CrossRef]
10. Gremião, I.D.F.; Oliveira, M.M.E.; De Miranda, L.H.M.; Freitas, D.F.S.; Pereira, S.A. Geographic Expansion of Sporotrichosis, Brazil. *Emerg. Infect. Dis.* **2020**, *26*, 621–624. [CrossRef]
11. Brilhante, R.S.N.; Rodrigues, A.M.; Sidrim, J.J.C.; Rocha, M.F.G.; Pereira, S.A.; Gremião, I.D.F.; Schubach, T.M.P.; de Camargo, Z.P. In vitrosusceptibility of antifungal drugs againstSporothrix brasiliensisrecovered from cats with sporotrichosis in Brazil: Table 1. *Med. Mycol.* **2016**, *54*, 275–279. [CrossRef]
12. Nakasu, C.C.T.; Waller, S.B.; Ripoll, M.K.; Ferreira, M.R.A.; Conceição, F.R.; Gomes, A.D.R.; Osório, L.D.G.; de Faria, R.O.; Cleff, M.B. Feline sporotrichosis: A case series of itraconazole-resistant Sporothrix brasiliensis infection. *Braz. J. Microbiol.* **2021**, *52*, 163–171. [CrossRef]
13. Pereira, S.A.; Schubach, T.M.P.; Gremião, I.D.F.; Da Silva, D.T.; Figueiredo, F.B.; De Assis, N.V.; Passos, S.R.L. Aspectos terapêuticos da esporotricose felina. *Acta Sci. Vet.* **2018**, *37*, 311–321. [CrossRef]
14. Waller, S.B.; Ripoll, M.K.; Madrid, I.M.; Acunha, T.; Cleff, M.B.; Chaves, F.C.; de Mello, J.R.B.; de Faria, R.O.; Meireles, M.C.A. Susceptibility and resistance of Sporothrix brasiliensis to branded and compounded itraconazole formulations. *Braz. J. Microbiol.* **2021**, *52*, 155–162. [CrossRef]

15. Waller, S.B.; Lana, D.F.D.; Quatrin, P.M.; Ferreira, M.R.A.; Fuentefria, A.M.; Mezzari, A. Antifungal resistance on Sporothrix species: An overview. *Braz. J. Microbiol.* **2021**, *52*, 73–80. [CrossRef]
16. Borba-Santos, L.P.; Rodrigues, A.M.; Gagini, T.B.; Fernandes, G.F.; Castro, R.; de Carmargo, Z.P.; Nucci, M.; Lopes-Bezerra, L.M.; Ishida, K.; Rozental, S. Susceptibility of Sporothrix brasiliensis isolates to amphotericin B, azoles, and terbinafine. *Med. Mycol.* **2015**, *53*, 178–188. [CrossRef]
17. Macêdo-Sales, P.A.; Souza, L.O.P.; Della-Terra, P.P.; Lozoya-Pérez, N.E.; Machado, R.L.D.; Rocha, E.M.D.S.D.; Lopês-Bezerra, L.M.; Guimarães, A.J.; Rodrigues, A.M.; Mora-Montes, H.M.; et al. Coinfection of domestic felines by distinct Sporothrix brasiliensis in the Brazilian sporotrichosis hyperendemic area. *Fungal Genet. Biol.* **2020**, *140*, 103397. [CrossRef]
18. Rodrigues, A.M.; de Hoog, G.S.; de Cássia Pires, D.; Brihante, R.S.N.; da Costa Sidrim, J.J.; Gadelha, M.F.; Colombo, A.L.; de Camargo, Z.P. Genetic diversity and antifungal susceptibility profiles in causative agents of sporotrichosis. *BMC Infect. Dis.* **2014**, *14*, 219. [CrossRef] [PubMed]
19. Sanchotene, K.O.; Brandolt, T.M.; Klafke, G.B.; Poester, V.R.; Xavier, M.O. In vitro susceptibility of Sporothrix brasiliensis: Comparison of yeast and mycelial phases. *Med. Mycol.* **2017**, *55*, 869–876. [CrossRef]
20. Forezi, L.; Borba-Santos, L.P.; Cardoso, M.F.C.; Ferreira, V.F.; Rozental, S.; Silva, F.D.C.D. Synthesis and Antifungal Activity of Coumarins Derivatives against *Sporothrix* spp. *Curr. Top. Med. Chem.* **2018**, *18*, 164–171. [CrossRef]
21. Garcia Ferreira, P.; Pereira Borba-Santos, L.; Noronha, L.L.; Deckman Nicoletti, C.; de Sá Haddad Queiroz, M.; de Carvalho da Silva, F.; Rozental, S.; Omena Futuro, D.; Francisco Ferreira, V. Synthesis, Stability Studies, and Antifungal Evaluation of Substituted α- and β-2,3-Dihydrofuranaphthoquinones against Sporothrix brasiliensis and Sporothrix schenckii. *Molecules* **2019**, *24*, 930. [CrossRef]
22. Mathias, L.; Almeida, J.; Passoni, L.; Gossani, C.; Taveira, G.; Gomes, V.; Motta, O. Antifungal activity of silver salts of Keggin-type heteropolyacids against *Sporothrix* spp. *J. Microbiol. Biotechnol.* **2020**, *30*, 540–551. [CrossRef]
23. Espinel-Ingroff, A.; Abreu, D.P.B.; Almeida-Paes, R.; Brilhante, R.S.N.; Chakrabarti, A.; Chowdhary, A.; Hagen, F.; Córdoba, S.; Gonzalez, G.M.; Govender, N.P.; et al. Multicenter, International Study of MIC/MEC Distributions for Definition of Epidemiological Cutoff Values for Sporothrix Species Identified by Molecular Methods. *Antimicrob. Agents Chemother.* **2017**, *61*, e01057-17. [CrossRef]
24. NCCLS. *Performance Standards for Antimicrobial Disk Susceptibility Tests*; NCCLS document M2-A8; Approved Standard—Eighth ed.; NCCLS: Wayne, PA, USA, 2003; ISBN 1-56238-485-6.
25. Almeida-Paes, R.; Brito-Santos, F.; Figueiredo-Carvalho, M.H.G.; Machado, A.C.S.; Oliveira, M.M.E.; Pereira, S.A.; Gutierrez-Galhardo, M.C.; Zancopé-Oliveira, R.M. Minimal inhibitory concentration distributions and epidemiological cutoff values of five antifungal agents against Sporothrix brasiliensis. *Mem. Inst. Oswaldo Cruz* **2017**, *112*, 376–381. [CrossRef]
26. Riffel, A.; Medina, L.; Stefani, V.; Santos, R.; Bizani, D.; Brandelli, A. In vitro antimicrobial activity of a new series of 1,4-naphthoquinones. *Braz. J. Med Biol. Res.* **2002**, *35*, 811–818. [CrossRef] [PubMed]
27. Tandon, V.K.; Chhor, R.B.; Singh, R.V.; Rai, S.; Yadav, D.B. Design, synthesis and evaluation of novel 1,4-naphthoquinone derivatives as antifungal and anticancer agents. *Bioorganic Med. Chem. Lett.* **2004**, *14*, 1079–1083. [CrossRef] [PubMed]
28. Louvis, A.D.R.; Silva, N.A.A.; Semaan, F.S.; Silva, F.D.C.D.; Saramago, G.; de Souza, L.C.S.V.; Ferreira, B.L.A.; Castro, H.C.; Salles, J.P.; Souza, A.L.A.; et al. Synthesis, characterization and biological activities of 3-aryl-1,4-naphthoquinones—Green palladium-catalysed Suzuki cross coupling. *New J. Chem.* **2016**, *40*, 7643–7656. [CrossRef]
29. Tandon, V.K.; Maurya, H.K.; Mishra, N.N.; Shukla, P.K. Micelles catalyzed chemoselective synthesis 'in water' and biological evaluation of oxygen containing hetero-1,4-naphthoquinones as potential antifungal agents. *Bioorganic Med. Chem. Lett.* **2011**, *21*, 6398–6403. [CrossRef]
30. Tandon, V.K.; Yadav, D.B.; Maurya, H.K.; Chaturvedi, A.K.; Shukla, P.K. Design, synthesis, and biological evaluation of 1,2,3-trisubstituted-1,4-dihydrobenzo[g]quinoxaline-5,10-diones and related compounds as antifungal and antibacterial agents. *Bioorg. Med. Chem.* **2006**, *14*, 6120–6126. [CrossRef]
31. Janeczko, M.; Kubiński, K.; Martyna, A.; Muzyczka, A.; Boguszewska-Czubara, A.; Czernik, S.; Tokarska-Rodak, M.; Chwedczuk, M.; Demchuk, O.M.; Golczyk, H.; et al. 1,4-Naphthoquinone derivatives potently suppress Candida Albicans growth, inhibit formation of hyphae and show no toxicity toward zebrafish embryos. *J. Med. Microbiol.* **2018**, *67*, 598–609. [CrossRef]
32. Chandra, S.; Vandana; Kumar, S. Synthesis, spectroscopic, anticancer, antibacterial and antifungal studies of Ni(II) and Cu(II) complexes with hydrazine carboxamide, 2-[3-methyl-2-thienyl methylene]. *Spectrochim. Acta Part A Mol. Biomol. Spectrosc.* **2015**, *135*, 356–363. [CrossRef]
33. Casanova, B.B.; Muniz, M.N.; De Oliveira, T.; De Oliveira, L.F.; Machado, M.M.; Fuentefria, A.M.; Gosmann, G.; Gnoatto, S.C.B. Synthesis and Biological Evaluation of Hydrazone Derivatives as Antifungal Agents. *Molecules* **2015**, *20*, 9229–9241. [CrossRef] [PubMed]
34. Cordeiro, R.D.A.; de Melo, C.V.S.; Marques, F.J.D.F.; Serpa, R.; Evangelista, A.J.D.J.; Caetano, E.P.; Mafezoli, J.; Oliveira, M.D.C.F.D.; da Silva, M.R.; Bandeira, T.D.J.P.G.; et al. Synthesis and in vitro antifungal activity of isoniazid-derived hydrazones against Coccidioides posadasii. *Microb. Pathog.* **2016**, *98*, 1–5. [CrossRef]
35. Secci, D.; Bizzarri, B.; Bolasco, A.; Carradori, S.; D'Ascenzio, M.; Rivanera, D.; Mari, E.; Polletta, L.; Zicari, A. Synthesis, anti-Candida activity, and cytotoxicity of new (4-(4-iodophenyl)thiazol-2-yl)hydrazine derivatives. *Eur. J. Med. Chem.* **2012**, *53*, 246–253. [CrossRef]

36. Carvalho, P.H.D.A.; Duval, A.R.; Leite, F.R.M.; Nedel, F.; Cunico, W.; Lund, R.G. (7-Chloroquinolin-4-yl)arylhydrazones: C andida albicans enzymatic repression and cytotoxicity evaluation, Part 2. *J. Enzym. Inhib. Med. Chem.* **2015**, *31*, 126–131. [CrossRef]
37. Abu-Melha, S.; Gomha, S.; Abouzied, A.; Edrees, M.; Dena, A.A.; Muhammad, Z. Microwave-Assisted One Pot Three-Component Synthesis of Novel Bioactive Thiazolyl-Pyridazinediones as Potential Antimicrobial Agents against Antibiotic-Resistant Bacteria. *Molecules* **2021**, *26*, 4260. [CrossRef]
38. Hadni, H.; Elhallaoui, M. 3D-QSAR, docking and ADMET properties of aurone analogues as antimalarial agents. *Heliyon* **2020**, *6*, e03580. [CrossRef]
39. Ortiz, C.L.; Completo, G.C.; Nacario, R.C.; Nellas, R.B. Potential Inhibitors of Galactofuranosyltransferase 2 (GlfT2): Molecular Docking, 3D-QSAR, and In Silico ADMETox Studies. *Sci. Rep.* **2019**, *9*, 17096. [CrossRef] [PubMed]
40. Larregieu, C.A.; Benet, L.Z. Drug Discovery and Regulatory Considerations for Improving In Silico and In Vitro Predictions that Use Caco-2 as a Surrogate for Human Intestinal Permeability Measurements. *AAPS J.* **2013**, *15*, 483–497. [CrossRef]
41. Deb, P.K.; Al-Attraqchi, O.; Prasad, M.R.; Tekade, R.K. Protein and Tissue Binding: Implication on Pharmacokinetic Parameters. In *Dosage Form Design Considerations*; Elsevier: Amsterdam, The Netherlands, 2018; Volume I, ISBN 9780128144244.
42. Mikov, M.; Đanić, M.; Pavlović, N.; Stanimirov, B.; Golocorbin-Kon, S.; Stankov, K.; Al-Salami, H. The Role of Drug Metabolites in the Inhibition of Cytochrome P450 Enzymes. *Eur. J. Drug Metab. Pharmacokinet.* **2017**, *42*, 881–890. [CrossRef]
43. Correia, M.A.; Monteflano, P.R.O. Inhibition of cytochrome P450 enzymes. In *Cytochrome P450*; Ortiz de Montellano, P.R., Ed.; Springer: Cham, Switzerland, 2015; ISBN 978-3-319-12107-9.
44. de Oliveira Moraes, A.D.T.; de Miranda, M.D.S.; Jacob, Í.T.T.; da Cruz Amorim, C.A.; de Moura, R.O.; da Silva, S.Â.S.; Soares, M.B.P.; de Almeida, S.M.V.; de Lima Souza, T.R.C.; de Oliveira, J.F.; et al. Synthesis, in vitro and in vivo biological evaluation, COX-1/2 inhibition and molecular docking study of indole-N-acylhydrazone derivatives. *Bioorg. Med. Chem.* **2018**, *26*, 5388–5396. [CrossRef]
45. Nakajima, M.; Yoshida, R.; Shimada, N.; Yamazaki, H.; Yokoi, T. Inhibition and inactivation of human cytochrome P450 isoforms by phenethyl isothiocyanate. *Drug Metab. Dispos.* **2001**, *29*, 1110–1113.
46. Hu, B.; Joseph, J.; Geng, X.; Wu, Y.; Suleiman, M.R.; Liu, X.; Shi, J.; Wang, X.; He, Z.; Wang, J.; et al. Refined pharmacophore features for virtual screening of human thromboxane A2 receptor antagonists. *Comput. Biol. Chem.* **2020**, *86*, 107249. [CrossRef]
47. Björnsson, E.S. Hepatotoxicity by Drugs: The Most Common Implicated Agents. *Int. J. Mol. Sci.* **2016**, *17*, 224. [CrossRef]
48. Vedani, A.; Smieško, M. In Silico Toxicology in Drug Discovery—Concepts Based on Three-dimensional Models. *Altern. Lab. Anim.* **2009**, *37*, 477–496. [CrossRef] [PubMed]
49. Pires, D.E.V.; Blundell, T.L.; Ascher, D.B. pkCSM: Predicting Small-Molecule Pharmacokinetic and Toxicity Properties Using Graph-Based Signatures. *J. Med. Chem.* **2015**, *58*, 4066–4072. [CrossRef] [PubMed]
50. Drwal, M.N.; Banerjee, P.; Dunkel, M.; Wettig, M.R.; Preissner, R. ProTox: A web server for the in silico prediction of rodent oral toxicity. *Nucleic Acids Res.* **2014**, *42*, W53–W58. [CrossRef]
51. Banerjee, P.; Siramshetty, V.B.; Drwal, M.N.; Preissner, R. Computational methods for prediction of in vitro effects of new chemical structures. *J. Cheminform.* **2016**, *8*, 51. [CrossRef] [PubMed]
52. CLSI. *Reference Method for Broth Dilution Antifungal Susceptibility Testing of Yeasts*, 3rd ed.; Clinical and Laboratory Standards Institute: Wayne, PA, USA, 2008.
53. CLSI. *Reference Method for Broth Dilution Antifungal Susceptibility Testing of Filamentous Fungi*, 2nd ed.; Clinical and Laboratory Standards Institute: Wayne, PA, USA, 2008.
54. Oliveira, D.S. de Avaliação Clínico-Epidemiológica e Perfil de Sensibilidade a Antifúngicos de *Sporothrix brasiliensis* Isolados a Partir de Felinos Domésticos do Estado do Rio de Janeiro. Master's Thesis, Universidade Federal Fluminense, Faculdade de Medicina Veterinária, Rio de Janerio, Brazil, 2016.
55. Novais, J.S.; Rosandiski, A.C.; De Carvalho, C.M.; Silva, L.S.D.S.; Souza, L.C.D.S.V.D.; Santana, M.V.; Martins, N.R.C.; Castro, H.C.; Ferreira, V.F.; Gonzaga, D.T.G.; et al. Efficient Synthesis and Antibacterial Profile of Bis(2-hydroxynaphthalene- 1,4-dione). *Curr. Top. Med. Chem.* **2020**, *20*, 121–131. [CrossRef] [PubMed]
56. Sander, T.; Freyss, J.; Von Korff, M.; Reich, J.R.; Rufener, C. OSIRIS, an Entirely in-House Developed Drug Discovery Informatics System. *J. Chem. Inf. Model.* **2009**, *49*, 232–246. [CrossRef]
57. Daina, A.; Michielin, O.; Zoete, V. SwissADME: A free web tool to evaluate pharmacokinetics, drug-likeness and medicinal chemistry friendliness of small molecules. *Sci. Rep.* **2017**, *7*, 42717. [CrossRef]

pharmaceuticals

Review

Drug Repurposing in Medical Mycology: Identification of Compounds as Potential Antifungals to Overcome the Emergence of Multidrug-Resistant Fungi

Lucie Peyclit [1,2,†], Hanane Yousfi [1,2,†], Jean-Marc Rolain [1,2] and Fadi Bittar [1,2,*]

1   Aix Marseille Univ, IRD, APHM, MEPHI, 13005 Marseille, France; lucie.peyclit@ap-hm.fr (L.P.);
    hanane.yousfi@univ-evry.fr (H.Y.); jean-marc.rolain@univ-amu.fr (J.-M.R.)
2   IHU Méditerranée Infection, 13005 Marseille, France
*   Correspondence: fadi.bittar@univ-amu.fr
†   The authors contributed equally.

Abstract: Immunodepression, whether due to HIV infection or organ transplantation, has increased human vulnerability to fungal infections. These conditions have created an optimal environment for the emergence of opportunistic infections, which is concomitant to the increase in antifungal resistance. The use of conventional antifungal drugs as azoles and polyenes can lead to clinical failure, particularly in immunocompromised individuals. Difficulties related to treating fungal infections combined with the time required to develop new drugs, require urgent consideration of other therapeutic alternatives. Drug repurposing is one of the most promising and rapid solutions that the scientific and medical community can turn to, with low costs and safety advantages. To treat life-threatening resistant fungal infections, drug repurposing has led to the consideration of well-known and potential molecules as a last-line therapy. The aim of this review is to provide a summary of current antifungal compounds and their main resistance mechanisms, following by an overview of the antifungal activity of non-traditional antimicrobial drugs. We provide their eventual mechanisms of action and the synergistic combinations that improve the activity of current antifungal treatments. Finally, we discuss drug repurposing for the main emerging multidrug resistant (MDR) fungus, including the *Candida auris*, *Aspergillus* or *Cryptococcus* species.

Keywords: drug repurposing; antifungals; repositioning; yeasts; emerging fungi; multidrug resistance; therapeutic alternatives; new targets; *Candida auris*; *Aspergillus* spp.

Citation: Peyclit, L.; Yousfi, H.; Rolain, J.-M.; Bittar, F. Drug Repurposing in Medical Mycology: Identification of Compounds as Potential Antifungals to Overcome the Emergence of Multidrug-Resistant Fungi. *Pharmaceuticals* **2021**, *14*, 488. https://doi.org/10.3390/ph14050488

Academic Editors: Jong Heon Kim, Luisa W. Cheng and Kirkwood Land

Received: 12 March 2021
Accepted: 18 May 2021
Published: 20 May 2021

Publisher's Note: MDPI stays neutral with regard to jurisdictional claims in published maps and institutional affiliations.

1. Introduction

Most public health organizations, including the World Health Organization, do not have a fungal infection surveillance program, despite the fact that invasive fungal infections present a high mortality rate worldwide, often exceeding 50% [1,2]. Fungal infections have long been poorly documented and recognized, perhaps due to the need to treat other severe and serious bacterial and viral infections. However, mycoses should no longer be ignored.

The signs and symptoms of fungal infection appear during antibiotic therapy, particularly due to opportunistic fungal agents. In particular, invasive fungal infections affect patients with compromised immune systems, such as those with hematologic malignancies, HIV infection, chemotherapy treatments, etc. [3,4]. In addition, other factors such as the ageing of the population, which is susceptible to these opportunistic infections and improvements in diagnostic methods have led to their increasing prevalence in hospitals. A significant number of fungal agents, including yeast and yeast-like species such as *Candida* spp., *Cryptococcus* spp. and molds such as the *Aspergillus* species, complicate clinical management, with a variety of symptoms, prevalence and clinical outcomes [5]. Moreover, fungal infections can also occur in healthy people, so it is difficult to control their spread [6]. Therefore, greater consideration should be given to monitoring fungal infections [7–9].

To deal with these fungal infections, there are only four main therapeutic classes currently used in clinical practice, namely polyenes, azoles, echinocandins and flucytosine. Although these drugs remain active, they display several limitations that complicate their routine use including off-target toxicity, drugs interaction, clinical failure and long-term treatment [10]. Furthermore, the emerging resistance to antifungals and the poor clinical response of many isolates to antifungal therapy make this an even greater public-health concern. For example, previous exposure to an antifungal agent such as fluconazole has been shown to increase the risk of fluconazole-resistant *Candida* infections in immunocompromised patients [11]. Antifungal resistance remains a critical global problem, although it may vary depending on the species, geography and available therapeutic alternatives [5]. Some species are known to be more resistant than others, leading to treatment failure; these include *Candida* pathogens (*Candida glabrata*, *Candida krusei*, *Candida lusitaniae* and the very newly emerging yeast: *Candida auris*), some cryptococcal species and opportunistic *Aspergillus* or *Fusarium* species associated with immunocompromised hosts [5]. The aforementioned species commonly exhibit high intrinsic antifungal resistance profiles, sometimes to different classes of antifungals as is the case with almost all *Fusarium* spp. to triazoles, 5-fluorocytosine and echinocandins [5].

Despite great efforts made internationally to deal with antibiotic resistance by reducing the inappropriate consumption of antimicrobials, this does not yet extend to the use of antifungal agents. In reality, antifungal resistance may arise from fungicide use in agriculture, as described in a recent study in 2017, where plant bulbs were found to be positive for triazole-resistant *Aspergillus fumigatus*, rather than from clinical use [12]. Furthermore, antifungal therapies are mostly given to immunocompromised patients or those in intensive care units (ICUs), when treatment is unavoidable, rather than in preventive use in community medicine. Demers et al. described the impact of the heterogeneity of a single gene (MRR1), found in different *C. lusitaniae* subpopulations in an azole-naïve cystic fibrosis patient, on the level of fluconazole resistance, highlighting other indirect factors involved, such as the host immune system and coinfecting bacteria [13]. Therefore, reducing the consumption of antifungals may not be the only solution to improving the difficult management of invasive fungal diseases.

The development of new antifungal drugs represents a major challenge for the pharmaceutical industry, since fungi are eukaryotic organisms and have a close evolutionary relationship with their human hosts [14]. In addition, the pharmaceutical industry is no longer interested in developing and marketing new antimicrobials, including antifungals [1,15]. Indeed, the drug development process remains very expensive, time-consuming and risky due to many factors [16]. Given the difficulties of treating invasive fungal infections, consideration should be made to use efficient alternative strategies to implement immediate and appropriate measures. New targets involving enzymes and other metabolic pathways, or new formulations/generations are under development to broaden the spectrum of antifungal activities and to potentially overcome current resistances [17–20]. New antifungals can also be found in natural products or plant extracts, inspired by traditional medicine and aromatherapy [21,22]. Indeed, various essential oils showed in vitro efficiency against clinical yeast or fungi [23–25]. Preclinical studies to determine human toxicities, pharmacodynamics and clinical trials must continue [21,25]. At the same time, the strategy of drug repurposing (also called drug repositioning) consists of identifying drugs known to be effective for another indication than that for which they are marketed [15]. This strategy has gained in popularity in recent years and has already proven to be effective, particularly in oncology, cardiology and Alzheimer's disease. These reinvestigated drugs have already completed preclinical trials, main human toxicities are well known, and the research and development process can be, considerably, reduced allowing for lower investment costs and faster potential clinical use [15,26]. However, the setting up of this strategy is not a straightforward matter as it does not exclude carrying out further clinical trials before the simple repositioning of a given drug, but it must also initially deal with the intellectual property rights, regulatory/authority process, license grant, pricing, patient's acceptance, marketing strategy and commercialization in order to avoid any failure at the development stage [27].

The aim of this review is to report on the non-antifungal drugs that may be active against the most common emerging multidrug-resistant (MDR) fungi in human pathology. We first summarize the antifungal compounds currently used for clinical therapies and their main mechanisms of resistance and then report on alternative drugs used to treat fungal infections that have been reported in the literature so far. Finally, we address different drugs that have a potential for repurposing to treat the main difficult to treat fungi.

2. Current Antifungal Agents and Their Mechanisms of Resistance

Since their first and progressive discovery in the mid-20th century, systemic antifungals have improved the management of many invasive fungal infections. The spectrum of antifungals and their mechanisms of action are diverse, since they act on different structures in the fungal cell; we describe them below (Figure 1).

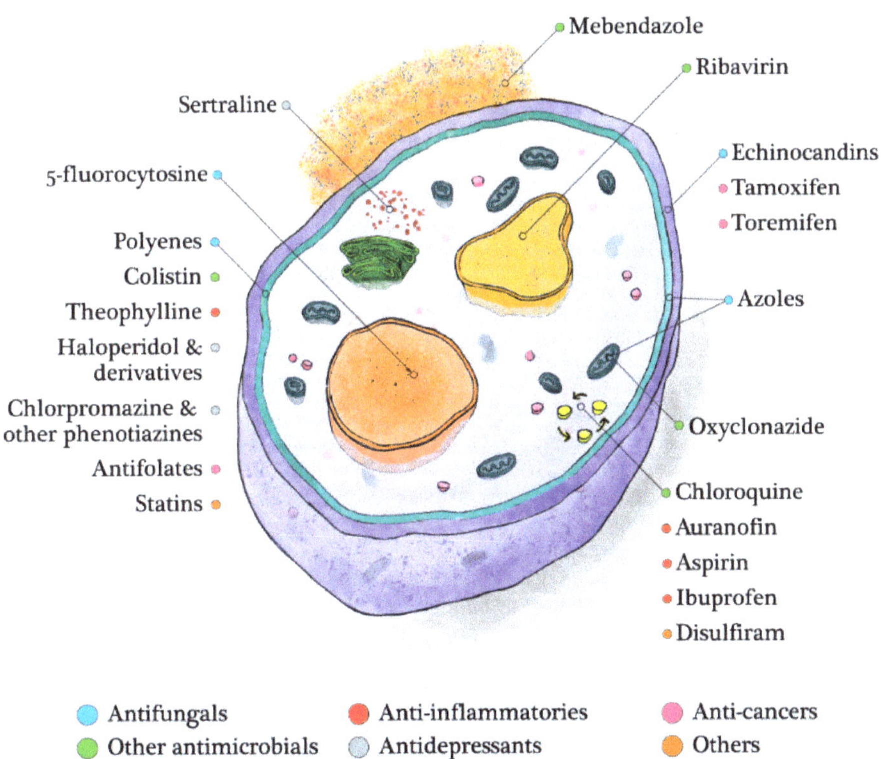

Figure 1. Therapeutic targets in the fungal cell of the compounds listed in the review: known antifungals and molecules that have a potential for repurposing as antifungals.

The first antifungal agents, polyenes, were developed in 1950s for clinical use and have a large spectrum of activity against yeast and filamentous fungi. Polyenes are macrocyclic organic molecules from a soil actinomycete, *Streptomyces nodosus* [28]. The two most clinically relevant members of this class are topical nystatin and intravenous amphotericin B. Both bind to a sterol moiety, ergosterol, on the fungal cell membrane, disrupting cell permeability and leading to cellular lysis. Nystatin is not effective against dermatophytes but is effective against the *Candida* species [29]. Liposomal amphotericin B has been developed to allow the administration of higher doses with less nephrotoxicity (fewer side effects) to mammalian cells.

Resistance to polyenes is mainly related to changes in the lipid structure of the membrane and subsequently a modification in its fluidity and absorbency. The principal altered effects concern enzymes that are involved in the synthesis of ergosterol. Deficiencies in the *ERG2* and *ERG3* genes, which code for the isomerase of C-8 sterol and delta-5,6-desaturase, induce modifications in membrane sterols. The quantity of ergosterol is modified, consequently affecting polyene activity. Boosted activity of catalase, an antioxidant enzyme that decreases oxidative injury, represents another polyene-resistance mechanism. However, intrinsic amphotericin B resistance frequently described in some *Aspergillus* spp. strains not only include the alteration of the ergosterol pathway but the signaling pathways, such as those described in *A. terreus* [30] or increased enzymatic activity of the peroxidase and superoxide dismutase in *A. flavus* [31] (Table 1).

Discovered in 1980s, azoles now represent the best conventional antifungal agents for medical treatment. Azole compounds have had a major impact on the treatment of invasive fungal infections over the last 35 years. The use of the first available azoles, imidazoles including ketoconazole, miconazole and clotrimazole, was primarily restricted to the treatment of superficial fungal infections. These compounds were substituted by the first-generation of triazoles, such as fluconazole and itraconazole, to broaden the range of application. Later, the search for new antifungals was intensified to overcome some efficiency limitations and to prevent emerging resistant pathogens. The second generation of triazoles (voriconazole, posaconazole, efinaconazole and isavuconazole) were developed with an extended spectrum of activity [32,33]. The final target of these drugs is ergosterol from the cell membrane. They inhibit lanosterol-14α-demethylase in mitochondria, which interferes with the synthesis of the membrane ergosterol. Each azole has a different affinity in its inhibition of lanosterol, which may explains the differences in spectrum of activity among azole agents [34], but all have a strong inhibitor effect on the CYP450 enzyme system, which is responsible for many drug interactions [35]. The global HIV epidemic led to the widespread and significant use of fluconazole to treat oro-oesophageal candidiasis in HIV-infected patients and fluconazole-resistant *Candida* strains were later widely reported in these patients [34,36].

Four mechanisms of resistance have been demonstrated so far: (i) activation of the efflux pumps due to an overexpression of membrane-associated transporters encoded by the gene families of transporters (CDR and MDR) reduces azole plasmatic concentrations [30,34]; (ii) azole agents cannot, qualitatively, bind to their enzymatic target to target changes induced by a mutation in *ERG11* gene, which encodes for the lanosterol-14α-demethylase [33,34]; (iii) some strains induce the overexpression of ERG11 as a compensatory mechanism and increase the intracellular concentration of this protein. Thus, an increasing concentration of a given azole agent is needed to remain efficient. This resistance mechanism involves quantitative changes by upregulating the target enzyme [30,34]; (iv) some strains, with a mutation of the ERG3 gene, developed a pathway bypassing fungal membrane biosynthesis by replacing ergosterol with 14α-methyl-fecosterol and preventing the accumulation of a toxic product: 14α-methyl-3,6-diol. This alternative route of ergosterol biosynthesis maintains both the function of the fungal membrane and the resistance to azoles [33] (Table 1).

Mannan, chitin and α- and β-glucans are the main compounds of the fungal cell wall. Drugs belonging to the echinocandins target one of these components and act as antifungals, inhibiting β-(1,3)-glucan synthetase leading to a depletion of β-(1,3)-glucan, an essential component for the structuring and function of the cell wall [33]. Echinocandins (caspofungin, anidulafungin and micafungin) are semisynthetic cyclic lipopeptides derived from natural products. They present a reserve supplement to the arsenal of drugs available to treat invasive fungal diseases with a fungicidal action against the *Candida* species and a fungistatic activity against the *Aspergillus* species [37].

Table 1. Current main antifungal agents: mechanisms of action, clinical indications, side effects and mechanisms of antifungal resistance. FCZ: fluconazole, ITZ: itraconazole, VRZ: voriconazole.

Antifungal Classes	Mechanisms of Action	Clinical Indications	Side Effects	Mechanisms of Resistance	Common Resistant Species
Polyenes *Amphotericin B* *Nystatin*	Ergosterol binding (membrane) permeabilization by ion channel formation Cell content leakage	Invasive fungal infection Topical *Candida* infections	Renal toxicity Hypokalemia Phlebitis Immunoallergic reaction	Deficiencies in *ERG2* and *ERG3* genes Ergosterol synthesis alteration Modifications in membrane sterols Changes of enzymatic activity or signaling pathways	*Scedosporium* spp., *Candida lusitaniae*, *Aspergillus terreus*
Azoles *Fluconazole* *Itraconazole* *Voriconazole* *Posaconazole* *Efinaconazole* *Isavuconazole*	Inhibition of lanosterol Ergosterol synthesis inhibition Alteration of fungal membrane fluidity and agility	All invasive candidiasis Cryptococcal meningitis *Aspergillus* spp. infections	Digestive disturbancesCephalgias Hepatotoxicity Drug interactions (CYPP450)	Over expression of efflux pump's function *ERG11* gene mutations inducing blockage in azoles binding Up-regulation of enzyme target Bypass pathway development by *ERG3* gene mutation	FCZ: *Candida krusei*, *Aspergillus* spp., *Scedosporium* spp., *Fusarium* spp., Mucorales ITZ: *Fusarium* spp. VRZ: Mucorales
Echinocandins *Micafungin* *Caspofungin* *Anidulafungin*	Inhibition of β-1,3-glucan synthase (β-GS) Formation of a defective cell wall	Invasive candidiasis Invasive aspergillosis (2nd intention)	Good overall tolerance	Mutations on *FKS1* gene (encoding for a subunit of β-GS) Decrease of affinity between drug and target	*Cryptococcus* spp., *Fusarium* spp., *Scedosporium* spp., Mucorales
5-fluorocytosine	Nucleoside analogue Disruption of protein synthesis Inhibition of DNA synthesis	Cryptococcosis Invasive candidiasis if treatment failure Always in association	Gastrointestinal troubles Hepatotoxicity Hematotoxicity	Mutations on *FUR1* gene (encoding uracil phosphoribosyl transferase) Mutations on *FCY1* gene (encoding cytosine deaminase enzyme)	Ineffective against many filamentous fungi

Candida resistance to echinocandins has been related to several mutations in a hot-spot region of the *FKS1* gene, which encodes for a subunit of echinocandin target, resulting in a lower affinity between the antifungal and its target [5,30] (Table 1).

A pyrimidine analogue, 5-fluorocytosine (5-FC), was developed in 1957 as an antimetabolite. Without any potential use as an anticancer treatment, it was used to treat fungal infections. Once 5-fluorocytosine enters the fungal cell, enzymes convert it in compounds to be incorporated in the synthesized RNA. This disrupts the protein synthesis of the affected fungi. It is also converted as a potent inhibitor of thymidylate synthase, which interferes in DNA synthesis and nuclear division [33]. It is always combined with other drugs, in association with azoles or amphotericin B, due to the high prevalence of intrinsic resistance in many fungal species and to the rapid development of resistance in yeast [33,34]. The primary resistance was about 10% in the *Candida albicans* strains [30]. Resistance to 5-FC may be due to mutations in *FUR1* or in the genes *FCY1* and *FCY2* leading to defects in flucytosine metabolism [30] (Table 1).

Finally, it is notable that, besides the limited number of available antimycotic agents and the burden of antifungal resistance (described above), toxicities and drug–drug interactions of antifungals, the treatment failure due to clinical resistance is frequently reported in invasive mycoses [34]. In fact, in vitro antifungal susceptibility testing does not guarantee the success of in vivo treatment. Many factors affecting the infected patient, the responsible fungal strain and the prescribed antifungal agent, which have been well reviewed elsewhere [34], may explain this clinical resistance.

3. Non-Antifungal Drugs Identified as Having a Potential Antifungal Activity against Invasive Fungal Strains

Given the rapid evolution of resistance to antifungal drugs and the high prevalence of mycoses in clinical settings due to the increasing number of human immunodeficiency cases and/or as the result of improved fungal diagnosis, there is an urgent need to improve the efficacy of current treatments and to develop new therapeutic strategies. It is worth considering innovative approaches [5] or, simply, associations of existing drugs, which could be a promising approach to extending the use of current antifungal agents. Indeed, drug combination resulting in a synergistic activity has the potential to impede the evolution of drug resistance through employing several mechanisms or targets [15]. However, this should not be considered as the only "miracle" solution because interactions can be indifferent and combinations can be unsuccessful [38].

Drug repositioning or repurposing allows for new indications for previously approved drugs that are already marketed for other medical reasons. This approach offers many benefits over de novo drug development. Previously established pharmacokinetic and pharmacodynamic profiles and toxicity data allow for faster and cheaper development of repositioned molecules. Consequently, clinical use may be considered to overcome the rapid emergence of resistant fungi and outbreaks [15,39]. In this paper, we reported on molecules that have been found to be active in vitro or even in vivo against fungal agents, according to their initial therapeutic class (Figure 1 and Table 2).

3.1. Antimicrobials Apart from Antifungals

Polymyxins including colistin and polymyxin B are peptide antibiotics and target the outer membrane of Gram-negative bacteria [40]. Tested against MDR yeasts and molds, polymyxins showed an antifungal activity with minimum inhibitory concentrations (MICs) ranging from 16 to 128 μg/mL [41]. In particular, a fungicidal effect was described with colistin against *C. albicans*, *Cryptococcus neoformans* and *Rhodotorula mucilaginosa*. The mechanism of action was similar to the bacterial mechanism, by inducing membrane damage to MDR-*C. albicans*, as observed under fluorescent microscopy [41]. Using a checkerboard microdilution assay, synergistic activity was revealed with colistin-amphotericin B and colistin-itraconazole, against MDR *C. albicans* and *Lichtheimia corymbifera* strains [41]. The colistin-azoles combination has also been reported more recently in strains showing low susceptibilities to fluconazole. Bibi M. et al. confirm that colistin binds to the lipids of fungal membranes and works in relation to an ergosterol depletion level due to the previous action of azoles [42].

Table 2. Drugs with reported in vitro antifungal activities. NR: not reported.

	Drug	First Indication	Antifungal Activity	Activity Range	Antifungal Mechanism of Action	References
Antimicrobials	**Polymyxins** *Colistin* *Polymyxin B*	Gram-negative bacterial infections	*C. albicans* *C. neoformans* *R. mucilaginosa* *S. apiospermum* *L. prolificans* *F. oxysporum* *F. solani* *R. oryzae*	16–128 μg/mL	Membrane damages on *Candida albicans*	[41]
	Ribavirin	Hepatitis C	*C. albicans* *C. tropicalis* *C. parapsilosis*	0.37–3.02 μg/mL	Disruption of vacuoles function of *C. albicans* strains	[41,43]
	Oxyclozanide	Animal parasitosis	*C. albicans*	16–32 μg/mL	Uncoupling the mitochondrial electron transport from phosphorylation and changing the mitochondrial membrane potential	[44]

Table 2. *Cont.*

	Drug	First Indication	Antifungal Activity	Activity Range	Antifungal Mechanism of Action	References
	Chloroquine	Malaria	*C. neoformans*	3.19 µg/mL (10 µM)	Iron deprivation	[45]
			C. albicans	31.2–250 µg/mL	Inhibition of ergosterol biosynthesis &	[46]
			S. cerevisiae	NR	Growth inhibition via blocking thiamine transportation	[47]
	Mebendazole	Helminthiasis	*C. neoformans*	92.5 ng/mL	Morphological alterations by reducing capsular dimension	[48]
			C. gatti	(0.3125 µM)		
Anti-inflammatory	**Auranofin**	Rheumatoid arthritis	*C. albicans* *A. fumigatus* *S. apiospermum* *L. prolificans* *C. neoformans*	0.25–16 µg/mL	Action on reactive-oxygen-mediated cell death	[49,50]
	Aspirin Ibuprofen	Inflammation	*Cryptococcus* spp. *Candida* spp.	1–10 mg/mL	Stress induction via ROS-mediated damage	[51,52]
	Theophylline	Asthma, COPD	*Candida* spp.	1.4–1.8 mg/mL	Membrane damages by ionic and ergosterol modifications	[53]
	Haloperidol **Trifluperidol**	Psychosis	*C. albicans* *C. neoformans*	<4 µg/mL	Possible action on GPCRs, mediators of signals across the cell membrane	[54,55]
Antipsychotics	**Sertraline**	Depression	*C. neoformans* *Lomentospora prolificans* *Scedosporium* spp., *Fusarium* spp. *Paecilomyces* spp., *Alternaria* spp. and *Curvularia* spp.	2–6 µg/mL 8–32 µg/mL	Inhibition of protein synthesis	[56,57]
	Chlorpromazine	Schizophrenia	*Candida* spp. *C. neoformans* Filamentous fungi: *Aspergillus* spp., *Scedosporium* spp., *Pseudallescheria* spp. and Zygomycetes	1–16 µg/mL	Possible modifications of membrane	[58,59]
Anticancers	**Tamoxifen** **Toremifene**	Breast cancer	*Candida* spp. *C. neoformans*	8–64 µg/mL	Prevention of proteins calmodulin from binding to calcineurin, cell lysis and alteration of fungal development Disturb the cell wall integrity via interaction with Ccr1	[60–62]
Others	**Disulfiram**	Alcoholism	*Candida* spp. *C. neoformans* *Aspergillus* spp.	1–16 µg/mL	Chelating metals Inhibition of multidrug transporter implicated in drug resistance	[63,64]

Other non-antifungal agents with bacterial activity have demonstrated a synergy with antifungals, as described by Rossato et al. [65]. For example, erythromycin with amphotericin B showed no toxic in vivo effect and could be a promising novel combination in invasive antifungal therapy [66].

Ribavirin, a purine nucleoside analogue, displays a broad-spectrum activity against many RNA and DNA viruses. Ribavirin is used to treat hepatitis C virus in combination with interferon-α [67]. Tournu et al. identified ribavirin as a potential *C. albicans* disrupting agent of vacuole, which is essential to yeast pathogenicity [43]. Based on this, we demonstrated the fungistatic activity of ribavirin against MDR *C. albicans* and fungicidal activity against *C. parapsilosis*. MICs largely ranged from 1.56 to 12.5 µmol/L (0.37–3.02 µg/mL) against *C. albicans*, *C. parapsilosis* and *C. tropicalis*. Synergistic activity was also reported when the antiviral agent was combined with either amphotericin B, fluconazole or itraconazole, against MDR *C. albicans* and was thus proposed to be further investigated for clinical use [68].

In terms of antiparasitic (anthelmintic) drugs, the activity of oxyclozanide, a halogenated salicylanilide, was demonstrated against *C. albicans* isolates, including ones that were resistant to azole and echinocandin. This anthelmintic agent seems to alter/disturb the mitochondrial oxidative phosphorylation function and thus its ability to use the non-fermentable carbon sources by disrupting the mitochondrial membrane potential [44]. Oxyclozanide is widely used as an antiparasitic veterinary drug against the liver fluke *Fasciola hepatica* [69], and has been studied for antibacterial properties against colistin-resistant Gram-negative bacilli infections [70]. Pic et al. showed that oxyclozanide inhibited, at 58% and 99%, the growth of *C. albicans* at a concentration of 10 and 100 μM, respectively. These concentrations are comparable to the therapeutic dose used for an ovine weighing 45 kg [71], however its repurposing strategy in human therapies requires further and careful data on pharmacokinetics/pharmacodynamics and/or in vivo therapeutic assays, as this drug has not yet been used in humans.

The first and main antimalarial drug, chloroquine, is able to be used in different indications as it has many effects on inflammatory responses, metabolic process, the immune system and infections [72,73]. Weber et al. described that chloroquine treatment of macrophages infected with cryptococcal cells led to the formation of iron complexes inducing the death of *C. neoformans* [45]. Chloroquine has also been shown to inhibit thiamine transporters in the yeast *Saccharomyces cerevisiae*, linked to glucose metabolism [47]. It can also damage fungal morphogenesis due to an anormal synthesis of ergosterol in drug-resistant *C. albicans* strains [74].

Benzimidazoles such as mebendazole, albendazole, flubendazole and triclabendazole are broad-spectrum anthelmintic drugs. Joffe et al. demonstrated the efficacy of benzimidazoles in inhibiting the growth of *C. neoformans*, especially mebendazole and flubendazole [48]. Mebendazole has been suggested to be repurposed as an anticryptococcal drug because it can efficiently penetrate the blood–brain barrier in animal models [48,75].

3.2. Anti-Inflammatory Drugs

Recently, Ogundeji et al. reported that aspirin and ibuprofen can control the growth of cryptococcal cells, with a high susceptibility of *C. neoformans* strains. Ibuprofen had a greater inhibitory effect than aspirin on all 10 *Cryptococcus* spp. tested strains at various drug concentrations [51]. The effects of ibuprofen seem to be dose-dependent; at high concentration (10 mg/mL), *Candida* cells are killed whereas at lower concentration (5 mg/mL), the drug was fungistatic [52]. In addition, synergistic outcomes were observed between ibuprofen and fluconazole or amphotericin B in *Cryptococcus* spp. and *Candida* spp., with fractional inhibitory concentration (FIC) < 0.5 [51,52,76].

Auranofin inhibits several inflammatory pathways and has been used since 1985 as an antirheumatic drug. It has already been found to be effective to treat bacterial infections [77] but Wiederhold et al. showed that auranofin also displayed an activity against various yeast and molds such as *A. fumigatus*, *Scedosporium apiospermum* and *Lomentospora prolificans* [49]. Although auranofin MICs were sometimes higher than those of the reference treatment (i.e., fluconazole), these concentrations could be easily achieved in patients' blood treated with the usual therapeutic dose [49]. Therefore, auranofin, along with its activity on biofilms [78], could be used as a promising antifungal treatment [54,66]. However, due to its immunosuppressive action, this drug should be carefully investigated prior to administration as an antifungal agent, given that fungi-infected patients are commonly immunocompromised.

Finally, theophylline is generally used for asthma or chronic obstructive pulmonary disease (COPD) and has been proposed to treat candidiasis as it had effects on cell membrane integrity [53].

3.3. Antipsychotic Drugs

Oral antipsychotic drugs are often used in routine clinical practice, so their side effects and toxicity are now well known. Recently, haloperidol and trifluperidol were described for their antifungal activity against *C. albicans* (with MICs values < 4 μg/mL) or against *C. neoformans* [55,79]. The authors demonstrated that the two antipsychotics had a similar effect to fluconazole, acting on the yeast membrane [54]. The combination of a haloperidol-derivative with the antifungal posaconazole displayed a 16-fold reduction in MIC values for both the azole agent and the antipsychotic, compared with each drug alone (from >32 and >128 to 2 and 8 μg/mL respectively) with a fractional inhibitory concentration index (FICI) of 0.13 [79]. Strong synergies were also observed against *C. glabrata* and *Aspergillus terreus* [79].

Sertraline, the most frequently prescribed antidepressant, has been reported to be fungicidal against *C. neoformans*, with MICs ranging between 2 and 6 μg/mL [56,80,81] but also against emerging fungi [57,82]. Moreover, in vivo sertraline antifungal activity was tested, in murine models of cryptococcosis, showing a reduction in the fungal burden [56]. The combination of this compound and azoles or amphotericin B showed efficiency against various *Cryptococcus* spp. strains [83,84]. Of the 53 tested isolates, 31 were affected by the synergistic combination sertraline-fluconazole (FICI ≤ 0.5) [83]. However, sertraline has shown some antagonist effects with fluconazole against *Candida* strains [56].

Despite many in vitro, in vivo and human studies showing the efficiency of sertraline as antifungal, disparities appear in the clinicals studies [80,85]. In 2019, a clinical trial of 486 participants testing sertraline as an adjunctive treatment to IV amphotericin B and oral fluconazole to treat HIV-associated cryptococcal meningitis did not significatively improve survival, but did reveal a similar fungal clearance rate between groups [85]. In 2020, another randomized trial testing sertraline as pre-emptive therapy was stopped without a final conclusion due to the severe side effects of sertraline that were observed [86].

Chlorpromazine and trifluoperazine, dopamine antagonists, are used to treat schizophrenia. They are largely reported as having antibacterial and antifungal effects [58,59,87] and Vitale et al. confirmed the in vitro antifungal activity against difficult-to-treat filamentous fungi such as *Aspergillus* species (*A. fumigatus*, *A. ustus* and *A. terreus*), zygomycetes (*Absidia corymbifera*, *Rhizopus oryzae* and *R. microspores*) and *Scedosporium* species (*S. apiospermum* and *S. prolificans*) [58]. Phenothiazines including chlorpromazine appear to act on the fungal membrane but this needs further consideration [58]. The use of the combination must be advised to avoid resistance, to have greater efficacy and a less toxic effect: for example, chlorpromazine-amphotericin B against *C. neoformans* [84] or *Candida* species [59] demonstrated an interesting synergism profile.

3.4. Anticancer Drugs

Anticancer drugs may represent an important source of potential repurposing drugs. Indeed, as they often work on the basic metabolism pathways of eucaryotic and human cells such as the DNA replicating pathway, yeast cells may also be affected. Various anticancer drugs have been seen to be efficient in vitro against yeast growth [54,88]. Butts et al. demonstrated the fungicidal activity of tamoxifen and toremifene, two estrogenic receptor antagonists, against *C. neoformans* within macrophages, where the main pathogenesis of this organism happens [60]. Tamoxifen is described as an inhibitor of calmodulin. In this way, tamoxifen-treated yeasts showed cell lysis and an alteration of fungal development [61,89]. Recently, the interaction between tamoxifen and its target Ccr1 has also been described as causing the disruption of cell wall integrity [62].

However, administration of anticancer molecules to patients who are often immuno-compromised should only be taken after careful consideration due to the numerous side effects that these compounds can cause, including immunosuppression [90]. A bypass solution via a synergistic association could circumvent this limitation. Interestingly, both compounds were synergistic in vitro with amphotericin B and fluconazole against cryptococcal cells [60]. In addition, in vivo candidiasis was cured, in a murine model, by the administration of 200 mg/kg of body weight per day of tamoxifen [61]. Since then, a randomized phase II clinical trial of tamoxifen as an adjuvant to the gold standard therapy for cryptococcal meningitis is in progress [91].

Antifolates as inhibitors of purine synthesis have also been reported to be effective agents against yeast development by reducing the quantity of ergosterol [65].

3.5. Other Approved Drugs

Finally, we reported on a few other well-known and old drugs that could be repurposed as their indications, doses and side effects are known and thus can be easily managed.

Fluvastatin, rosuvastatin, atorvastatin and simvastatin are statins used to lower the synthesis of cholesterol by inhibiting HMG-CoA. Statins may be an important adjuvant for the treatment of fungal infections because their efficiency has been reported against fungi, even against azole-resistant yeasts [65,92]. Indeed, Macreadie et al. demonstrated a strong inhibition of the growth of *Candida* spp. (with the exception of *C. krusei* on YEPD media containing 100 μM of statins) and *A. fumigatus* [93]. On the fungal cell, statins might work on the pathway of mevalonate synthesis causing a decrease in ergosterol quantity of the cell membrane [93]. Statins showed in vitro synergy with various azoles and may display beneficial outcomes on candidiasis according to one cohort study in 2013 [65,81,94].

Disulfiram is an alcohol antagonist drug that has been used in clinical practice for many years. Khan et al. reported its antifungal potential in 2007 and provided an MIC range from 1 to 16 μg/mL for fluconazole-sensitive and resistant yeast strains, with an inhibiting effect on biofilm formation [95,96]. It also had a fungicidal activity on *Aspergillus* spp. [95]. Earlier, Shukla et al. demonstrated that disulfiram could reverse Cdr1p-mediated drug resistance so it could be used in combination to sensitize resistant strains [63,64]. Broadly, side effects of disulfiram are uncommon, contraindications include pregnancy and unstable cardiovascular disease and close hepatic monitoring is required [97]. Given all this information, in vitro, in vivo and clinical trials should be pursued to fully justify the repositioning of this molecule.

4. Some Emerging Multidrug-Resistant Fungi and Their Compounds with Repurpose Potential Identified through Phenotypic Screening

Resistance to at least one class of antifungal agents is a concern of most existing fungal species. However, some major pathogens have a relatively high resistance rate and constitute a serious public health burden, especially *C. albicans*, *Cryptococcus* spp. and *Aspergillus* spp. In addition, other emerging and potentially life-threatening pathogens are increasingly being reported [98]. Some are still not well characterized and are opportunistic and MDR, such as *C. auris* [99], *Scedosporium* spp. and *Fusarium* spp. [100]. We present a non-exhaustive list of compounds below that could be effective as an alternative therapeutic strategy, depending on these MDR species (Figure 2).

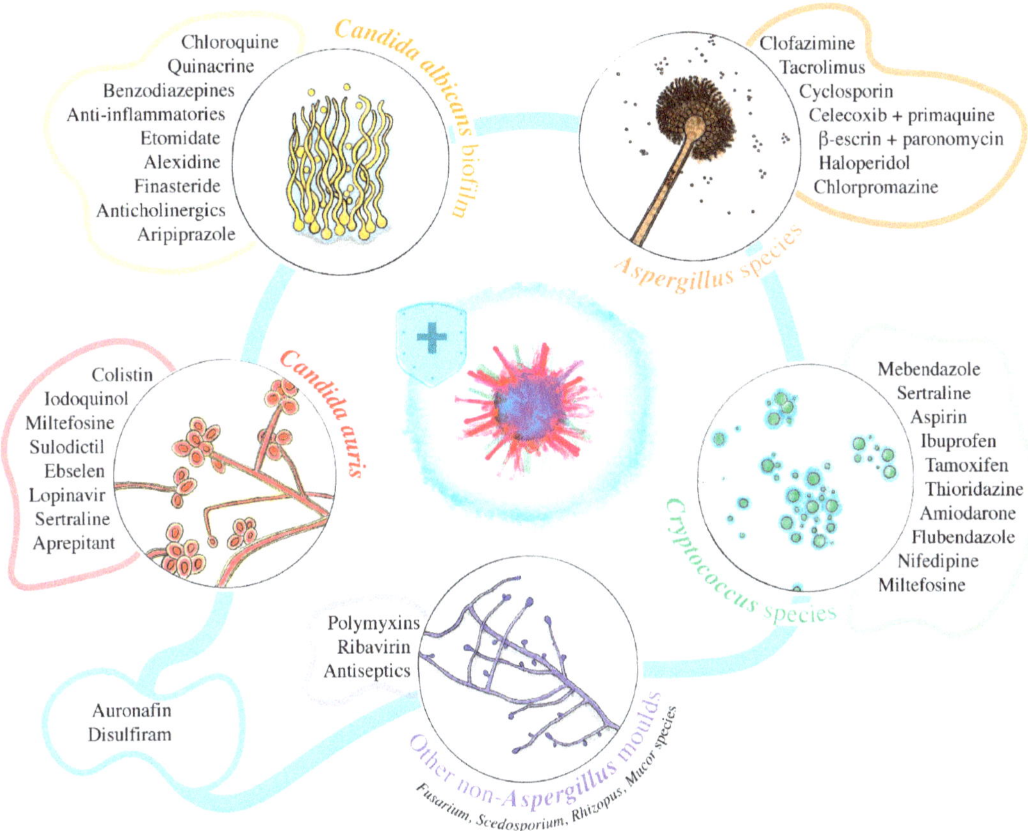

Figure 2. Examples of approved and candidate drugs with activity against emerging MDR fungi.

4.1. C. albicans Biofilms

The pathogenicity of the *Candida* species resides in their ability to form biofilms, thereby protecting them from external elements such as antifungal agents or the host's immune system. The common implantation of a *Candida* biofilm on a medical device such as a catheter can lead to candidemia and severe systemic infections. In the absence or failure of treatment, the health and economic consequences are significant [101]. Some drugs alone or in combination with a currently used antifungal are effective in vitro to reduce or inhibit the formation of a biofilm [102,103].

As reported above, chloroquine reverts the azole resistance in biofilms [46], and aspirin [104], disulfiram [96] or auranofin [78] decrease in vitro the biofilm formation of many *Candida* species. Quinacrine is an antimalarial, which was used during WWII but remains available to treat giardiasis or cutaneous leishmaniasis. It has been effective at preventing and treating *C. albicans* biofilms with MIC ranges of 64–256 µg/mL [93]. In addition, both amphotericin B and caspofungin have synergy with quinacrine with an FICI equal to 0.37 and 0.31, respectively [105]. Benzodiazepines such as midazolam or diazepam have also been proposed to be repurposed against the biofilm formation of yeast, due to interaction with its virulence factors [106,107]. Anti-inflammatory compounds appear to be very efficient, such as celecoxib, etodolac, meloxicam, etc. [81]. In addition, flufenamic acid (a nonsteroidal anti-inflammatory drug, NSAID) showed an excellent action in the prevention and treatment of biofilms from echinocandin-resistant strains [108]. The hypnotic agent, etomidate, works against the biofilm of fluconazole-resistant *Candida* spp. strains [109]. The antiseptic alexidine [103] and the finasteride used to treat the prostatic hyperplasia [110] were highly active in vitro at preventing *C. albicans* biofilms. Nile et al. highlighted the cholinergic receptor, which slowed down the biofilm-mediated virulence of *C. albicans* while having a boosting effect on the host immune response using pilocarpine [111] or tropicamide in a study by Machado et al. [112]. The antipsychotic drug aripiprazole was as effective as azoles by acting on the early formation of the pseudo hyphal [113]. As mentioned, many drugs from different therapeutic classes can be effective as an antibiofilm agent. In vivo studies first, followed by clinical trials remain important to confirm those in vitro options.

4.2. C. auris

C. auris is considered to be a serious global health threat due to its association with nosocomial invasive infections, its high mortality rate and its multidrug resistant profile [114]. High rates of antifungal resistance have been reported for fluconazole and amphotericin B, which is considered to be an intrinsic resistance, but some acquired resistances to echinocandins have been described in some countries [115,116]. The need for effective agents must be a priority in order to address future outbreaks, which, given the current situation, are highly likely [117]. To quickly identify candidate drugs, screening molecule libraries can offer various solutions [49,66,118]. Cheng et al. found six novel anti-*C. auris* compounds among more than 4300 approved drugs with 13 possible different drug associations [118]. The amebicide iodoquinol and leishmanicide miltefosine were reported in those screenings as being potential repositionable compounds due to their inhibition of *C. auris* growth [49]. Sulodictil [66], ebselen [66], antiemetic aprepitant [119] and lopinavir [120] were active against *C. auris*, either alone or in association with currently used antifungals. Finally, most of the drugs described above as being effective on yeast are also effective on *C. auris*, such as sertraline [121], oxyclonazide [44], colistin [112], etc. Indeed, these molecules are generally active via novel targets, which are still naïve in antifungal treatments and therefore in yeast adaptation.

4.3. Aspergillus Species

Invasive aspergillosis threatens the lives of millions of immunocompromised patients every year, with a mortality rate of 50/60% [98]. These species come from the environment and some patients are infected via contaminated foods [122,123]. Used for agricultural production, certain fungicides display a similar mechanism as azoles used in clinical routine. The evolution of resistance by selective pressure is attributed to their widespread use and there is an urgent need to find alternative strategies to current antifungals [124–126]. As a result, different drug screenings have been performed and have found that clozafimine, tacrolimus, cyclosporin [127], haloperidol [79], disulfiram [95], chlorpromazine [58] and auranofin [122] as reported above, could be repurposed as an anti-*Aspergillus* spp. treatment. Interesting synergies were also obtained between celecoxib and primaquine and paromomycin and β-escin with a FICI of <0.27 or <0.38, respectively, against *A. fumigatus* growth [128].

4.4. Cryptococcus Species

Occurring predominantly during the course of an immunocompromised patient's disease, cryptococcosis causes pneumonia or meningoencephalitis, due to an inhalation of *Cryptococcus* cells from the environment [129]. In addition to their worldwide spread, treatment of these infections remains challenging for clinicians who deal with only three classes of antifungal agents, as echinocandins are not effective against them [89]. Several molecules were found to be active against *Cryptococcus* cell growth, as mentioned above, such as auranofin, aspirin, ibuprofen [122], tamoxifen [129], mebendazole [48], sertraline [86,121] and disulfiram [95], and thioridazine, amiodarone [122], miltefosine [130] and calcium channel blockers such as nifedipine [131] have also been described. The antiparasitic drug, flubendazole, was very active, with MICs ranging from 0.039 to 0.156 μg/mL, even against fluconazole-resistant strains [131]. Despite this, the perfect anticryptococcal treatment would exhibit a low toxicity for polymedicated, immunosuppressed patients would be well-distributed around the body to eradicate all cryptococcal niches including the cerebrospinal fluid [129].

4.5. Other Non-Aspergillus Molds

More rarely, other MDR fungal pathogens can cause aggressive and disseminated infections associated with poor prognosis such as Hyalohyphomycetes, including the genera *Fusarium*, *Scedosporium*, etc., or the Mucormycetes with *Mucor* and *Rhizomucor* group. Early and effective treatment is required to prevent the progression of the infection and to limit outbreaks [132,133]. However, this infection remains difficult to treat because the aforementioned fungi are resistant to most current antifungal drugs (Table 1) [132,134]. We previously screened a library of 1280 drugs against six of these filamentous fungi including *Fusarium*, *Scedosporium*, *Rhizopus* and *Lichtheimia* species [135]. The main hits found were antifungals, antiseptics and some antineoplastics against a few strains, and polymyxins [41], disulfiram [95], auranofin [122] and ribavirin [68], as mentioned above. These fungi have fewer hits and further investigations are warranted in order not to reach fatal therapeutic impasses. Some combinations of antibiotics and antifungals, however, displayed strong synergies [65].

5. Further Assessment and Prioritization of Repurpose Potential

It is important to highlight in this effort that preliminary in vitro studies, however, encouraging, do not automatically imply repurpose potential. In addition, not all compounds identified above possess equal repurpose potential while others may be altogether inappropriate for the clinical use in question. Given that safety data for marketed drugs and advanced clinical candidates are available, further assessment should be undertaken to evaluate and prioritize each molecule with respect to its repurpose potential. For example, effective auranofin concentrations that inhibited yeasts and molds' growth, ranged from 0.25 to 16 µg/mL [50], but human pharmacokinetics displayed a *Cmax* of 0.025 µg/mL after 6 mg per day [136]. Aspirin on the other hand was effective against yeasts [44,69] in accordance with human doses and used for a long time now. For analgesic dosages, the *Cmax* was reported on average at 50 µM (9 mg/L) and were higher for anti-inflammatory use [137]. Another example, quinacrine alone, was effective against *C. albicans* biofilms with MICs between 64 and 256 µg/mL [63]. However, quinacrine pharmacokinetics from an intrapleural dose displayed a *Cmax* below 1 µg/mL for a 600 mg dose [138] and the therapeutic dose is initially 100 mg, which makes the use of this molecule in monotherapy inappropriate. In spite of that, its use in synergic combinations led to a reduce in the initial dose, from 64 µg/mL alone to 4 µg/mL with caspofungine [105] and approaching acceptable human concentrations. At the current stage, the use of quinacrine as a monotherapy does not reach the reported human concentrations, but the study of its combinations with other molecules is still interesting in view of the reported synergisms. Plasma concentrations of synergistic combinations of azoles or amphotericin B and colistin can be above some previously reported MICs, where human doses of colistin have resulted in serum concentrations up to 32 µg/mL [41]. In such cases, the renal condition of the patient must be considered, although nebulized doses of colistin could avoid this toxicity [139]. Therefore, an analysis process must be applied to all molecules before their repurposing after in vitro, in vivo and other assays. In summary, strict and careful analysis is essential before administration, concerning the consistency between the effective dose and the serum concentration/toxicity dose, but also covering all other parameters related or not to the molecule, such as potential drug interactions, bioavailability of the molecule, strain susceptibility, patient condition and consent, approval of the patient's care panel, etc.

6. Conclusions

The emergence of new mycotic agents and the increase in antifungal resistance has led to the need to find new and/or alternative drugs. We are currently seeing that non-traditional antimicrobial agents, previously prescribed to treat non-infectious conditions, may display antimicrobial properties, and it would be a worthwhile investment to further explore these compounds before being repurposed. Otherwise, these reported drugs could serve as a starting point for the reinnovation of a new molecule. We noted compounds ranging from anti-inflammatory to antipsychotic drugs, which have been documented to control fungal growth and to be repositionable. However, their clinical application may be limited to treat life-threatening fungi, due to drug toxicity, especially with anticancer drugs that generally target eukaryotic organisms or with drugs inducing an immuno-suppressive state. Susceptibility testing on each fungus, careful analysis of pharmacokinetic/pharmacodynamic (PK/PD) data and human achievable and tolerable concentrations must be confirmed before administration to assess the benefit/risk. All effective treatments should be taken into consideration as a last-line therapy, even if they have side effects, if they could save the patient [140]. Fortunately, in in vitro assays some of these compounds act synergistically with currently used antifungal agents. Thus, combinations enable the use of low concentrations with the advantage of minimizing any possible undesired physiological effects. This also enhances the efficiency of traditional antimicrobial drugs, which are fungistatic when used alone under normal conditions.

Assuming that obstacles may be overcome, drug repurposing is a promising alternative strategy into which further research and clinical trials are essential to combat the increase in invasive fungal infections.

Author Contributions: L.P., H.Y., J.-M.R. and F.B. drafted and revised the manuscript. All authors read and approved the final manuscript.

Funding: This work was supported by the French Government under the "Investissements d'avenir" (Investments for the Future) program managed by the Agence Nationale de la Recherche (ANR, fr: National Agency for Research), (reference: Méditerranée Infection 10-IAHU-03). This work was supported by Région Provence-Alpes-Côte d'Azur and European funding (FEDER (Fonds européen de développement régional) PRIMMI (Plateformes de Recherche et d'Innovation Mutualisées Méditerranée Infection)).

Institutional Review Board Statement: Not applicable.

Informed Consent Statement: Not applicable.

Data Availability Statement: No new data were created or analyzed in this study. Data sharing is not applicable to this article.

Acknowledgments: We thank our designer Juan Manuel Osorio Rozo for Figures 1 and 2 (https://www.behance.net/juanmanuelosoriorozo, accessed on 18 May 2021).

Conflicts of Interest: The authors declare no conflict of interest.

References

1. Brown, G.D.; Denning, D.W.; Gow, N.A.R.; Levitz, S.M.; Netea, M.G.; White, T.C. Hidden Killers: Human Fungal Infections. *Sci. Transl. Med.* **2012**, *4*, 165rv13. [CrossRef]
2. Denning, D.W. Calling upon all public health mycologists: To accompany the country burden papers from 14 countries. *Eur. J. Clin. Microbiol. Infect. Dis.* **2017**, *36*, 923–924. [CrossRef]
3. Livio, P.; Morena, C.; Anna, C.; Massimo, O.; Luana, F.; Bruno, M.; Domenico, P.; Marco, P.; Alessandro, B.; Anna, C.; et al. The epidemiology of fungal infections in patients with hematologic malignancies: The SEIFEM-2004 study. *Haematologica* **2006**, *91*, 1068–1075.
4. Armstrong-James, D.; Meintjes, G.; Brown, G.D. A neglected epidemic: Fungal infections in HIV/AIDS. *Trends Microbiol.* **2014**, *22*, 120–127. [CrossRef] [PubMed]
5. Geddes-McAlister, J.; Shapiro, R.S. New pathogens, new tricks: Emerging, drug-resistant fungal pathogens and future prospects for antifungal therapeutics. *Ann. N. Y. Acad. Sci.* **2019**, *1435*, 57–78. [CrossRef] [PubMed]

6. Gupta, R.; Goel, N.; Gupta, A.; Gupta, K.B.; Chaudhary, U.; Sood, S. A rare fungal infiltration of lungs in a healthy young girl. *Case Rep. Pulmonol.* **2011**, *2011*, 1–3. [CrossRef]

7. Srinivasan, A.; Lopez-Ribot, J.L.; Ramasubramanian, A.K. Overcoming antifungal resistance. *Drug Discov. Today Technol.* **2014**, *11*, 65–71. [CrossRef]

8. Alcazar-Fuoli, L.; Mellado, E. Current status of antifungal resistance and its impact on clinical practice. *Br. J. Haematol.* **2014**, *166*, 471–484. [CrossRef]

9. Meurman, J.H.; Hämäläinen, P.; Janket, S.J. Oral infections in older adults. *Aging Health* **2006**, *2*, 1013–1023. [CrossRef]

10. Perfect, J.R. The antifungal pipeline: A reality check. *Nat. Rev. Drug Discov.* **2017**, *16*, 603–616. [CrossRef] [PubMed]

11. Sanguinetti, M.; Posteraro, B.; Lass-Flörl, C. Antifungal drug resistance among *Candida* species: Mechanisms and clinical impact. *Mycoses* **2015**, *58*, 2–13. [CrossRef]

12. Barber, A.E.; Riedel, J.; Sae-Ong, T.; Kang, K.; Brabetz, W.; Panagiotou, G.; Deising, H.B.; Kurzai, O. Effects of agricultural fungicide use on *Aspergillus fumigatus*: Abundance, antifungal susceptibility, and population structure. *mBio* **2020**, *11*. [CrossRef] [PubMed]

13. Demers, E.G.; Biermann, A.R.; Masonjones, S.; Crocker, A.W.; Ashare, A.; Stajich, J.E.; Hogan, D.A. Evolution of drug resistance in an antifungal-naive chronic *Candida lusitaniae* infection. *Proc. Natl. Acad. Sci. USA* **2018**, *115*, 12040–12045. [CrossRef]

14. Hall, R.A.; Noverr, M.C. Fungal interactions with the human host: Exploring the spectrum of symbiosis. *Curr. Opin. Microbiol.* **2017**, *40*, 58–64. [CrossRef]

15. Peyclit, L.; Baron, S.A.; Rolain, J.-M. Drug repurposing to fight colistin and carbapenem-resistant bacteria. *Front. Cell. Infect. Microbiol.* **2019**, *9*, 193. [CrossRef] [PubMed]

16. Dimasi, J.A.; Feldman, L.; Seckler, A.; Wilson, A. Trends in risks associated with new drug development: Success rates for investigational drugs. *Clin. Pharmacol. Ther.* **2010**, *87*, 272–277. [CrossRef]

17. Houšt', J.; Spížek, J.; Havlícek, V. Antifungal Drugs. *Metabolites* **2020**, *10*. [CrossRef]

18. Wiederhold, N.P. The antifungal arsenal: Alternative drugs and future targets. *Int. J. Antimicrob. Agents* **2018**, *51*, 333–339. [CrossRef]

19. Rauseo, A.M.; Coler-Reilly, A.; Larson, L.; Spec, A. Hope on the horizon: Novel fungal treatments in development. *Open Forum Infect. Dis.* **2020**, *7*, 1–19. [CrossRef]

20. Su, H.; Han, L.; Huang, X. Potential targets for the development of new antifungal drugs. *J. Antibiot.* **2018**, *71*, 978–991. [CrossRef]

21. Mahboubi, M. *Artemisia sieberi* Besser essential oil and treatment of fungal infections. *Biomed. Pharmacother.* **2017**, *89*, 1422–1430. [CrossRef]

22. D'agostino, M.; Tesse, N.; Frippiat, J.P.; Machouart, M.; Debourgogne, A. Essential oils and their natural active compounds presenting antifungal properties. *Molecules* **2019**, *24*, 3713. [CrossRef] [PubMed]

23. Córdoba, S.; Vivot, W.; Szusz, W.; Albo, G. Antifungal activity of essential oils against *Candida* species isolated from clinical samples. *Mycopathologia* **2019**, *184*, 615–623. [CrossRef]

24. Barac, A.; Donadu, M.; Usai, D.; Spiric, V.T.; Mazzarello, V.; Zanetti, S.; Aleksic, E.; Stevanovic, G.; Nikolic, N.; Rubino, S. Antifungal activity of *Myrtus communis* against *Malassezia* sp. isolated from the skin of patients with pityriasis versicolor. *Infection* **2018**, *46*, 253–257. [CrossRef]

25. González-Burgos, E.; Gómez-Serranillos, M.P. Natural Products for Vulvovaginal Candidiasis Treatment: Evidence from Clinical Trials. *Curr. Top. Med. Chem.* **2018**, *18*, 1324–1332. [CrossRef] [PubMed]

26. Pushpakom, S.; Iorio, F.; Eyers, P.A.; Escott, K.J.; Hopper, S.; Wells, A.; Doig, A.; Guilliams, T.; Latimer, J.; McNamee, C.; et al. Drug repurposing: Progress, challenges and recommendations. *Nat. Rev. Drug Discov.* **2018**, *18*, 41–58. [CrossRef]

27. Gill, R.; Amberkar Mohan Babu, V.; Meena Kumari, K. Assets and liabilities of drug repositioning. *Int. J. Pharma Bio Sci.* **2016**, *7*, 47–53. [CrossRef]

28. Nett, J.E.; Andes, D.R. Antifungal Agents: Spectrum of activity, pharmacology, and clinical indications. *Infect. Dis. Clin. N. Am.* **2016**, *30*, 51–83. [CrossRef]

29. Kuriyama, T.; Karasawa, T.; Williams, D.W. Antimicrobial Chemotherapy: Significance to Healthcare. In *Biofilms in Infection Prevention and Control: A Healthcare Handbook*; Elsevier Inc.: Amsterdam, The Netherlands, 2014; pp. 209–244. ISBN 9780123970435.

30. Scorzoni, L.; de Paula e Silva, A.C.A.; Marcos, C.M.; Assato, P.A.; de Melo, W.C.M.A.; de Oliveira, H.C.; Costa-Orlandi, C.B.; Mendes-Giannini, M.J.S.; Fusco-Almeida, A.M. Antifungal therapy: New advances in the understanding and treatment of mycosis. *Front. Microbiol.* **2017**, *8*. [CrossRef] [PubMed]

31. Rudramurthy, S.M.; Paul, R.A.; Chakrabarti, A.; Mouton, J.W.; Meis, J.F. Invasive aspergillosis by *Aspergillus flavus*: Epidemiology, diagnosis, antifungal resistance, and management. *J. Fungi* **2019**, *5*, 55. [CrossRef]

32. Allen, D.; Wilson, D.; Drew, R.; Perfect, J. Azole antifungals: 35 years of invasive fungal infection management. *Expert Rev. Anti Infect. Ther.* **2015**, *13*, 787–798. [CrossRef]

33. Ghannoum, M.A.; Rice, L.B. Antifungal agents: Mode of action, mechanisms of resistance, and correlation of these mechanisms with bacterial resistance. *Clin. Microbiol. Rev.* **1999**, *12*, 501–517. [CrossRef]

34. Kanafani, Z.A.; Perfect, J.R. Resistance to antifungal agents: Mechanisms and clinical impact. *Clin. Infect. Dis.* **2008**, *46*, 120–128. [CrossRef]

35. Yu, D.T.; Peterson, J.F.; Seger, D.L.; Gerth, W.C.; Bates, D.W. Frequency of potential azole drug-drug interactions and consequences of potential fluconazole drug interactions. *Pharmacoepidemiol. Drug Saf.* **2005**, *14*, 755–767. [CrossRef] [PubMed]

36. Canuto, M.M.; Rodero, F.G. Antifungal drug resistance to azoles and polyenes. *Lancet Infect. Dis.* **2002**, *2*, 550–563. [CrossRef]
37. Aguilar-Zapata, D.; Petraitiene, R.; Petraitis, V. Echinocandins: The expanding antifungal armamentarium. *Clin. Infect. Dis.* **2015**, *61*, S604–S611. [CrossRef]
38. Chassot, F.; Venturini, T.P.; Piasentin, F.B.; Santurio, J.M.; Svidzinski, T.I.E.; Alves, S.H. Activity of antifungal agents alone and in combination against echinocandin-susceptible and -resistant *Candida parapsilosis* strains. *Rev. Iberoam. Micol.* **2019**, *36*, 44–47. [CrossRef] [PubMed]
39. Ciliberto, G.; Mancini, R.; Paggi, M.G. Drug repurposing against COVID-19: Focus on anticancer agents. *J. Exp. Clin. Cancer Res.* **2020**, *39*. [CrossRef] [PubMed]
40. Baron, S.; Hadjadj, L.; Rolain, J.-M.; Olaitan, A.O. Molecular mechanisms of polymyxin resistance: Knowns and unknowns. *Int. J. Antimicrob. Agents* **2016**, *48*, 583–591. [CrossRef]
41. Yousfi, H.; Ranque, S.; Rolain, J.M.; Bittar, F. *In vitro* polymyxin activity against clinical multidrug-resistant fungi. *Antimicrob. Resist. Infect. Control* **2019**, *8*. [CrossRef] [PubMed]
42. Bibi, M.; Murphy, S.; Benhamou, R.I.; Rosenberg, A.; Ulman, A.; Bicanic, T.; Fridman, M.; Berman, J. Combining Colistin and Fluconazole Synergistically Increases Fungal Membrane Permeability and Antifungal Cidality. *ACS Infect. Dis.* **2021**, *7*. [CrossRef]
43. Tournu, H.; Carroll, J.; Latimer, B.; Dragoi, A.M.; Dykes, S.; Cardelli, J.; Peters, T.L.; Eberle, K.E.; Palmer, G.E. Identification of small molecules that disrupt vacuolar function in the pathogen *Candida albicans*. *PLoS ONE* **2017**, *12*. [CrossRef]
44. Pic, E.; Burgain, A.; Sellam, A. Repurposing the anthelminthic salicylanilide oxyclozanide against susceptible and clinical resistant *Candida albicans* strains. *Med. Mycol.* **2019**, *57*, 387–390. [CrossRef]
45. Weber, S.M.; Levitz, S.M.; Harrison, T.S. Chloroquine and the fungal phagosome. *Curr. Opin. Microbiol.* **2000**, *3*, 349–353. [CrossRef]
46. Shinde, R.B.; Raut, J.S.; Chauhan, N.M.; Karuppayil, S.M. Chloroquine sensitizes biofilms of *Candida albicans* to antifungal azoles. *Braz. J. Infect. Dis.* **2013**, *17*, 395–400. [CrossRef]
47. Huang, Z.; Srinivasan, S.; Zhang, J.; Chen, K.; Li, Y.; Li, W.; Quiocho, F.A.; Pan, X. Discovering thiamine transporters as targets of chloroquine using a novel functional genomics strategy. *PLoS Genet.* **2012**, *8*. [CrossRef] [PubMed]
48. Joffe, L.S.; Schneider, R.; Lopes, W.; Azevedo, R.; Staats, C.C.; Kmetzsch, L.; Schrank, A.; Del Poeta, M.; Vainstein, M.H.; Rodrigues, M.L. The anti-helminthic compound mebendazole has multiple antifungal effects against *Cryptococcus neoformans*. *Front. Microbiol.* **2017**, *8*. [CrossRef]
49. Wall, G.; Chaturvedi, A.K.; Wormley, F.L.; Wiederhold, N.P.; Patterson, H.P.; Patterson, T.F.; Lopez-Ribot, J.L. Screening a repurposing library for inhibitors of multidrug-resistant *Candida auris* identifies ebselen as a repositionable candidate for antifungal drug development. *Antimicrob. Agents Chemother.* **2018**, *62*. [CrossRef] [PubMed]
50. Wiederhold, N.P.; Patterson, T.F.; Srinivasan, A.; Chaturvedi, A.K.; Fothergill, A.W.; Wormley, F.L.; Ramasubramanian, A.K.; Lopez-Ribot, J.L. Repurposing auranofin as an antifungal: *In vitro* activity against a variety of medically important fungi. *Virulence* **2017**, *8*, 138–142. [CrossRef] [PubMed]
51. Ogundeji, A.O.; Pohl, C.H.; Sebolai, O.M. Repurposing of aspirin and ibuprofen as candidate anti-Cryptococcus drugs. *Antimicrob. Agents Chemother.* **2016**, *60*, 4799–4808. [CrossRef]
52. Pina-Vaz, C.; Sansonetty, F.; Rodrigues, A.G.; Martinez-De-Oliveira, J.; Fonseca, A.F.; Mårdh, P.A. Antifungal activity of ibuprofen alone and in combination with fluconazole against Candida species. *J. Med. Microbiol.* **2000**, *49*, 831–840. [CrossRef] [PubMed]
53. Singh, S.; Fatima, Z.; Ahmad, K.; Hameed, S. Repurposing of respiratory drug theophylline against *Candida albicans*: Mechanistic insights unveil alterations in membrane properties and metabolic fitness. *J. Appl. Microbiol.* **2020**, *129*, 860–875. [CrossRef]
54. Stylianou, M.; Kulesskiy, E.; Lopes, J.P.; Granlund, M.; Wennerberg, K.; Urban, C.F. Antifungal application of nonantifungal drugs. *Antimicrob. Agents Chemother.* **2014**, *58*, 1055–1062. [CrossRef] [PubMed]
55. Ji, C.; Liu, N.; Tu, J.; Li, Z.; Han, G.; Li, J.; Sheng, C. Drug Repurposing of Haloperidol: Discovery of New Benzocyclane Derivatives as Potent Antifungal Agents against Cryptococcosis and Candidiasis. *ACS Infect. Dis.* **2019**, *6*. [CrossRef]
56. Zhai, B.; Wu, C.; Wang, L.; Sachs, M.S.; Lin, X. The antidepressant sertraline provides a promising therapeutic option for neurotropic cryptococcal infections. *Antimicrob. Agents Chemother.* **2012**, *56*, 3758–3766. [CrossRef]
57. Villanueva-Lozano, H.; González, G.M.; Espinosa-Mora, J.E.; Bodden-Mendoza, B.A.; Andrade, A.; Martínez-Reséndez, M.F.; Treviño-Rangel, R.D.J. Evaluation of the expanding spectrum of sertraline against uncommon fungal pathogens. *J. Infect. Chemother.* **2020**, *26*, 309–311. [CrossRef]
58. Vitale, R.G.; Afeltra, J.; Meis, J.F.G.M.; Verweij, P.E. Activity and post antifungal effect of chlorpromazine and trifluopherazine against *Aspergillus*, *Scedosporium* and zygomycetes. *Mycoses* **2007**, *50*, 270–276. [CrossRef] [PubMed]
59. Galgóczy, L.; Bácsi, A.; Homa, M.; Virágh, M.; Papp, T.; Vágvölgyi, C. *In vitro* antifungal activity of phenothiazines and their combination with amphotericin B against different Candida species. *Mycoses* **2011**, *54*. [CrossRef] [PubMed]
60. Butts, A.; Koselny, K.; Chabrier-Roselló, Y.; Semighini, C.P.; Brown, J.C.S.; Wang, X.; Annadurai, S.; DiDone, L.; Tabroff, J.; Childers, W.E.; et al. Estrogen receptor antagonists are anti-cryptococcal agents that directly bind EF hand proteins and synergize with fluconazole *in vivo*. *mBio* **2014**, *5*. [CrossRef]
61. Dolan, K.; Montgomery, S.; Buchheit, B.; DiDone, L.; Wellington, M.; Krysan, D.J. Antifungal activity of tamoxifen: In vitro and in vivo activities and mechanistic characterization. *Antimicrob. Agents Chemother.* **2009**, *53*, 3337–3346. [CrossRef] [PubMed]

62. Liu, Q.; Guo, X.; Jiang, G.; Wu, G.; Miao, H.; Liu, K.; Chen, S.; Sakamoto, N.; Kuno, T.; Yao, F.; et al. NADPH-cytochrome P450 reductase ccr1 is a target of tamoxifen and participates in its antifungal activity via regulating cell wall integrity in fission yeast. *Antimicrob. Agents Chemother.* **2020**, *64*. [CrossRef]

63. Shukla, S.; Sauna, Z.E.; Prasad, R.; Ambudkar, S.V. Disulfiram is a potent modulator of multidrug transporter Cdr1p of *Candida albicans*. *Biochem. Biophys. Res. Commun.* **2004**, *322*, 520–525. [CrossRef]

64. Sauna, Z.E.; Shukla, S.; Ambudkar, S.V. Disulfiram, an old drug with new potential therapeutic uses for human cancers and fungal infections. *Mol. Biosyst.* **2005**, *1*, 127–134. [CrossRef]

65. Rossato, L.; Camargo dos Santos, M.; Vitale, R.G.; de Hoog, S.; Ishida, K. Alternative treatment of fungal infections: Synergy with non-antifungal agents. *Mycoses* **2020**. [CrossRef]

66. Rossi, S.A.; De Oliveira, H.C.; Agreda-Mellon, D.; Lucio, J.; Soares Mendes-Giannini, M.J.; García-Cambero, J.P.; Zaragoza, O. Identification of off-patent drugs that show synergism with amphotericin B or that present antifungal action against *Cryptococcus neoformans* and *Candida spp.*. *Antimicrob. Agents Chemother.* **2020**, *64*. [CrossRef]

67. Te, H.S.; Randall, G.; Jensen, D.M. Mechanism of action of ribavirin in the treatment of chronic hepatitis C. *Gastroenterol. Hepatol.* **2007**, *3*, 218–225.

68. Yousfi, H.; Cassagne, C.; Ranque, S.; Rolain, J.M.; Bittar, F. Repurposing of ribavirin as an adjunct therapy against invasive *Candida* strains in an *in vitro* study. *Antimicrob. Agents Chemother.* **2019**, *63*. [CrossRef] [PubMed]

69. Swan, G.E. The pharmacology of halogenated salicylanilides and their anthelmintic use in animals. *J. S. Afr. Vet. Assoc.* **1999**, *70*, 61–70. [CrossRef]

70. Ayerbe-Algaba, R.; Gil-Marqués, M.L.; Miró-Canturri, A.; Parra-Millán, R.; Pachón-Ibáñez, M.E.; Jiménez-Mejías, M.E.; Pachón, J.; Smani, Y. The anthelmintic oxyclozanide restores the activity of colistin against colistin-resistant Gram-negative bacilli. *Int. J. Antimicrob. Agents* **2019**, *54*, 507–512. [CrossRef]

71. ANSES RCP—DOUVISTOME. Available online: http://www.ircp.anmv.anses.fr/rcp.aspx?NomMedicament=DOUVISTOME (accessed on 17 February 2021).

72. Rolain, J.M.; Colson, P.; Raoult, D. Recycling of chloroquine and its hydroxyl analogue to face bacterial, fungal and viral infections in the 21st century. *Int. J. Antimicrob. Agents* **2007**, *30*, 297–308. [CrossRef]

73. Plantone, D.; Koudriavtseva, T. Current and Future Use of Chloroquine and Hydroxychloroquine in Infectious, Immune, Neoplastic, and Neurological Diseases: A Mini-Review. *Clin. Drug Investig.* **2018**, *38*, 653–671. [CrossRef]

74. Shinde, R.B.; Rajput, S.B.; Raut, J.S.; Mohan Karuppayil, S. An in vitro repositioning study reveals antifungal potential of chloroquine to inhibit growth and morphogenesis in *Candida albicans*. *J. Gen. Appl. Microbiol.* **2013**, *59*, 167–170. [CrossRef]

75. Bai, R.Y.; Staedtke, V.; Wanjiku, T.; Rudek, M.A.; Joshi, A.; Gallia, G.L.; Riggins, G.J. Brain penetration and efficacy of different mebendazole polymorphs in a mouse brain tumor model. *Clin. Cancer Res.* **2015**, *21*, 3462–3470. [CrossRef] [PubMed]

76. Scott, E.M.; Tariq, V.N.; McCrory, R.M. Demonstration of synergy with fluconazole and either ibuprofen, sodium salicylate, or propylparaben against *Candida albicans in vitro*. *Antimicrob. Agents Chemother.* **1995**, *39*, 2610–2614. [CrossRef]

77. Torres, N.S.; Abercrombie, J.J.; Srinivasan, A.; Lopez-Ribot, J.L.; Ramasubramanian, A.K.; Leung, K.P. Screening a commercial library of pharmacologically active small molecules against *Staphylococcus aureus* biofilms. *Antimicrob. Agents Chemother.* **2016**, *60*, 5663–5672. [CrossRef]

78. She, P.; Liu, Y.; Wang, Y.; Tan, F.; Luo, Z.; Wu, Y. Antibiofilm efficacy of the gold compound auranofin on dual species biofilms of *Staphylococcus aureus* and *Candida* spp. *J. Appl. Microbiol.* **2020**, *128*, 88–101. [CrossRef]

79. Holbrook, S.Y.L.; Garzan, A.; Dennis, E.K.; Shrestha, S.K.; Garneau-Tsodikova, S. Repurposing antipsychotic drugs into antifungal agents: Synergistic combinations of azoles and bromperidol derivatives in the treatment of various fungal infections. *Eur. J. Med. Chem.* **2017**, *139*, 12–21. [CrossRef]

80. Rhein, J.; Nielsen, K.; Boulware, D.R.; Meya, D.B. Sertraline for HIV-associated cryptococcal meningitis—Authors' reply. *Lancet Infect. Dis.* **2016**, *16*, 1111–1112. [CrossRef]

81. Moraes, D.C.; Ferreira-Pereira, A. Insights on the anticandidal activity of non-antifungal drugs. *J. Mycol. Med.* **2019**, *29*, 253–259. [CrossRef] [PubMed]

82. da Rosa, T.F.; Machado, C.; de, S.; Serafin, M.B.; Bottega, A.; Foletto, V.S.; Coelho, S.S.; Hörner, R. Repositioning or Redirection of Antidepressant Drugs in the Treatment of Bacterial and Fungal Infections. *Am. J. Ther.* **2020**, *27*, e528–e532. [CrossRef]

83. Nayak, R.; Xu, J. Effects of sertraline hydrochloride and fluconazole combinations on *Cryptococcus neoformans* and *Cryptococcus gattii*. *Mycology* **2010**, *1*, 99–105. [CrossRef]

84. Rossato, L.; Loreto, É.S.; Zanette, R.A.; Chassot, F.; Santurio, J.M.; Alves, S.H. *In vitro* synergistic effects of chlorpromazine and sertraline in combination with amphotericin B against *Cryptococcus neoformans var. grubii*. *Folia Microbiol. (Praha)* **2016**, *61*, 399–403. [CrossRef]

85. Rhein, J.; Huppler Hullsiek, K.; Tugume, L.; Nuwagira, E.; Mpoza, E.; Evans, E.E.; Kiggundu, R.; Pastick, K.A.; Ssebambulidde, K.; Akampurira, A.; et al. Adjunctive sertraline for HIV-associated cryptococcal meningitis: A randomised, placebo-controlled, double-blind phase 3 trial. *Lancet Infect. Dis.* **2019**, *19*, 843–851. [CrossRef]

86. Boulware, D.R.; Nalintya, E.; Rajasingham, R.; Kirumira, P.; Naluyima, R.; Turya, F.; Namanda, S.; Rutakingirwa, M.K.; Skipper, C.P.; Nikweri, Y.; et al. Adjunctive sertraline for asymptomatic cryptococcal antigenemia: A randomized clinical trial. *Med. Mycol.* **2020**, *58*, 1037–1043. [CrossRef]

87. Homa, M.; Galgóczy, L.; Tóth, E.; Tóth, L.; Papp, T.; Chandrasekaran, M.; Kadaikunnan, S.; Alharbi, N.S.; Vágvölgyi, C. In vitro antifungal activity of antipsychotic drugs and their combinations with conventional antifungals against *Scedosporium* and *Pseudallescheria* isolates. *Med. Mycol.* **2015**, *53*, 890–895. [CrossRef]

88. Montoya, M.C.; Krysan, D.J. Repurposing estrogen receptor antagonists for the treatment of infectious disease. *mBio* **2018**, *9*. [CrossRef]

89. Butts, A.; Martin, J.A.; DiDone, L.; Bradley, E.K.; Mutz, M.; Krysan, D.J. Structure-Activity Relationships for the Antifungal Activity of Selective Estrogen Receptor Antagonists Related to Tamoxifen. *PLoS ONE* **2015**, *10*. [CrossRef]

90. Quezada, H.; Martínez-Vázquez, M.; López-Jácome, E.; González-Pedrajo, B.; Andrade, Á.; Fernández-Presas, A.M.; Tovar-García, A.; García-Contreras, R. Repurposed anti-cancer drugs: The future for anti-infective therapy? *Expert Rev. Anti Infect. Ther.* **2020**, *18*, 609–612. [CrossRef]

91. Ngan, N.T.T.; Mai, N.T.H.; Tung, N.L.N.; Lan, N.P.H.; Tai, L.T.H.; Phu, N.H.; Chau, N.V.V.; Binh, T.Q.; Hung, L.Q.; Beardsley, J.; et al. A randomized open label trial of tamoxifen combined with amphotericin B and fluconazole for cryptococcal meningitis. *Wellcome Open Res.* **2019**, *4*. [CrossRef]

92. de Oliveira Neto, A.S.; Souza, I.L.A.; Amorim, M.E.S.; de Freitas Souza, T.; Rocha, V.N.; do Couto, R.O.; Fabri, R.L.; de Freitas Araújo, M.G. Antifungal efficacy of atorvastatin-containing emulgel in the treatment of oral and vulvovaginal candidiasis. *Med. Mycol.* **2020**. [CrossRef]

93. Macreadie, I.G.; Johnson, G.; Schlosser, T.; Macreadie, P.I. Growth inhibition of *Candida* species and *Aspergillus fumigatus* by statins. *FEMS Microbiol. Lett.* **2006**, *262*, 9–13. [CrossRef]

94. de Queiroz Ribeiro, N.; Costa, M.C.; Magalhães, T.F.F.; Carneiro, H.C.S.; Oliveira, L.V.; Fontes, A.C.L.; Santos, J.R.A.; Ferreira, G.F.; de Sousa Araujo, G.R.; Alves, V.; et al. Atorvastatin as a promising anticryptococcal agent. *Int. J. Antimicrob. Agents* **2017**, *49*, 695–702. [CrossRef] [PubMed]

95. Khan, S.; Singhal, S.; Mathur, T.; Upadhyay, D.J.; Rattan, A. Antifungal potential of disulfiram. *Jpn. J. Med. Mycol.* **2007**, *48*, 109–113. [CrossRef] [PubMed]

96. Hao, W.; Qiao, D.; Han, Y.; Du, N.; Li, X.; Fan, Y.; Ge, X.; Zhang, H. Identification of disulfiram as a potential antifungal drug by screening small molecular libraries. *J. Infect. Chemother.* **2020**. [CrossRef]

97. Blanc, M.; Daeppen, J.-B. Does disulfiram still have a role in alcoholism treatment? *Rev. Med. Suisse* **2005**, *1*, 1728–1730.

98. Talbot, G.H.; Bradley, J.; Edwards, J.E.; Gilbert, D.; Scheid, M.; Bartlett, J.G. Bad bugs need drugs: An update on the development pipeline from the Antimicrobial Availability Task Force of the Infectious Diseases Society of America. *Clin. Infect. Dis.* **2006**, *42*, 657–668. [CrossRef]

99. Osei Sekyere, J. *Candida auris*: A systematic review and meta-analysis of current updates on an emerging multidrug-resistant pathogen. *Microbiologyopen* **2018**, *7*. [CrossRef]

100. Ramirez-Garcia, A.; Pellon, A.; Rementeria, A.; Buldain, I.; Barreto-Bergter, E.; Rollin-Pinheiro, R.; De Meirelles, J.V.; Xisto, M.I.D.S.; Ranque, S.; Havlicek, V.; et al. *Scedosporium* and *Lomentospora*: An updated overview of underrated opportunists. *Med. Mycol.* **2018**, *56*, S102–S125. [CrossRef]

101. Nobile, C.J.; Johnson, A.D. *Candida albicans* Biofilms and Human Disease. *Annu. Rev. Microbiol.* **2015**, *69*, 71–92. [CrossRef]

102. Kathwate, G.H.; Shinde, R.B.; Mohan Karuppayil, S. Non-antifungal drugs inhibit growth, morphogenesis and biofilm formation in *Candida albicans*. *J. Antibiot.* **2021**, 1–8. [CrossRef]

103. Kim, J.H.; Cheng, L.W.; Chan, K.L.; Tam, C.C.; Mahoney, N.; Friedman, M.; Shilman, M.M.; Land, K.M. Antifungal Drug Repurposing. *Antibiotics* **2020**, *9*, 812. [CrossRef] [PubMed]

104. Stepanović, S.; Vuković, D.; Ješić, M.; Ranin, L. Influence of acetylsalicylic acid (Aspirin) on biofilm production by *Candida* species. *J. Chemother.* **2004**, *16*, 134–138. [CrossRef] [PubMed]

105. Kulkarny, V.V.; Chavez-Dozal, A.; Rane, H.S.; Jahng, M.; Bernardo, S.M.; Parra, K.J.; Lee, S.A. Quinacrine inhibits *Candida albicans* growth and filamentation at neutral pH. *Antimicrob. Agents Chemother.* **2014**, *58*, 7501–7509. [CrossRef] [PubMed]

106. Holanda, M.A.V.; Da Silva, C.R.; De A Neto, J.B.; Do Av Sá, L.G.; Do Nascimento, F.B.S.A.; Barroso, D.D.; Da Silva, L.J.; Cândido, T.M.; Leitão, A.C.; Barbosa, A.D.; et al. Evaluation of the antifungal activity in vitro of midazolam against fluconazole-resistant *Candida* spp. isolates. *Future Microbiol.* **2021**, *16*, 71–81. [CrossRef]

107. Juvêncio da Silva, L.; Dias Barroso, F.D.; Vieira, L.S.; Carlos Mota, D.R.; da Silva Firmino, B.K.; Rocha da Silva, C.; de Farias Cabral, V.P.; Cândido, T.M.; Sá, L.G.D.A.V.; Barbosa da Silva, W.M.; et al. Diazepam's antifungal activity in fluconazole-resistant *Candida* spp. and biofilm inhibition in *C. albicans*: Evaluation of the relationship with the proteins ALS3 and SAP5. *J. Med. Microbiol.* **2021**. [CrossRef]

108. Chavez-Dozal, A.A.; Jahng, M.; Rane, H.S.; Asare, K.; Kulkarny, V.V.; Bernardo, S.M.; Lee, S.A. In vitro analysis of flufenamic acid activity against *Candida albicans* biofilms. *Int. J. Antimicrob. Agents* **2014**, *43*, 86–91. [CrossRef]

109. Do Amaral Valente Sá, L.G.; Da Silva, C.R.; De Andrade Neto, J.B.; Do Nascimento, F.B.S.A.; Barroso, F.D.D.; Da Silva, L.J.; De Farias Cabral, V.P.; Barbosa, A.D.; Silva, J.; Marinho, E.S.; et al. Antifungal activity of etomidate against growing biofilms of fluconazole-resistant *Candida* spp. strains, binding to mannoproteins and molecular docking with the ALS3 protein. *J. Med. Microbiol.* **2020**, *69*, 1221–1227. [CrossRef]

110. Chavez-Dozal, A.A.; Lown, L.; Jahng, M.; Walraven, C.J.; Lee, S.A. *In vitro* analysis of finasteride activity against Candida albicans urinary biofilm formation and filamentation. *Antimicrob. Agents Chemother.* **2014**, *58*, 5855–5862. [CrossRef]

111. Nile, C.; Falleni, M.; Cirasola, D.; Alghamdi, A.; Anderson, O.F.; Delaney, C.; Ramage, G.; Ottaviano, E.; Tosi, D.; Bulfamante, G.; et al. Repurposing Pilocarpine Hydrochloride for Treatment of *Candida albicans* Infections. *mSphere* **2019**, *4*. [CrossRef]

112. Machado, C.B.; da Silva, C.R.; Barroso, F.D.; Campos, R.D.S.; Sá, L.V.; do Nascimento, F.A.; Cavalcanti, B.C.; Júnior, H.V.N.; Neto, J.A. In vitro evaluation of anti-fungal activity of tropicamide against strains of *Candida* spp. resistant to fluconazole in planktonic and biofilm form. *J. Mycol. Med.* **2021**, *31*. [CrossRef]

113. Rajasekharan, S.K.; Lee, J.H.; Lee, J. Aripiprazole repurposed as an inhibitor of biofilm formation and sterol biosynthesis in multidrug-resistant Candida albicans. *Int. J. Antimicrob. Agents* **2019**, *54*, 518–523. [CrossRef] [PubMed]

114. Lone, S.A.; Ahmad, A. *Candida auris*—the growing menace to global health. *Mycoses* **2019**, *62*, 620–637. [CrossRef] [PubMed]

115. Lockhart, S.R. *Candida auris* and multidrug resistance: Defining the new normal. *Fungal Genet. Biol.* **2019**, *131*. [CrossRef]

116. Kean, R.; Ramage, G. Combined Antifungal Resistance and Biofilm Tolerance: The Global Threat of *Candida auris*. *mSphere* **2019**, *4*. [CrossRef] [PubMed]

117. Allaw, F.; Kara Zahreddine, N.; Ibrahim, A.; Tannous, J.; Taleb, H.; Bizri, A.R.; Dbaibo, G.; Kanj, S.S. First *Candida auris* Outbreak during a COVID-19 Pandemic in a Tertiary-Care Center in Lebanon. *Pathogens* **2021**, *10*, 157. [CrossRef] [PubMed]

118. Cheng, Y.-S.; Roma, J.S.; Shen, M.; Fernandes, C.M.; Tsang, P.S.; Forbes, H.E.; Boshoff, H.; Lazzarini, C.; Del Poeta, M.; Zheng, W.; et al. Identification of Antifungal Compounds against Multidrug Resistant *Candida auris* Utilizing a High Throughput Drug Repurposing Screen. *Antimicrob. Agents Chemother.* **2021**. [CrossRef]

119. Eldesouky, H.E.; Lanman, N.A.; Hazbun, T.R.; Seleem, M.N. Aprepitant, an antiemetic agent, interferes with metal ion homeostasis of *Candida auris* and displays potent synergistic interactions with azole drugs. *Virulence* **2020**, *11*, 1466–1481. [CrossRef]

120. Eldesouky, H.E.; Salama, E.A.; Lanman, N.A.; Hazbun, T.R.; Seleem, M.N. Potent synergistic interactions between lopinavir and azole antifungal drugs against emerging multidrug-resistant *Candida auris*. *Antimicrob. Agents Chemother.* **2021**, *65*. [CrossRef]

121. Gowri, M.; Jayashree, B.; Jeyakanthan, J.; Girija, E.K. Sertraline as a promising antifungal agent: Inhibition of growth and biofilm of Candida auris with special focus on the mechanism of action in vitro. *J. Appl. Microbiol.* **2020**, *128*, 426–437. [CrossRef]

122. Kim, J.H.; Chan, K.L.; Cheng, L.W.; Tell, L.A.; Byrne, B.A.; Clothier, K.; Land, K.M. High efficiency drug repurposing design for new antifungal agents. *Methods Protoc.* **2019**, *2*, 31. [CrossRef]

123. Vermorel-Faure, O.; Lebeau, B.; Mallaret, M.R.; Michallet, M.; Brut, A.; Ambroise-Thomas, P.; Grillot, R. Food-related fungal infection risk in agranulocytosis. Mycological control of 273 food items offered to patients hospitalized in sterile units. *Press. Med.* **1993**, *22*, 157–160.

124. Meis, J.F.; Chowdhary, A.; Rhodes, J.L.; Fisher, M.C.; Verweij, P.E. Clinical implications of globally emerging azole resistance in Aspergillus fumigatus. *Philos. Trans. R. Soc. B Biol. Sci.* **2016**, *371*. [CrossRef] [PubMed]

125. De Lucca, A.J. Harmful fungi in both agriculture and medicine. *Rev. Iberoam. Micol.* **2007**, *24*, 3–13. [CrossRef]

126. Perlin, D.S.; Rautemaa-Richardson, R.; Alastruey-Izquierdo, A. The global problem of antifungal resistance: Prevalence, mechanisms, and management. *Lancet Infect. Dis.* **2017**, *17*, e383–e392. [CrossRef]

127. Osherov, N.; Kontoyiannis, D.P. The anti- *Aspergillus* drug pipeline: Is the glass half full or empty? *Med. Mycol.* **2017**, *55*, 118–124. [CrossRef] [PubMed]

128. Vallières, C.; Singh, N.; Alexander, C.; Avery, S.V. Repurposing Nonantifungal Approved Drugs for Synergistic Targeting of Fungal Pathogens. *ACS Infect. Dis.* **2020**, *6*, 2950–2958. [CrossRef] [PubMed]

129. May, R.C.; Stone, N.R.H.; Wiesner, D.L.; Bicanic, T.; Nielsen, K. *Cryptococcus*: From environmental saprophyte to global pathogen. *Nat. Rev. Microbiol.* **2016**, *14*, 106–117. [CrossRef]

130. Spadari, C.D.C.; Wirth, F.; Lopes, L.B.; Ishida, K. New approaches for cryptococcosis treatment. *Microorganisms* **2020**, *8*, 613. [CrossRef]

131. Truong, M.; Monahan, L.G.; Carter, D.A.; Charles, I.G. Repurposing drugs to fast-track therapeutic agents for the treatment of cryptococcosis. *PeerJ* **2018**, *2018*. [CrossRef]

132. Tortorano, A.M.; Richardson, M.; Roilides, E.; van Diepeningen, A.; Caira, M.; Munoz, P.; Johnson, E.; Meletiadis, J.; Pana, Z.D.; Lackner, M.; et al. ESCMID and ECMM joint guidelines on diagnosis and management of hyalohyphomycosis: Fusarium spp., Scedosporium spp. and others. *Clin. Microbiol. Infect.* **2014**, *20*, 27–46. [CrossRef]

133. Douglas, A.P.; Chen, C.-A.; Slavin, M.A. Emerging infections caused by non-*Aspergillus* filamentous fungi. *Clin. Microbiol. Infect.* **2016**, *22*, 670–680. [CrossRef] [PubMed]

134. Shinohara, M.M.; George, E. *Scedosporium apiospermum*: An emerging opportunistic pathogen that must be distinguished from *Aspergillus* and other hyalohyphomycetes. *J. Cutan. Pathol.* **2009**, *36*, 39–41. [CrossRef] [PubMed]

135. Yousfi, H.; Ranque, S.; Cassagne, C.; Rolain, J.M.; Bittar, F. Identification of repositionable drugs with novel antimycotic activity by screening the Prestwick Chemical Library against emerging invasive moulds. *J. Glob. Antimicrob. Resist.* **2020**, *21*, 314–317. [CrossRef] [PubMed]

136. Torres, N.S.; Montelongo-Jauregui, D.; Abercrombie, J.J.; Srinivasan, A.; Lopez-Ribot, J.L.; Ramasubramanian, A.K.; Leung, K.P. Antimicrobial and antibiofilm activity of synergistic combinations of a commercially available small compound library with colistin against *Pseudomonas aeruginosa*. *Front. Microbiol.* **2018**, *9*. [CrossRef] [PubMed]

137. Dovizio, M.; Bruno, A.; Tacconelli, S.; Patrignani, P. Mode of action of aspirin as a chemopreventive agent. *Recent Results Cancer Res.* **2013**, *191*, 39–65. [CrossRef] [PubMed]

138. Bjorkman, S.; Ove Elisson, L.; Gabrielsson, J. Pharmacokinetics of Quinacrine after Intrapleural Instillation in Rabbits and Man. *J. Pharm. Pharmacol.* **1989**, *41*, 160–163. [CrossRef]
139. Steinfort, D.P.; Steinfort, C. Effect of long-term nebulized colistin on lung function and quality of life in patients with chronic bronchial sepsis. *Intern. Med. J.* **2007**, *37*, 495–498. [CrossRef]
140. Rolain, J.M.; Baquero, F. The refusal of the Society to accept antibiotic toxicity: Missing opportunities for therapy of severe infections. *Clin. Microbiol. Infect.* **2016**, *22*, 423–427. [CrossRef]

pharmaceuticals

Review

Repurposing Antifungals for Host-Directed Antiviral Therapy?

Sebastian Schloer [1,2,*], Jonas Goretzko [1] and Ursula Rescher [1,*]

[1] Institute-Associated Research Group "Regulatory Mechanisms of Inflammation",
    Institute of Medical Biochemistry, Center for Molecular Biology of Inflammation,
    and "Cells in Motion" Interfaculty Centre, University of Muenster, Von-Esmarch-Str. 56,
    D-48149 Muenster, Germany; j_gore01@uni-muenster.de
[2] Leibniz Institute for Experimental Virology, D-20251 Hamburg, Germany
*   Correspondence: SebastianMaximilian.Schloer@ukmuenster.de (S.S.); rescher@uni-muenster.de (U.R.);
    Tel.: +49-2518352118 (S.S. & U.R.)

Abstract: Because of their epidemic and pandemic potential, emerging viruses are a major threat to global healthcare systems. While vaccination is in general a straightforward approach to prevent viral infections, immunization can also cause escape mutants that hide from immune cell and antibody detection. Thus, other approaches than immunization are critical for the management and control of viral infections. Viruses are prone to mutations leading to the rapid emergence of resistant strains upon treatment with direct antivirals. In contrast to the direct interference with pathogen components, host-directed therapies aim to target host factors that are essential for the pathogenic replication cycle or to improve the host defense mechanisms, thus circumventing resistance. These relatively new approaches are often based on the repurposing of drugs which are already licensed for the treatment of other unrelated diseases. Here, we summarize what is known about the mechanisms and modes of action for a potential use of antifungals as repurposed host-directed anti-infectives for the therapeutic intervention to control viral infections.

Keywords: antifungals; host-directed drug therapy; drug repurposing; azoles; polyenes; echinocandins; viral infections

Citation: Schloer, S.; Goretzko, J.; Rescher, U. Repurposing Antifungals for Host-Directed Antiviral Therapy? *Pharmaceuticals* 2022, 15, 212. https://doi.org/10.3390/ph15020212

Academic Editors: Jong Heon Kim, Luisa W. Cheng, Kirkwood Land and Daniela De Vita

Received: 16 December 2021
Accepted: 8 February 2022
Published: 10 February 2022

Publisher's Note: MDPI stays neutral with regard to jurisdictional claims in published maps and institutional affiliations.

1. Introduction

Throughout history, the world's population has been impacted by virus outbreaks, epidemics and pandemics. The recently emerged severe acute respiratory syndrome coronavirus 2 (SARS-CoV-2) and the resulting coronavirus disease 2019 (COVID-19) have impressively shown how quickly the spread of a novel virus becomes a major threat for public health worldwide [1]. The ongoing pandemic with high mortality and morbidity rates highlights the urgent need for safe and fast development of new pharmaceutics to combat newly emerging viral infections [2,3]. While the ultrafast development of vaccines to fight SARS-CoV-2 has been an unprecedented success story, the high mutation rates leading to rapidly changing viral genomes also require a continuous update of vaccines [4–6]. Moreover, antigenic shift, i.e., the recombination of viral genomes and the appearance of new viral subtypes, is a great public health concern [5,6]. Thus, antiviral drugs are essential for the management and control of viral infections to fill the gap until virus-specific routine immunization strategies are available.

While drugs that directly act on viral components (direct antivirals) usually offer safe and effective treatment options, they also come with the risk of emerging resistances, as seen by the tremendously fast and effective adaptation of influenza A virus (IAV) to the viral neuraminidase inhibitor oseltamivir [7,8]. Although viruses come in all shapes and sizes and have confusingly diverse replication cycles, the common denominator is the absolute dependence on a host cell for their propagation. As a consequence, all viruses exploit fundamental cellular processes to gain access to the cellular replication and transport machinery for the biosynthesis of the viral genome, virus proteins, and the assembly and release of virions. The common basic steps in infection cycles of viruses are shown in Figure 1. Notably, virus replication depends on (i) the interaction with cellular membranes (during virus internalization, assembly, release), (ii) translation and modification of viral proteins, and (iii) manipulation of cellular signaling pathways to suppress virus detection and destruction and promote viral assembly and release. Drugs that target these cellular functions will most likely also affect the propagation of a whole range of otherwise unrelated viruses. Thus, such host-directed therapeutic strategies could be advantageous as new antiviral approaches, especially for the threat posed by new emerging viruses [9]. Repurposing clinically approved drugs that have been developed for other indications but also target such cellular processes provides a particularly attractive strategy to move such treatments into the clinic faster and safer, as these drugs are already in use. Indeed, this repurposing strategy is currently being increasingly pursued in the fight against COVID-19 [10]. Here, we present an overview of the commonly used antifungals and review what is known about their antiviral potential and the putative molecular targets and mechanisms/modes of action in mammalian cells.

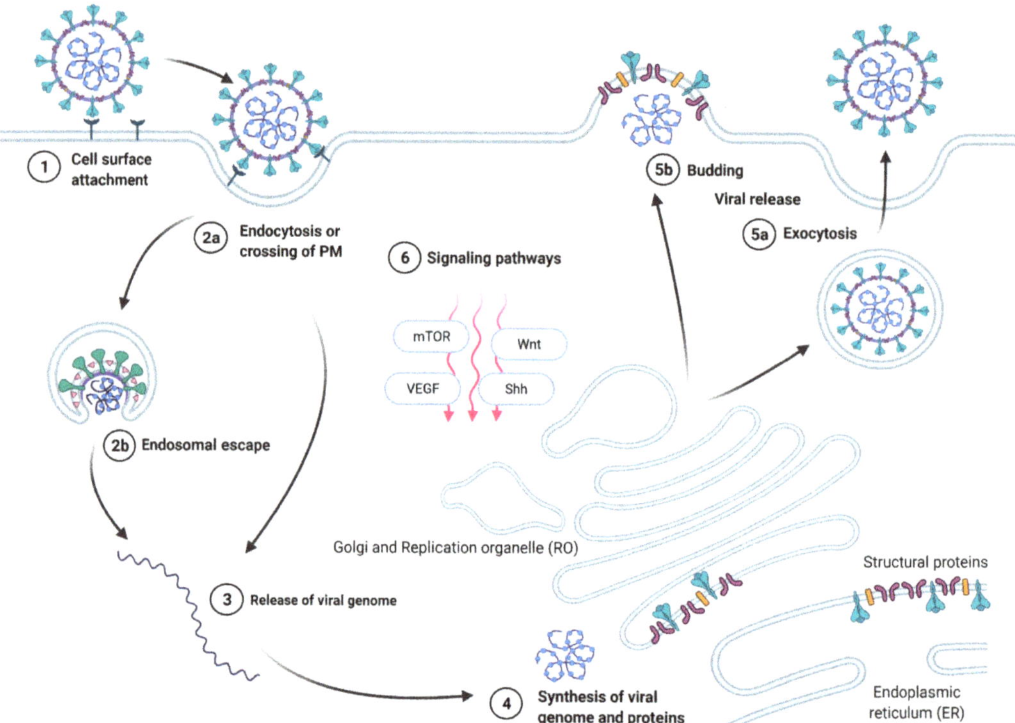

Figure 1. Basic steps of the replication cycle of viruses in mammalian cells. Upon attachment of the virus to receptors on the host cell surface (1), viruses cross cellular membranes to gain access to the host cell, either by penetration (for nonenveloped viruses) of or fusion (for enveloped viruses) with the plasma membrane (2a) or endosomes (2b), and the viral genome is released into the host cell (3). Viral replication depends on the synthesis of viral components using the existing or modified host cell organelles (4) and the release of the newly assembled virions from the host cell via either exocytosis of virion-containing vesicles (5a) or budding (5b). To support the viral replication, viruses modulate cellular signaling pathways, such as the Wnt, Shh, VEGF, and mTORC1 signal transduction

pathways (6). Repurposed drugs might target (i) viral interaction with cellular membranes (during virus internalization, assembly, release), (ii) translation and modification of viral proteins, and (iii) cellular signaling pathways hijacked by the virus to suppress virus detection and destruction and promote viral assembly and release. Adapted from "Coronavirus Replication Cycle", by BioRender.com (2020). Retrieved from https://app.biorender.com/biorender-templates (accessed date on 15 December 2021).

2. Antifungal Drugs

The synthetic small molecule antifungals are classified according to their targets and mechanisms of action, and comprise four major classes, the polyenes, pyrimidine analogs (flucytosine), azoles, and echinocandins (Table 1).

Table 1. Overview of antifungal drugs, their mechanism of action, their clinical use, and a potential use as repurposed antivirals.

Antifungal Drug Family	Mechanism of Action	Clinical Use as Antifungals	Antiviral Potential
Polyenes	Bind sterol components and form pores, resulting in a compromised fungal plasma membrane [11].	Aspergillosis, cryptococcosis, candidiasis, zygomycosis, fusariosis, coccidioidomycosis, paracoccidioidomycosis, histoplasmosis, blastomycosis, mucormycosis, penicilliosis, and phaeohyphomycosis [11].	Japanese encephalitis virus [12], herpes simplex virus (HSV) [13], human immunodeficiency virus (HIV) [14], rubella virus [12], vesicular stomatitis virus (VSV) [15]
Flucytosine	Interferes with fungal nucleic acid synthesis [16].	Candidiasis, and cryptococcosis [16].	Not known
Echinocandins	Inhibit the fungal enzyme β1,3-glucan synthase, leading to incomplete fungal cell wall formation [17].	Aspergillosis, and candidiasis [17].	Chikungunya virus (CHIKV) [18], enteroviruses [19], dengue virus [20], SARS-CoV-2 [21], Sindbis virus (SINV) and Semliki Forest virus (SFV) [18]
Azoles	Primarily inhibit the fungal sterol biosynthesis, leading to compromised fungal membranes [22].	Aspergillosis, candidiasis, and cryptococcosis [22].	SARS-CoV-2 [23,24], influenza virus [25], Ebola virus [26–28], Parechovirus A3 [29], dengue virus [30], enteroviruses [31], human cytomegalovirus [32]

2.1. Polyenes—Disruptors of Fungal Membrane

Polyenes exert their antifungal activity by binding ergosterol in the fungal cell membrane. The resulting membrane disintegration by pore formation increases the permeability, and the membrane leakage leads to the subsequent death of the fungal cell. A well-known, highly effective polyene is amphotericin B, which is widely used in the clinics against invasive fungal infections. Amphotericin B was shown to also induce oxidative stress to fungal cells and modulate the immune system [33].

A potential antiviral capacity of amphotericin B has been reported against a variety of viruses, including human immunodeficiency virus (HIV) [14], Japanese encephalitis virus (JEV) [12], herpes simplex virus (HSV) [13], rubella virus [12] and vesicular stomatitis virus (VSV) [15]. Amphotericin B inhibits the infectivity of JEV in a concentration-dependent manner up to 200-fold at a postinfection step likely at viral replication and/or synthesis of viral proteins without affecting virus adsorption to host cell surfaces [12].

In vitro results highlight the possibility of using amphotericin B against HIV infections, providing a dual effect, against the virus itself and the opportunistic fungal infections that often accompany HIV infections due to the patients' compromised immune system [14]. Furthermore, amphotericin B potentiated the antiviral efficacy of acyclovir against pseudorabies virus (PRV) without a direct effect on PRV replication in the absence of acyclovir [34]. The mechanism of action remains unclear; however, a recent high-throughput virtual screening approach showed amphotericin B, among other antiparasitic drugs, to possess potential inhibitory features against 10 SARS-CoV-2 molecular targets including the RNA-dependent RNA polymerase [35].

2.2. Flucytosine—A Selective Inhibitor of Fungal Nuclear Acid Synthesis

Flucytosine as such has no antifungal capacity but is converted by the fungal enzyme cytosine deaminase (which is not present in mammalian cells) into 5-fluorouracil, which is further metabolized. The incorporation of 5-fluorouracil and its metabolites into fungal DNA and RNA then causes aberrant fungal RNA and DNA synthesis [36,37]. To our knowledge, there are no studies reporting antiviral properties of flucytosine yet.

2.3. Echinocandins—Noncompetitive Inhibitors of (1,3)-β-D-Glucan Synthase

Echinocandins attack the fungal cell wall by inhibiting the (1,3)-β-D-glucan synthesis and thereby triggering osmotic stress and subsequent cell lysis. They are fungicidal against molds and yeast (most *Candida* species) and are generally considered well-tolerable due to little adverse effects and drug–drug interactions. Therefore, they are the preferred treatment option for invasive candidiasis [38,39]. Micafungin is an FDA-approved echinocandin that exhibits broad antifungal activity against a variety of *Candida* species. Of note, a potential antiviral capacity against chikungunya virus (CHIKV) [18], enteroviruses [19] and dengue virus [20], among others, has been reported in recent years. Mosquito-borne CHIKV belongs to the alphaviruses and is a global health problem. Micafungin was able to attenuate the cytopathic effects of CHIKV, reduce viral replication, release and spread, and impair viral stability. Micafungin also had antiviral effects against the alphaviruses Sindbis virus (SINV) and Semliki Forest virus (SFV) [18]. Enterovirus 71 (EV71), the major causative agent of hand-foot-and-mouth disease (HFMD), was shown sensitive to micafungin treatment as well. Micafungin effectively diminished EV71 proliferation and replication already at a micromolar dose [19]. Against other enteroviruses like Coxsackievirus group B type 3 (CVB3) and human rhinovirus (HRV), its antiviral capacity was only moderate. The authors proposed a virion-independent mechanism of action targeting intracellular processes such as translation, polyprotein processing or replication [19]. Recently, micafungin and its analogs, caspofungin and anidulafungin, were suggested for treatment of dengue virus (DENV) infection. In this case, the mechanism of action depends on the direct binding to the envelope protein DENV-2, thereby destabilizing and destroying the virion [20]. Of note, recent in silico studies argued for a binding of the echinocandins micafungin and pneumocandin B0 to the 3C-like protease (3CLpro) of porcine epidemic diarrhea virus (PEDV), as well as to the severe acute respiratory syndrome coronavirus 2 (SARS-CoV-2) main protease (Mpro), with micafungin having a higher calculated binding affinity towards 3CLpro and pneumocandin B0 binding preferably to Mpro, warranting further investigation into the use of echinocandin-type antifungal drugs as antiviral agents that act via binding to defined viral molecular targets [21].

2.4. Azoles—Inhibitors of Ergosterol Biosynthesis

The antifungal azoles are classified by the numbers of nitrogen atoms in the azole ring and include imidazoles (e.g., ketoconazole, miconazole, and clotrimazole) and triazoles (e.g., itraconazole, posaconazole, and fluconazole) [40]. Their efficacy and relative safety have led to their widespread clinical use for the antifungal therapy against aspergillosis, candidiasis, and cryptococcosis [41]. While the clinical use of imidazoles (the only exception is ketoconazole) is limited to the treatment of superficial mycoses, triazoles are used for superficial and systemic fungal infections [42]. Azoles exert their antifungal activities through multiple modes of action. By inhibiting the two cytochrome P450 enzymes that function in the ergosterol biosynthesis (CYP51, lanosterol 14α-demethylase) and the conversion of ergostatrienol into ergostatetraenol, (CYP61, 22-desaturase), azoles lead to a depletion of ergosterol. The resulting accumulation of toxic sterol precursors impairs membrane fluidity, asymmetry, and membrane integrity in fungal cells [43].

Interestingly, some of the azoles (itraconazole, posaconazole, voriconazole, and ketoconazole) also impact the mammalian cholesterol metabolism at higher concentrations [44]. Itraconazole and posaconazole influence cellular cholesterol levels by impairing different pivotal steps in the cholesterol homeostasis. Both antifungals include (i) the inhibition of the lanosterol 14α-demethylase and thereby the homeostasis and de novo synthesis of cholesterol [45,46], (ii) the inhibition of the cholesterol-transferring membrane protein Niemann–Pick C1 (NPC1) that results in the accumulation of cholesterol in the endolysosomal system [47], and they further (iii) interact with the oxysterol-binding protein (OSBP), which blocks the shuttling of cholesterol and phosphatidylinositol-4-phosphate (PI4P) between membranes [31]. In the following we will exemplarily discuss the interaction of itraconazole with cellular proteins.

2.4.1. Itraconazole Directly Interacts with the Endolysosomal Cholesterol Transporter NPC1

Niemann–Pick disease, type C1 (NPC1) protein NPC1 was first identified and characterized as a membrane protein that when mutated causes Niemann–Pick disease, type C1, a rare autosomal neurovisceral lipid storage disorder [48]. NPC1 protein is an endolysosomal integral membrane protein and mediates endolysosomal cholesterol transport [49]. A dysfunctional NPC1 protein, which is found mutated in 95% of the NPC patients, disturbs the intracellular lipid transport, leading to the excessive accumulation of lipid products including cholesterol in the endolysosomal compartment [48]. Interestingly, a possible function of NPC1 as a drug target in antiviral strategies has been explored in several recent publications [50,51]. The Ebola virus entry has been shown to directly rely on NPC1 function via binding of the Ebola virus glycoprotein (GP) to NPC1 [52]. The Ebola virus GP is cleaved by endosomal proteases to unmask the NPC1 binding site, and GP–NPC1 engagement within lysosomes promotes the viral escape into the host cytoplasm [53]. Consistent with this vital dependence of the Ebola virus replication on NPC1 protein, cells lacking NPC1 are nonpermissive for the virus entry and NPC1 knockout mice are protected from lethal Ebola virus infection [52]. Blocking NPC1 has also been reported to cause accumulation of the HIV-1 viral gag protein in the endolysosomal compartment [54,55], resulting in a profound suppression of virion release [55]. The finding that the HIV-1 accessory protein Nef induces host cell genes involved in cholesterol biosynthesis and homeostasis [56] emphasizes the strong dependency of HIV-1 on the host cell cholesterol levels, suggesting that NPC1 is a candidate drug target in the treatment of HIV-1 infections. Interestingly, NPC1 also emerged as a candidate drug target for other enveloped viruses, namely IAV and SARS-CoV-2. Viral replication rates were decreased in cells in which NPC1 was functionally blocked, and the increased endolysosomal cholesterol levels were suggested to interfere with the proper insertion of the fusogenic IAV hemagglutinin domains and the SARS-CoV-2 spike protein, thus affecting virus uncoating [23–25].

In many of these and other in vitro studies exploring the importance of NPC1 in diverse cellular functions, the cell-permeable hydrophobic polyamine U18666A, a small-molecule NPC1 inhibitor, was used [57,58]. However, the substantial toxicity of this compound limits a clinical use [59,60]. Of note, itraconazole has been shown to also directly bind and inhibit NPC1 [47], and thus might serve as an attractive candidate for NPC1 targeting strategies via drug repurposing. In favor of this notion, itraconazole-treated cells generated lesser IAV and Ebola virus progeny [25,28], and a beneficial treatment outcome was indeed confirmed in a mouse IAV infection model in vivo [25]. Itraconazole-mediated induction of type I interferons (IFNs), which is considered a fundamental step in establishing antiviral immunity, might also contribute to the observed antiviral effects [25]. While itraconazole also proved its antiviral potential in a 3D cell culture model for SARS-CoV-2 infection [23], an antiviral effect was not seen in the hamster infection model [61].

2.4.2. Itraconazole Interferes with OSBP and OSBP-Related Proteins (ORP) Functionality

Azoles also impair cellular lipid metabolism via an inhibitory effect on oxysterol-binding protein 1 (OSBP) and on other proteins that belong to the OSBP-related proteins (ORP) family, and this property might add to their antiviral use. OSBP was first identified as an intracellular protein that binds cytosolic 25-hydroxycholesterol [62]. Beside its capacity to bind 25-hydroxycholesterol, OSBP orchestrates the formation of endoplasmic reticulum (ER)–Golgi complex membrane contact sites and thereby shuttles sterols into the Golgi and phosphatidylinositol-4-phosphate (PI4P) back to the ER [63]. OSBP and the family of OSBP-related proteins (ORP) share a lipid-binding domain that binds either a sterol or a nonsterol ligand as well as a PI4P-binding N-terminal pleckstrin homology (PH) domain [64,65]. Another binding motif found in many ORPs, including OSBP, is the FFAT-motif which interacts with the ER-resident VAMP-associated proteins (VAP) receptors [64,65]. Both motifs are involved in shaping the ER–Golgi or, in the case of some viruses, the ER–replication organelle (RO) contact sites and are considerably engaged in lipid transport through different organelles.

Enterovirus, dengue virus, and hepatitis C virus replication are reported to depend on ORP and OSBP [30,31]. Pharmacologic inhibition, siRNA knockdown, and rescue of replication by overexpression have demonstrated the importance of ORPs and OSBP for enterovirus replication and propagation [31]. The virus-induced accumulation of PI4P lipids drives the recruitment of OSBP to these contact sides, and the OSBP-mediated transport of cholesterol and PI4P is pivotal for the formation and functionality of the enterovirus RO [31,66–68]. In line, OSBP knockdown and treatment with 25-hydroxycholesterol, an inhibitor of the cholesterol-PI4P exchange, negatively affects virus replication [69]; however, the precise molecular mechanism remains unclear. Itraconazole directly binds OSBP [31], leading to increased PI4P levels at the Golgi (in uninfected cells) or the RO (in infected cells), while the accumulation of cholesterol at the RO is blocked [31].

2.4.3. Targeting mTOR Signaling via Itraconazole

Itraconazole not only impairs lipid homeostasis, but also affects different signaling pathways, including mammalian target of rapamycin (mTOR), hedgehog, and Wnt signaling pathways that are hijacked by a broad range of viruses to drive the production of infectious particles.

The mammalian target of rapamycin (mTOR) signaling cascade is a pivotal signaling pathway that regulates apoptosis and counteracts stress-induced autophagy (such as, e.g., that elicited by viruses). Although different cellular locations for mTOR complex 1 (mTORC1) and mTORC2 have been reported, mTORC1 lysosomal localization appears critical for its ability to sense and respond to cell starvation [70]. Cholesterol was recently identified to promote the recruitment of mTORC1 to the lysosomal membrane [71] and the mTOR signaling cascade is regulated in a cholesterol-dependent manner [72].

Several viruses have evolved strategies to subvert the mTORC1 signaling network to drive their replication and propagation [73–81]. The Semliki Forest virus (SFV), Sindbis virus (SINV), and Chikungunya virus (CHIKV), members of the alphavirus family, cause different diseases but have in common that they encode nonstructural proteins (nsP) [82]. The activation of the PI3K/Akt/mTOR pathway is mediated through the phosphorylated and membrane-attached protein nsP3, which forms the viral replication complex upon virus internalization [83–85]. The activation of mTOR signaling is also fundamental for infection with the *Flaviviridae* West Nile virus (WNV), Japanese encephalitis virus (JEV) and dengue virus (DENV) [86–88]. *Flaviviridae* infection increases mTOR activity through a PI3K-dependent mechanism to maintain translation of its positive-sense RNA genome and also delays WNV-induced apoptosis [76,89,90]. The hepatitis C virus, another *Flaviviridae* member, increases phosphorylation of mTOR through the nonstructural protein 5A (NS5A) [91]. NS5A seems to activate PI3K/Akt signaling by directly binding PI3K [92,93]. The activation of the mTORC1 pathway by HCV has been linked to antiapoptotic signals that ensure cell survival and maintain persistence by promoting steady-state levels of virus replication [94,95]. Among the β-herpesvirus, the human cytomegalovirus (HCMV) maintains mTORC1 activation [96,97] through the expression of the two HCMV immediate early proteins, IEP72 and IEP86 [98]. Another persistent virus that tightly regulates mTOR signaling pathways is the human immunodeficiency virus type 1 (HIV-1). In dendritic cells, the HIV-1 envelope glycoprotein activates mTOR to prevent autophagy and to increase virus infection. Pharmacological treatment with rapamycin decreased viral spreading [99]. Another study implied that the HIV-1 protein Nef initiates mTOR activation which can be blocked by inhibitors of mTOR or PI3K [100,101], suggesting that drugs that modify the mTORC1 signaling pathway could act as anti-HIV-1 agents [102,103].

Well-balanced mTOR signaling is vitally important for IAV infection [104,105]. Thus, the pharmacological inhibition of the mTOR signaling axis might serve as potential antiviral target. In contrast to other azoles, itraconazole additionally inhibits mTOR signaling through affecting the upstream 5′-AMP-dependent protein kinase (AMPK) [106], which is activated upon an increased AMP/ATP ratio and serves as a regulator of cellular energy levels [107]. Once activated, AMPK inhibits mTOR signaling [108]. The activation of AMPK through itraconazole is a result of direct binding and inhibition of the mitochondrial Voltage-Dependent Anion Channel 1 (VDAC1), a critical regulator of mitochondrial metabolism, resulting in a drop in cellular energy levels [27]. Itraconazole treatment also impairs vascular endothelial growth factor receptor 2 (VEGFR2) functionality in endothelial cells, which is mostly due to altered VEGFR2 glycosylation, trafficking, and signaling [109]. As some viruses like human papillomaviruses or hepatitis viruses promote angiogenesis to facilitate optimal supply by nutrients [110–112], this might be an additional beneficial effect of itraconazole in antiviral strategy.

2.4.4. Itraconazole, a Modulator of Hedgehog Signaling

Another signaling axis that is hijacked by viruses to promote their own replication and spreading is the hedgehog (Hh) signaling pathway. Some viruses, e.g., influenza viruses, interfere with the expression of hedgehog by directly modulating the specific activity of the transcriptional effector, glioma-associated oncogene homolog (GLI) [113]. The GLI family of zinc-finger transcription factors and Smoothened (Smo) are the signal transducers of the Sonic hedgehog (Shh) pathway [114]. Shh is secreted from cells and binds to the Patched 1 (Ptch1) receptor, which in the unbound state inhibits the activity of the transmembrane protein Smo [114]. Smo activation and GLI1 nuclear relocation drive the expression of genes involved in proliferation and apoptosis [114]. A recent study showed that the IAV nonstructural protein 1 (NS1) alters the expression of Hh target genes by directly modulating the specific activity of the transcriptional effector GLI [113]. Smelkinson et al. identified a point mutation (A122V) in the NS1 protein, which led to significantly accelerated lethality when incorporated into a mouse-adapted influenza-A virus [113].

Hh signaling is also associated with hepatitis B virus (HBV) and HCV infection. Patients suffering from chronic HBV and HCV infection display increased hepatocyte production of Hh ligands [115]. Hh pathway activation often occurs as a response to fibrogenic repair of liver damage due to chronic viral hepatitis [115,116]. The HBV protein HBx stimulates the GLI activation through protein stabilization and nuclear localization in liver cancer cells while the exact mechanism is not fully understood [117]. These data clearly showed the importance of Hh signaling for the outcome of viral infections.

Earlier studies proposed an inhibitory effect of itraconazole on Hh signaling by direct action on Smo [118,119]; however, recent research showed that itraconazole inhibits the expression of Shh and GLI1 proteins without affecting the expression of Ptch1 and Smo [120]. In the case of viral infection, itraconazole could subvert the ability of the virus to increase the host cell permissiveness for viral replication.

2.4.5. Itraconazole and Its Inhibitory Effect on Wnt Signaling

The ancient and evolutionary-conserved Wnt signaling network contains two arms, the β-catenin-dependent, and the β-catenin-independent pathway [121]. While in the "off" state, β-catenin is degraded in the proteasome; the "on" state is initiated by binding of Wnt to a receptor called Frizzled (FZD) [122], leading to increased cytosolic β-catenin levels, and β-catenin translocation into the nucleus, where it orchestrates, together with other transcription factors, the expression of genes involved in differentiation and proliferation [121]. Depending on the receptor and ligand combinations, Wnt signaling can also activate signaling pathways independently of β-catenin, e.g., the calcium-dependent activation of PKC and Ca^{2+}/calmodulin-dependent protein kinases II (Ca^{2+}/CAMKII), resulting in changes in cell adhesion.

Viruses intervene with Wnt signaling by either epigenetic modification of Wnt gene expression or through interaction with specific Wnt pathway members, often resulting in the nuclear translocation of β-catenin and activation of Wnt signaling [122]. Both arms of Wnt signaling modulate the expression of genes that are required for the maintenance of viral pathogenesis such as for adenovirus and coxsackievirus B3 [122]. Furthermore, Wnt signaling is also involved in viruses-induced cancer development [123–127]. Given that diverse viruses affect this signaling pathway, Wnt signaling might be modulated in an antiviral manner. Indeed, itraconazole was found to negatively affect the expression of Wnt growth factor protein Wnt3A and to downregulate β-catenin, while increasing the levels of the endogenous Wnt inhibitor Axin-1 [128,129].

3. Conclusions

In most of the cases, the reported antiviral activities of the different antifungal agents are rather descriptive, thus it is not yet possible to formulate general principles of their antiviral activities (except for the azoles, which seem to act mostly via interference with the host cell cholesterol homeostasis, including the direct inhibitory binding to the endolysosomal cholesterol transporter NPC1). For most of the antifungals, the precise molecular targets, a prerequisite to an understanding of a drug's mechanism(s) of action (MOA), remain to be uncovered. Moreover, an antifungal's antiviral activity might most probably be associated with several host cell targets (see, for example, itraconazole), or might affect viral components (see, for example, micafungin). In this regard, the above-listed reports can only serve as starting points to accelerate the ultimate goal of developing tailored pharmaceuticals that recognize and interact with the respective molecular structure in the desired manner. However, the fact that a drug MOA is unknown or unclear does not mean that the drug holds no therapeutic potential. Rather, we would like to draw the readers' attention to their antiviral modes of action (MoA), i.e., the observed lowered levels of viral replication, for the following reasons:

Repurposing clinically licensed drugs with well-known safety profiles (in this case, antifungals) that additionally target host cell factors required for virus entry, replication, or propagation might be a promising starting point for the development of novel prophylaxis and treatment approaches of viral infections. As they interfere with cellular metabolism and processes, such compounds are thought to cause fewer resistances, a key issue with direct antivirals. The combinatory use of such host-directed drugs with common antivirals could strengthen the antiviral effect and help to overcome viral infections. While drugs that directly target viral components are much more efficient to eliminate the pathogens, drugs that act on essential host cell factors are considered to circumvent the risk of resistance emergence. Combination therapy using two or more drugs to simultaneously hit multiple targets is, therefore, considered a key strategy to achieve therapeutic success at lower doses and a reduced likelihood of drug resistance development [130–132]. Targeting the virus/host interface via repurposing of azoles in combination with direct antivirals might thus provide a superior antiviral strategy [23,133]. Indeed, several studies have shown the benefit of combining azoles with antivirals [23,32,133]. In vitro studies revealed an additive effect of combinatory treatments with itraconazole or posaconazole and oseltamivir, a well-tolerated inhibitor of IAV neuraminidase. A synergistic antiviral effect was observed in SARS-CoV-2 infection models in vitro, when itraconazole was administered in combination with the antiviral remdesivir, a viral RNA-dependent RNA-polymerase inhibitor [23]. Similarly, a synergistic antiviral effect against HCMV was obtained with the combination of posaconazole and the anti-HCMV drug ganciclovir [32]. These data strongly argue for azoles as a useful element in combinatory treatments to combat certain viral infections. As already stated, itraconazole is a direct inhibitor of the endolysosomal cholesterol transporter NPC1 [47]. Because host cell cholesterol balance has been observed to exert a pivotal role in the infection cycles of several enveloped viruses including Ebola and influenza viruses [25,50,51], analyzing the relationships between the chemical structure of itraconazole and its biological activity (i.e., blocking NPC1 functionality) appears the path forward, and by understanding the structure–activity relationships (SAR), molecular docking simulation can then be used to rationally design novel antiviral derivatives.

Nevertheless, although beneficial effects were already observed as stated above, there are several issues to be considered:

Bioavailability—The low water solubility of the highly lipophilic azoles, including itraconazole, and the resulting poor bioavailability after oral application is a major disadvantage [134,135]. Itraconazole absorption critically depends on low pH, thus reduced gastric acidity caused by fasting conditions or medications (e.g., proton-pump inhibitors such as the widely used drug omeprazole) can considerably reduce the bioavailability and the absorption from the gastrointestinal tract [136], leading to high variability in the plasma levels of patients [134,137], and intake of acidic beverages is known to improve itraconazole absorption [138]. Indeed, a failure to reach the adequate serum concentrations intended in patients receiving systemic antifungal treatment has been observed [139].

Safety—Ideally, the therapeutically effective dose of a drug is much lower than the dose that leads to unwanted adverse effects. Nevertheless, several well-known and widely used drugs have a narrow therapeutic index and thus require frequent monitoring. Drug-induced organ injury is a considerable safety risk, with liver, heart, and kidney damage being the most common reasons for stopping the medication. Hepatotoxicity has been reported as a main adverse effect of antifungal treatment, more frequently in patients treated with azoles [140], and therapeutic drug monitoring has been recommended [141]. Again, SAR analyses might accelerate the development of engineered derivatives with enhanced bioavailability and safety profiles.

Author Contributions: Conceptualization, U.R. and S.S.; writing—original draft preparation, S.S.; writing—review and editing, J.G., U.R. and S.S.; supervision, U.R. and S.S. All authors have read and agreed to the published version of the manuscript.

Funding: This research was funded by grants from the German Research Foundation (DFG), CRC1009 "Breaking Barriers", Project A06 (to U.R.), CRC 1348 "Dynamic Cellular Interfaces", Project A11 (to U.R.), the Interdisciplinary Center for Clinical Research (IZKF) of the Münster Medical School, grant number Re2/022/20 (to U.R.), the Innovative Medizinische Forschung (IMF) of the Münster Medical School, grant number SC121912 (to S.S.)

Institutional Review Board Statement: Not applicable.

Informed Consent Statement: Not applicable.

Data Availability Statement: Data sharing is not applicable to this article.

Conflicts of Interest: The authors declare no conflict of interest. The funders had no role in the design of the study; in the collection, analyses, or interpretation of data; in the writing of the manuscript, or in the decision to publish the results.

References

1. Cascella, M.; Rajnik, M.; Cuomo, A.; Dulebohn, S.C.; Di Napoli, R. *Features, Evaluation and Treatment Coronavirus (COVID-19)*; StatPearls Publishing: Treasure Island, FL, USA, 2020.
2. Kumar, S.; Çalışkan, D.M.; Janowski, J.; Faist, A.; Conrad, B.C.G.; Lange, J.; Ludwig, S.; Brunotte, L. Beyond Vaccines: Clinical Status of Prospective COVID-19 Therapeutics. *Front. Immunol.* **2021**, *12*, 752227. [CrossRef]
3. Huang, C.; Wang, Y.; Li, X.; Ren, L.; Zhao, J.; Hu, Y.; Zhang, L.; Fan, G.; Xu, J.; Gu, X.; et al. Clinical features of patients infected with 2019 novel coronavirus in Wuhan, China. *Lancet* **2020**, *395*, 497–506. [CrossRef]
4. Anand, U.; Jakhmola, S.; Indari, O.; Jha, H.C.; Chen, Z.S.; Tripathi, V.; Pérez de la Lastra, J.M. Potential Therapeutic Targets and Vaccine Development for SARS-CoV-2/COVID-19 Pandemic Management: A Review on the Recent Update. *Front. Immunol.* **2021**, *12*, 658519. [CrossRef]
5. Martin, D.P.; Weaver, S.; Tegally, H.; San, J.E.; Shank, S.D.; Wilkinson, E.; Lucaci, A.G.; Giandhari, J.; Naidoo, S.; Pillay, Y.; et al. The emergence and ongoing convergent evolution of the SARS-CoV-2 N501Y lineages. *Cell* **2021**, *184*, 5189–5200.e7. [CrossRef]
6. de Souza, U.J.B.; Dos Santos, R.N.; Campos, F.S.; Lourenço, K.L.; da Fonseca, F.G.; Spilki, F.R. High rate of mutational events in sars-cov-2 genomes across brazilian geographical regions, february 2020 to june 2021. *Viruses* **2021**, *13*, 1806. [CrossRef]
7. Hurt, A.C.; Holien, J.K.; Parker, M.W.; Barr, I.G. Oseltamivir resistance and the H274Y neuraminidase mutation in seasonal, pandemic and highly pathogenic influenza viruses. *Drugs* **2009**, *69*, 2523–2531. [CrossRef]
8. Trebbien, R.; Pedersen, S.S.; Vorborg, K.; Franck, K.T.; Fischer, T.K. Development of oseltamivir and zanamivir resistance in influenza A(H1N1)pdm09 virus, Denmark, 2014. *Eurosurveillance* **2017**, *22*, 30445. [CrossRef]
9. Kaufmann, S.H.E.; Dorhoi, A.; Hotchkiss, R.S.; Bartenschlager, R. Host-directed therapies for bacterial and viral infections. *Nat. Rev. Drug Discov.* **2018**, *17*, 35–56. [CrossRef]
10. Gordon, D.E.; Jang, G.M.; Bouhaddou, M.; Xu, J.; Obernier, K.; White, K.M.; O'Meara, M.J.; Rezelj, V.V.; Guo, J.Z.; Swaney, D.L.; et al. A SARS-CoV-2 protein interaction map reveals targets for drug repurposing. *Nature* **2020**, *583*, 459–468. [CrossRef]
11. Carolus, H.; Pierson, S.; Lagrou, K.; Van Dijck, P. Amphotericin b and other polyenes—Discovery, clinical use, mode of action and drug resistance. *J. Fungi* **2020**, *6*, 321. [CrossRef]
12. Kim, H.; Kim, S.J.; Park, S.N.; Oh, J.W. Antiviral effect of amphotericin B on Japanese encephalitis virus replication. *J. Microbiol. Biotechnol.* **2004**, *14*, 121–127.
13. Shiota, H.; Jones, B.R.; Schaffner, C.P. Anti-Herpes Simplex Virus (HSV) Effect of Amphotericin B Methyl Ester In Vivo. In *Parasites, Fungi, and Viruses*; Springer US: Boston, MA, USA, 1976; pp. 339–346.
14. Konopka, K.; Guo, L.S.S.; Düzgüneş, N. Anti-HIV activity of amphotericin B-cholesteryl sulfate colloidal dispersion in vitro. *Antiviral Res.* **1999**, *42*, 197–209. [CrossRef]
15. Jordan, G.W.; Humphreys, S.; Zee, Y.C. Effect of amphotericin B methyl ester on vesicular stomatitis virus morphology. *Antimicrob. Agents Chemother.* **1978**, *13*, 340–341. [CrossRef]
16. Padda, I.S.; Parmar, M. *Flucytosine*; StatPearls Publishing: Treasure Island, FL, USA, 2021.
17. Hüttel, W. Echinocandins: Structural diversity, biosynthesis, and development of antimycotics. *Appl. Microbiol. Biotechnol.* **2021**, *105*, 55–66. [CrossRef] [PubMed]
18. Ho, Y.-J.; Liu, F.-C.; Yeh, C.-T.; Yang, C.M.; Lin, C.-C.; Lin, T.-Y.; Hsieh, P.-S.; Hu, M.-K.; Gong, Z.; Lu, J.-W. Micafungin is a novel anti-viral agent of chikungunya virus through multiple mechanisms. *Antiviral Res.* **2018**, *159*, 134–142. [CrossRef]
19. Kim, C.; Kang, H.; Kim, D.E.; Song, J.H.; Choi, M.; Kang, M.; Lee, K.; Kim, H.S.; Shin, J.S.; Jeong, H.; et al. Antiviral activity of micafungin against enterovirus 71. *Virol. J.* **2016**, *13*, 1–9. [CrossRef]
20. Chen, Y.-C.; Lu, J.-W.; Yeh, C.-T.; Lin, T.-Y.; Liu, F.-C.; Ho, Y.-J. Micafungin Inhibits Dengue Virus Infection through the Disruption of Virus Binding, Entry, and Stability. *Pharmaceuticals* **2021**, *14*, 338. [CrossRef]
21. Vergoten, G.; Bailly, C. In silico analysis of echinocandins binding to the main proteases of coronaviruses PEDV (3CLpro) and SARS-CoV-2 (Mpro). *Silico Pharmacol.* **2021**, *9*, 1–10. [CrossRef]
22. Shafiei, M.; Peyton, L.; Hashemzadeh, M.; Foroumadi, A. History of the development of antifungal azoles: A review on structures, SAR, and mechanism of action. *Bioorg. Chem.* **2020**, *104*, 104240. [CrossRef]
23. Schloer, S.; Brunotte, L.; Mecate-Zambrano, A.; Zheng, S.; Tang, J.; Ludwig, S.; Rescher, U. Drug synergy of combinatory treatment with remdesivir and the repurposed drugs fluoxetine and itraconazole effectively impairs SARS-CoV-2 infection in vitro. *Br. J. Pharmacol.* **2021**, *178*, 2339–2350. [CrossRef]
24. Van Damme, E.; De Meyer, S.; Bojkova, D.; Ciesek, S.; Cinatl, J.; De Jonghe, S.; Jochmans, D.; Leyssen, P.; Buyck, C.; Neyts, J.; et al. In vitro activity of itraconazole against SARS-CoV-2. *J. Med. Virol.* **2021**, *93*, 4454–4460. [CrossRef] [PubMed]
25. Schloer, S.; Goretzko, J.; Kühnl, A.; Brunotte, L.; Ludwig, S.; Rescher, U. The clinically licensed antifungal drug itraconazole inhibits influenza virus in vitro and in vivo. *Emerg. Microbes Infect.* **2019**, *8*, 80–93. [CrossRef] [PubMed]
26. Flemming, A. Antivirals: Achilles heel of Ebola viral entry. *Nat. Rev. Drug Discov.* **2011**, *10*, 731. [CrossRef]
27. Head, S.A.; Shi, W.Q.; Yang, E.J.; Nacev, B.A.; Hong, S.Y.; Pasunooti, K.K.; Li, R.J.; Shim, J.S.; Liu, J.O. Simultaneous targeting of NPC1 and VDAC1 by itraconazole leads to synergistic inhibition of MTOR signaling and angiogenesis. *ACS Chem. Biol.* **2017**, *12*, 174–182. [CrossRef] [PubMed]
28. Kummer, S.; Lander, A.; Goretzko, J.; Kirchoff, N.; Rescher, U.; Schloer, S. Pharmacologically induced endolysosomal cholesterol imbalance through clinically licensed drugs itraconazole and fluoxetine impairs Ebola virus infection in vitro. *Emerg. Microbes Infect.* **2022**, *11*, 195–207. [CrossRef]

29. Rhoden, E.; Ng, T.F.F.; Campagnoli, R.; Nix, W.A.; Konopka-Anstadt, J.; Selvarangan, R.; Briesach, L.; Oberste, M.S.; Weldon, W.C. Antifungal triazole posaconazole targets an early stage of the parechovirus A3 life cycle. *Antimicrob. Agents Chemother.* **2020**, *64*, e02372-19. [CrossRef]

30. Meutiawati, F.; Bezemer, B.; Strating, J.R.P.M.; Overheul, G.J.; Žusinaite, E.; van Kuppeveld, F.J.M.; van Cleef, K.W.R.; van Rij, R.P. Posaconazole inhibits dengue virus replication by targeting oxysterol-binding protein. *Antiviral Res.* **2018**, *157*, 68–79. [CrossRef]

31. Strating, J.R.P.M.; van der Linden, L.; Albulescu, L.; Bigay, J.; Arita, M.; Delang, L.; Leyssen, P.; van der Schaar, H.M.; Lanke, K.H.W.; Thibaut, H.J.; et al. Itraconazole Inhibits Enterovirus Replication by Targeting the Oxysterol-Binding Protein. *Cell Rep.* **2015**, *10*, 600–615. [CrossRef]

32. Mercorelli, B.; Luganini, A.; Celegato, M.; Palù, G.; Gribaudo, G.; Lepesheva, G.I.; Loregian, A. The clinically approved antifungal drug posaconazole inhibits human cytomegalovirus replication. *Antimicrob. Agents Chemother.* **2020**, *64*, e00056-20. [CrossRef]

33. Mesa-Arango, A.C.; Scorzoni, L.; Zaragoza, O. It only takes one to do many jobs: Amphotericin B as antifungal and immunomodulatory drug. *Front. Microbiol.* **2012**, *3*, 286. [CrossRef]

34. Malewicz, B.; Momsen, M.; Jenkin, H.M.; Borowski, E. Potentiation of antiviral activity of acyclovir by polyene macrolide antibiotics. *Antimicrob. Agents Chemother.* **1984**, *25*, 772–774. [CrossRef] [PubMed]

35. Marak, B.N.; Dowarah, J.; Khiangte, L.; Singh, V.P. Step toward repurposing drug discovery for COVID-19 therapeutics through in silico approach. *Drug Dev. Res.* **2021**, *82*, 374–392. [CrossRef] [PubMed]

36. Di Mambro, T.; Guerriero, I.; Aurisicchio, L.; Magnani, M.; Marra, E. The yin and yang of current antifungal therapeutic strategies: How can we harness our natural defenses? *Front. Pharmacol.* **2019**, *10*, 80. [CrossRef] [PubMed]

37. Carmona, E.M.; Limper, A.H. Overview of Treatment Approaches for Fungal Infections. *Clin. Chest Med.* **2017**, *38*, 393–402. [CrossRef]

38. Gintjee, T.J.; Donnelley, M.A.; Thompson, G.R. Aspiring Antifungals: Review of Current Antifungal Pipeline Developments. *J. Fungi* **2020**, *6*, 28. [CrossRef] [PubMed]

39. Van Daele, R.; Spriet, I.; Wauters, J.; Maertens, J.; Mercier, T.; Van Hecke, S.; Brüggemann, R. Antifungal drugs: What brings the future? *Med. Mycol.* **2019**, *57*, S328–S343. [CrossRef]

40. Sheehan, D.J.; Hitchcock, C.A.; Sibley, C.M. Current and emerging azole antifungal agents. *Clin. Microbiol. Rev.* **1999**, *12*, 40–79. [CrossRef]

41. Francois, I.; Cammue, B.; Borgers, M.; Ausma, J.; Dispersyn, G.; Thevissen, K. Azoles: Mode of Antifungal Action and Resistance Development. Effect of Miconazole on Endogenous Reactive Oxygen Species Production in Candida albicans. *Antiinfect. Agents Med. Chem.* **2006**, *5*, 3–13. [CrossRef]

42. Lass-Flörl, C. Triazole antifungal agents in invasive fungal infections: A comparative review. *Drugs* **2011**, *71*, 2405–2419. [CrossRef]

43. Kelly, S.L.; Lamb, D.C.; Baldwin, B.C.; Corran, A.J.; Kelly, D.E. Characterization of Saccharomyces cerevisiae CYP61, Sterol Δ22-Desaturase, and Inhibition by Azole Antifungal Agents. *J. Biol. Chem.* **1997**, *272*, 9986–9988. [CrossRef]

44. Minnebruggen, G.V.; Francois, I.E.J.A.; Cammue, B.P.A.; Thevissen, K.; Vroome, V.; Borgers, M.; Shroot, B. A General Overview on Past, Present and Future Antimycotics. *Open Mycol. J.* **2010**, *4*, 22–32. [CrossRef]

45. Warrilow, A.G.; Parker, J.E.; Kelly, D.E.; Kelly, S.L. Azole Affinity of Sterol 14-Demethylase (CYP51) Enzymes from Candida albicans and Homo sapiens. *Antimicrob. Agents Chemother.* **2013**, *57*, 1352–1360. [CrossRef]

46. Munayyer, H.K.; Mann, P.A.; Chau, A.S.; Yarosh-Tomaine, T.; Greene, J.R.; Hare, R.S.; Heimark, L.; Palermo, R.E.; Loebenberg, D.; McNicholas, P.M. Posaconazole is a potent inhibitor of sterol 14α-demethylation in yeasts and molds. *Antimicrob. Agents Chemother.* **2004**, *48*, 3690–3696. [CrossRef] [PubMed]

47. Trinh, M.N.; Lu, F.; Li, X.; Das, A.; Liang, Q.; De Brabander, J.K.; Brown, M.S.; Goldstein, J.L.; Chang, T.-Y.; Maxfield, F.R. Triazoles inhibit cholesterol export from lysosomes by binding to NPC1. *Proc. Natl. Acad. Sci. USA* **2017**, *114*, 89–94. [CrossRef] [PubMed]

48. Vanier, M.; Millat, G. Niemann-Pick disease type C. *Clin. Genet.* **2003**, *64*, 269–281. [CrossRef] [PubMed]

49. Pfeffer, S.R. NPC intracellular cholesterol transporter 1 (NPC1)-mediated cholesterol export from lysosomes. *J. Biol. Chem.* **2019**, *294*, 1706–1709. [CrossRef]

50. Herbert, A.S.; Davidson, C.; Kuehne, A.I.; Bakken, R.; Braigen, S.Z.; Gunn, K.E.; Whelan, S.P.; Brummelkamp, T.R.; Twenhafel, N.A.; Chandran, K.; et al. Niemann-pick C1 is essential for ebolavirus replication and pathogenesis in vivo. *MBio* **2015**, *6*, 1–12. [CrossRef]

51. Côté, M.; Misasi, J.; Ren, T.; Bruchez, A.; Lee, K.; Filone, C.M.; Hensley, L.; Li, Q.; Ory, D.; Chandran, K.; et al. Small molecule inhibitors reveal Niemann-Pick C1 is essential for Ebola virus infection. *Nature* **2011**, *477*, 344–348. [CrossRef]

52. Carette, J.E.; Raaben, M.; Wong, A.C.; Herbert, A.S.; Obernosterer, G.; Mulherkar, N.; Kuehne, A.I.; Kranzusch, P.J.; Griffin, A.M.; Ruthel, G.; et al. Ebola virus entry requires the cholesterol transporter Niemann-Pick C1. *Nature* **2011**, *477*, 340–343. [CrossRef]

53. Miller, E.H.; Obernosterer, G.; Raaben, M.; Herbert, A.S.; Deffieu, M.S.; Krishnan, A.; Ndungo, E.; Sandesara, R.G.; Carette, J.E.; Kuehne, A.I.; et al. Ebola virus entry requires the host-programmed recognition of an intracellular receptor. *EMBO J.* **2012**, *31*, 1947–1960. [CrossRef]

54. Coleman, E.M.; Walker, T.N.; Hildreth, J.E.K. Loss of Niemann Pick type C proteins 1 and 2 greatly enhances HIV infectivity and is associated with accumulation of HIV Gag and cholesterol in late endosomes/lysosomes. *Virol. J.* **2012**, *9*, 31. [CrossRef]

55. Tang, Y.; Leao, I.C.; Coleman, E.M.; Broughton, R.S.; Hildreth, J.E.K. Deficiency of Niemann-Pick Type C-1 Protein Impairs Release of Human Immunodeficiency Virus Type 1 and Results in Gag Accumulation in Late Endosomal/Lysosomal Compartments. *J. Virol.* **2009**, *83*, 7982–7995. [CrossRef]

56. van 't Wout, A.B.; Swain, J.V.; Schindler, M.; Rao, U.; Pathmajeyan, M.S.; Mullins, J.I.; Kirchhoff, F. Nef Induces Multiple Genes Involved in Cholesterol Synthesis and Uptake in Human Immunodeficiency Virus Type 1-Infected T Cells. *J. Virol.* **2005**, *79*, 10053–10058. [CrossRef] [PubMed]

57. Lu, F.; Liang, Q.; Abi-Mosleh, L.; Das, A.; de Brabander, J.K.; Goldstein, J.L.; Brown, M.S. Identification of NPC1 as the target of U18666A, an inhibitor of lysosomal cholesterol export and Ebola infection. *Elife* **2015**, *4*, 1–16. [CrossRef] [PubMed]

58. Liscum, L.; Faust, J.R. The intracellular transport of low density lipoprotein-derived cholesterol is inhibited in Chinese hamster ovary cells cultured with 3-β-[2-(diethylamino)ethoxy]androst-5-en-17-one. *J. Biol. Chem.* **1989**, *264*, 11796–11806. [CrossRef]

59. Lloyd-Evans, E.; Morgan, A.J.; He, X.; Smith, D.A.; Elliot-Smith, E.; Sillence, D.J.; Churchill, G.C.; Schuchman, E.H.; Galione, A.; Platt, F.M. Niemann-Pick disease type C1 is a sphingosine storage disease that causes deregulation of lysosomal calcium. *Nat. Med.* **2008**, *14*, 1247–1255. [CrossRef] [PubMed]

60. Cenedella, R.J.; Jacob, R.; Borchman, D.; Tang, D.; Neely, A.R.; Samadi, A.; Mason, R.P.; Sexton, P. Direct perturbation of lens membrane structure may contribute to cataracts caused by U18666A, an oxidosqualene cyclase inhibitor. *J. Lipid Res.* **2004**, *45*, 1232–1241. [CrossRef]

61. Liesenborghs, L.; Spriet, I.; Jochmans, D.; Belmans, A.; Gyselinck, I.; Teuwen, L.-A.; ter Horst, S.; Dreesen, E.; Geukens, T.; Engelen, M.M.; et al. Itraconazole for COVID-19: Preclinical studies and a proof-of-concept randomized clinical trial. *EBioMedicine* **2021**, *66*, 103288. [CrossRef]

62. Ridgway, N.D.; Dawson, P.A.; Ho, Y.K.; Brown, M.S.; Goldstein, J.L. Translocation of oxysterol binding protein to Golgi apparatus triggered by ligand binding. *J. Cell Biol.* **1992**, *116*, 307–319. [CrossRef]

63. Mesmin, B.; Bigay, J.; Moser von Filseck, J.; Lacas-Gervais, S.; Drin, G.; Antonny, B. A Four-Step Cycle Driven by PI(4)P Hydrolysis Directs Sterol/PI(4)P Exchange by the ER-Golgi Tether OSBP. *Cell* **2013**, *155*, 830–843. [CrossRef]

64. Weber-Boyvat, M.; Zhong, W.; Yan, D.; Olkkonen, V.M. Oxysterol-binding proteins: Functions in cell regulation beyond lipid metabolism. *Biochem. Pharmacol.* **2013**, *86*, 89–95. [CrossRef]

65. Raychaudhuri, S.; Prinz, W.A. The diverse functions of oxysterol-binding proteins. *Annu. Rev. Cell Dev. Biol.* **2010**, *26*, 157–177. [CrossRef]

66. Hsu, N.Y.; Ilnytska, O.; Belov, G.; Santiana, M.; Chen, Y.H.; Takvorian, P.M.; Pau, C.; van der Schaar, H.; Kaushik-Basu, N.; Balla, T.; et al. Viral reorganization of the secretory pathway generates distinct organelles for RNA replication. *Cell* **2010**, *141*, 799–811. [CrossRef] [PubMed]

67. Arita, M.; Kojima, H.; Nagano, T.; Okabe, T.; Wakita, T.; Shimizu, H. Phosphatidylinositol 4-Kinase III Beta Is a Target of Enviroxime-Like Compounds for Antipoliovirus Activity. *J. Virol.* **2011**, *85*, 2364–2372. [CrossRef] [PubMed]

68. Arita, M. Phosphatidylinositol-4 kinase III beta and oxysterol-binding protein accumulate unesterified cholesterol on poliovirus-induced membrane structure. *Microbiol. Immunol.* **2014**, *58*, 239–256. [CrossRef]

69. Roulin, P.S.; Lötzerich, M.; Torta, F.; Tanner, L.B.; Van Kuppeveld, F.J.M.; Wenk, M.R.; Greber, U.F. Rhinovirus uses a phosphatidylinositol 4-phosphate/cholesterol counter-current for the formation of replication compartments at the ER-Golgi interface. *Cell Host Microbe* **2014**, *16*, 677–690. [CrossRef] [PubMed]

70. Betz, C.; Hall, M.N. Where is mTOR and what is it doing there? *J. Cell Biol.* **2013**, *203*, 563–574. [CrossRef]

71. Castellano, B.M.; Thelen, A.M.; Moldavski, O.; Feltes, M.; Van Der Welle, R.E.N.; Mydock-McGrane, L.; Jiang, X.; Van Eijkeren, R.J.; Davis, O.B.; Louie, S.M.; et al. Lysosomal cholesterol activates mTORC1 via an SLC38A9-Niemann-Pick C1 signaling complex. *Science* **2017**, *355*, 1306–1311. [CrossRef]

72. Lim, C.Y.; Davis, O.B.; Shin, H.R.; Zhang, J.; Berdan, C.A.; Jiang, X.; Counihan, J.L.; Ory, D.S.; Nomura, D.K.; Zoncu, R. ER-lysosome contacts enable cholesterol sensing by mTORC1 and drive aberrant growth signalling in Niemann–Pick type C. *Nat. Cell Biol.* **2019**, *21*, 1206–1218. [CrossRef]

73. Karam, B.S.; Morris, R.S.; Bramante, C.T.; Puskarich, M.; Zolfaghari, E.J.; Lotfi-Emran, S.; Ingraham, N.E.; Charles, A.; Odde, D.J.; Tignanelli, C.J. mTOR inhibition in COVID-19: A commentary and review of efficacy in RNA viruses. *J. Med. Virol.* **2021**, *93*, 1843–1846. [CrossRef]

74. Shin, Y.K.; Liu, Q.; Tikoo, S.K.; Babiuk, L.A.; Zhou, Y. Effect of the phosphatidylinositol 3-kinase/Akt pathway on influenza A virus propagation. *J. Gen. Virol.* **2007**, *88*, 942–950. [CrossRef] [PubMed]

75. Hale, B.G.; Jackson, D.; Chen, Y.H.; Lamb, R.A.; Randall, R.E. Influenza A virus NS1 protein binds p85β and activates phosphatidylinositol-3-kinase signaling. *Proc. Natl. Acad. Sci. USA* **2006**, *103*, 14194–14199. [CrossRef] [PubMed]

76. Shives, K.D.; Beatman, E.L.; Chamanian, M.; O'Brien, C.; Hobson-Peters, J.; Beckham, J.D. West Nile Virus-Induced Activation of Mammalian Target of Rapamycin Complex 1 Supports Viral Growth and Viral Protein Expression. *J. Virol.* **2014**, *88*, 9458–9471. [CrossRef] [PubMed]

77. Bose, S.K.; Shrivastava, S.; Meyer, K.; Ray, R.B.; Ray, R. Hepatitis C Virus Activates the mTOR/S6K1 Signaling Pathway in Inhibiting IRS-1 Function for Insulin Resistance. *J. Virol.* **2012**, *86*, 6315–6322. [CrossRef] [PubMed]

78. Joubert, P.E.; Stapleford, K.; Guivel-Benhassine, F.; Vignuzzi, M.; Schwartz, O.; Albert, M.L. Inhibition of mTORC1 Enhances the Translation of Chikungunya Proteins via the Activation of the MnK/eIF4E Pathway. *PLoS Pathog.* **2015**, *11*, e1005091. [CrossRef]

79. Das, I.; Basantray, I.; Mamidi, P.; Nayak, T.K.; Pratheek, B.M.; Chattopadhyay, S.; Chattopadhyay, S. Heat shock protein 90 positively regulates Chikungunya virus replication by stabilizing viral non-structural protein nsP2 during infection. *PLoS One* **2014**, *9*, e100531. [CrossRef] [PubMed]

80. Shi, Y.; He, X.; Zhu, G.; Tu, H.; Liu, Z.; Li, W.; Han, S.; Yin, J.; Peng, B.; Liu, W. Coxsackievirus A16 elicits incomplete autophagy involving the mTOR and ERK pathways. *PLoS One* **2015**, *10*, e0122109. [CrossRef] [PubMed]

81. Kuss-Duerkop, S.K.; Wang, J.; Mena, I.; White, K.; Metreveli, G.; Sakthivel, R.; Mata, M.A.; Muñoz-Moreno, R.; Chen, X.; Krammer, F.; et al. Influenza Virus Differentially Activates mTORC1 and mTORC2 Signaling to Maximize Late Stage Replication. *PLoS Pathog.* **2017**, *13*, e1006635. [CrossRef] [PubMed]

82. Bakar, F.A.; Ng, L.F.P. Nonstructural proteins of alphavirus—potential targets for drug development. *Viruses* **2018**, *10*, 71. [CrossRef]

83. Mohankumar, V.; Dhanushkodi, N.R.; Raju, R. Sindbis virus replication, is insensitive to rapamycin and torin1, and suppresses Akt/mTOR pathway late during infection in HEK cells. *Biochem. Biophys. Res. Commun.* **2011**, *406*, 262–267. [CrossRef]

84. Joubert, P.E.; Werneke, S.W.; de la Calle, C.; Guivel-Benhassine, F.; Giodini, A.; Peduto, L.; Levine, B.; Schwartz, O.; Lenschow, D.J.; Albert, M.L. Chikungunya virus-induced autophagy delays caspase-dependent cell death. *J. Exp. Med.* **2012**, *209*, 1029–1047. [CrossRef]

85. Thaa, B.; Biasiotto, R.; Eng, K.; Neuvonen, M.; Götte, B.; Rheinemann, L.; Mutso, M.; Utt, A.; Varghese, F.; Balistreri, G.; et al. Differential Phosphatidylinositol-3-Kinase-Akt-mTOR Activation by Semliki Forest and Chikungunya Viruses Is Dependent on nsP3 and Connected to Replication Complex Internalization. *J. Virol.* **2015**, *89*, 11420–11437. [CrossRef]

86. Klaitong, P.; Smith, D.R. Roles of non-structural protein 4a in flavivirus infection. *Viruses* **2021**, *13*, 2077. [CrossRef]

87. Lahon, A.; Arya, R.P.; Banerjea, A.C. Dengue Virus Dysregulates Master Transcription Factors and PI3K/AKT/mTOR Signaling Pathway in Megakaryocytes. *Front. Cell. Infect. Microbiol.* **2021**, *11*, 715208. [CrossRef]

88. Albentosa-González, L.; de Oya, N.J.; Arias, A.; Clemente-Casares, P.; Martin-Acebes, M.Á.; Saiz, J.C.; Sabariegos, R.; Mas, A. Akt kinase intervenes in flavivirus replication by interacting with viral protein ns5. *Viruses* **2021**, *13*, 896. [CrossRef]

89. Urbanowski, M.D.; Hobman, T.C. The West Nile Virus Capsid Protein Blocks Apoptosis through a Phosphatidylinositol 3-Kinase-Dependent Mechanism. *J. Virol.* **2013**, *87*, 872–881. [CrossRef] [PubMed]

90. Lee, C.-J.; Liao, C.-L.; Lin, Y.-L. Flavivirus Activates Phosphatidylinositol 3-Kinase Signaling To Block Caspase-Dependent Apoptotic Cell Death at the Early Stage of Virus Infection. *J. Virol.* **2005**, *79*, 8388–8399. [CrossRef] [PubMed]

91. George, A.; Panda, S.; Kudmulwar, D.; Chhatbar, S.P.; Nayak, S.C.; Krishnan, H.H. Hepatitis C virus NS5A binds to the mRNA cap-binding eukaryotic translation initiation 4F (eIF4F) complex and up-regulates host translation initiation machinery through eIF4E-binding protein 1 inactivation. *J. Biol. Chem.* **2012**, *287*, 5042–5058. [CrossRef]

92. He, Y.; Nakao, H.; Tan, S.-L.; Polyak, S.J.; Neddermann, P.; Vijaysri, S.; Jacobs, B.L.; Katze, M.G. Subversion of Cell Signaling Pathways by Hepatitis C Virus Nonstructural 5A Protein via Interaction with Grb2 and P85 Phosphatidylinositol 3-Kinase. *J. Virol.* **2002**, *76*, 9207–9217. [CrossRef] [PubMed]

93. Street, A.; Macdonald, A.; Crowder, K.; Harris, M. The Hepatitis C Virus NS5A Protein Activates a Phosphoinositide 3-Kinase-dependent Survival Signaling Cascade. *J. Biol. Chem.* **2004**, *279*, 12232–12241. [CrossRef] [PubMed]

94. Le Sage, V.; Cinti, A.; Amorim, R.; Mouland, A.J. Adapting the stress response: Viral subversion of the mTOR signaling pathway. *Viruses* **2016**, *8*, 152. [CrossRef]

95. Mannová, P.; Beretta, L. Activation of the N-Ras–PI3K–Akt-mTOR Pathway by Hepatitis C Virus: Control of Cell Survival and Viral Replication. *J. Virol.* **2005**, *79*, 8742–8749. [CrossRef]

96. Clippinger, A.J.; Maguire, T.G.; Alwine, J.C. Human Cytomegalovirus Infection Maintains mTOR Activity and Its Perinuclear Localization during Amino Acid Deprivation. *J. Virol.* **2011**, *85*, 9369–9376. [CrossRef] [PubMed]

97. Kudchodkar, S.B.; Del Prete, G.Q.; Maguire, T.G.; Alwine, J.C. AMPK-Mediated Inhibition of mTOR Kinase Is Circumvented during Immediate-Early Times of Human Cytomegalovirus Infection. *J. Virol.* **2007**, *81*, 3649–3651. [CrossRef] [PubMed]

98. Yu, Y.; Alwine, J.C. Human Cytomegalovirus Major Immediate-Early Proteins and Simian Virus 40 Large T Antigen Can Inhibit Apoptosis through Activation of the Phosphatidylinositide 3′-OH Kinase Pathway and the Cellular Kinase Akt. *J. Virol.* **2002**, *76*, 3731–3738. [CrossRef] [PubMed]

99. Blanchet, F.P.; Moris, A.; Nikolic, D.S.; Lehmann, M.; Cardinaud, S.; Stalder, R.; Garcia, E.; Dinkins, C.; Leuba, F.; Wu, L.; et al. Human immunodeficiency virus-1 inhibition of immunoamphisomes in dendritic cells impairs early innate and adaptive immune responses. *Immunity* **2010**, *32*, 654–669. [CrossRef]

100. Campbell, G.R.; Rawat, P.; Bruckman, R.S.; Spector, S.A. Human Immunodeficiency Virus Type 1 Nef Inhibits Autophagy through Transcription Factor EB Sequestration. *PLoS Pathog.* **2015**, *11*, e1005018. [CrossRef]

101. Cinti, A.; Le Sage, V.; Milev, M.P.; Valiente-Echeverría, F.; Crossie, C.; Miron, M.-J.; Panté, N.; Olivier, M.; Mouland, A.J. HIV-1 enhances mTORC1 activity and repositions lysosomes to the periphery by co-opting Rag GTPases. *Sci. Rep.* **2017**, *7*, 5515. [CrossRef]

102. Heredia, A.; Le, N.; Gartenhaus, R.B.; Sausville, E.; Medina-Moreno, S.; Zapata, J.C.; Davis, C.; Gallo, R.C.; Redfield, R.R. Targeting of mTOR catalytic site inhibits multiple steps of the HIV-1 lifecycle and suppresses HIV-1 viremia in humanized mice. *Proc. Natl. Acad. Sci. USA* **2015**, *112*, 9412–9417. [CrossRef]

103. Donia, M.; McCubrey, J.A.; Bendtzen, K.; Nicoletti, F. Potential use of rapamycin in HIV infection. *Br. J. Clin. Pharmacol.* **2010**, *70*, 784–793. [CrossRef]

104. Zhou, Z.; Jiang, X.; Liu, D.; Fan, Z.; Hu, X.; Yan, J.; Wang, M.; Gao, G.F. Autophagy is involved in influenza A virus replication. *Autophagy* **2009**, *5*, 321–328. [CrossRef]

105. Zhirnov, O.P.; Konakova, T.E.; Garten, W.; Klenk, H.-D. Caspase-Dependent N-Terminal Cleavage of Influenza Virus Nucleocapsid Protein in Infected Cells. *J. Virol.* **1999**, *73*, 10158–10163. [CrossRef]
106. Tsubamoto, H.; Inoue, K.; Sakata, K.; Ueda, T.; Takeyama, R.; Shibahara, H.; Sonoda, T. Itraconazole inhibits AKT/mTOR signaling and proliferation in endometrial cancer cells. *Anticancer Res.* **2017**, *37*, 515–520. [CrossRef]
107. Garcia, D.; Shaw, R.J. AMPK: Mechanisms of Cellular Energy Sensing and Restoration of Metabolic Balance. *Mol. Cell* **2017**, *66*, 789–800. [CrossRef]
108. Cork, G.K.; Thompson, J.; Slawson, C. Real Talk: The Inter-play Between the mTOR, AMPK, and Hexosamine Biosynthetic Pathways in Cell Signaling. *Front. Endocrinol. (Lausanne)* **2018**, *9*, 522. [CrossRef] [PubMed]
109. Nacev, B.A.; Grassi, P.; Dell, A.; Haslam, S.M.; Liu, J.O. The antifungal drug itraconazole inhibits Vascular Endothelial Growth Factor Receptor 2 (VEGFR2) glycosylation, trafficking, and signaling in endothelial cells. *J. Biol. Chem.* **2011**, *286*, 44045–44056. [CrossRef] [PubMed]
110. Vrancken, K.; Paeshuyse, J.; Liekens, S. Angiogenic activity of hepatitis B and C viruses. *Antivir. Chem. Chemother.* **2012**, *22*, 159–170. [CrossRef] [PubMed]
111. Li, G.; He, L.; Zhang, E.; Shi, J.; Zhang, Q.; Le, A.D.; Zhou, K.; Tang, X. Overexpression of human papillomavirus (HPV) type 16 oncoproteins promotes angiogenesis via enhancing HIF-1α and VEGF expression in non-small cell lung cancer cells. *Cancer Lett.* **2011**, *311*, 160–170. [CrossRef]
112. Hassan, M.; Selimovic, D.; El-Khattouti, A.; Soell, M.; Ghozlan, H.; Haikel, Y.; Abdelkader, O.; Megahed, M. Hepatitis C virus-mediated angiogenesis: Molecular mechanisms and therapeutic strategies. *World J. Gastroenterol.* **2014**, *20*, 15467–15475. [CrossRef]
113. Smelkinson, M.G.; Guichard, A.; Teijaro, J.R.; Malur, M.; Loureiro, M.E.; Jain, P.; Ganesan, S.; Zúñiga, E.I.; Krug, R.M.; Oldstone, M.B.; et al. Influenza NS1 directly modulates Hedgehog signaling during infection. *PLoS Pathog.* **2017**, *13*, 1–24. [CrossRef]
114. Rimkus, T.K.; Carpenter, R.L.; Qasem, S.; Chan, M.; Lo, H.W. Targeting the sonic hedgehog signaling pathway: Review of smoothened and GLI inhibitors. *Cancers* **2016**, *8*, 22. [CrossRef] [PubMed]
115. Pereira, T.D.A.; Witek, R.P.; Syn, W.K.; Choi, S.S.; Bradrick, S.; Karaca, G.F.; Agboola, K.M.; Jung, Y.; Omenetti, A.; Moylan, C.A.; et al. Viral factors induce Hedgehog pathway activation in humans with viral hepatitis, cirrhosis, and hepatocellular carcinoma. *Lab. Investig.* **2010**, *90*, 1690–1703. [CrossRef] [PubMed]
116. Granato, M.; Zompetta, C.; Vescarelli, E.; Rizzello, C.; Cardi, A.; Valia, S.; Antonelli, G.; Marchese, C.; Torrisi, M.R.; Faggioni, A.; et al. HCV derived from sera of HCVinfected patients induces profibrotic effects in human primary fibroblasts by activating GLI2. *Sci. Rep.* **2016**, *6*, 1–11. [CrossRef] [PubMed]
117. Kim, H.Y.; Cho, H.K.; Hong, S.P.; Cheong, J.H. Hepatitis B virus X protein stimulates the Hedgehog-Gli activation through protein stabilization and nuclear localization of Gli1 in liver cancer cells. *Cancer Lett.* **2011**, *309*, 176–184. [CrossRef] [PubMed]
118. Kim, J.; Tang, J.Y.; Gong, R.; Kim, J.; Lee, J.J.; Clemons, K.V.; Chong, C.R.; Chang, K.S.; Fereshteh, M.; Gardner, D.; et al. Itraconazole, a Commonly Used Antifungal that Inhibits Hedgehog Pathway Activity and Cancer Growth. *Cancer Cell* **2010**, *17*, 388–399. [CrossRef]
119. Dirix, L. Discovery and exploitation of novel targets by approved drugs. *J. Clin. Oncol.* **2014**, *32*, 720–721. [CrossRef]
120. Deng, H.; Huang, L.; Liao, Z.; Liu, M.; Li, Q.; Xu, R. Itraconazole inhibits the Hedgehog signaling pathway thereby inducing autophagy-mediated apoptosis of colon cancer cells. *Cell Death Dis.* **2020**, *11*, 1–15. [CrossRef] [PubMed]
121. Clevers, H.; Nusse, R. Wnt/β-catenin signaling and disease. *Cell* **2012**, *149*, 1192–1205. [CrossRef]
122. van Zuylen, W.J.; Rawlinson, W.D.; Ford, C.E. The Wnt pathway: A key network in cell signalling dysregulated by viruses. *Rev. Med. Virol.* **2016**, *26*, 340–355. [CrossRef]
123. Shackelford, J.; Maier, C.; Pagano, J.S. Epstein-Barr virus activates β-catenin in type III latently infected B lymphocyte lines: Association with deubiquitinating enzymes. *Proc. Natl. Acad. Sci. USA* **2003**, *100*, 15572–15576. [CrossRef]
124. Anastas, J.N.; Moon, R.T. WNT signalling pathways as therapeutic targets in cancer. *Nat. Rev. Cancer* **2013**, *13*, 11–26. [CrossRef] [PubMed]
125. Hsieh, A.; Kim, H.S.; Lim, S.O.; Yu, D.Y.; Jung, G. Hepatitis B viral X protein interacts with tumor suppressor adenomatous polyposis coli to activate Wnt/β-catenin signaling. *Cancer Lett.* **2011**, *300*, 162–172. [CrossRef]
126. Xie, Q.; Chen, L.; Shan, X.; Shan, X.; Tang, J.; Zhou, F.; Chen, Q.; Quan, H.; Nie, D.; Zhang, W.; et al. Epigenetic silencing of SFRP1 and SFRP5 by hepatitis B virus X protein enhances hepatoma cell tumorigenicity through Wnt signaling pathway. *Int. J. Cancer* **2014**, *135*, 635–646. [CrossRef] [PubMed]
127. Umer, M.; Qureshi, S.A.; Hashmi, Z.Y.; Raza, A.; Ahmad, J.; Rahman, M.; Iqbal, M. Promoter hypermethylation of Wnt pathway inhibitors in hepatitis C virus - Induced multistep hepatocarcinogenesis. *Virol. J.* **2014**, *11*, 117. [CrossRef]
128. Popova, S.A.; Buczacki, S.J.A. Itraconazole perturbs colorectal cancer dormancy through SUFU-mediated WNT inhibition. *Mol. Cell. Oncol.* **2018**, *5*, e1494950. [CrossRef]
129. Liang, G.; Liu, M.; Wang, Q.; Shen, Y.; Mei, H.; Li, D.; Liu, W. Itraconazole exerts its anti-melanoma effect by suppressing Hedgehog, Wnt, and PI3K/mTOR signaling pathways. *Oncotarget* **2017**, *8*, 28510–28525. [CrossRef]
130. Shyr, Z.A.; Cheng, Y.S.; Lo, D.C.; Zheng, W. Drug combination therapy for emerging viral diseases. *Drug Discov. Today* **2021**, *26*, 2367–2376. [CrossRef]
131. Govorkova, E.A.; Webster, R.G. Combination chemotherapy for influenza. *Viruses* **2010**, *2*, 1510–1529. [CrossRef] [PubMed]
132. Reeves, J.D.; Piefer, A.J. Emerging drug targets for antiretroviral therapy. *Drugs* **2005**, *65*, 1747–1766. [CrossRef]

133. Schloer, S.; Goretzko, J.; Pleschka, S.; Ludwig, S.; Rescher, U. Combinatory Treatment with Oseltamivir and Itraconazole Targeting Both Virus and Host Factors in Influenza A Virus Infection. *Viruses* **2020**, *12*, 703. [CrossRef]
134. Domínguez-Gil Hurlé, A.; Sánchez Navarro, A.; García Sánchez, M.J. Therapeutic drug monitoring of itraconazole and the relevance of pharmacokinetic interactions. *Clin. Microbiol. Infect.* **2006**, *12*, 97–106. [CrossRef]
135. Prentice, A.G.; Glasmacher, A. Making sense of itraconazole pharmacokinetics. *J. Antimicrob. Chemother.* **2005**, *56*, i17–i22. [CrossRef]
136. Jaruratanasirikul, S.; Sriwiriyajan, S. Effect of omeprazole on the pharmacokinetics of itraconazole. *Eur. J. Clin. Pharmacol.* **1998**, *54*, 159–161. [CrossRef] [PubMed]
137. Lestner, J.; Hope, W.W. Itraconazole: An update on pharmacology and clinical use for treatment of invasive and allergic fungal infections. *Expert Opin. Drug Metab. Toxicol.* **2013**, *9*, 911–926. [CrossRef] [PubMed]
138. Bae, S.K.; Park, S.J.; Shim, E.J.; Mun, J.H.; Kim, E.Y.; Shin, J.G.; Shon, J.H. Increased oral bioavailability of itraconazole and its active metabolite, 7-hydroxyitraconazole, when coadministered with a vitamin C beverage in healthy participants. *J. Clin. Pharmacol.* **2011**, *51*, 444–451. [CrossRef]
139. Vena, A.; Muñoz, P.; Mateos, M.; Guinea, J.; Galar, A.; Pea, F.; Alvarez-Uria, A.; Escribano, P.; Bouza, E. Therapeutic Drug Monitoring of Antifungal Drugs: Another Tool to Improve Patient Outcome? *Infect. Dis. Ther.* **2020**, *9*, 137–149. [CrossRef]
140. Tverdek, F.P.; Kofteridis, D.; Kontoyiannis, D.P. Antifungal agents and liver toxicity: A complex interaction. *Expert Rev. Anti. Infect. Ther.* **2016**, *14*, 765–776. [CrossRef] [PubMed]
141. Ashbee, H.R.; Barnes, R.A.; Johnson, E.M.; Richardson, M.D.; Gorton, R.; Hope, W.W. Therapeutic drug monitoring (TDM) of antifungal agents: Guidelines from the british society for medical mycology. *J. Antimicrob. Chemother.* **2014**, *69*, 1162–1176. [CrossRef]

pharmaceuticals

Article

Evaluation of Amebicidal and Cysticidal Activities of Antifungal Drug Isavuconazonium Sulfate against *Acanthamoeba* T4 Strains

Brian Shing [1,2], Mina Balen [2,3] and Anjan Debnath [2,*]

1   Biomedical Sciences Graduate Division, University of California San Diego, 9500 Gilman Drive, MC 0685, La Jolla, CA 92093-0756, USA; bjshing@health.ucsd.edu
2   Center for Discovery and Innovation in Parasitic Diseases, Skaggs School of Pharmacy and Pharmaceutical Sciences, University of California San Diego, 9500 Gilman Drive, MC 0756, La Jolla, CA 92093-0756, USA; mbalen@ucsd.edu
3   Division of Biological Sciences, University of California San Diego, 9500 Gilman Drive, MC 0346, La Jolla, CA 92093-0756, USA
*   Correspondence: adebnath@health.ucsd.edu; Tel.: +1-858-822-5265

Abstract: *Acanthamoeba* species of amebae are often associated with *Acanthamoeba* keratitis, a severe corneal infection. Isavuconazonium sulfate is an FDA-approved drug for the treatment of invasive aspergillosis and mucormycosis. This prodrug is metabolized into the active isavuconazole moiety. Isavuconazole was previously identified to have amebicidal and cysticidal activity against *Acanthamoeba* T4 strains, but the activity of its prodrug, isavuconazonium sulfate, against trophozoites and cysts remains unknown. Since it is not known if isavuconazonium can be metabolized into isavuconazole in the human eye, we evaluated the activities of isavuconazonium sulfate against trophozoites and cysts of three T4 genotype strains of *Acanthamoeba*. Isavuconazonium displayed amebicidal activity at nanomolar concentrations as low as 1.4 nM and prevented excystation of cysts at concentrations as low as 136 µM. We also investigated the cysticidal activity of isavuconazonium sulfate in combination with a currently used amebicidal drug polyhexamethylene biguanide (PHMB). Although combination of isavuconazonium with PHMB did not elicit an obvious synergistic cysticidal activity, the combination did not cause an antagonistic effect on the cysts of *Acanthamoeba* T4 strains. Collectively, these findings suggest isavuconazonium retains potency against *Acanthamoeba* T4 strains and could be adapted for *Acanthamoeba* keratitis treatment.

Keywords: *Acanthamoeba*; free-living ameba; *Acanthamoeba* keratitis; isavuconazonium sulfate; cyst; drug

Citation: Shing, B.; Balen, M.; Debnath, A. Evaluation of Amebicidal and Cysticidal Activities of Antifungal Drug Isavuconazonium Sulfate against *Acanthamoeba* T4 Strains. *Pharmaceuticals* **2021**, *14*, 1294. https://doi.org/10.3390/ph14121294

Academic Editor: Christophe Dardonville

Received: 23 November 2021
Accepted: 8 December 2021
Published: 11 December 2021

Publisher's Note: MDPI stays neutral with regard to jurisdictional claims in published maps and institutional affiliations.

1. Introduction

Acanthamoeba castellanii is a causative agent of *Acanthamoeba* keratitis (AK). It is a serious infection of the eye that causes inflammation in the cornea and can result in permanent visual impairment or blindness. *Acanthamoeba* is common in nature and can be found in soil, air and water, including insufficiently chlorinated pools, hot tubs, tap and shower water. In unfavorable environments, the ameboid form of the organism called a 'trophozoite' transforms into a drug-resistant double-walled cyst. Cyst resistance to therapeutic agents, and recurrence of infection due to *Acanthamoeba* excystment, remain challenges for disease prevention and cures. Infection recurrence occurs in approximately 10% of cases [1], due possibly to excystment. No single drug has yet been shown effective at therapeutic concentrations against both the trophozoite and cyst stages of *Acanthamoeba*. Current treatment of AK involves an aggressive disinfectant chlorhexidine, in combination with diamidines, polyhexamethylene biguanide (PHMB) and neomycin. Combination therapies have proven more successful than monotherapies [2–4]. The most aggressive and severe cases of AK require corneal grafts or surgical removal of the eye [5]. Despite advances in combination therapies and surgery, the resistance of cysts to therapeutic agents poses challenges that are yet to be addressed [6]. Therefore, discovering and identifying therapeutics that are effective against both stages of the parasite would be critical to reducing AK recurrence and improving existing therapies.

Earlier, we identified isavuconazole as amebicidal and cysticidal [7], but clinically the prodrug isavuconazonium sulfate is administered orally or intravenously for the treatment of fungal infections. Isavuconazonium sulfate is metabolized by plasma esterase enzymes, specifically butyrylcholinesterase, into isavuconazole [8]. However, it is unknown if isavuconazonium can be metabolized into isavuconazole in the human eye. As such, we evaluated the activity of isavuconazonium sulfate activity against the trophozoites and cysts of three separate T4 genotype strains of *Acanthamoeba*.

2. Results

2.1. Determination of Amebicidal Activity

The amebicidal activity of isavuconazonium sulfate was tested against three T4 genotype strains of *Acanthamoeba* (strains Ma, CDC:V240, and MEEI 0184). The trophozoites were exposed to serial dilutions of isavuconazonium sulfate with final concentrations ranging from 50 μM to 0.006 nM. All three strains of *Acanthamoeba* trophozoite appeared to be highly susceptible to isavuconazonium sulfate. Isavuconazonium displayed an EC_{50} of 0.001 μM against *Acanthamoeba* strain Ma (Figure 1A), which was about 1700- to 5000-fold more potent than the current standards of care chlorhexidine and PHMB. The EC_{50} of isavuconazonium against *Acanthamoeba* strain CDC:V240 was 0.037 μM (Figure 1B), which was about 30- to 300-fold more potent than chlorhexidine and PHMB, respectively. Isavuconazonium exhibited an EC_{50} of 0.024 μM against clinical strain MEEI 0184 (Figure 1C). This EC_{50} was about 1.5-fold better than the EC_{50} demonstrated against the CDC:V240 strain (Table 1). Overall, these nanomolar potencies demonstrate that isavuconazonium retains its potency against trophozoites of *Acanthamoeba* T4 strains.

2.2. Determination of Cysticidal Activity

Acanthamoeba T4 cysts are also clinically relevant, as they are often more difficult to treat than trophozoites and require higher concentrations of antimicrobial compounds for efficacy. In order to determine the activity of isavuconazonium sulfate against cysts of *Acanthamoeba* T4 strains, cysts from all three strains were tested against higher concentrations of isavuconazonium sulfate than for evaluating trophozoites. Final concentrations of isavuconazonium ranged from 200 μM to 100 μM in increments of 10 μM, and a treatment was considered to be cysticidal if there was no evidence of trophozoite proliferation or excystation by day 7, which is a commonly used end point for cysticidal assays [9–13].

Table 1. EC_{50} values of isavuconazonium sulfate against trophozoites of *Acanthamoeba* T4 strains.

Inhibitor	Strain	Mean (μM)	95% Lower CL (μM) [a]	95% Upper CL (μM) [a]
Isavuconazonium sulfate	Ma	0.001	0.001	0.002
	CDC:V240	0.037	0.027	0.049
	MEEI 0184	0.024	0.021	0.027
Standards of care				
Chlorhexidine [7]	Ma	1.7	1.4	1.9
	CDC:V240	1.1	1.0	1.2
	MEEI 0184	1	0.9	1.1
PHMB [7]	Ma	7.2	6.6	8.0
	CDC:V240	11.8	10.5	13.4
	MEEI 0184	4.6	3.0	7.1

[a] CL, confidence limit.

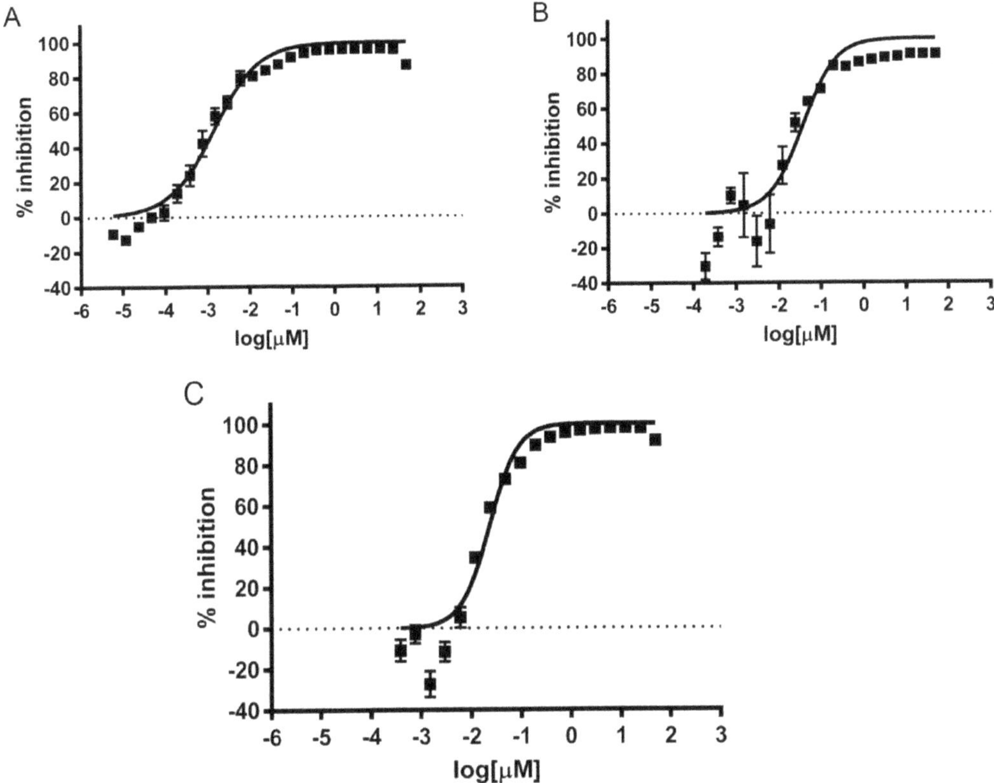

Figure 1. Concentration—dependent inhibition of growth of *Acanthamoeba* trophozoites by isavuconazonium sulfate. Different concentrations of isavuconazonium were tested in triplicate for activity against trophozoites of *Acanthamoeba* T4 strains. The data points represent mean percentage growth inhibition of (**A**) Ma strain, (**B**) CDC:V240, and (**C**) MEEI 0184 of different concentrations of isavuconazonium. EC_{50} curves were generated from mean values of percentage growth inhibition of isavuconazonium against *Acanthamoeba*.

Acanthamoeba T4 cysts were treated with isavuconazonium and allowed to recover in PYG growth media for 7 days and evaluated for cysticidal activity. Isavuconazonium displayed cysticidal activity against all three tested strains, as no excystation was observed at day 7 (Figure 2A,D,G). Isavuconazonium exhibited an average minimum cysticidal concentration (MCC) of 167.1 $\pm$ 23.6 µM against *Acanthamoeba* Ma strain. *Acanthamoeba* strain CDC:V240 had an average MCC of 136.0 $\pm$ 11.4 µM. *Acanthamoeba* strain MEEI 0184 displayed an average MCC of 187.5 $\pm$ 5.0 µM.

2.3. Effect of Combination of Isavuconazonium and PHMB on Cysts

While isavuconazonium has low nanomolar potency against trophozoites, it appears to display cysticidal activity only at high micromolar concentrations. Since cysts require higher isavuconazonium concentrations to prevent excystation, we wanted to evaluate if isavuconazonium in combination with other currently used drugs can display synergy to reduce the isavuconazonium concentration required to treat cysts.

Isavuconazonium was combined with PHMB and qualitatively assessed for excystation after 7 days of incubation in growth media (Figure 3). Isavuconazonium displayed an MCC of 167.1 µM. Combined with 40.41 µM PHMB, isavuconazonium was able to be lowered to 20.52 µM and still have minimal excystation (Figure 4A). Monotherapy of 20.52 µM isavuconazonium was confluent with trophozoites by day 7 (Figure 4B). Monotherapy of 40.41 µM PHMB had minimal to no excystation (Figure 4C). Since the combination of 20.52 µM of isavuconazonium with 40.41 µM of PHMB elicited a similar effect to what was caused by 40.41 µM of PHMB alone, it is apparent that the combination of two compounds did not have a synergistic effect able to suppress excystation or kill cysts. It was clear that the combination of these two compounds at this concentration did not cause an antagonistic effect on the cysts of Ma strain of *Acanthamoeba*.

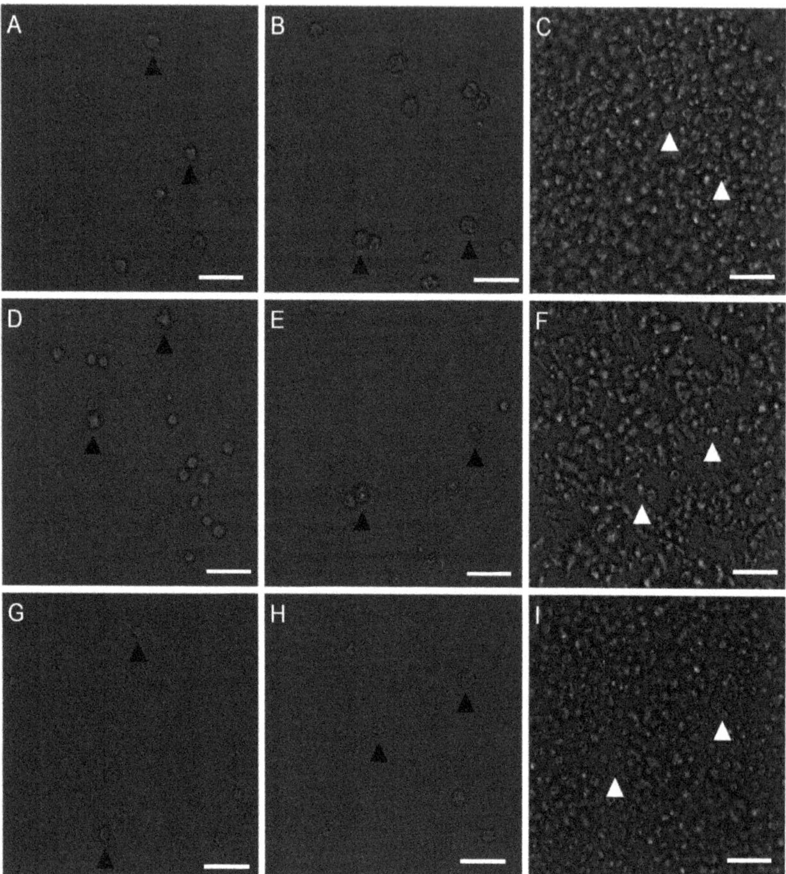

Figure 2. Effect of isavuconazonium sulfate on the morphology of *Acanthamoeba* Ma, CDC:V240 and MEEI 0184 cysts. Ma cysts were treated with (**A**) 170 µM isavuconazonium, (**B**) 461.85 µM PHMB, (**C**) 0.5% (*v/v*) DMSO. CDC:V240 strain cysts were treated with (**D**) 140 µM isavuconazonium, (**E**) 461.85 µM PHMB, (**F**) 0.5% (*v/v*) DMSO. MEEI 0184 strain cysts were treated with (**G**) 190 µM isavuconazonium, (**H**) 461.85 µM PHMB, (**I**) 0.5% (*v/v*) DMSO. Morphology and excystation of *Acanthamoeba* T4 cysts after 7 days of incubation in PYG growth media are displayed. Black arrowheads: cysts. White arrowheads: trophozoites. Magnification: 200×. Scale bar: 50 µm.

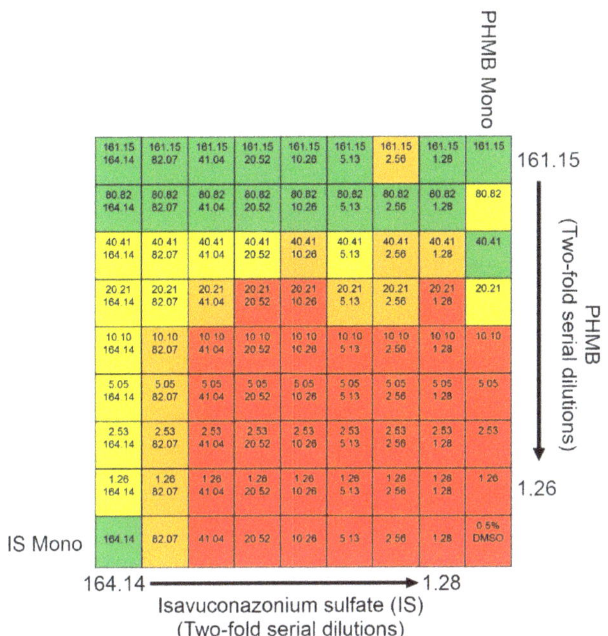

Figure 3. Isavuconazonium-PHMB combination excystation heatmap. Heatmap displaying qualitative scoring of excystation for different isavuconazonium sulfate (IS)–PHMB combination treatments. Top number in grids represents PHMB concentration in µM, bottom number in grids represents isavuconazonium sulfate concentration in µM. Green: 0% trophozoite plate coverage; yellow: 0–1% trophozoite plate coverage; orange: 50–80% trophozoite plate coverage; red: 100% trophozoite plate coverage.

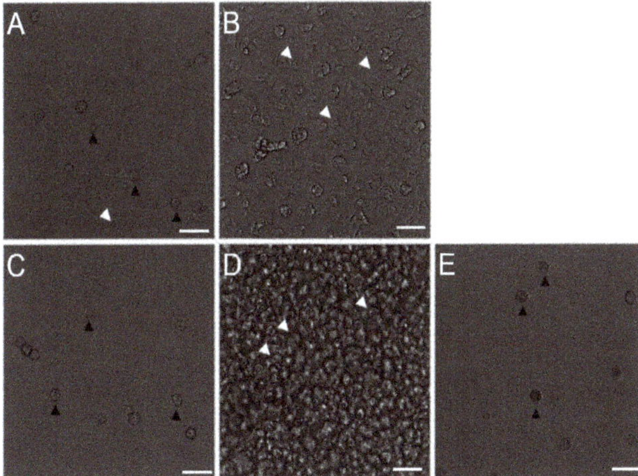

Figure 4. Effect of combination of isavuconazonium sulfate and PHMB on the morphology of *Acanthamoeba* Ma cysts. Ma cysts were treated with (**A**) 20.52 µM isavuconazonium and 40.41 µM PHMB, (**B**) 20.52 µM isavuconazonium monotherapy, (**C**) 40.41 µM PHMB monotherapy, (**D**) 0.5% (*v/v*) DMSO, (**E**) 461.85 µM PHMB. Morphology and excystation of *Acanthamoeba* cysts were evaluated after 7 days of incubation in PYG growth media. Black arrowheads: cysts. White arrowheads: trophozoites. Magnification: 200×. Scale bar: 50 µm.

3. Discussion

Isavuconazole has previously been evaluated and demonstrated potent amebicidal and cysticidal activity [7]. Since isavuconazole is typically administered as the prodrug isavuconazonium sulfate, there is a possibility that it may not be metabolized in the human eye and that *Acanthamoeba* T4 strains would not be susceptible to isavuconazole. Previously, isavuconazonium was identified as effective against *Acanthamoeba* T4 trophozoites [14], but its effect was not investigated against multiple strains and, more importantly, on the cysts of *Acanthamoeba* T4 strains. In this work, we evaluated the amebicidal and cysticidal activity of isavuconazonium against multiple strains of *Acanthamoeba* T4, and also explored the possibility of combining isavuconazonium with PHMB against the cysts of *Acanthamoeba*.

Isavuconazonium had mean EC_{50} values against trophozoites ranging from 0.001 μM (strain Ma) to 0.037 μM (strain CDC:V240) (Table 1). Our reported values are lower than those reported by Rice et al. against *Acanthamoeba* strain Ma (EC_{50} of 0.09 ± 0.02 μM [14]. This could be due to the differences in experimental conditions. We also previously evaluated the active drug isavuconazole against trophozoites and reported an EC_{50} of <0.001 μM (strain CDC:V240), 0.005 μM (strain Ma), and 0.026 μM (strain MEEI 0184) [7]. The EC_{50} values of isavuconazonium for strains Ma and MEEI 0184 are comparable to those previously reported values for isavuconazole. Interestingly, the EC_{50} of isavuconazonium against CDC:V240 was approximately 40× higher than that of the previously reported isavuconazole EC_{50} value. Taken together, this suggests Acanthamoeba T4 trophozoites are still susceptible to the prodrug isavuconazonium.

To our knowledge, this is the first reported evaluation of the cysticidal activity of isavuconazonium sulfate against *Acanthamoeba* T4 strains. We previously reported that isavuconazole, the active form of isavuconazonium, displayed cysticidal activity against *Acanthamoeba* Ma at 70 μM [7]. In this work, we determined the MCC of isavuconazonium against various *Acanthamoeba* T4 strains. The MCC of isavuconazonium varies from 1.9× (CDC:V240) to 2.6× (MEEI 0184) higher than previously reported about isavuconazole [7]. Since isavuconazonium is a prodrug that must be metabolized to active isavuconazole, it is possible that higher concentrations of the prodrug are required as not all of the isavuconazonium is metabolized into isavuconazole.

AK therapies frequently rely on chlorhexidine or PHMB as monotherapy, or in combination with propamidine isethionate and hexamidine. Commonly used combinations include PHMB with propamidine, chlorhexidine with propamidine, chlorhexidine with PHMB, and PHMB with propamidine and neomycin [15]. Since combination therapies were found to be more successful than monotherapies, we investigated the effect of the combination of isavuconazonium and PHMB on cysts of an *Acanthamoeba* T4 strain. In spite of the challenges associated with the excystation-based cysticidal assay that depends on a "cysticidal-or-not" readout rather than percentage inhibition [16], we identified that the combination of isavuconazonium and PHMB did not cause antagonistic or synergistic cysticidal effects on the cysts of the Ma strain of *Acanthamoeba*. Future studies will require confirmation of the effect of the combination of isavuconazonium and PHMB on trophozoites and cysts of different strains. Whether isavuconazonium can be combined with other commonly used drugs will require further investigation.

Isavuconazonium sulfate is FDA-approved for the treatment of invasive aspergillosis and mucormycosis [17]. In terms of fungal infections, isavuconazonium sulfate inhibits lanosterol 14α-demethylase [8], which prevents the biosynthesis of ergosterol and results in its eventual depletion. It is available in both oral and intravenous formulations and, following administration, it is rapidly cleaved to the active isavuconazole. The tissue distribution of isavuconazole was evaluated in animals after oral and intravenous administrations of isavuconazonium sulfate, and a low concentration of isavuconazole was detected in the eye lens [17,18]. This low concentration of isavuconazole may not be sufficient to kill *Acanthamoeba* T4 cysts if the drug is administered orally or intravenously. The preferable route of administration of drugs for the treatment of AK is topical, but the distribution of isavuconazole in the eye has not been evaluated when administered topically. Therefore, it was important to determine the effect of the prodrug isavuconazonium sulfate in case topical administration of prodrug isavuconazonium sulfate does not lead to the formation of active isavuconazole in human eyes. The potent amebicidal and cysticidal activities of the prodrug isavuconazonium sulfate against multiple T4 strains of *Acanthamoeba* provide confidence that the FDA-approved isavuconazonium sulfate is a promising lead for the treatment of AK.

4. Materials and Methods

4.1. Cell Culture

Acanthamoeba strains Ma (American Type Culture Collection #50370, Manassas, VA, USA), CDC:V240 (CDC, Atlanta, GA, USA), and MEEI 0184 (Tufts University, Medford, OR, USA), belonging to T4 genotype, were cultured as described by Shing et al. [7]. Trophozoites were cultured at 28 °C and 5% CO_2 in peptone yeast glucose (PYG) medium supplemented with 100 U/mL penicillin and 100 µg/mL streptomycin.

4.2. Determination of Amebicidal Activity

Stock 10 mM isavuconazonium sulfate (Cayman Chemical, Ann Arbor, MI, USA) and 10 mM chlorhexidine were prepared in DMSO. Isavuconazonium sulfate was serially diluted two-fold to generate solutions ranging in concentration from 10 mM to 1.2 nM. Next, 0.5 µL of each of these isavuconazonium sulfate dilutions was added to 96-well white, flat bottom microplates (Greiner Bio-One, Kremsmünster, Austria). This was followed by the addition of 5×10^3 trophozoites in 99.5 µL of PYG media to each well, giving final isavuconazonium sulfate concentrations ranging from 50 µM to 5.96 pM. Additionally, 0.5 µL of DMSO was added as a negative control (0.5% (v/v) DMSO), while 0.5 µL of chlorhexidine (50 µM) was added as a positive control. The plates were incubated for 48 h at 28 °C and 5% CO_2. At 48 h, viability measurements were taken using the CellTiter-Glo luminescent cell viability assay (Promega, Madison, WI, USA) [7]. For the measurements, 25 µL of CellTiter-Glo was added to each well and shaken on an orbital shaker at 360 RPM for 10 min prior to luminescence readings on an EnVision 2104 Multilabel Reader (PerkinElmer, Waltham, MA, USA). Data from a minimum of three independent experiments (biological replicates) conducted in triplicate were analyzed on GraphPad Prism 6 to determine EC_{50} values.

4.3. Cyst Generation

Encystment of *Acanthamoeba* T4 strains was induced by culturing trophozoites in an encystation media (95 mM NaCl; 5 mM KCl; 8 mM $MgSO_4$; 0.4 mM $CaCl_2$; 1 mM $NaHCO_3$; 20 mM Tris-HCl, pH 9.0) [19]. The trophozoite harvesting and encystation protocols were conducted as previously described by Shing et al. [7]. Briefly, trophozoites of *Acanthamoeba* T4 strains were centrifuged at $200 \times g$ for 5 min and washed in phosphate-buffered saline (PBS) three times prior to resuspension in encystation media. Then, 5×10^3 cells in 99.5 µL were added to each well of a 96 well clear-bottom plate (Corning, Corning, NY, USA). The cells were incubated in encystation media to facilitate the encystation of trophozoites into cysts for 48 h prior to any cysticidal or combination experiments.

4.4. Determination of Cysticidal Activity

Cyst plate generation was conducted as previously described in Section 4.3 [7]. After 48 h, 0.5 µL of isavuconazonium sulfate solution was added to a final concentration of 200, 190, 180, 170, 160, 150, 140, 130, 120, 110, or 100 µM. To serve as negative and positive controls, 0.5% (v/v) DMSO and 461.85 µM PHMB were used, respectively. The cysts were incubated for 48 h. Afterwards, the wells were washed four times with 100 µL of PBS before the addition of 100 µL of PYG medium. The cysts were then incubated for one week and imaged by an ImageXpress Micro XLS (Molecular Devices, San Jose, CA, USA) at $200\times$ magnification. The PYG growth media were exchanged for fresh media on day 3 and day 5. The images were manually reviewed for excysted trophozoites, and cysticidal activity was defined as having no trophozoites by day 7. Image brightness and contrast were adjusted by ImageJ. All experiments were conducted in triplicate and images were analyzed from a minimum of four independent experiments.

4.5. Effect of Combination of Isavuconazonium and PHMB on Cysts

Isavuconazonium was evaluated in combination with PHMB to assess potential synergy. After generating a plate of cysts using the methods described in Section 4.3, the cysts were treated with a checkerboard dilution scheme.

Isavuconazonium sulfate was serially diluted two-fold to generate solutions ranging in concentration from 65.7 mM to 512 μM. PHMB was serially diluted two-fold to generate solutions ranging in concentration from 64.7 mM to 504 μM. Then, 0.25 μL of each of the isavuconazonium sulfate and PHMB dilutions were added to each well of the plate to generate various concentration combinations. Isavuconazonium's dilution gradient varied horizontally across the plate to give final concentrations ranging from 164.14 to 1.28 μM. PHMB's dilution gradient varied vertically across the plate to give final concentrations ranging from 161.15 to 1.26 μM. As a negative control, 0.5% (*v/v*) DMSO was used, while 461.85 μM PHMB served as a positive control. Additionally, isavuconazonium sulfate and PHMB were also tested in monotherapy using the same final concentrations (164.14 to 1.28 μM and 161.15 to 1.26 μM, respectively) in separate columns.

After treatment for 48 h, the plate was washed with PBS and imaged for one week on the ImageXpress Micro XLS as described in Section 4.4. Images from three independent experiments were manually reviewed for excystation.

5. Conclusions

In this work, we tested isavuconazonium sulfate against *Acanthamoeba* T4 trophozoites and cysts to evaluate its potential as an anti-*Acanthamoeba* treatment. We found multiple T4 strains of *Acanthamoeba* to be susceptible to isavuconazonium, with appreciable activity against trophozoites even at low nanomolar concentrations. Cysts required significantly higher micromolar concentrations to prevent excystation. These findings suggest isavuconazonium to be potentially useful for clinical treatment of *Acanthamoeba* keratitis. Future studies should focus on in vivo animal models to validate isavuconazonium as a treatment option.

Author Contributions: Conceptualization, A.D. and B.S.; methodology, B.S., A.D. and M.B.; formal analysis, B.S. and M.B.; investigation, B.S. and M.B.; writing—original draft preparation, B.S. and M.B.; writing—review and editing, A.D., B.S. and M.B.; visualization, B.S. and M.B.; supervision, A.D. and B.S.; project administration, A.D. and B.S. All authors have read and agreed to the published version of the manuscript.

Funding: A.D. was supported by the National Eye Institute of the National Institutes of Health under Award Number R21EY032601.

Institutional Review Board Statement: Not applicable.

Informed Consent Statement: Not applicable.

Data Availability Statement: Data is contained within the article.

Acknowledgments: We thank Ibne Karim M. Ali and Shantanu Roy of CDC for providing CDC:V240 strain of *Acanthamoeba*. We are grateful to Noorjahan Panjwani of Tufts University for providing MEEI 0184 strain of *Acanthamoeba*. M.B. acknowledges the support of Undergraduate Research Scholarship Ledell Family Research Award for Science and Engineering.

Conflicts of Interest: The authors declare no conflict of interest.

References

1. Siddiqui, R.; Aqeel, Y.; Khan, N.A. The Development of Drugs against Acanthamoeba Infections. *Antimicrob. Agents Chemother.* **2016**, *60*, 6441–6450. [CrossRef] [PubMed]
2. Gupta, D.; Panda, G.S.; Bakhshi, S. Successful treatment of acanthamoeba meningoencephalitis during induction therapy of childhood acute lymphoblastic leukemia. *Pediatr. Blood Cancer* **2008**, *50*, 1292–1293. [CrossRef] [PubMed]

3.   Singhal, T.; Bajpai, A.; Kalra, V.; Kabra, S.K.; Samantaray, J.C.; Satpathy, G.; Gupta, A.K. Successful treatment of Acanthamoeba meningitis with combination oral antimicrobials. *Pediatr. Infect. Dis. J.* **2001**, *20*, 623–627. [CrossRef]

4.   Visvesvara, G.S. Amebic meningoencephalitides and keratitis: Challenges in diagnosis and treatment. *Curr. Opin. Infect. Dis.* **2010**, *23*, 590–594. [CrossRef]

5.   Schuster, F.L.; Visvesvara, G.S. Free-living amoebae as opportunistic and non-opportunistic pathogens of humans and animals. *Int. J. Parasitol* **2004**, *34*, 1001–1027. [CrossRef] [PubMed]

6.   Ledee, D.R.; Iovieno, A.; Miller, D.; Mandal, N.; Diaz, M.; Fell, J.; Fini, M.E.; Alfonso, E.C. Molecular identification of t4 and t5 genotypes in isolates from acanthamoeba keratitis patients. *J. Clin. Microbiol.* **2009**, *47*, 1458–1462. [CrossRef]

7.   Shing, B.; Singh, S.; Podust, L.M.; McKerrow, J.H.; Debnath, A. The Antifungal Drug Isavuconazole Is both Amebicidal and Cysticidal against Acanthamoeba castellanii. *Antimicrob. Agents Chemother.* **2020**, *64*, e02223-19. [CrossRef] [PubMed]

8.   Murrell, D.; Bossaer, J.B.; Carico, R.; Harirforoosh, S.; Cluck, D. Isavuconazonium sulfate: A triazole prodrug for invasive fungal infections. *Int. J. Pharm. Pract.* **2017**, *25*, 18–30. [CrossRef] [PubMed]

9.   Heaselgrave, W.; Hamad, A.; Coles, S.; Hau, S. In Vitro Evaluation of the Inhibitory Effect of Topical Ophthalmic Agents on Acanthamoeba Viability. *Transl. Vis. Sci. Technol.* **2019**, *8*, 17. [CrossRef] [PubMed]

10. Iovieno, A.; Miller, D.; Ledee, D.R.; Alfonso, E.C. Cysticidal activity of antifungals against different genotypes of Acanthamoeba. *Antimicrob. Agents Chemother.* **2014**, *58*, 5626–5628. [CrossRef] [PubMed]

11. Lamy, R.; Chan, E.; Good, S.D.; Cevallos, V.; Porco, T.C.; Stewart, J.M. Riboflavin and ultraviolet A as adjuvant treatment against Acanthamoeba cysts. *Clin. Exp. Ophthalmol.* **2016**, *44*, 181–187. [CrossRef] [PubMed]

12. Lee, J.E.; Oum, B.S.; Choi, H.Y.; Yu, H.S.; Lee, J.S. Cysticidal effect on acanthamoeba and toxicity on human keratocytes by polyhexamethylene biguanide and chlorhexidine. *Cornea* **2007**, *26*, 736–741. [CrossRef] [PubMed]

13. Talbott, M.; Cevallos, V.; Chen, M.C.; Chin, S.A.; Lalitha, P.; Seitzman, G.D.; Lietman, T.M.; Keenan, J.D. Synergy Testing of Antiamoebic Agents for Acanthamoeba: Antagonistic Effect of Voriconazole. *Cornea* **2019**, *38*, 1309–1313. [CrossRef] [PubMed]

14. Rice, C.A.; Troth, E.V.; Russell, A.C.; Kyle, D.E. Discovery of Anti-Amoebic Inhibitors from Screening the MMV Pandemic Response Box on Balamuthia mandrillaris, Naegleria fowleri, and Acanthamoeba castellanii. *Pathogens* **2020**, *9*, 476. [CrossRef] [PubMed]

15. Oldenburg, C.E.; Acharya, N.R.; Tu, E.Y.; Zegans, M.E.; Mannis, M.J.; Gaynor, B.D.; Whitcher, J.P.; Lietman, T.M.; Keenan, J.D. Practice patterns and opinions in the treatment of acanthamoeba keratitis. *Cornea* **2011**, *30*, 1363–1368. [CrossRef] [PubMed]

16. Shing, B.; Balen, M.; McKerrow, J.H.; Debnath, A. Acanthamoeba Keratitis: An update on amebicidal and cysticidal drug screening methodologies and potential treatment with azole drugs. *Expert. Rev. Anti. Infect. Ther.* **2021**, *19*, 1427–1441. [CrossRef] [PubMed]

17. Schmitt-Hoffmann, A.H.; Kato, K.; Townsend, R.; Potchoiba, M.J.; Hope, W.W.; Andes, D.; Spickermann, J.; Schneidkraut, M.J. Tissue Distribution and Elimination of Isavuconazole following Single and Repeat Oral-Dose Administration of Isavuconazonium Sulfate to Rats. *Antimicrob. Agents Chemother.* **2017**, *61*, e01292-17. [CrossRef] [PubMed]

18. Natesan, S.K.; Chandrasekar, P.H. Isavuconazole for the treatment of invasive aspergillosis and mucormycosis: Current evidence, safety, efficacy, and clinical recommendations. *Infect. Drug Resist.* **2016**, *9*, 291–300. [CrossRef] [PubMed]

19. Sohn, H.J.; Kang, H.; Seo, G.E.; Kim, J.H.; Jung, S.Y.; Shin, H.J. Efficient Liquid Media for Encystation of Pathogenic Free-Living Amoebae. *Korean J. Parasitol.* **2017**, *55*, 233–238. [CrossRef] [PubMed]

MDPI

St. Alban-Anlage 66

4052 Basel

Switzerland

Tel. +41 61 683 77 34

Fax +41 61 302 89 18

www.mdpi.com

Pharmaceuticals Editorial Office

E-mail: pharmaceuticals@mdpi.com

www.mdpi.com/journal/pharmaceuticals

MDPI

St. Alban-Anlage 66

4052 Basel

Switzerland

Tel. +41 61 683 77 34

Fax +41 61 302 89 18

www.mdpi.com

Foods Editorial Office

E-mail: foods@mdpi.com

www.mdpi.com/journal/foods